ENVIRONMENTAL INORGANIC CHEMISTRY

© 1985 VCH Publishers, Inc. Deerfield Beach, Florida

Distribution:
VCH Verlagsgesellschaft mbH, P.O. Box 1260/1280, D-6940 Weinheim, Federal Republic of Germany

USA AND Canada: VCH Publishers, Inc., 303 N.W. 12th Avenue, Deerfield Beach, FL 33442-1705, USA

LIBRARY & INFO

ENVIRONMENTAL INORGANIC CHEMISTRY

Edited by
Kurt J. Irgolic
and
Arthur E. Martell

Kurt J. Irgolic
Associate Director
Center for Energy and Mineral Resources
Texas A&M University
College Station, Texas 77843

Arthur E. Martell
Distinguished Professor
Department of Chemistry
Texas A&M University
College Station, Texas 77843

Library of Congress Cataloging in Publication Data

Main entry under title:

Environmental inorganic chemistry.

　　Includes bibliographies and index.
　　Based on papers presented at the U.S.-Italy Joint Seminar and Workshop on
Environmental Inorganic Chemistry held at San Miniato, Italy from June 5-10, 1983.
　　1. Environmental chemistry—Congresses.　2. Chemistry, Inorganic—Congresses.
I. Irgolic, Kurt J.　II. Martell, Arthur Earl, 1916-　.　III. U.S.-Italy Joint Seminar and
Workshop on Environmental Inorganic Chemistry (1983 : San Miniato, Italy)
TD193.E56　1985　　574.5'222　　　85-5297
ISBN 0-89573-145-2

© 1985 VCH Publishers, Inc.

This work is subject to copyright.

All rights are reserved, whether the whole or part of the material is concerned, specifically those of translation, reprinting, re-use of illustrations, broadcasting, reproduction by photocopying machine or similar means, and storage in data banks.

Registered names, trademarks, etc, used in this book, even when not specifically marked as such, are not to be considered unprotected by law.

Printed in the United States of America.

ISBN 0-89573-145-2　VCH Publishers
ISBN 3-527-26194-X　VCH Verlagsgesellschaft

PREFACE

The U.S.-Italy Joint Seminar and Workshop on Environmental Inorganic Chemistry was held at the University Conference Center in the town of San Miniato al Monte thirty miles west of Florence.

The Workshop was attended by forty invited participants from China, England, Italy, Switzerland, Sweden, Wales, and the United States, and provided a much needed forum for the assessment of the major research trends, the current state-of-the-art, and the major concerns in environmental inorganic chemistry from the viewpoint of the inorganic chemists coming from the participating countries. The information presented during the workshop and the contacts made among the participants will lead to international cooperative efforts in environmental inorganic chemistry that will help promote rapid transfer of research results, joint use of specialized instrumentation and facilities, exchange of scientists, interchange of ideas, and the development of research priorities. Such scientific cooperation between countries will maximize the results of environmental research by planned exchange of results and by eliminating unnecessary duplication of effort.

Italy, surrounded by the Mediterranean Sea, is vitally interested in alleviating, preventing, and monitoring pollution in ocean waters. Concern about the Mediterranean Sea led to the establishment of the "Long-Term Program for Pollution Monitoring and Research (Med-Pol-Phase II)" of the Mediterranean Action Plan administered by the United Nations Environment Program. Other countries such as Sweden, England, Switzerland, and China are also concerned about environmental problems, and their scientists have contributed to the literature in this area. China, with one quarter of the World's population, is now moving toward large-scale industrialization. China's environmental pollution problems are expected to grow rapidly in the next decade. The Chinese delegation to this workshop made contact with environmental scientists from other countries. Technical information presented at the workshop, useful for the assessment and control of the environmental hazards that might develop in China, is now available to Chinese environmental scientists. Such exchange of information avoids duplication of work that has already been carried out in laboratories of the western countries.

Inorganic and organometallic compounds of main-group and transition elements are part of the environment and exert their beneficial or detrimental effects on biological systems through specific, but largely unknown chemical or physicochemical interactions with molecules essential for life. Biological processes are known to transform inorganic compounds through ligand-exchange, carbon-element bond formation, and redox reactions, to increase the variety of substances present in the environment, and to greatly enhance the complexity of the interactions that occur. Generally, environmental problems can be completely understood and solved in the most expeditious manner, or prevented before they occur, only when their chemical foundations have been elucidated and the nature of the metal compounds involved has been determined. To obtain the necessary knowledge, investigations of the toxic or beneficial effects of elements and determinations of total element concentrations in organisms and their biological compartments must be followed by studies identifying the various compounds, their transformations in biological systems, and their interactions with biomolecules. These principles apply to inorganic compounds in land, air, sea, and groundwaters, and surface waters.

The workshop addressed ongoing and planned future activities in the area of "molecular" environmental inorganic chemistry, examined the present state of knowledge, provided information on the reactions of inorganic and organometallic compounds on the molecular level, and evaluated hypotheses that have been advanced to explain the observed reactions. Many chemical disciplines are involved in this search for a fundamental understanding of the impact of inorganic and organometallic compounds on the environment: analytical chemists must provide the sensitive and compound-specific methods, without which advances will be almost impossible; chemists specializing in synthesis must make available sufficient quantities of important model compounds for detailed study; physical chemists must elucidate reaction mechanisms, determine formation constants, establish the nature of compounds in solution and ascertain molecular structures. This workshop brought together not only delegates from several countries to establish international cooperation, but also representatives of various chemical disciplines to assure an interdisciplinary approach to environmental inorganic chemistry.

A number of basic interdisciplinary issues emerged during

Preface

the workshop that provide challenging research topics to the inorganic chemist. The influence of this workshop will make itself felt as ideas generated during the workshop lead to research projects, that explore the relationship between inorganic and organometallic compounds and biota, and - in a cyclic fashion - refine the list of priority research topics of fundamental significance to inorganic chemistry and the environmental sciences.

Environmental scientists and personnel in regulatory agencies have traditionally relied on a limited data base obtained from organic chemistry and geochemistry to estimate the scope, reactivity, bioavailability, and control strategies for both essential and toxic metals and metalloids. This workshop with its interdisciplinary emphasis on the role of inorganic chemisty in the environment emphasizes the importance of revitalizing and reassessing approaches to world-wide needs of industrial and developing countries for biotechnological controls and health measures associated with inorganic and organometallic compounds.

The proceedings of the San Miniato Workshop consist of the Workshop Summary and thirty five papers. The Workshop Summary is based on the presentations and discussions at San Miniato. After the plenary sessions and general discussions, at which the participants presented their insights and concerns, six groups were formed to probe in detail important topics in environmental inorganic chemistry. The topics of these group discussions were

- the role of inorganic processes in the environment,
- species, functions, and toxicities of metals,
- environmental biochemistry and toxicology,
- radionuclides in the environment,
- seawater, and
- analytical aspects of environmental inorganic chemistry.

The reports and recommendations that emerged from these groups were organized and edited. The edited Workshop Summary was reviewed by all workshop participants. Corrections and changes recommended during this review were incorporated into the final version. The Workshop Summary consists of six sections: coordination compounds, organometallic and organometalloidal compounds, radionuclides, inorganic compounds as therapeutic agents, environmental biochemistry of inorganic and organometallic compounds, and analytical aspects of environmental inorganic chemistry.

Each section sketches current knowledge, highlights important problems, recommends areas of research, and provides a bibliography. Roman numerals in brackets refer to pertinent papers collected in the second part of the proceedings.

The papers delivered at the workshop provide information about the general state of development of environmental inorganic chemistry and report recent, significant advances in this area. All papers were reviewed by the participants' peers and thoroughly edited. Although the workshop was held from June 5 - June 10, 1983, the articles were updated and the latest references added during the review and editing process. Each paper contains a liberal number of references (including 1984 references and references to papers in press) that may serve as a convenient entry to the literature on environmental inorganic chemistry.

This workshop on environmental inorganic chemistry was a beginning. Additional international workshops with participants from disciplines not represented in San Miniato (organic chemistry, industry, agriculture, medicine, atmospheric sciences, regulatory agencies) should be organized and held annually. The proceedings of these workshops will provide guidelines for the proper development of environmental inorganic chemistry, serve as a source of information for funding agencies, regulatory agencies and political bodies, and facilitate the entry into this field by interested scientists.

A. E. Martell
Chairman, U.S. Committee

P. A. Vigato
Co-Chairman
Italian Committee

Ugo Croatto
Co-Chairman
Italian Committee

Kurt J. Irgolic
Editor

ACKNOWLEDGEMENT

Without the financial support of the U.S. National Science Foundation and the Consiglio Nazionale Delle Ricerche of Italy this workshop would not have become a reality. All participants appreciate the hospitality they experienced in San Miniato and are especially thankful to Professore Ivano Bertini, University of Florence, for his untiring efforts on behalf of the workshop.

The hard work of all participants and reviewers, who helped to bring this volume into its final form, is greatly appreciated. The editing process and the preparation of the camera-ready copy was materially aided by Texas A&M University's Center for Energy and Mineral Resources. Prompted by appreciation of the importance of environmental questions for the generation of energy and the production of minerals, the Center placed its facilities and its staff at the disposal of the editors. For this invaluable assistance we thank Dr. Spencer Baen, Director of the Center, and the Center's staff.

Kurt J. Irgolic

A. E. Martell

TABLE OF CONTENTS

Preface	III
Acknowledgements	VII
Environmental Inorganic Chemistry:	
Workshop Summary	1
K. J. Irgolic, A. E. Martell, Workshop Participants	
Coordination Compounds	4
The Problem of Species Distribution	5
Examples for the Relationship between Toxicity and Chemical Structure	6
Examples of Problems with Toxicity Testing	7
Determination of Species Distribution	8
Species Distribution in Homogeneous Systems	9
Species Distribution in Heterogenous Systems	10
Bibliography	11
Recommendations for Research	12
Organometallic and Organometalloidal Compounds	18
The Problem of Speciation	19
Structure-Toxicity Relationships	20
Determination of Species Distribution	20
Bibliography	21
Recommendations for Research	23
Radionuclides	27
Bibliography	29
Recommendations for Research	30
Inorganic Compounds as Therapeutic Agents	32
Bibliography	34
Recommendations for Research	35
Environmental Biochemistry of Inorganic and Organometallic Compounds	36
Bibliography	39
Recommendations for Research	41
Analytical Aspects of Environmental Inorganic Chemistry	42
Determination of Total Trace Element Concentrations and Concentrations of Trace Element Compounds	42
Separation and Analysis Techniques	43
Labile Metal Complexes and Speciation	44
Surface Analysis	46
Sampling	47

	Stability of Trace Element Compounds	48
	Standards for Environmental Analyses	50
	Reporting of Analytical Results	50
	Analytical Methods for Soils, Sediments and the Atmosphere	52
	Bibliography	52
	Recommendations for Research	55
I.	Inorganic Chemistry and the Environment U. Croatto	59
II.	Inorganic Chemistry of the Ocean S. Ahrland	65
III.	Speciation of Metal Complexes and Methods of Predicting Thermodynamics of Metal-Ligand Reactions A. E. Martell, R. J. Motekaitis, R. M. Smith	89
IV.	Factors Influencing the Coordination Chemistry of Metal Ions in Aqueous Solutions R. D. Hancock	117
V.	Environmental Factors in the Inorganic Chemistry of Natural Systems: The Estuarine Benthic Sediment Environment O. P. Bricker	135
VI.	Trace Metals in the Ligurian and Northern Tyrrhenian Seas: Results and Suggestions for Further Studies R. Capelli, V. Minganti	155
VII.	Chemistry of Metal Oxides in Natural Water: Catalysis of the Oxidation of Manganese(II) by γ-FeOOH and Reductive Dissolution of Manganese(III) and (IV) Oxides J. J. Morgan, W. Sung, A. Stone	167
VIII.	Secondary Minerals: Natural Ion Buffers P. A. Williams	185
IX.	Environmental Inorganic Chemistry of Main Group Elements with Special Emphasis on Their Occurrence as Methyl Derivatives F. E. Brinckman	195
X.	Rates of Methylation of Mercury(II) Species in Water by Organotin and Organosilicon Compounds J. M. Bellama, K. L. Jewett, J. D. Nies	239
XI.	The Environmental Chemistry of Metals with Examples from Studies of the Speciation of Cadmium P. J. Sadler, D. P. Higham, J. K. Nicholson	249

XII.	Design of Metal-Specific Ligands	273
	D. E. Fenton, U. Casellato, P. A. Vigato	
XIII.	Mixed-Metal Complexes of Biofunctional Ligands: Formation, Stability and Possible Role in Model Systems with Metal Ion Overloads . .	285
	P. Amico, G. Arena, P. G. Daniele, G. Ostacoli, E. Rizzarelli, S. Sammartano	
XIV.	Behavior of Radionuclides in the Marine Environment: Present State of Knowledge and Future Needs	299
	P. Scoppa, C. Myttenaere	
XV.	Speciation of Plutonium in Seawater and Freshwater	307
	G. R. Choppin	
XVI.	Bioturbation and Fate of Radionuclides in Benthic Marine Ecosystems	321
	E. H. Schulte	
XVII.	Specific Sequestering Agents for Iron and Actinides	331
	K. N. Raymond	
XVIII.	Mercury Accumulation in a Pelagic Foodchain	349
	M. Bernhard	
XIX.	Chemical Studies of Aquatic Pollution by Heavy Metals in China	359
	Liu Ching-I, Tang Hongxiao	
XX.	Determination of the Chemical Forms of Toxic Elements in Environmental Aquatic Samples from China	373
	Lu Zongpeng, Mao Meizhou, Wei Jinxi	
XXI.	Mercury in River Sediments	393
	Peng An, Wang Zijan	
XXII.	Deposition of Atmospheric and Metal Pollutants and Their Impact on the Freshwater Environment: Case Studies in Switzerland	401
	W. Stumm	
XXIII.	Atmospheric Acidity: Analytical Problems Related to Its Determination	419
	A. Liberti, I. Allegrini, A. Febo, M. Possanzini	
XXIV.	Recovery and Reuse of Inorganic Materials in the Tannery Industry	431
	M. Acampora, U. Croatto, G. Piovan, P. A. Vigato	
XXV.	Assessment of Exposure of Human Populations to Heavy Metals through Dietary Intakes . .	437
	G. Tomassi	

XXVI.	Metals as Genotoxic Agents: The Model of Chromium V. Bianchi, A. G. Levis	447
XXVII.	Urinary Chromium as an Estimator of Air Exposure to Different Types of Hexavalent Chromium-Containing Aerosols A. Mutti, C. Minoia, C. Pedroni, G. Arfini, G. Micoli, A. Cavalleri, I. Franchini	463
XXVIII.	Metabolism and Toxicity of Chromium Compounds I. Franchini, A. Mutti	473
XXIX.	Microbial Resistance to Heavy Metals J. M. Wood, Hong-Kang Wang	487
XXX.	Bacterial Transformations of and Resistances to Heavy Metals S. Silver	513
XXXI.	Environmental Hazards from the Genetic Toxicity of Transition Metal Ions E. H. Abbott	541
XXXII.	Environmental Inorganic Analytical Chemistry K. J. Irgolic	547
XXXIII.	Neutron Activation and Radiotracer Methods Applied to Research on Trace Metal Exposure and Health Effects E. Sabbioni, R. Pietra, F. Mousty, M. Castiglioni	565
XXXIV.	Trace Element-Organic Ligand Speciation in Oil Shale Wastewaters J. S. Stanley, R. E. Sievers	579
XXXV.	Environmental Conditions in the Bay of Augusta, Sicily: Applicability of Simultaneous Inductively Coupled Argon Plasma Emission Spectrometry for the Analysis of Seawater G. Magazzu, C. G. Pappalardo, K. J. Irgolic	601

Appendices 613
   List of Participants 615
   Author Index 621
   Subject Index 623
   List of Abbreviations 653

ENVIRONMENTAL INORGANIC CHEMISTRY
WORKSHOP SUMMARY

K. J. Irgolic*, A. E. Martell* and Workshop Participants

*Department of Chemistry, Texas A&M University
College Station, Texas 77843 USA

Environmental Inorganic Chemistry addresses the beneficial and inimical interactions of inorganic, organometallic, and organometalloidal compounds with living organisms. To properly describe these interactions and identify their effects, detailed knowledge about the interacting entities and their transformations on a molecular basis is indispensable.

The space accessible to life is only a very small fraction of the total volume of earth and her atmosphere. Most of the planet consists of inorganic substances that supply the raw materials for the biotic domain. Biota need certain essential elements to carry out their life functions such as growth, maintenance, and reproduction, but would probably function better in the absence of the other, nonessential elements. Figure 1 identifies the essential

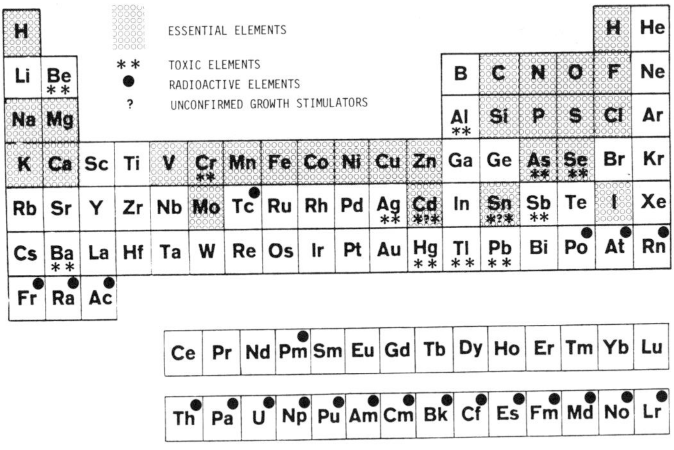

Figure 1. Radioactive elements, and essential and toxic elements in the environment.

© 1985 VCH Publishers, Inc.
Environmental Inorganic Chemistry

elements and the toxic elements. Essentiality and toxicity are not exclusive properties of an element or its compounds. Which of these two properties manifests itself is determined both by the concentration of the element [XXIX] and its specific molecular form [III, IX].

Elements are cycled in nature (Fig. 2). Most of the

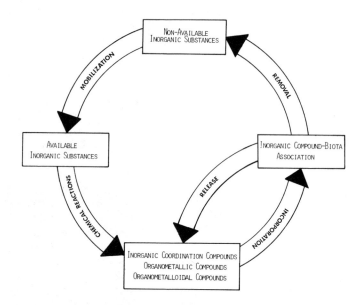

Figure 2. Schematic representation of the cycling of inorganic, organometallic and organometalloidal compounds in the environment.

elements are locked up in the form of insoluble inorganic compounds in rocks and sediments. These nonavailable inorganic substances must be mobilized and made soluble before biota can use them. This mobilization is effected by weathering processes involving abiotic chemical reactions and biologically mediated reactions. Metal ions can be released from sediments, for instance, as a consequence of the degradation of organic material and of redox changes [V, VII]. Among the biologically mediated reactions are those initiated by metal-tolerant bacteria and by man [IX, X]. Human activities release many substances into the environment in chemical forms capable of interacting with biota. After mobilization, which generally transforms insoluble substances to soluble substances, the primary products are changed by chemical reactions to inorganic

Workshop Summary

coordination compounds, organometallic compounds and organometalloidal compounds. Which types of compounds are formed is determined by the nature of the metal and by the reactants present in the environmental compartment considered. These "inorganic compounds" - a term which shall refer to organometallic, organometalloidal and purely inorganic compounds - may now interact with biota. Such interactions - inimical or beneficial - may proceed on the outer surfaces of cells or may require transport of compounds through membranes into cells [XXIX]. The biochemical apparatus operating on these "inorganic compounds" may produce new chemical species, which can be released through excretion into the environment and made available to other biota. Alternately, the "inorganic compounds" associated with cellular materials may be removed from circulation upon the death of the organism, for instance, by deposition in sediments. There, these compounds may join the pool of nonavailable substances awaiting mobilization.

Such a brief, general summary of the cycling of "inorganic compounds" cannot cover the many important details of the various steps within the cycles. However, the complexities of the systems, which are the domain of environmental inorganic chemistry, become apparent upon inspection of the major components of such systems.

- Eighty-six naturally occurring elements - among them approximately 20 essential trace elements and ten toxic elements - provide the building blocks for a large variety of compounds. Most of the elements are classified as metals. These metallic elements form, upon mobilization, simple ionic compounds and more complex coordination compounds [III]. A few of the metallic elements (Co, Pd, Pt, Au, Hg, Ge, Sn, Pb, Sb, Bi, Po) and all the metalloids form element-carbon bonds, which are sufficiently stable toward hydrolysis to survive in largely aqueous media [IX-XI].

- Man-made radioactive elements (isotopes of naturally existing elements and synthetic elements) add compounds, which - in addition to their potential chemical toxicity - emit damaging radiation [XIV-XVII].

- Aqueous solutions of various compositions provide media in which chemical and biochemical reactions proceed under the influence and control of other dissolved substances [II-IV, VII-XI].

- Solid matter – suspended in water or deposited in sediments – offers adsorption sites for dissolved substances and provides catalytically active surfaces to influence the rates of reactions [VII, XIX]. Solid matter may also serve as a buffer system for their constituent ions in water [VIII].

- Biota with their variability in size, function, and biochemistry assure, that a multitude of interactions with "inorganic compounds" occur.

Environmental processes are a complex mixture of biological and abiotic reactions. Environmental inorganic chemistry is far from a detailed understanding of these interdependent and perhaps synergistic interactions. To improve our understanding of environmental processes and assist us in predicting their directions and effects, appropriate experimental systems serving as models for environmental compartments must be studied in the laboratory under well-controlled conditions. The results of such studies will contribute to the data base, which must be available to successfully interpret observations made on "real" environmental samples.

Traditionally, studies of the transformations and fate of chemicals in the environment dealt almost exclusively with organic compounds. The true role of metals and metalloids in biogeochemical cycles has only recently begun to receive well-deserved attention of inorganic chemists. Not unexpectedly, many compounds of metals and metalloids have been found in the environment. It is reasonable to assume that many more new compounds will be identified and reaction pathways unique to environmental processes uncovered. In this manner the new field of "ENVIRONMENTAL INORGANIC CHEMISTRY" is created.

COORDINATION COMPOUNDS

Eighty-two percent of the 108 known elements are metals. The problem of determining the nature of the coordination compounds formed by metal ions in the environment is enormous, even when only the thirty most important metallic elements are considered. These metal ions compete for a large number of ligands [III].

Approximately 100,000 different organic chemicals are produced and used throughout the world. Production and use are concentrated in the industrialized countries. These chemicals and an even larger number of byproducts of their manufacture find their way into the environment. Whereas only a fraction of these organic chemicals have functional groups, which allow them to act as ligands toward metal ions, most of these compounds may be modified in the environment by biological, chemical, or photochemical processes such as hydrolysis and oxidation, and converted to metal-binding agents. The number of organic substances and potential ligands increases annually by approximately two percent through development of new products.

The thirty essential and toxic elements might join with the available ligands to produce several million coordination compounds. A metal ion may form several complexes in solution and on surfaces with the same ligand and a much larger number of mixed-ligand complexes with several different ligands. Thus, the existence of five to ten million metal complexes in the environment is a possibility although only a small fraction of the possible complexes are expected to form. Their formation is governed by their thermodynamic stabilities and kinetic limitations under prevailing environmental conditions [III, IV, XIII].

The Problem of Species Distribution

It is important to know which complexes are actually present in the environment, because the toxicity of a metal is highly dependent on the nature of the ligand with which it is combined. Toxicity data are available for only a few complexes. The experimental determination of the nature of all potentially existing complexes and of their toxicities is physically and financially impossible. The determination of reliable toxicity data for a new compound costs approximately $100,000.

The problem of speciation plaguing environmentalists and inorganic chemists alike is largely unresolved for "inorganic compounds" in natural systems. Although progress has been made during the past few years in identifying "inorganic compounds" in environmental samples and laboratory systems mimicking environmental conditions, much more, well-targeted activity is urgently needed to identify "inorganic compounds" in the environment, determine their concentrations, and study model systems under well-defined conditions in the laboratory. These data can then be used to predict species distribution, lifetimes, and toxicities.

When analytical methods permit the identification of an "inorganic compound" and the determination of its concentration in an environmental sample, this particular compound might exist and be present at a certain concentration because thermodynamic equilibrium has been established. However, the system may not be at equilibrium, and the concentration of the compound may be determined by kinetic factors, which may or may not establish a steady state condition. Close attention must always be paid to the residence time, lifetime, or similar measure of the kinetic lability of a species, which is part of the dynamic flux in the environment. Because the state-of-the-art of speciation is rather undeveloped, it is in most cases not possible to state with certainty, that a result obtained with an environmental sample is characteristic of an equilibrium situation. Synthetic samples mimicking the environment are better defined than "real" samples and, therefore, have been used to establish a data base that will help to calculate thermodynamic species-distribution employing stability constants pertinent to environmental conditions, and to establish structure-activitiy relationships. The thermodynamically favored species distribution identifies the state, toward which kinetically controlled systems will move.

Examples for the Relationship Between Toxicity and Chemical Structure

The toxicity and reactivity of a metal complex depends on the nature of the ligands. Very stable complexes may not interact with biological systems and may be essentially inert and nontoxic. However, when such complexes are formed by essential elements, which as a result of complex formation become unavailable, the ligands would cause metal deficiencies. Labile, nonessential metal ion complexes may be highly toxic through their competition with essential metal ions for important biological receptors. Thus, the binding affinities of metal ions and the ligands, with which they form complexes, profoundly influence the biological, chemical, and toxicological properties of an element.

It is well known that lead(II) acetate is extremely toxic. In contrast lead(II) ethylenediaminetetraacetate is nontoxic. Acute lead poisoning is combatted by intravenous injection of calcium ethylenediaminetetraacetate. By metal ion exchange in the body the lead complex is formed, which is then eliminated unmetabolized in the urine. Administration of sodium ethylenediaminetetraacetate would cause calcium deficiency, because of rapid and quantitative removal of essential calcium ion from the blood serum.

It has been observed that certain essential trace elements become toxic at moderate to high concentrations. Toxicity at higher levels might be a common characteristic of all essential metal ions. This toxicity is caused by the combination of the metal ions with biomolecules other than their natural receptors. Even iron, which is required at moderate levels, is toxic when the normal iron-carriers are overloaded. Toxic excesses of iron are associated with diseases such as β-thalassemia and haemosiderosis.

Dramatic changes in biological activity and toxicity occur when the oxidation state of the metal or the ligands are changed. Bakers yeast is thought to contain a chromium(III)-nicotinic acid-glutathione complex, which has specific receptors on cell surfaces and regulates insulin metabolism. Hexavalent chromium as chromate or dichromate is mutagenic [XXVI-XXVIII]. It passes through cell membranes more readily than chromium(III). Nickel tetracarbonyl, an intermediate in the Mond process for nickel refining, is instantly lethal to man at a concentration of 30 mg/L in air. Trinickel disulfide, an intermediate in the nickel refining process, induced tumors when injected into rats. Nickel compounds do not seem to be carcinogenic, however, when injested orally.

Examples of Problems with Toxicity Testing

Speciation methods that permit inorganic chemists to track environmental molecules in biotic and abiotic natural systems are generally developed in the laboratory in systems devoid of the natural matrix with concentrations of reactants much higher than found in nature. At nanomolar to picomolar environmental concentrations, reactions unimportant in the laboratory systems may become predominant. For instance, bimolecular processes (ligand exchange, dimerization) may be replaced by a pseudo-first order process (hydrolysis). The relative influence of surfaces (container walls, particulate materials) increases with decreasing concentrations. Because of disregard of the changes in the chemical nature of labile metal complexes with concentration, pH, redox intensity, and composition of the medium, the results of many toxicity tests carried out with metal ions and ligands are being misinterpreted and lead to incorrect conclusions. Numerous examples are available to show, that crucial dose-response studies on animals and humans were erroneously based on formulas given on the reagent labels or established at concentrations much higher than found in the environment.

Nitrilotriacetic acid, a substitute for polyphosphate in detergents, was tested extensively to assess its environmental impact [III]. Animals were fed high doses of the highly alkaline and completely ionic trisodium salt and not the calcium, magnesium, or zinc complexes, which are the principal compounds in waste water from commercial and domestic use. Not surprisingly, the tests produced results not at all applicable to environmental situations. Completely erroneous conclusions were drawn about the toxicity and carcinogenicity of nitrilotriacetic acid and its complexes in the environment.

Many metal ions form binuclear species in the concentration ranges most often used in experimental work. At lower concentrations most of these species dissociate and form mononuclear complexes. For instance, plutonium(IV) and plutonium(VI) form polymeric hydroxides at plutonium concentrations higher than 10^{-8} M [XV]. At concentrations of 10^{-10} to 10^{-15} M the plutonium hydroxides are monomeric and possess properties very different from those of the polymeric compounds. It is very important that investigators involved in testing metal complexes and ligands become aware of the reactions that are induced by dilution. Results obtained at high concentrations cannot be linearly extrapolated to the low concentration ranges encountered in the environment.

Determination of Species Distribution

The experimental determination of the nature and concentrations of all metal complexes potentially present in the environment is practically impossible. Reasonable estimates of the stabilities of metal complexes can be obtained even in multi-component systems provided that stability constants for the individual complexes are available. High-speed digital computers facilitate the required calculations. Currently, accurate stability constants are available for approximately 200 organic and inorganic ligands with many of the metal ions present in environmental systems. Additional constants for a few metal ions with several hundred ligands have also been determined. The limited data base for the calculation of species distribution thus consists of several thousand stability constants. New ligands are generated in the environment mainly from organic compounds as a result of chemical, photochemical, and biological processes. Information about these ligand transformations is inadequate for a comprehensive assessment of their influence on species distribution. The expansion of these data bases is urgently needed.

Species Distribution in Homogeneous Systems

Theoretical and empirical approaches have been employed for the estimation of stability constants in homogeneous systems in the absence of experimental data. The state-of-the-art for such estimations is far from satisfactory. For complexes of alkaline earth ions, first-row divalent transition metal ions, and a few lanthanide ions with several types of ligands including several multidentate synthetic ligands, empirical linear correlations for stability trends have been established. Limited extrapolations and interpolations should produce reasonable estimates for such systems. With few exceptions, such estimates have not yet been made.

Developments of stability relationships from first principles based on modern concepts of coordinate bonding and ionic interactions have been far less successful. Metal-ligand stabilities may be predicted for simple monodentate donors by equations requiring six parameters [III, IV]. The extension of this approach to bidentate donors may be possible. More complicated donors require the use of additional empirical parameters for structural variations. The resulting mathematical relationships are complex and cumbersome.

Changes in the composition of the solutions including pH, in which metal complexes are formed, influence the species distribution. The effects of concentration, ionic strength, and temperature are fairly well understood and may be estimated from available data. Pitzer's specific ionic interaction model can, for instance, be used to complement the ion association model. Although calorimetric data for heats of complex formation are not as numerous as equilibrium constants, sufficient values are available to estimate temperature coefficients for many equilibrium systems. However, this data base is also in need of expansion.

Species Distribution in Heterogeneous Systems

The environment is not a homogeneous system. Solution phases are in contact and interact with suspended solids, sediments, and biota. According to generally accepted estimates, ninety percent of the metals in "solution" are adsorbed on solid surfaces or locked up in solid matrices. The characterization of heterogeneous systems is much more complicated than that of homogeneous systems.

Solubility products are generally well known for metal ions, most inorganic ligands, and a few organic ligands. For nearly all of the organic ligands that may be found in the environment solid-solution equilibrium data are not available. In natural systems equilibria are established between dissolved metal complexes and metal complexes adsorbed on colloidal materials, soil particles, and clays. In addition to adsorption, metal complexes may exchange ligands, which they bind in solution, for donor atoms in the solid material upon contact with such phases. The lack of data for such heterogeneous systems causes serious problems, because chemical processes at the particle-solution interface play a dominant role in regulating the concentration of most reactive elements in natural waters. The concentration of metals is typically much higher in the solid phase than in the solution phase. The higher the affinity of a metal for coordinating sites on a particle, the more completely will the metal be removed from the aqueous phase. Present concepts in surface and colloid chemistry need to be further developed and extended to gain better insights into the coordination chemistry of solid-solution interfaces. To achieve better understanding of such interactions, the properties of surfaces must be determined - if at all possible - at molecular and atomic levels. The chemical nature of metal complexes on surfaces must be determined. Presently available analytical techniques such as sequential extraction procedures are usually not sufficiently selective for speciation.

BIBLIOGRAPHY

Albert, A. **"Selective Toxicity"**, 5th Ed.; Chapman and Hall: London, 1973.

Goldberg, E. Ed.; **"The Nature of Seawater"**, Dahlem Workshop Report, Dahlem Konferenzen, Wallotstraße 19, D-1000 Berlin 33, Germany, 1975.

Jenne, E.A., **"Chemical Modeling in Aqueous Systems"**, <u>ACS Symposium Series 93</u>, American Chemical Society, Washington, D.C., 1979.

Martell, A. E.; Smith, R. M. **"Critical Stability Constants"**, 5 Volumes, Supplements, Plenum Publishing Corporation: New York, N.Y., 1972-1982.

Singer, P. D. Ed.; **"Trace Metals and Metal-Organic Interactions in Natural Waters"**, Ann Arbor Science Publishers, Inc.: Ann Arbor, Michigan, 1973.

Stevenson, F. J. **"Humus Chemistry"**, John Wiley & Sons: New York, N.Y., 1982.

Stumm, W.; Morgan, J. J. **"Aquatic Chemistry"**, 2nd Ed. John Wiley & Sons: New York, N.Y., 1981.

Underwood, E. J., **"Trace Elements in Human and Animal Nutrition"**, 4th Ed., Academic Press: New York, 1977.

RECOMMENDATIONS FOR RESEARCH

Environmental Chemistry of Coordination Compounds

Determination of stability constants for metal complexes in homogeneous systems: The current data base consisting of accurately measured stability constants for metal complexes and related equilibrium constants should be expanded to include metals and ligands of importance to the environment. Quantitative data on representative metal-ligand interactions are essential to provide the base for reliable estimation of stability constants for the large number of complexes, which for practical reasons cannot be investigated experimentally.

Quantitative data are needed for the following ligands: carbohydrates, cellulose, starch, carbohydrate derivatives; proteins, peptides, amino acids; carboxylic acids, polycarboxylic acids, hydroxycarboxylic acids; degradation products of natural organic compounds such as humic and fulvic acids; phenols, phenolic acids, sulfonates, phthalates, organophosphates, lignins, and similar substances discarded as industrial wastes; and organic materials at cell walls and membranes such as glycophosphates.

Much more attention must be given to metal ions, that may be encountered as contaminants in the environment, such as As, Cd, Hg, Sb, Sn, Pb and actinides. Accurate data seem to be restricted to those metals (alkali metals, alkaline earth metals, first row transition elements, and lanthanides) that are easy to handle experimentally.

So called "interactive" effects between metal and organometallic ions may be explicable by improved speciation. For example, molybdenum interferes with copper metabolism. This interaction is thought to be caused by tetrathiomolybdate. Ruminants are very susceptible to this interference, because they reduce sulfate to sulfide, which may convert molybdate to tetrathiomolybdate.

The data base needs to be extended to include solvent systems and oxidation states of metal ions, which have not yet been extensively investigated. For example, in the anoxic regions of the oceans copper(I) is probably stable. Few stability constants pertaining to such conditions have been measured. The extreme pH conditions in various biological systems, e.g., pH 1-3 in the stomach, pH 5 in some lysosomes, may give rise to new complexes, which might be difficult to predict. Biological systems contain non-aqueous compartments and therefore the variation of metal-ligand affinities with increasing hydrophobicity of solvents needs to be studied for a wide range of ligands. Even plasma has hydrophobic regions. Albumin is coated with fatty acids. Lipids are stored in chylomicrons. Oil slicks and other organic films on natural water are other examples of hydrophobic regions in a largely aqueous medium.

Estimation of species distributions: The development of semi-empirical equations for the prediction of stability constants should be continued. Although the results of these endeavors will be of only limited usefulness for environmental scientists for the next ten to twenty years, continued efforts may lead to success as the understanding of metal-ligand bonding becomes more sophisticated, and computer techniques are developed for the quantitative treatment of metal-ligand interactions. Accurate calculations based on fundamental atomic properties and physical principles are not yet possible. The most reasonable and productive approach appears to be the use of comparisons between new complexes and known constants of analogous ligands for the prediction of thermodynamic equilibrium parameters. Effects of ionic strength and temperature may be estimated on the basis of the known behavior of various types of ligands. The likelihood of success for these estimation procedures increases with the size of the data base of reliable stability constants and associated protonation and hydrolysis constants.

Reliable estimates of equilibrium parameters should extend the data base to cover eventually all the metal ions and ligands of importance in environmental systems. The species distribution in an environmental system can then be estimated and the biological activities and toxicities of the complexes evaluated.

Identification of metal complexes and ligands released by organisms: Microorganisms such as bacteria, fungi, and algae are known to release metal complexes and ligands to the environment. Under extreme conditions of metal deficiency, powerful ionophores, for instance for iron(III), are excreted, which could also bind other metal ions such as chromium(III) and aluminum(III), and mobilize these ions for entrance into the food chain. Microorganisms will lyse at the end of their growth phase and release metal compounds into the environment. Cadmium-resistant bacteria release, for instance, a cadmium-cysteine complex. Such complexes may be more toxic than the cadmium compound to which the organism was originally exposed. The metal complexes and ligands released by biota must be identified and studied.

Development of convenient and inexpensive toxicity tests: To provide guidance to researchers exploring structure-toxicity relationships, convenient and inexpensive methods for the identification of toxic substances must be developed. Well-defined systems for in vitro toxicological testing of metal complexes must be devised. Such tests might be based on LD-50 (lethal dose 50%) of cells in culture, DNA binding, fidelity of DNA replication, or chromosomal damage in cultured mammalian cells. For each test the distribution of the metal complex under scrutiny among the components of the medium should be computed. The chosen buffer system must include all substances necessary for cell growth.

Testing of very insoluble inorganic compounds for toxicity requires special considerations. A test for the reactivity of mineral surfaces using cell or protein adhesion might be useful. Perhaps specific solubility tests for classes of compounds such as oxides and sulfides should be carried out as a step preliminary to determinations of toxicities. The test for chromosomal damage in cultured mammalian cells allows the detection of the genotoxicity of insoluble metal compounds, provided they are taken up by the cells via endocytosis and reach the genetic target.

Metal complexes are generally dissolved in dimethyl sulfoxide, detergents such as Tween 80, or saline before toxicity tests are carried out. This dissolution procedure should be standardized and the ligand-exchange reactions likely to occur during dissolution and in solution should be computed.

Every effort should be made to make toxicologists involved in toxicity testing aware of the fact that the natural complexes formed by a metal will change with the concentrations of ligands and metal. Conclusions based on results from feeding artificially high doses cannot be extrapolated to dilute solutions common in the environment.

Investigations of particulates and solution-particle interactions: The elucidation of the chemical and mineralogical nature of particles in natural aqueous systems such as mineral and rock fragments, carbonates, clays, humic and fulvic acids, semi-polymerized oxyhydroxydes of silicon, aluminum, iron(III) and manganese(III, IV), is of special importance. Humic (fulvic) acid-clay composites may, for instance, be of central importance to natural transport processes of metal ions. More knowledge about humic and fulvic materials at the molecular and not just the functional level is urgently needed. The dimensions of these particulates and their degrees of crystallinity influence their activity and must, therefore, be investigated. A first step toward a better characterization of particles by size and crystallinity is the collection and critical review of literature data.

To achieve a well-founded understanding of the behavior of metals in the environment, the detailed nature of the interactions of metal ions and complexes with the surfaces of particles must be known. The type of interaction will vary with the properties of the metal and the particle. Initially, the primary coordination sphere of the metal on the surface should be investigated. There is no apparent reason why this information should not become available quickly through concerted application of the extensive array of sophisticated, investigative tools of coordination chemistry.

Extensive new studies need to be undertaken to identify particles and their compositions important for metal transport. Thermodynamic investigations of carefully selected coordination reactions occurring at interfaces appear to be the next logical step. Particular attention ought to be paid to the possibility that ternary complexes with organic and inorganic components might have special stability. The simultaneous interactions of organic and inorganic ligands with metal ions, which may be complexed to surfaces, must be important. Such interactions should be explored with model systems. Information generated by research workers in other areas should be checked for usefulness with regard to the problem of ternary associations. An important process related to the formation of ternary complexes and worthy of detailed study is the transfer of metal species across humic (fulvic) acid films on clay/colloid particles. This process may remove metal ions from solution.

Exploration of kinetic aspects in heterogeneous systems: Although retention times in the environment frequently allow the application of thermodynamic models, kinetic factors are also needed to explain observations. Kinetic factors may even be of seminal importance in heterogeneous systems. It is highly probable, for instance, that various metal compounds claimed to be present in solution are really chemisorbed on and carried by suspended particulates at different rates. Perhaps of even greater significance are rate constants for the dissociation of metal complexes in heterogeneous and homogeneous systems. It is important to know the rates at which equilibria are attained under environmental conditions.

Development of particle standards: Standards for particulate matter are urgently needed. The adoption of a standard clay should be seriously considered. The currently available standard bentonite might be appropriate. This bentonite has been extensively studied by agricultural scientists. Standards for sediments from different environments are also needed. The cooperation of professional organizations is required for the establishment of particulate standards. Their implementation might be administered and monitored by appropriate international associations. Particularly pressing is the need for standards of humic and fulvic acids. Materials currently taken from defined sites cannot be useful for much longer. Synthetic standards must be introduced for all studies on humic and fulvic materials. Sufficient information is now available to prepare synthetic standards containing appropriate donor sites of defined chirality. Other standards prepared by coating standard clays with standard organic macromolecules according to a standardized procedure would also be useful. Additional chemical studies with natural humic and fulvic substances are still needed. However useful information is more likely to be generated by systematic "continuous-variation" studies with standard substances than by work with natural materials.

Collection of natural materials should continue. However, more material than is required should be obtained and the excess stored at a central depository. The availability of such properly preserved materials allows comparative studies to be made in the future with improved techniques.

Compilation of data for solid-solution interactions of metal complexes: Kinetic aspects and equilibria of solid-solution interactions have been studied extensively by earth scientists and soil chemists. Although such studies generally did not address the distribution of chemical species containing metal ions, other useful information may be available from these sources. Therefore, a literature survey of solid-solution interactions should be undertaken. Even knowledge about gross effects would be helpful as a guide for more detailed studies of solid-solution interactions of metal complexes.

ORGANOMETALLIC AND ORGANOMETALLOIDAL COMPOUNDS

Several metallic elements (Co, Pd, Pt, Au, Hg) and all metalloids form element-carbon bonds that are hydrolytically stable. These organometallic compounds, a term which shall stand also for organometalloidal compounds, may therefore be present in a largely aqueous environment. Methylation, a very common reaction in biological systems, may also proceed abiotically and generate methyl-element compounds. Examples of such compounds, which have been detected in environmental samples, are dimethyl selenide, selenoamino acids, methylarsonic acid, dimethylarsinic acid, dimethyltin species, methyl- and dimethylmercury, and tetramethyllead. However, compounds with methyl groups as the organic moieties are not the only organometallic derivatives found in nature. Compounds with more complex organic groups have been identified in biota. The arsenic-cycle shown in Fig. 3 may serve to convey an idea about the diversity of compounds formed by arsenic. The discovery of additional compounds of arsenic and of other elements in the environment is a certainty. Although the potential number of different organometallic compounds is expected to be much smaller than that of the coordination compounds, their identification and detection in the various environmental compartments is still a formidable task that is far from completion.

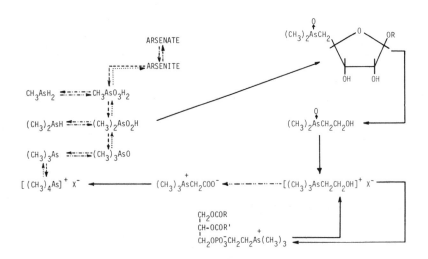

Figure 3. Arsenic compounds in the environment and some of their transformations (---- methylation, -.-.- reduction, ... demethylation, -..-..- oxidation, ____ other reactions).

The Problem of Speciation

Speciation appears to be easier with organometallic compounds than with labile coordination compounds. The organic group-element bond is in most cases sufficiently stable over time and in different media to allow the separation of organometallic compounds from their matrix and concentrate them to levels required by analytical techniques appropriate for detection and identification. A variety of concentration techniques and analytical methods, including element-specific detection systems for chromatographs, have been developed for this purpose.

Most of the elements forming organometallic compounds of environmental interest have various valence states. These compounds - especially those with a central atom in a high valence state - have available coordination sites for binding ligands. Relatively little is known about the coordination chemistry of such organometallic compounds under environmental conditions. The problems associated with the determination of coordination compounds outlined earlier apply equally to the determination of organometallic coordination compounds. The large number of potentially available ligands, the labile character of many of the complexes, the dependence of the structure of the complexes on concentration and the composition of the matrix, the largely unknown interactions of the complexes with solid phases, and their structure-dependent effects on biota make this rather undeveloped area of environmental inorganic chemistry a prime and promising target for important research.

Of special importance to organometallic coordination chemistry are a group of ligands that have chalcogen atoms as donors. The ubiquitous sulfur atom is the most important member of the chalcogen group. Sulfur atoms are incorporated into many important biomolecules (thioamino acids, enzymes) forming sulfides (R-S-R), disulfides (R-SS-R) and thiols (R-SH). All of these sulfur derivatives may serve as ligands. Thiols may combine with organometallic compounds to form element-sulfur bonds through condensation reactions. Most of the elements that have stable carbon-element bonds form also strong element-sulfur bonds. The well-established stability and insolubility of inorganic arsenic sulfides and mercury sulfide may serve as examples. Such sulfur-element interactions between, for instance, thiol groups in enzymes and trivalent arsenic compounds were postulated to be the cause of the toxicity of trivalent arsenic. Thiols may also serve as reducing agents for high-valent organometallic compounds. Condensation reactions initially produce a derivative with at least two organylthio groups, which are subsequently eliminated as disulfide with concomitant reduction of the valence state of the central atom by two units. Much remains to be done before the environmental chemistry of organometallic compounds can be claimed to have a solid foundation.

Structure-Toxicity Relationships

The toxicities of organometallic compounds and of organometallic coordination compounds are determined by their structures and stabilities in the same way as discussed for coordination compounds. There is also recent evidence, that toxic actions of organometals on organisms are encoded in the basic metal-carbon skeletal backbone [IX]. Mercury sulfide and methylmercury compounds are celebrated, though infamous, examples of such relationships. Whereas the very insoluble mercury sulfide (HgS) is non-toxic because of its inertness, the lipid-soluble methylmercury derivatives are insidious and potent neurotoxins. Biota have unfortunately found a way for the methylation of mercury even in the form of its sulfides [XXIX].

Not all methylation reactions increase the toxicity of an element or a compound. Arsenite, which is rather toxic, is converted through methylation to methylarsonic acid and dimethylarsinic acid that are less toxic than arsenite. The arsenic-containing carbohydrates and lipids may cause further reduction of the inimical effects of arsenic.

In the past, the evaluation of environmental hazards of the elements that form organometallic compounds was largely based on "total element" concentrations [XXXII]. Elements forming coordination compounds were treated similarly. Considering the widely diverging toxic properties of organometallic compounds, the description, evaluation, remediation and regulation of environmental hazards as well as nutritional needs purely in terms of "total element" concentrations are completely unacceptable.

Determination of Species Distribution

The experimental determination of the nature and concentrations of organometallic compounds and their non-labile coordination compounds is within the capabilities of analytical techniques. Most of these techniques are still in the development stage and have not yet found the wide application they deserve. It also would be very useful to increase the sensitivities of these techniques by one or two orders of magnitude in order to obviate preconcentration procedures.

The determination of labile coordination compounds of organometallic derivatives is plagued by the same problems as the determination of inorganic coordination compounds. Techniques for the in-situ characterization of such compounds at very low concentrations are urgently needed but are not available. The large number of potential ligands makes the experimental determination of the complexes a Herculean task [III]. Therefore, theoretical and empirical relationships must be employed to predict species distributions based on stability constants and enthalpies of formation as discussed in more detail in the section on coordination compounds. Although the literature is not a bounteous source of values for such predictions for organometallic compounds, sufficient information is available to make reasonable estimates for homogeneous systems. Experimental techniques do exist - mainly developed for coordination compounds - to determine the necessary data for organometallic compounds.

Heterogeneous systems containing organometallic entities suffer from the same lack of data and experimental techniques needed to obtain these data as systems with coordination compounds.

BIBLIOGRAPHY

Branica, M., Konrad, Z., Eds.; **"Lead in the Marine Environment"**, Pergamon Press, New York, N.Y., 1980.

Brinckman, F. E., Bellama, J. M., Eds.; **"Organometals and Organometalloids - Occurrence and Fate in the Environment"**. ACS Symposium Series, No. 82, 1978.

Brinckman, F. E., Fish, R. H., Eds.; **"Environmental Speciation and Monitoring Needs for Trace Metal-Containing Substances from Energy-Related Process"**, U.S. National Bureau of Standards Special Publication No. 618, November 1981, 324 pp.

Committee on Medical and Biological Effects of Environmental Pollutants, National Research Council, National Academy of Sciences, Washington, D.C., **"Mercury"**, 1978, **"Arsenic"**, 1977, **"Selenium"**, 1976, **"Lead"**, 1972.

Lederer, W. H., Fensterheim, R. J., Eds.; **"Arsenic: Industrial, Biomedical, Environmental Perspectives"**. Van Nostrand Reinhold Company, New York, N.Y., 1983.

Leppard, G. G., Ed.; **"Trace Element Speciation in Surface Waters and Its Ecological Implications"**. **NATO Conference Series. I. Ecology, Vol. 6,** Plenum Press, New York, N.Y., 1983.

Shamberger, R. J., **"Biochemistry of Selenium"**. Plenum Press, New York, N.Y., 1983.

Thayer, J. S., Brinckman, F. E., **"The Biological Methylation of Metals and Metalloids"**, in **"Advances in Organometallic Chemistry"**, **Vol. 20,** Stone, F. G. A., West, R., Eds.; Academic Press, New York, N.Y., 1982, p. 313.

Wong, C. S., Boyle, E., Bruland, K. W., Burton, J. D., Goldberg, E. D., Eds.; **"Trace Metals in Sea Water"**. **NATO Conference Series. IV: Marine Sciences, Vol. 9,** Plenum Press, New York, N.Y., 1983.

RECOMMENDATIONS FOR RESEARCH
Organometallic and Organometalloidal Compounds

The recommendations for research in environmental organometallic and organometalloidal compounds (collectively referred to as organometallic compounds) apply also to coordination compounds. There is no fundamental difference between compounds with hydrolytically stable carbon-element bonds and stable metal complexes.

Identification of naturally occurring organometallic compounds: The determination of environmental organometallic compounds of biotic and abiotic origin is in its infancy. Knowledge about the nature of these molecules, probably present in all environmental compartments, often at subnanomolar concentration under either steady state, transient, or equilibrium conditions, is crucial to progress in the field of inorganic environmental chemistry. It is difficult, if not impossible, to predict the types of compounds that organisms form. Therefore, a variety of organisms needs to be surveyed for metal and metalloid content. The most promising biota must then be subjected to procedures for concentrating, purifying, and identifying the organometallic species. This procedure, which is laborious, slow, and labor-intensive, was successfully applied to the identification of arsenic compounds in marine organisms and should also lead to success with other organometallic compounds.

Synthesis of organometallic compounds found in the environment: An organometallic compound is generally isolated from biota in extremely small quantities. Kilogram quantities of an organism yield milligrams of a compound, just enough to determine its chemical composition and perhaps its structure. There is no possibility of conducting toxicity tests, to study the chemical and physical behavior, and to develop analytical techniques for the detection of the compound in crude extracts. It is very desirable to develop a synthesis for the naturally-occurring compound. The synthetic material can then be produced in large quantities at much lower cost than by isolation from biota. Examples of compounds that are known or expected to be formed by biota are dimethylarsinoylribose derivatives and arsenolecithins. The organometallic synthetic chemist must be prepared to apply his preparative skills toward producing newly discovered compounds.

Exploration of the chemistry of organometallic compounds at very low concentrations: Organometallic compounds occur in biota or in various media outside biota at extremely low concentrations often at levels lower than nanomolar. At these low concentrations reaction pathways and reaction rates are probably different from those encountered at millimolar and higher concentrations. For example, methylarsines when prepared in the laboratory are easily and sometimes rapidly oxidized by atmospheric oxygen. However, at very low concentration the arsines have a surprisingly long halflife. A similar behavior is characteristic of methyltin hydrides in estuaries. The chemistry of organometallic compounds at extremely low concentrations should be investigated with the most sensitive techniques available. Should experimental difficulties prevent work at concentrations encountered in the environment, then the lowest feasible concentration should be used.

Laboratory studies of the transformation of inorganic and organometallic compounds: The natural environment is difficult to control. Too many variables defeat almost all attempts to establish cause-effect relationships. Much insight can be gained by modeling an environmental compartment in the laboratory. Organisms can be grown in a well-defined medium. Their exposure to inorganic or organometallic compounds can be strictly controlled. The time of exposure to compounds can be varied as desired. The organisms can be harvested and analyzed at various times. Experiments of this kind can hardly be carried out outside the laboratory. For these experiments to succeed, biologists, who are much better than chemists in raising organisms, and analytical and inorganic chemists, who can contribute their chemical intuition, must closely collaborate. Surprisingly, very little work has been done in this area. The potential scientific rewards from such studies are considerable. A funding mechanism coupled with patience of the funding agency must be found and maintained to bring biologists and chemists together to learn more about the transformation of trace element compounds by organisms.

Coordination chemistry of organometallic compounds: Linking organic groups with an element by abiotic or biotic means through formation of carbon-element bonds does generally not saturate the bonding capacity of a central atom. Most of the metalloids have several valence states and many have high coordination numbers. The coordination chemistry of metalloids and organometalloidal compounds, however, is not as varied as that of metals and organometallic compounds. For ligands to become linked to metalloid atoms rather strongly-bonded groups such as -OH or =O must be replaced, whereas in metal complexes a looser association between central atom and ligands is often encountered. Central atom-ligand interactions, especially those with ligands containing sulfur donor atoms, are nevertheless very important. The toxicity of arsenite is, for instance, attributed to the reaction of sulfhydryl groups in enzymes with the hydroxyl groups on the arsenic atom. The coordination chemistry of environmentally important organometallic and organometalloidal compounds must be investigated in much more detail than has been done thus far. Many of the research recommendations presented in the section on coordination compounds are also applicable to organometallic compounds.

Investigation of the interactions of cell membranes with organometallic compounds: The interactions of a compound with a cell wall is the first step in exposing an organism to an element. This first interaction, although important for any compound, is specially interesting with organometallic compounds that have hydrophobic and hydrophilic regions in their molecules. The organic moiety in the organometallic compounds may provide a convenient probe for the study of their interactions with cell walls and other membranes. Without information about these interactions, the uptake, permeation, and transformation mechanisms cannot be established. A concerted effort to shed light on these processes promises to lead to a much better understanding of fundamental aspects of environmental inorganic chemistry such as bioavailability.

Development of models predicting uptake and membrane transmission of organometallic compounds: After experimental studies have created a sufficiently large molecular data base of naturally occurring organometallic (and inorganic) substances and appropriate laboratory systems have been used to evaluate lifetimes, chemical, and physical properties, and likely reaction pathways, a predictive model must be developed for the uptake of compounds by organisms and the transmission of such compounds through membranes. The model must be based on molecular topologies. The change in the properties of organometallic compounds crossing from an aqueous to a nonaqueous medium should be studied.

Establishment of a data base for organometallic compounds of environmental importance: To provide a firm foundation, on which to develop much needed research in environmental organometallic (and inorganic) chemistry, pertinent data not now available must be collected, evaluated, and organized on an international scale. Such data include physical and chemical constants, structures, equilibrium constants, and rate constants. These data must be made available as reports or books similar to the series of volumes dedicated to stability constants for complexes of metal ions, transmitted at an annual clearing house held in tandem with suitable internal meetings and communicated through a monthly news bulletin. These activities require the sponsorship of at least one but preferably several international science-supporting agencies.

RADIONUCLIDES

For over three decades the environmental behavior of radionuclides has been studied by national and international organizations. The results of these studies were reported in a series of international meetings, of which those sponsored by the International Atomic Energy Agency (IAEA) are worthy of special note. The proceedings of these meetings published by IAEA should be consulted by anyone interested in the environmental behavior of radionuclides. Several general references are given at the end of this section.

Radionuclides are produced in fission reactors operated for research purposes and for the generation of power. In addition, radionuclides are released as byproducts of nonnuclear processes such as combustion of coal, mining of uranium ore, and processing of phosphate rock. Industry and medicine uses radionuclides for special purposes. Some of these radionuclides are also released into the environment.

To assess the relative impact of different radionuclides on the environment including man, the sources of radionuclides must be identified, and the mechanism of their dispersion in the atmosphere, in soil, and in water must be delineated. The dispersion must be studied on a local as well as a global scale. Mobilization processes, and kinetic and equilibrium factors are crucial aspects in these mechanistic considerations [XIV-XVI].

The present state of knowledge concerning the environmental behavior of radionuclides from the nuclear fuel cycle have been assessed in the following manner:

- A large amount of data exists for the equilibrium distributions of most fission and activation products in freshwater, seawater, soil, sediments, and marine organisms.

- Mathematical models have been developed that are satisfactory for evaluating the radiological consequences of the release of radionuclides during the normal operation of the nuclear fuel cycle.

- These models describe the effects of a steady state operation of the nuclear fuel cycle, but are inadequate for the evaluation of the results of large-scale, accidental releases of radionuclides.

- Uncertainties remain about the ability to predict the long-term consequences of the placement of radionuclides in permanent, high-level, waste disposal sites.

Far less attention has been given to questions related to the dispersion and migration of radionuclides released by operations not producing nuclear power. These radionuclides increase the natural radiation background in an as yet poorly defined manner. This little studied topic must be more fully investigated in regard to transfer of radionuclides to and through the food chain, the degree of exposure of biota to these radionuclides, and potential methods to prevent or reduce their release to the environment.

Research on radionuclides in the environment should address the long-term behavior of long-lived radionuclides, which is important for the evaluation of cummulative dose commitments; should explore the kinetic aspects of the environmental behavior of radionuclides, which is important in relation to accidental releases; should determine the uncertainties associated with the results provided by models, which are necessary for the calculation of dose optimization; should clarify the long-range transport of mobile radionuclides in the environment; and should assess the bioavailability of various physicochemical forms of radionuclides. Radionuclides are generally metallic in nature and thus form coordination compounds. Radionuclides derived from elements capable of forming hydrolytically stable organometallic or organometalloidal compounds will undergo the same reactions as their nonradioactive isotopes. Therefore, the statements and suggestions with regard to the determination of coordination compounds and organometallic and organometalloidal compounds apply equally to radionuclides.

Such research should produce a satisfactory base for the evaluation of the effects of regular and accidental release of radionuclides to the environment. Knowledge gained from the study of radionuclides may provide valuable and often unique information about the environmental behavior of stable elements at trace levels.

BIBLIOGRAPHY

The following publications provide general and detailed information about radionuclides and their interactions with the environment, and extensive references to pertinent literature:

Ionizing Radiation: Sources and Biological Effects. UNSCEAR Report for 1982, United Nations, New York, N.Y., 1982.

Radioecology: Nuclear Energy and the Environment. F. W. Whicker and V. Schult, CRC Press, Boca Raton, Florida, 1982.

Impacts of Radionuclide Releases into the Environment. International Atomic Energy Agency, Vienna, Austria, 1981.

International Commission on Radiation Protection. Publication ICRP-26, Pergamon Press, New York, N.Y., 1977.

Transuranium Nuclides in the Environment. International Atomic Energy Agency, STI/PUB/410, Vienna, Austria, 1975.

RECOMMENDATIONS FOR RESEARCH
Radionuclides

Identification of radionuclide compounds in the environment: Better, more sensitive methods for the identification and quantitation of radionuclide compounds in freshwater, seawater, estuarine waters, sediments, and soils are needed. Particular attention should be given to sediment-water interfaces.

Assessment of the importance of equilibrium and kinetic factors for the distribution of radionuclide compounds: Most studies are currently concerned with the characterization of radionuclides in systems at equilibrium. In nature, steady state conditions may be more important than equilibrium conditions. Reaction rates pertinent to the transformation of radionuclide compounds at the ultratrace concentrations, at which radionuclides occur in nature, must be determined before the role of kinetics can be assessed in the distribution of species in the environment.

Bioavailability of radionuclides: The chemical nature of compounds formed by radionuclides in the environment determines their bioavailability. Bioavailability and biomagnification are dependent on the physical and chemical properties of radionuclide compounds. Before these biological properties of radionuclides can be understood, the nature of the radionuclide compounds must be established. The efforts of the microflora and microfauna on the transformation of radionuclide species are not well understood and must be elucidated.

Long-term behavior of long-lived radionuclides: Transuranic elements and technetium are man-made radionuclides, which have been introduced into the environment for the first time only a few decades ago. Direct information on the long-term behavior of these radionuclides is not yet available. Extrapolations from short-term experiments are unreliable, because the species present in the short-term experiments might not be the species characteristic of an equilibrium state that might be reached after long periods. However, naturally-occurring elements with properties similar to the long-lived radionuclides should be studied. These "analogous elements" that have been in the environment long enough to be in a biogeochemical equilibrium may serve as models. This "analogous element" approach should be pursued vigorously.

Development of models for the behavior of radionuclides in the environment: The laboratory and field data pertaining to radionuclides must be carefully analyzed and used to develop models that satisfactorily explain and predict the behavior of radionuclides at trace concentrations in the ecosystem under various conditions of temperature, pressure, ionic strength, and pH in the presence of biota. Proper weight must be given to equilibrium and kinetic factors to make the models useful for understanding the short- and long-term effects of the addition of radionuclides to ecosystems.

Disposal of high-level radioactive wastes in the deep sea: The deep sea is a promising disposal site for high-level radioactive wastes. In addition to economical and technical difficulties associated with this disposal method, lack of information about the physics, chemistry, and biology of the deep sea is a troublesome problem. Most of the knowledge of the various pathways exposing man to radionuclides and of the impact of radionuclides on the marine environment is derived from studies in shallow water. Extrapolations to deep-sea conditions may be misleading and unreliable. A better understanding of vertical and horizontal dispersion processes in the sea is required. The interactions between waste components and abyssal sediments must be known. Little information exists about uptake, accumulation, and mobilization of radionuclides by deep-sea biota. The evaluation of collective dose commitments from waste disposal operations requires a better understanding of the biogeochemical processes involved in the long-term behavior of radiologically important nuclides. Pertinent studies that provide the information needed for the assessment of the deep sea as a disposal site for high-level radioactive wastes are urgently needed and should be begun without delay.

INORGANIC COMPOUNDS AS THERAPEUTIC AGENTS

Inorganic and organometallic compounds should not be viewed only as providing elements essential or inimical to life, but also as agents useful for combatting disease. Many organic compounds were synthesized that are now used by the medical profession in its fight against illness. There is an equally enormous scope for the design and the use of inorganic drugs. Pharmacological activities depend - just as do toxic effects - on the exact structure of a chemical compound.

Several inorganic substances are now employed as pharmaceuticals. Platinum complexes such as cis-diaminoplatinum(II) dichloride serve as anticancer drugs. Mercury compounds such as phenyl mercury(II) nitrate are potent bacteriocides. Silver sulfadiazine is effective against *Pseudomonas* in burns. Antimony drugs have cured many people from disease caused by infectious parasites. Gold(I) thiolates are used for treatment of rheumatoid arthritis. Lithium carbonate is given prophylactically for manic depressive psychoses.

These few examples indicate the potential importance of inorganic drugs. Most of the known coordination compounds and organometallic compounds await testing for pharmacological activity. The US National Cancer Institute has data for only 11,000 metal and metalloid compounds of 55 elements. Thirty-five of these elements have at least one active compound. Among the enormously large number of coordination derivatives formed by organometallic substances and purely inorganic compounds are certainly valuable pharmaceuticals waiting to be discovered. The activities of such compounds can be altered by variation of valence states of the central atoms, by modification of ligands, and by ligand exchange. The interplay between the thermodynamics and kinetics of ligand exchange has enormous potential in pharmacology.

The design of inorganic pharmaceuticals is not the only concern of environmental inorganic chemistry in the area of inorganic pharmacology. Metal ions and organometallic compounds that have found their way into an organism to cause interferences with essential life processes must be removed or made inactive. This task can often be accomplished by providing to an organism one or more appropriately structured ligands that will coordinate with the offending inorganic compounds. This "ligand therapy" is also an important and productive area of environmental inorganic chemistry. More attention should be devoted by synthetic chemists to the design and synthesis of new agents for the complexation and removal of toxic metal ions. The treatment of lead and plutonium poisoning should serve as examples [III, XI, XVII].

The present treatment of lead poisoning is relatively nonspecific. Calcium ethylenediaminetetraacetate and occasionally 1-hydroxy-2,3-propanedithiol (British Anti-Lewisite, BAL) are administered for complexation of lead. These agents react, however, with many other elements. The design of a nontoxic chelating agent of high specificity for lead would be of great value for the improvement of lead therapy.

Plutonium (Pu), a major byproduct of the nuclear power program, is present in the environment at very low concentrations with a limited geographical distribution [XIV-XVI]. The long half-life of plutonium and the high biological hazard when incorporated into an organism are the reasons for special environmental concerns. Plutonium is present in higher animals as Pu(IV), an ion with properties remarkably similar to iron(III). Pu(IV) is coordinated by transferrin, the human iron-transport protein in serum and thus becomes irreversibly incorporated unless chelation therapy is administered [XVII]. Presently, calcium or zinc diethylenetriaminepentaacetate is administered in the U.S. for complexation of plutonium. Relatively large amounts of the calcium or zinc chelate are needed to remove a small fraction of Pu(IV). The agent complexes strongly with many other metal ions and may cause metal deficiencies.

During the last few years much more specific complexing agents for Pu(IV) were prepared based on the similarity between Pu(IV) and Fe(III) and naturally-occurring complexing agents for Fe(III) of microbial origin called siderophores. Incorporation of catecholate groups – found in certain siderophores – into octadentate ligands that can completely encapsulate Pu(IV) produced compounds, which removed at low doses at least 75 percent of Pu(IV) from test animals. These unique properties of compounds related to biogenic siderophores suggest that a critical survey of naturally-occurring cryptate or polydentate ligands might uncover useful agents for the treatment of metal poisoning [XII, XVII]. Such a survey might also uncover microbiological metabolites that are capable of endo- and exocellular complexation of common and uncommon technological metals and metalloids. These compounds could mobilize toxic substances and create a health hazard.

The rational design of metal-ion-specific complexing agents will have a great impact in the treatment of metal ion poisoning and contamination and deserves the full attention of environmental inorganic chemists.

BIBLIOGRAPHY

Birch, N. J. and Sadler, P. J., **"Specialist Periodical Report: Inorganic Biochemistry"**, Hill, H. A. O., Ed., **Vol. 1-3,** Royal Society London, 1979, 1980, 1981.

F. Fenchel and T. H. Blackburn, **"Bacteria and Mineral Cycling"**, Academic Press, London, 1979.

Sadler, P. J., **"Inorganic Pharmacology"**, Chemistry in **Britain,** March 1982, p. 182.

Siegel, H., Ed., **"Metal Ions in Biological Systems"**, Vol. 11, **"Metal Complexes as Anitcancer Agents"**, Marcel Dekker, New York, N.Y., 1980.

RECOMMENDATIONS FOR RESEARCH
Inorganic Compounds as Therapeutic Agents

Design and synthesis of ligands for the treatment of metal poisoning: Only recently have chemists begun to design biocompatible complexing agents for metals and metalloids. The properties of the ligands such as geometry, ligating groups, and basicity must restrict strong complexation to specific metal ions and thus target a particular metal ion. The expansion and further development of such efforts will lead to dramatic improvements in complexing and controlling metal and organometal ions that pose environmental hazards.

Design and synthesis of inorganic therapeutic agents: There is enormous scope for the design of inorganic drugs. Platinum compounds are used to treat cancer, gold derivatives help fight arthritis, and arsenic compounds were widely used to combat venereal disease before the advent of antibiotics. Arsenic derivatives are still part of the pharmacopoeia of the veterinarian. Inorganic compounds can be designed to produce specific biological effects. Pharmacological activity depends intimately and often subtly on the exact structure containing the element. Concerted efforts to design and synthesize inorganic drugs containing elements such as tin, gold, silicon, platinum, vanadium, gallium, and others require close collaboration with pharmacologists and clinicians and will certainly lead to a rich harvest of medically useful drugs.

ENVIRONMENTAL BIOCHEMISTRY OF INORGANIC AND ORGANOMETALLIC COMPOUNDS

Living organisms have influenced the steady state levels of elements in the atmosphere, in freshwater systems, in the soil, and in the oceans through geological time. Inorganic and organometallic compounds have been synthesized, changed, and decomposed through biological activities for five billion years. Only during the past 200 years have industrial activities dramatically changed the distribution of the elements at the surface of the earth. Organisms have been and are adapting to these changes.

The fates of pollutants in the biosphere have received much attention during past decades. The cycling of heavy metals, radioactive wastes, and atmospheric pollutants have been studied in terrestrial and marine systems. Although much of this work is analytically significant, it has been performed without a full appreciation of the impact of metabolic capabilities of the organisms that live in the regions stressed by industrial pollution. The exploration of the biochemical bases for toxicity and for resistance to pollutants are emerging, fertile endeavors for environmental scientists. The principles that govern the transport and modification of pollutants in living cells must be understood, before the molecular basis for toxicity and resistance can be determined.

Microorganisms are evolving strategies to maintain low intracellular concentrations of toxic pollutants in response to imposed stresses. This adaptation to resist toxic substances has arisen in two distinct patterns. First, the ability to resist toxins may have been inherited from organisms that lived under extreme environmental conditions. It is not unusual to find, for example, naturally-occurring populations of bacteria and algae that tolerate high concentrations of metal ions at low pH and high temperatures. Hot springs, volcanic lakes, and deep sea vents provide such extreme conditions that are more representative of the state of the world billions of years ago than the general conditions found now. Secondly, some bacteria developed resistance rather easily through the acquisition of extrachromosomal DNA molecules (plasmids) [XXIX-XXXI].

The emerging biochemical basis for resistance to toxic metals indicates the existence of various mechanisms.

- Energy-driven efflux pumps keep the concentrations of toxic elements in the interior of the cell at low levels. Such mechanisms were described for cadmium and arsenate.

- Oxidation or reduction converts a more toxic form of an element enzymatically to a less toxic form. Examples for such mechanisms are the oxidation of arsenite to arsenate and the reduction of mercury(II) ions to elemental mercury.

- Intracellular polymers synthesized by the organisms trap and remove metal ions from solution. Such traps were described for cadmium, calcium, nickel and copper.

- The cell surface binds metal ions and thus prevents their uptake. Nickel was found to be concentrated 3000-fold over the concentration in the medium at the cell surface of cyanobacteria and green algae.

- Metal compounds of low solubility are precipitated on cell surfaces. *Cyanidium caldarium*, a thermophilic alga, precipitates metal sulfides and oxides in this manner.

- Metals and metalloids are methylated and their methyl derivatives transported through cell membranes by diffusion controlled processes. Examples of elements that can be methylated are mercury, tin, selenium, and arsenic.

Each of these mechanisms for resistance requires cellular energy and moves the distribution of elements at the earth's surface farther away from equilibrium. Therefore, the kinetic aspects of these processes are especially important to the field of environmental health and toxicity. The dynamic aspects of the interactions of biota with pollutants have not been and are still not sufficiently appreciated.

Resistance mechanisms in higher organisms including man seem to be quite variable. The rat responds, for instance, to an arsenate dose by binding arsenic selectively to vicinal thiol groups on a serum protein. The rabbit, however, reduces arsenate to arsenite and methylates arsenite to dimethylarsinic acid with 5-adenosylmethionine as the methyl donor. The rather nonpolar dimethylarsinic acid is rapidly excreted through the urinary system

[XXXIII].

The interactions between essential and nonessential elements are very complex. For example, selenium was shown to effectively decrease the toxicity of ionic mercury and of methylmercury. Very little, molecularly-based knowledge about such fascinating interactions is available, although many more are certain to exist.

Basic to environmental biochemistry is an understanding of the processes involved in the transport of elements through cell membranes. Physical and chemical parameters such as pH, temperature, redox conditions, the presence of competing elements, and the nature of the compound to be transported have a profound effect on the selection of a species for transport and on toxicity. Knowledge of the chemical species present in a system is, therefore, very important.

The cycling of metals in the biosphere not only influences the availability of these metals but also controls the biochemical fate of organic compounds. For example, enzymes that dehalogenate chlorinated hydrocarbons require transition metals such as iron(III), copper(II), and manganese(II) for their active sites. The uptake of metals by organisms is, therefore, intimately connected with the rates of biodegradation of natural and anthropogenic toxic organic compounds. Much work is urgently needed to identify the molecular basis of element transport through cell membranes, of the chemical changes forced upon inorganic and organometallic compounds inside the cell and of the genetic apparatus that makes these changes possible.

The protection of man and his environment from harmful and possibly irreversible effects of inorganic and organometallic compounds is the ultimate goal of environmental inorganic chemistry. The effects of such compounds on man cannot be as easily assessed as those on other organisms. Valuable knowledge can, however, be gained by well-planned and well-executed monitoring and investigation of population groups exposed to metals and metalloids under special circumstances established by nature or human activities. Exposure to mercury in regions of mercury deposits [XVIII, XXV], contact with chromium compounds in areas with many tanneries [XXVI-XXVIII], and exposure to arsenic close to smelters are examples of such special circumstances.

To assure that the results of monitoring programs can be interpreted in a meaningful manner, biological data and chemical data must be obtained. Dose-response relationships are not very meaningful unless the nature of the compounds and their concentrations in the environment are known. Control groups not exposed to the particular pollutant to be studied above the natural background will provide information about normal (baseline) reactions.

Man reacts to the influx of pollutants by mobilizing his biochemical armory, which is expected not to be drastically different from that available to other organisms. Careful epidemiological studies, with concomitant speciation of environmental pollutants and coupled with investigations of the biochemical pathways through which pollutants are processed in animals and plants, will assist in protecting man from harm by metallic and organometallic pollutants.

BIBLIOGRAPHY

Brock, T. D. **"Thermophilic Microorganisms and Life at High Temperatures"**. Springer Verlag, Heidelberg, 1978.

Edmond, J. M.; Von Damm, K. **"Hot Springs on the Ocean Floor"**, **Scientific American 248** (1983) 78.

Ganther, H. E.; Goudi, C.; Sunde, M. L.; Kopecky, M. J.; Wagner, P.; Oh, Sang-Hwan; Hoekstra, W. G. **"Selenium: Relation to Decreased Toxicity of Methylmercury added to Diets Containing Tuna"**. Science 175 (1972) 1122.

Leonzio, C.; Focardi, S; Bacci, E. **"Complementary Accumulation of Selenium and Mercury in Fish Muscle"**. Sci. Total Environ. 24 (1982) 249.

Lovelock, J. E. **"Gaia, A New Look at Life on Earth"**. Oxford University Press, Oxford, England, 1979.

Ridley, W. P.; Dizikes, L. J.; Wood, J. M. **"Biomethylation of Toxic Elements in the Environment"**, **Science 197** (1977) 329.

Silver, S. **"Bacterial Transformation and Resistances to Heavy Metals"**, in **"Changing Biogeochemical Cycles of Metals and Human Health"**, Nriagu, J. O., Ed.; Springer Verlag,

Berlin, 1983.

Summers, A. O.; Silver, S. **"Microbial Transformations of Metals", Ann. Rev. Microbiol. 32** (1978) 637.

Swallow, K. C.; Westall, J. C.; McKnight, D. M.; Morel, W. M. L.; Morel, F. M. M. **"Potentiometric Determination of Copper Complexation by Phytoplankton Exudates", Limnol. Oceanogr. 23** (1978) 538.

Wood, J. M. **"Selective Biochemical Reactions of Environmental Significance"**. Chemica Scripta, Nobel Symposium on Inorganic Biochemistry. Karlskoga, Sweden, 21 (1982) 155.

Wood, J. M.; Hong-Kang Wang; **"Microbial Resistance to Metal Ion Toxicity", Environ. Sci. and Technol. 17** (1983).

Wood, J. M. **"Chlorinated Hydrocarbons: Oxidation in the Biosphere", Environ. Sci. and Technol. 16** (1982) 291A.

RECOMMENDATIONS FOR RESEARCH

Environmental Biochemistry of Inorganic and Organometallic Compounds

Investigation of the transport of metals through cell membranes: The mechanisms for transport of essential elements through cell membranes in micro- and higher organisms are largely unknown. The elucidation of these mechanisms is a first step toward an understanding of the interactions between organisms and metal compounds and might even provide molecular information about the carcinogenicity of metals such as nickel and chromium. The kinetics for transport of toxic metal ions has only been studied in depth for copper and mercury. The effect of environmental parameters and competing cations and anions, and the transport kinetics of other metal compounds needs exploration. The rates of transport into cells is crucial to an understanding of toxicology. A concerted research effort in this area is certain to be beneficial to the advancement of the field of environmental inorganic chemistry.

Exploration of the molecular basis of metal-ion resistance: The molecular basis for heavy metal ion resistance by bacteria and algae is just emerging. More basic information on the genetics of resistance through the acquisition of plasmids and other strategies is needed. The mechanisms for toxicity and resistance to highly reactive metal ions such as Cu^{2+} and Hg^{2+} deserve attention.

Elucidation of the neurotoxicity of metalalkyls: Methyl-mercury, methyltin and methyllead compounds are known to be neurotoxic. A common side-effect of the use of platinum anticancer drugs is peripheral neuropathy. The molecular basis of this toxicity is unknown and needs to be identified.

ANALYTICAL ASPECTS OF ENVIRONMENTAL INORGANIC CHEMISTRY

Without analytical chemistry most branches of chemistry would have been severely retarded in their development or would not have developed at all. Environmental chemistry would certainly be only of minor importance today without the accurate, precise, sensitive, and often sophisticated analytical methods that call attention to the presence of detrimental substances and the absence of essential elements in our environment. Analytical chemistry delivers the basic information that is necessary to assess an environmental situation, identify its causes, suggest preliminary action, and point toward a permanent solution. Aaron J. Ihde in his book "The Development of Modern Chemistry" assesses the importance of analytical chemistry in the following manner: "It is not incorrect to ascribe to analytical chemistry a position of primary importance, since only through chemical analysis can matter in its variety of forms be dealt with intelligently. The stimulus given to chemistry by new analytical approaches, either qualitative or quantitative, has been repeatedly observed." It is certainly true that problems of environmental inorganic chemistry cannot be dealt with intelligently without appropriate analytical techniques and their judicious applications. The array of analytical instruments and methods applicable to environmental inorganic chemistry is remarkable in its variety and sensitivity. Development of new instruments and instrument combinations promise a bright future.

Determination of Total Trace Element Concentrations and Concentrations of Trace Element Compounds

The art and science of the determinations of total trace elements have a long history and have achieved great success. When total trace element concentrations are determined, no attention is paid to the chemical nature of the compounds that contain the element of interest. A very large fraction of the available data describing the presence of inorganic and organometallic compounds in environmental samples relates to total element concentrations. The detection limits of these methods have reached for many elements levels approaching or exceeding natural background. Although analytical methods used in environmental work should have detection limits one to two orders of magnitude below natural background concentrations, further efforts to increase sensitivity will benefit environmental chemistry only marginally [XXXII].

The commercial availability of instruments for the simultaneous determination of many elements at microgram per liter levels allows the analytical chemist to provide the environmental scientist with a wealth of data that otherwise could not be obtained because of time and personnel limitations. These multi-element techniques, the most common of which are inductively coupled plasma emission spectrometry [XXXV], DC-plasma emission spectroscopy, plasma atomic fluorescence spectrometry, and inductively coupled plasma - mass spectrometry [XXXII], extract much more information from limited samples than any other technique perhaps with exception of neutron activation analysis, which, however, requires a nuclear reactor [XXXIII].

The determination of total trace element conentrations is a first step toward complete characterization of an environmental sample. The next step is speciation, the determination of the various chemical compounds that are responsible for the inimical and beneficial effects of inorganic and organometallic compounds. Various methods are available for speciation [XIX-XXI, XXXII-XXXIV]. Volatile compounds can be determined by gas chromatography with element- or compound-specific detectors such as mass spectrometers, and atomic absorption or emission spectrometers [XXXIV]. Non-volatile compounds can be separated by high pressure liquid chromatography or ion chromatography and detected by graphite furnace atomic spectrometers or plasma emission spectrometers [XXXII]. The systems with element-specific detectors are not as widely used as they should be. The availability of techniques for speciation make the description of environmental hazards and their regulations in terms of total element concentration unacceptable. The crude use of total element criteria will become unnecessary as appropriate research employing techniques for speciation is performed. Strenuous efforts to speciate and correlate reactivities of environmentally active metal and metalloid compounds are urgently needed.

Separation and Analysis Techniques

New techniques to separate very complex mixtures, hopefully without changing the chemical structure of components, and new methods to determine the composition of high molecular mass compounds of low volatility are also urgently needed. Although liquid chromatography with element-specific detection systems can establish the presence of a series of compounds, it cannot identify them without standards. These systems require advance knowledge of the species being investigated. Some of the new methods of ionization and separation based on mass spectrometric principles and chromatography deserve special attention. For example, fast atom bombardment mass spectrometry (FAB-MS) may provide important information about the composition of metal complexes that are not volatile enough for conventional electron impact mass spectrometry. MIT chemists have recently shown that nonvolatile complexes such as chelates of diethylenetriaminepentaacetic acid can be vaporized intact and ionized producing informative mass spectra. New mass spectrometers with mass ranges to 5000 Daltons make many metal complexes accessible to mass spectroscopic study. The examination of metallic and metalloidal compounds present in plants and animals by FAB-MS should provide valuable information about the structures of such compounds.

Ligands and metal complexes may be examined by "hyphenated" techniques such as gas chromatography-mass spectrometry-mass spectrometry (GC-MS-MS) and liquid chromatography-mass spectrometry-mass spectrometry (LC-MS-MS). Mass spectrometry followed by collisionally-induced dissociation to form secondary daughter ions holds promise for screening very complex environmental samples for metal complexes. The importance of gas and liquid chromatography coupled with mass spectrometry in the analysis, management, and cleanup of hazardous situations cannot be over-emphasized.

Labile Metal Complexes and Speciation

Information about the chemical nature of metal complexes in the environment, pertinent accurate stability constants, and thermodynamic functions are needed for very complex systems. Many of these systems contain labile complexes. It is desirable to know the structures of such labile complexes and their concentrations. The measurement process must not change the system. Many indirect techniques have been used with success in simple systems. The pH titration method is a good example. The hydrogen ion concentration of a sample is determined over a wide range of conditions and is then related to the presumed equilibria by adjusting equilibrium constants until the experimental results are reproduced within experimental error. In many systems of

Analytical Aspects

environmental interest indirect methods do not adequately prove the existence of the presumed equilibria.

In certain cases of sufficiently high concentration of an element with appropriate nuclear properties, nuclear magnetic resonance spectroscopy may be able to identify complexes and provide data leading to equilibrium constants [XI]. Unfortunately, the sensitivity of nuclear magnetic resonance spectroscopy is usually inadequate for the investigation of metal complexes at environmental concentrations. Paramagnetic or rapidly reacting systems may not be amenable to analysis by this technique.

Resonance Raman spectrometry is a promising technique which already can detect compounds at the mg/L level. The application of CW lasers, pulsed lasers, and Fourier Transform signal averaging might make this technique even more sensitive.

Many other methods available for the identification of labile complexes, such as UV-visible spectrophotometry, are not applicable to the low concentrations predominant in the environment. Separation and enrichment techniques are expected to change the equilibria, and the results obtained with enriched samples are not representative of the original material and are misleading at best.

It should be realized that the constants used to calculate the distribution of a metal at low concentrations among several complexes in thermodynamic equilibrium need not be determined at these low concentrations. If only the concentrations of the ligands are known, and the medium and the temperature are defined, the constants are independent of the actual metal concentration. The real difficulty with tracer concentrations of metals is the possibility that the metal is sequestered as an inert complex not in thermodynamic equilibrium, for instance, with chloride ions. Mercury in seawater presents such a case. Part of the mercury is present as kinetically inert methylmercury. Constants for the Hg(II)-chloride system at tracer concentrations of mercury are of little help in this situation. Techniques must be found to distinguish between labile and inert mercury complexes.

It is of course always satisfying when some calculated quantity can be independently checked. This would be possible for the concentration of free mercury(II) ions in seawater, provided that the mercury electrode would respond properly to a total equilibrium concentration of labile complexes of 10^{-11} M. It is doubtful that electrodes ever can be made specific at such low concentrations.

Much lower concentrations of free mercury ion can be easily determined at higher concentrations of labile mercury complexes. In a 10^{-3} M solution of labile mercury complexes containing 0.5 moles of chloride per liter practically all mercury is present as $HgCl_4^{2-}$. With a formation constant of 10^{30} M^{-4}, the concentration of free mercury ions is 10^{-31} M. In such cases, the exchange reaction at the electrode evidently does not involve Hg^{2+} at all. The potential is established via the complexes present. Free mercury ion is a thermodynamic fiction corresponding to one ion in 2×10^7 liters. Because a labile equilibrium exists, this quantity could nevertheless be considered a certain probability rather than an actual concentration.

A great need exists for techniques that can identify labile metal complexes and determine their very low concentrations in complex environmental systems and do not influence the equilibria, steady state conditions, or dynamic situation in the sample.

Surface Analysis

Much of environmental inorganic chemistry occurs on surfaces, and at interfaces between solutions and particulate matter [VII]. Without a clear understanding about the nature of species on surfaces and detailed knowledge about the reactions occurring at interfaces, the development of environmental inorganic chemistry will remain incomplete.

During the past fifteen years many instruments for the analysis of surfaces were developed and commercialized. The various electron spectroscopic techniques permit the study of the structures and distributions of molecular species or fragments on surfaces. Unfortunately, all these techniques require, that the sample be maintained in an ultrahigh vacuum. Substances that volatilize or decompose are not easily analyzed because of elevated background signals. Such instrumental restrictions prevent the application of surface analytical techniques to most problems of environmental interest. It is very important for environmental chemists to have techniques for the determination of surface structure and local composition.

Among the many specific problems, for which such techniques are needed, are the study of equilibria in heterogeneous systems, and the investigation of the interactions of dissolved substances with biological structures such as cell surfaces and membranes. Environmental chemistry is largely the study of heterogeneous systems, because nature is heterogeneous on a macroscopic and microscopic scale. The development of instruments and techniques for the characterization of surfaces and the elucidation of reactions proceeding at interfaces without altering the system is of highest priority. Perhaps nuclear magnetic resonance spectroscopy could be used in such studies. Small angle X-ray techniques require relatively high concentrations, but it should be possible to increase the sensitivity of this method.

Sampling

The efforts to develop analytical techniques with high sensitivity and selectivity have been rather successful. Constant striving for even more sensitive analytical methods and better instrumentation is a laudable endeavor. However, the proper collection and preservation of samples that are truly representative of the in-situ chemistry of the location, from which they were taken, generally attracts little attention. It appears that the available analytical capabilities have already outstripped our ability to obtain truly representative samples from many environments. A sample, the integrity of which was compromised during collection or storage, is worthless and may provide misleading information, even though it was analyzed by the most elegant, precise, and accurate technique. Many of the serious pollution problems will be solved only by the judicious marriage of sophisticated laboratory investigations to environmental sampling techniques that provide a true picture of the in-situ composition of the natural system.

The proper methodology to be used is determined by the environmental system to be sampled. Sampling of the atmosphere, which contains gases and particles, is almost always associated with preconcentration processes that might introduce artifacts and alter the collected material. Samples taken for gas and aerosol analysis are specially prone to alteration. Denuders are recommended to avoid such interferences. Particulate matter should be collected under constant flow conditions [XXII, XXIII].

The sampling methodology for the aquatic environment is well developed. Any investigation of chemical parameters has to take into account the hydrodynamics of the sampling site, a requirement of special significance for sites, at which the concentrations of chemical species change with time. For every sample the time and date of collection, the meteorological conditions, the location of the site, the depth at which the sample was taken, detailed information about the equipment and procedure used for sampling, and the measures employed for sample preservation must be recorded [VI].

When sediments are sampled, alteration of the sample and mixing with other materials must be avoided or at least minimized. Special care must be exercized when the surface of sediments is sampled. The collected samples must be stored to maintain the initial conditions.

The sampling methodology for living organisms must be based on the proper selection of the type of organism and assure that the sample will be representative. Organisms are to be selected consonant with the goal of the investigation, e.g., food organisms for human exposure studies, and bio-indicator organisms for general environmental work. When large organisms are investigated, the various tissues and organs must be analyzed, whereas a number of individuals of small organisms will provide representative results. Often, organisms of the same species and same size, caught at the same time and place, may contain significantly different metal concentrations [VI].

Stability of Trace Element Compounds

The desirability and necessity to identify and determine trace element compounds in samples taken for the purpose of characterizing the chemical composition of water, air, soil, or biota is well established. Just as samples collected for the determination of total trace element concentrations are suitably preserved to prevent adsorption of trace constituents on the walls of the container, to avoid losses through volatilization, and to counteract precipitation, samples collected to determine trace element compounds must be treated to prevent adsorption, losses, and inclusion in precipitates. In addition, such a treatment must assure that trace element compounds are not changed during the period between collection of the sample and analysis.

Analytical Aspects

Analysts are aware of the difficulties associated with the preservation of samples for total trace element analysis, and have at their disposal various methods to assure stability of their samples. Very little is known about the nature of trace element compounds in environmental samples and even less about methods for the successful preservation of these compounds. Many of the compounds suspected to be present in the environment are relatively unstable and their concentrations might be maintained by a steady state condition that might be disturbed in the samples collected for analysis. Unless such samples can be preserved to prevent reactions from occurring, even the most painstaking analysis will give erroneous results unusable for the evaluation of the state of the environmental compartment under investigation.

The determination of arsenite and arsenate may serve as an example. Because of the different toxicities of these two compounds their separate determination is often requested. No good preservation agent is yet available that would prevent the interconversion of these two arsenic compounds. The arsenite concentration was observed to increase initially and then decrease, or decrease monotonically, depending on the - in most cases unknown - other components of the sample.

It is therefore of vital importance that the concentration of already identified trace element compounds be studied as a function of time. Should changes occur, appropriate agents for preservation must be found. Such investigations must be conducted for each newly discovered trace element compound. Unless the integrity of a sample can be guaranteed, the time and money invested in an analysis is wasted.

Now is the time to collect, evaluate, and establish appropriate sampling and preservation methodologies for the investigation of natural systems and communicate recommended methodologies to scientists involved in environmental work. Although the United Nations Environmental Programme, Regional Seas Programme Activity Center, publishes a series "Reference Methods for Marine Pollution Studies", the international community of environmental inorganic chemists should consider establishing an international series of reports or a journal devoted to the dissemination of the best sampling methodologies and sponsoring "round-robin" interlaboratory comparisons of materials and methods.

Standards for Environmental Analyses

To establish that an analytical procedure gives reproducible results is generally easy. To ascertain that these reproducible results correspond to the true values is not easy. Standardization of instruments with materials of known composition, which differ from the matrix of the sample to be analyzed, does not assure that true values are obtained. To overcome the difficulties caused by matrix interferences with the analytical procedures, certified standards in various matrixes are needed. Such standards for total element concentrations are available commercially.

Environmental studies attempt now with increasing frequency to speciate, i.e., determine trace element compounds. For these endeavors, standards for trace element compounds are needed. Such standards are not yet available. The production of certified standards for compounds is complicated by the requirement, that not only the total element concentration but also the concentrations of the various compounds present must not change on storage.

In the absence of certified standards, "unofficial" standards should be developed, checked for stability, and made available to laboratories engaged in speciation work. The pooled experiences with such temporary standards might facilitate the development of certified standards.

Analytical procedures for the analysis of environmental samples should be standardized, revised as needed, and closely adhered to whenever possible. Under these circumstances data from different laboratories and different countries can be compared more meaningfully and the information generated is more useful.

Reporting of Analytical Results

Every number obtained through the application of an analytical technique to a sample should in the ideal case have associated with it an indication of its accuracy and precision. In many publications this vital information is not given. It is not admissible to use zeros or dashes when reporting results as being below the detection limit. The detection limit with the "less than" (<) sign should be used in such cases.

Analytical Aspects

The desirability of replicate analyses of the same sample is evident but often meets with practical difficulties such as time, cost, personnel constraints, and limited sample size. Replicate analysis of samples by the same method or instrumental technique is, however, not sufficient. The reliability of an analytical procedure should be checked by at least one other independent method, specially when the results are approaching detection limits. This protocol will help in uncovering systematic errors and perhaps will show ways to avoid them.

The concentrations of elements and compounds are now almost exclusively reported in the dimensionless units of ppm, ppb, and ppt. These units unless clearly defined are ambiguous, because mass per unit volume or mass per unit mass could be meant. An additional complication arises from the fact that a concentration expressed, for instance, in ppm selenium present as dimethyl selenide is different from that expressed in ppm dimethyl diselenide. When concentrations in homogeneous solutions are expressed in molarity or molality complications do not arise even if the molecular forms of the species are unknown. Although the presence of an element in a polymeric compound would make molarity or molality concentrations based on the element too high, such situations are not expected to be common in the dilute solutions encountered in the environment. The use of molarity or molality is recommended in preference to the "part per million" convention. Where concentrations of trace element compounds are reported, their chemical formulas must be specified.

Many scientists measure many different metal ion concentrations in a variety of matrices and attempt to interpret data intuitively usually on the basis of comparing concentrations by examining only two or three variables at a time. The more widespread use of statistical pattern recognition techniques such as factor analysis ought to greatly improve the information obtainable from collections of data. Computer-based programs are now available for the examination of associations, similarities, and differences in multidimensional space. It appears to be wise, to make fewer measurements more carefully and spend more effort critically examining and interpreting the data.

Quality control in analytical laboratories is of great importance and is discussed in many publications. Such quality control assuring accuracy and precision of results becomes even more important with the availability of computerized multi-element instruments, which produce data in such profusion, that it is often difficult to recognize incorrect analyses. A quality control program must be instituted and strictly adhered to. The measures taken to achieve accuracy and precision must be clearly described when data are reported. The use of standard samples matching as closely as possible the samples to be analyzed is very important.

Analytical methods for soils, sediments and the atmosphere: Most of the topics addressed in this section apply at least to some extent to all the major compartments of the environment. The most easily investigated systems are homogeneous phases, which are generally limited to aqueous and gaseous samples. The introduction of a solid as particulate matter into water or air immediately complicates analytical work. Soils and sediments are highly heterogeneous and require in most instances the application of separation procedures before analyses can be undertaken. Our knowledge about a system generally decreases with increasing inhomogeneity. Every effort must be made to learn more about these inhomogeneous systems, which are at least as important as the more easily handled atmosphere and aqueous solutions. The chemists' preference for well-defined, pure, homogeneous systems must be overcome for the sake of progress in the environmental inorganic chemistry of complex, inhomogeneous materials such as soils, sediments, and suspended particulates.

BIBLIOGRAPHY

Barnes, R. M., **"Development in Atomic Plasma Spectrochemical Analysis"**, Heydon & Son, Ltd.: London, 1981.

Ewing, G. W., Ed., **"Environmental Analysis"**, Academic Press, Inc.: New York, N.Y., 1977.

Gibb, T. R. P., Jr., Ed.; **"Analytical Methods in Oceanography"**, **Advances in Chemistry Series, No. 147,** American Chemical Society: Washington, D.C., 1975.

Jewett, K. L., Brinckman, F. E.; **"The Use of Element-Specific Detectors Coupled with High-Performance Liquid Chromatography"**, in "Detectors in Liquid Chromatography", Vickrey, T. M. Ed., Marcel Dekker: New York, N.Y., 1983, p.

205.

Kateman, G.; Pijpers, F. W.; "Quality Control in Analytical Chemistry", (Vol. 60 in Chemical Analysis, a Series of Monographs on Analytical Chemistry and its Applications), John Wiley & Sons: New York, N.Y., 1981.

Keith, L. H., Crummet, W., Deegan, J. Jr., Libby, R. A., Taylor, J. K., Wentler, G.; "Principles of Environmental Analysis", Anal. Chem., 55 (1983) 2210.

LaFleur, P. D., Ed.; "Accuracy in Trace Analysis: Sampling, Sample Handling, Analysis", Vol. 1 & 2; U.S. National Bureau of Standards Special Publication, No. 422, 1976.

Leppard, G. G., Ed.; "Trace Element Speciation in Surface Waters and Its Ecological Implications", Nato Conference Series. I: Ecology, Vol. VI, Plenum Press: New York, N.Y., 1983.

Luc Massart, D., Kaufman, L.; "The Interpretation of Analytical Chemical Data by the Use of Cluster Analysis", Vol. 65 in "Chemical Analysis", Elving, P.J.; Winefordner, J. D., Eds.; John Wiley & Sons: New York, N.Y., 1983.

Minear, R. A.; Keith, L. H.; "Water Analysis: Inorganic Species", Vol. 1, Academic Press: New York, N.Y., 1982.

Minczewski, J.; Chwastowska, J.; Dybczynski, R.; "Separation and Preconcentration Methods in Inorganic Trace Analysis", Halsted Press: New York, N.Y., 1982.

Mulick, J. D.; Sawicki, E., Ed.; "Ion Chromatographic Analysis of Environmental Pollutants", Vol. 2, Ann Arbor Science Publishers, Inc.: Ann Arbor, Michigan, 1979.

Nürnberg, H. W.; Valenta, P. "Potentialities and Applications of Voltammetry in Chemical Speciation of Trace Metals in the Sea", in "Trace Metals in Sea Water", Wong, C. S. Ed.; Plenum Press: New York, N.Y., 1983, p. 671.

Nürnberg, H. W. "Polarography and Voltammetry in Studies of Toxic Metals in Man and His Environment", Sci. Total Environ., 12 (1979) 35.

Pinta, M. "Detection and Determination of Trace Elements", Ann Arbor Science Publishers, Inc.: Ann Arbor, Michigan, 1975.

Pinta, M. **"Modern Methods for Trace Analysis"**, Ann Arbor Science Publishers, Inc.: Ann Arbor, Michigan, 1978.

Sawicki, E.; Mulick, J. D.; Wittgenstein, E. Eds.; **"Ion Chromatographic Analysis of Environmental Pollutants"**, Ann Arbor Science Publishers, Inc.: Ann Arbor, Michigan, 1978.

Schwedt, G.; **"Chromatographic Methods in Inorganic Analysis"**, Dr. Alfred Huthig Verlag: Heidelberg, 1981.

Vermuza, K.; **"Pattern Recognition in Chemistry"**, Vol. 21, Lecture Notes in Chemistry, Springer Verlag: Berlin, 1980.

Wong, C. S., Ed.; **"Trace Elements in Sea Water"**, **Nato Conference Series, IV: Marine Sciences, Vol. 9,** Plenum Press: New York, N.Y., 1983.

Zief, M.; Mitchell, J. W.; **"Contamination Control in Trace Element Analysis"**, (Vol. 47 in Chemical Analysis, a Series of Monographs on Analytical Chemistry and its Applications), John Wiley & Sons: New York, N.Y., 1976.

RECOMMENDATIONS FOR RESEARCH

Analytical Aspects of Environmental Inorganic Chemistry

Efforts to increase the sensitivity of total trace element determinations: It is now possible to determine many elements at 10^{-9} to 10^{-11} molar concentrations (low ppb levels) provided the matrix is not too complex. Many trace elements and trace element compounds are present in the environment at approximately these concentrations, which are close to the detection limits of many analytical techniques. To avoid cumbersome preconcentration procedures and achieve better precision and accuracy the detection limits of the analytical techniques should be improved by two orders of magnitude. Such an improvement would be specially beneficial for multi-element instruments.

Application of sensitive analytical techniques to complex matrices: Hardly any problems are encountered when samples of freshwater and similar solutions approaching distilled water in composition are analyzed. Insurmountable problems sometimes arise when concentrated brines, aqueous or organic extracts of biota, or other recalcitrant matrices are encountered. Presently, the only solution to these problems lies in separating the major solutes to allow the determination of minor constituents. Although various techniques exist for background correction and removal of spectral interferences, the analysis of trace constituents in a sample containing, for instance, 1 mole/L sodium chloride is not a routine task. Depending on the analytical technique employed, differences between the densities and viscosities of the standards and the sample might introduce serious errors. Matching the standards to the sample matrix is a possible solution. However, even the purest reagents of the major constituents of a sample might introduce larger amounts of trace elements than are present in the sample. Analytical techniques must be developed that permit the routine analysis of samples with complex matrices. It is imperative that detection limits are determined in the presence of the matrix. Detection limits obtained with distilled water solutions in the absence of the sample matrix are of little usefulness.

Development of speciation methods for organometallic compounds and nonlabile metal complexes: In principle, analytical techniques are available now for the determination of organometallic compounds and nonlabile metal complexes. Gas chromatographic (GC) separation of volatile compounds and liquid-chromatographic (LC) separation of nonvolatile compounds followed by element-specific detection, and electrochemical methods are the most promising techniques. Graphite furnace atomic absorption spectrometers (GFAA) and plasma emission spectrometers are examples of element-specific detectors. These techniques need to be applied more widely for speciation, and need to be refined and made applicable to samples with complex matrices.

Determination of labile metal complexes: Ideally, labile metal complexes should be identified and their concentrations determined by in-situ techniques that leave the sample undisturbed. This ideal has not been achieved yet, at least not for samples, in which the complexes are present at low environmental concentrations. In addition to the proven and reliable techniques of potentiometry, data relating to molecular properties can provide important information about the nature of a species in an environmental system. Such measurements are urgently needed to complement thermodynamic data that cannot provide information about composition and structure. In addition to spectrometric techniques, nuclear magnetic resonance and especially multinuclear magnetic resonance appears to be a promising tool. Current limits of sensitivity are now in the micromolar range for protons and the millimolar range for several metal ions. The development of more sensitive instruments and techniques is encouraged. Electrochemical techniques such as polarography and voltammetry are promising. Mercury in the form of Hg^{2+} can be determined with a Hg-sensitive electrode at 10^{-6} to 10^{-9} M. Ion-specific electrodes of similar sensitivity should be developed for other metals.

Development of mass spectrometric techniques for high molecular mass compounds: Several mass spectrometric (MS) techniques (Fourier Transform MS, fast ion bombardment MS and others) that are capable of examining rather involatile metallic and metalloidal compounds in the mass range of 5000 Dalton have recently become available. These techniques and chromatography-MS-MS must be explored for their usefulness to detect and identify metal complexes and organometallic compounds.

Investigations of kinetic aspects of environmental inorganic reactions: For a complete characterization and understanding of an environmental system, the kinetic parameters of the important reactions forming and degrading coordination compounds and organometallic compounds must be known. The determination of rate constants is much more complicated than speciation, which is a prerequisite for kinetic investigations. It is doubtful that kinetic data obtained at concentrations other than those found in the environment are of much use and reliability in assessing the chemical state of an environmental sample. Nevertheless, every effort should be made to accumulate kinetic information about environmentally important reactions. Such studies may not be possible with the available, strongly time-dependent analytical techniques (NMR with 10^{-3} s^{-1} spectrophotometry with 10^{-13} s^{-1} relaxation rates), but are of great importance to the assessment of the bioavailability of inorganic and organometallic species.

Development of in-situ surface analysis techniques: To learn more about the extremely important interactions between solutes in solution and the surfaces of suspended particulate matter, techniques for the characterization of surfaces without altering the easily changed particles must be available. Unfortunately, present instruments for surface analysis keep the sample in an ultrahigh vacuum that assures, that particles removed from water and placed in such a vacuum will be altered drastically. A breakthrough in instrument design must come before much progress can be made in this area. Perhaps methods based on spectrometry with variable wavelength lasers and epifluorescence might be useful.

Sampling and sample preservation: Much more attention must be paid to proper methodology that assures, that representative samples are taken from the atmosphere, the soil, water bodies, sediments, and biota. In addition, studies must be undertaken to identify the sampling methods that introduce the least amounts of contaminants. Of special importance to speciation is the ability to preserve the compounds in the sample in their original state until the analyses are completed. The task of finding suitable preservation methods, which will not be universally applicable but will very likely be compound-specific or at best suitable for a group of similar compounds, is at least as important as the development of analytical techniques for speciation. Much more work needs to be carried out in this area.

Development of standards for environmental analysis: Without widely available standards, preferably certified standards, the calibration of analytical techniques cannot be done properly. Many more standards are needed for "total trace element" techniques. With the obvious necessity for speciation, standards for trace element compounds in environmentally relevant matrices must also be developed.

Development of analytical methods for the atmospheric environment: Many of the preceding recommendations for research in analytical environmental inorganic chemistry apply equally to the water, soil, sediment, and atmospheric environment. However, several topics of special concern to atmospheric scientists are in need of further development. These topics are the measurement of the chemical composition of cloud water and of interstitial concentrations of gases and particles such as sulfur dioxide, sulfates, nitrogen oxides, nitric acid, hydrogen chloride, ammonia, hydrogen peroxide, and various aerosols; the aqueous phase chemistry of nitrogen oxides and nitric acid including the chemistry of nitrogen trioxide and dinitrogen pentoxide; development of improved methods for measuring free radicals; the formation and measurement of sulfur- and nitrogen-containing acids in atmospheric aerosols; and the determination of the contribution of naturally-produced compounds to acid rain.

I

INORGANIC CHEMISTRY AND THE ENVIRONMENT

Ugo Croatto

Istituto di Chimica Generale
Università di Padova
I-35100 Padova, Italy

The important environmental problems facing organisms on Earth are outlined and the necessity of better management of the environment is stressed. Examples of specific problems are given in the areas of energy production, industrial processes, extraction of metals, and agricultural activities. The role that inorganic chemistry can and must play in making and keeping Earth more habitable is emphasized.

In the past few decades our planet has experienced severe crises. Ores and nonrenewable energy sources such as oil, natural gas, and uranium have been consumed more rapidly than new reserves have been discovered. Our natural resources are being scattered over the planet increasing their entropy and decreasing their usefulness. This dilution of resources is specially noticable with metals. Water has become scarce for many urban, farming and industrial uses. Soils have been losing their productivity. The human population has grown exponentially and now doubles every 33 years. Food is not produced in sufficient quantities to provide even a minimal diet for millions of people. The environment is becoming increasingly polluted by the products of human activities and by wastes including radioactive materials and many biologically active pollutants. Our cultural heritage is threatened. Damage to buildings and to works of art is extensive. Even a cursory inspection of classical cities such as Rome and Venice plainly show the ravages of atmospheric pollution.

The environment must clearly be managed in a manner different from the past. The production and dispersion of wastes must be minimized. Materials must be recycled and reused as much as possible. Energy consumption must be reduced and waste-energy profitably employed. The pollution of air, water and soils must be minimized. Water must be used frugally and reused whenever possible. Depleted soils and barren soils must be reclaimed for agricultural production. The consumption of synthetic fertilizers should be reduced. Aquaculture must be developed. Because environmental issues are hardly ever isolated, a systems approach must be taken that evaluates the past, assesses the present, and predicts the future. The records of past conditions are preserved in materials such as sediments.

A small beginning has already been made in protecting the environment. Industrial processes are now being examined with the goal of reducing energy consumption, minimizing waste production, and thereby increasing profits. Biomass such as sludges from water purification, manure, and urban wastes is converted to biogas in a process that requires little energy. Humus is now prepared by aerobic fermentation of biomass properly balanced with respect to carbon, nitrogen, and phosphorus content. Soils unsuitable for farming need humus, macroelements, microelements, water-absorbing substances, root hormones, and sufficient populations of bacteria. The pH of soils should be maintained around seven, and additives are needed to improve chemico-physical properties and to increase productivity. Even soils in agricultural production require supplements, because monocultivation without crop rotation, heavy use of chemical fertilizers, and a drastic decrease in the use of animal manure have reduced the humus content and the chemical and microbiological qualities of these soils. Synthetic humus and industrial waste products such as calcium sulfate, iron sulfate, and loose particles from marble works that would have to be disposed can and should be used to improve soils and control their pH.

When materials must be dumped, waste disposal sites must be constructed that are capable of preventing contamination of the environment by leakage. Waste dumps should be managed to allow recovery of the wastes in the future, when our primary resources will have been depleted. A waste dump should be viewed as a future mine.

The energy demand should be met by increasing the quantities of renewable energy obtainable from biomass, solar radiation, wind, and geothermal fields. The waste heat from electric power generation stations using fossil or fissile fuels and from industrial steam plants can be employed to heat greenhouses and increase the productivity of aquaculture by raising fish, molluscs, and crustaceans in heated water basins. The energy in waste heat is almost twice the energy produced in the form of electricity.

Chemical pollutants belong to one of two classes. The first class consists of compounds of nitrogen and phosphorus that are essential for plants. Excessive amounts of these essential elements cause eutrophication of water bodies with immediate disastrous consequences. Eutrophication can be avoided by preventing excessive amounts of nutrients from reaching the aqueous environment. The second class of chemical pollutants comprises toxic substances such as poisons, mutagens, carcinogens, and teratogens. When toxicity is caused by a specific compound, conversion of this compound to a nontoxic substance eliminates the hazard. For example, the very toxic cyanide can be oxidized to the much less toxic cyanate. When the toxic effects are caused by metal ions, formation of organometallic derivatives, precipitation of insoluble salts, complexation with appropriate ligands, or change in oxidation states may be used for detoxification. Conversion of inorganic arsenic to methylarsonic acid, precipitation of barium as barium sulfate, combination of lead with ethylenediaminetetraacetic acid, and reduction of chromate to chromium(III) are examples of reactions producing substances with a toxicity lower than that of the starting materials. The dispersion of toxic chemicals all over Earth is a very serious problem that increases with time and affects the entire biosphere.

Counteracting this dispersion of chemical substances are natural anti-entropic systems that concentrate some elements by extraction from very dilute solutions. Organisms at various levels in the food chains sometimes contain very high concentrations of toxic and nontoxic elements. For example, the mussel *Mytilus galloprovincialis* filters in warm months 20 liters of water every hour and extracts remarkable quantities of dissolved elements from the sea. Plants in symbiosis with fungi can extract needed minerals from the soil from a circular area with a radius of 700 meters. Organisms may preferentially absorb vanadium, select strontium over calcium, distinguish sodium from potassium, or prefer calcium over magnesium.

Radioactive elements - and particularly alpha emitters - create radiation hazards and - in some cases - exert chemical toxicity. Long residence times of radioactive materials in organisms increase radiotoxicity. For instance, plutonium has a biological half-life of a few thousand years. An organism containing plutonium would have to live several thousand years to cut its plutonium burden in half. During a human lifespan plutonium is concentrated from the environment without the possibility of substantial elimination.

Physicochemical systems may also be very effective in concentrating substances. Surface films on water may cause nonuniform partitioning of dissolved materials between the bulk water and the film via microflotation. Tensioactive surfactants, which are now ubiquitous in the environment, are responsible for microflotation. Plutonium may, for instance, be present in surface films at concentrations 10,000-times higher than in the bulk water. The action of coast-bound winds and waves may bring the plutonium-rich surface films as sprays to the land. Manganese nodules are another example of a system that is capable of extracting metal ions from very dilute solutions. Soils can fix and accumulate toxic metal ions from water and other materials used in farming through ion exchange reactions with humus. The toxic metals are then transferred to plants and thus enter the food chain. Proper environmental management must limit pollution of soils by balancing the addition and removal of pollutants. To check the conditions of soils and organisms, the fluxes of pollutants through the system must be monitored.

Environmental inorganic chemistry should have a prominent role in the monitoring and managing of pollution by chemical plants, agricultural activities, and urban centers, and must provide chemical processes suitable for the protection of the environment from a variety of pollutants in air, water, and soils. Chemical pollutants in water are chromium, cobalt, nickel, copper, silver, barium, zinc, cadmium, mercury, arsenic, selenium, nitrogen, phosphorus, radioactive substances, and surfactants. Pollutants of soils are metals, industrial by-products such as sludges containing calcium carbonate, calcium sulfate, or iron sulfate, glass, and organic residues.

Of particular concern to inorganic chemistry should be processes for the recovery of metals from waste and very dilute solutions. For example, the fabrication of an item from scrap aluminum consumes only two percent of the energy required for making this item from aluminum extracted from bauxite. The recovery of expensive materials in a very dispersed state will become increasingly important, and the winning of metals from hydrometallurgical solutions must receive increased attention. Uranium is present in seawater at an average concentration of 3.4 µg/L. One cubic kilometer of seawater thus contains approximately three tons of uranium. It might be possible to extract this uranium from seawater and prevent the monopolization of an important source of nuclear energy. The extraction process must be economically viable and must consume less energy than a nuclear reactor can produce from the extracted uranium.

Inorganic chemistry has a very important role in the management of the environment. Particularly rewarding activities for inorganic chemists associated with environmental issues are to be found in the area of energy production from coal, petroleum, oil shale, and uranium, in the manufacture of inorganic pigments, electroplating, tannery operations, inorganic solvents and catalysts, and in the alkali industry, metal-working plants, sugar refineries, and synthetic fertilizer plants. Inorganic chemists might be able to find a way to halt the steadily progressing deterioration of architectural art works and sculptures. Inorganic analytical chemistry must have a major role in monitoring the state of the environment and in the development of sensitive and element- and compound-specific methods for the identification and quantitative determination of trace elements and trace element compounds at very low levels. Appropriate techniques include gas chromatography, mass spectrometry, radiochemical methods, solvent extraction, resin extraction, membrane extraction with special emphasis on supported liquid membranes, and many others. Thus, work on environmental issues by chemists with inorganic background is needed to help make our planet more habitable and to assure the availability of the material resources required to keep Earth a healthy place for all of its inhabitants.

DISCUSSION

K. N. Raymond (University of California, Berkeley): I believe, that the Japanese have begun trial studies of the industrial feasibility of Professor I. Tabushi's (Chemistry Dept., University of Kyoto) process for removing uranium from seawater. Extractors containing compounds with high affinity to uranium that are immobilized on a matrix are lowered into the sea. The strong currents off the eastern shore of Japan would assure the required flow of water through the extractors.

II

INORGANIC CHEMISTRY OF THE OCEAN

Sten Ahrland

Inorganic Chemistry 1, Chemical Center
University of Lund
S-220 07 Lund, Sweden

This survey presents some important aspects of the inorganic chemistry of "standard seawater", a solution of a composition characteristic of the bulk of the ocean. This water has a pH of 8.1 and a high oxidation potential. The source of the high oxidation potential is the oxygen of the air. The inert oxygen/water coupled cannot directly control the potential. The approaches tried to determine the actual potential and the responsible redox couples are presented. The pH is remarkably constant but rather far from the pH at which the carbonate system - the only protolytic system of appreciable concentration in seawater - has its maximal buffer capacity. To keep the pH constant, protons must be released on a large scale. The nature of potential proton-releasing reactions are discussed. The species distributions in seawater for alkali metal ions, alkaline earth ions, cobalt, nickel, copper, zinc, iron, manganese, cadmium, lead, mercury, uranium, and selenium are presented based on equilibrium calculations using the best available values for stability constants and solubility products. Several of these elements are not only involved in rapidly established equilibria but also in slow precipitation, complex formation and redox reactions.

Constituents of Seawater

Seawater is essentially a fairly concentrated salt solution. In most parts of the ocean the water is mixed very efficiently and the concentrations of the main constitutents are therefore quite uniform (Table 1). The ionic strength is fairly constant with Na^+, Mg^{2+}, Ca^{2+}, K^+,

Cl^- and SO_4^{2-} as the predominant ions [1]. Other ions contribute very little to the ionic medium. The minor ions are, however, of importance in other respects. For instance, the concentrations of bromide and carbonate, present as HCO_3^- and CO_3^{2-}, and perhaps of borate and fluoride (Table 1) are high enough to influence the chemical nature of metal ion species.

Table 1. Major and Some Minor Constituents of Seawater

Concentration (mM) of Main Constituents		Concentration (nM) of Trace Elements	
Na^+	479.	Mn	4.
Mg^{2+}	54.5	Fe	8.
Ca^{2+}	10.5	Co	0.1
K^+	10.4	Ni	5.
Cl^-	559.	Cu	4.
SO_4^{2-}	28.9	Zn	5.
HCO_3^-, CO_3^{2-}	2.35	Cd	0.1
Br^-	0.86	Hg	0.02
borate	0.4	Pb	0.05
F^-	0.075	U	14.
		Se	1.

Many of the metals and non-metals present at nanomolar or lower concentrations are important physiologically and might become sources of raw materials. Even a concentration as low as one nanomolar represents a large amount of approximately 10^{12} moles in a total volume of seawater of 10^{18} m^3. The problems associated with the exact determination of the concentrations of trace elements [2-4] and the enrichment of these elements from very dilute solutions are still formidable. Until recently, the concentrations of many trace elements in seawater were not known exactly. However, with improved analytical techniques a reasonable concensus about the true concentrations of most elements has now been reached. The values listed in Table 1 should be considered to be averages because the concentrations of trace elements are not as uniform as those of the main constituents. Even in the open sea, the concentrations of trace elements may vary five-fold [5]. In closed basins such as the Baltic Sea, concentrations significantly higher than those in the open ocean are often encountered [6].

Inorganic Chemistry of the Ocean   67

The formal ionic strength of seawater established by the concentrations of the main constitutents is 0.714 M corresponding to a salinity of 35‰. Because of excessive evaporation and restricted mixing with open-ocean water, the salinity may be markedly higher in some parts of the sea such as the Mediterranian, where the salinity reaches 40‰. Much lower salinities are found in regions where large quantities of freshwater are discharged and evaporation is relatively low. Such bodies of brackish water are found in estuaries and especially in basins that have very restricted communication with the open sea and receive freshwater from many rivers. The conditions in brackish waters are rather different from those prevailing in the open ocean.

pH of Seawater

The pH of ocean water varies with depth, temperature, and location [7, 8]. Under normal conditions, the pH is never far from 8. A value of 8.1 will be used as representative for the surface waters of the major parts of the ocean. With pK_w of 13.8 at the ionic strength of seawater [9] $p[OH^-]$ is 5.7. This value of $p[OH^-]$ is not far from one of the minima of the buffer capacity of the carbonate system, the only acid/base system of appreciable concentration in seawater. To account for the stable pH of the ocean water, other processes releasing protons on a huge scale must occur [10]. The first hypothesis for the release of protons postulated exchange of ions between seawater and the silicates transported into the ocean by rivers. In freshwater, silicates are relatively rich in protons that are exchanged for Na^+ and Mg^{2+} on contact with seawater [10, 11]. However, closer examination of this process showed, that such ion-exchange can certainly not provide protons at the required rate.

Recently a new source of protons was discovered. The many hot springs along the submarine ridges, where new crust is created from ascending magma, release protons through hydrothermal processes on a scale that might be large enough [12]. These protons come from water that penetrates the rifts and reacts with hot magma. Several processes are thought to contribute to the release of protons. Certainly very important are hydrolytic reactions involving basaltic silicates. Also, hydrogen sulfide formed by reduction of sulfate releases protons when metal sulfides are precipitated.

At the ionic strength of seawater, which is largely determined by salts such as sodium chloride presumed to be completely dissociated, the protonation constant for the carbonate ion (eqn. 1) is $10^{9.54}$ M^{-1} at 25° [1, 13]. The

$$K = \frac{[HCO_3^-]}{[H^+][CO_3^{2-}]} \tag{1}$$

conditional constant $K_c = 10^{8.95}$ M^{-1} is applicable to standard seawater under the assumption that the concentrations of carbonate and bicarbonate ion in the equilibrium expression represent the total concentrations of these ions in solution (free ions and ions bound in soluble complexes). If the difference is assumed to be caused by formation of complexes with Mg^{2+} and Ca^{2+}, the stabilities of these complexes may be estimated. Provided that only complexes with one ligand are formed at these low concentrations and that the magnesium and calcium complexes have the same stability, the stability constants, β, assume values of 45 M^{-1} for MCO_3 and 3.9 M^{-1} for $M(HCO_3)^+$. Certainly, β[$M(HCO_3)^+$], is much smaller than β[MCO_3]. The value for β[MCO_3] (45 M^{-1}) is somewhat higher than the value of 33 M^{-1} suggested for $MgCO_3$ in standard seawater [13].

In seawater at pH 8.1 with $K_c = 10^{8.95}$ M^{-1} the ratio of bicarbonate to carbonate is 7.1. At a total carbonate concentration of 2.35 mM, the carbonate concentration, i.e., free CO_3^{2-}, carbonate complexes, and $MgCO_3$ and $CaCO_3$ in solution, is 0.29 mM. With the reasonable values of 50 for β[MCO_3] and 1 for β[$M(HCO_3)^+$], the concentrations of free carbonate and bicarbonate ions are calculated to be 0.068 mM and 1.92 mM, respectively. The concentration of carbonate and bicarbonate bound in complexes is then 0.36 mM or fifteen percent of the total carbonate, a value similar to the twelve percent obtained by other calculations [1, 14]. These values refer to seawater at the surface at 25°. K_c varies considerably with temperature [5]. Therefore, the conditions are fairly different in deep water.

The calculated concentration of free CO_3^{2-} is close to that needed for precipitation of calcite or aragonite at the concentration of Ca^{2+} present in seawater. Also, the concentration of Mg^{2+} is high enough to make the precipitation of dolomite, $CaMg(CO_3)_2$, thermodynamically possible. Generally this reaction does not occur [15].

Oxidation Potential of Seawater

With exception of certain basins with very poor circulation and mixing, the oxidation potential of seawater is high. The exact value of this potential and the redox system determining it are matters of dispute [16, 17].

The high oxidation potential could be attributed to the oxygen/water system (eqn. 2), which has a standard oxidation potential (E^o) of 1229 mV [18]. Because this system is not

$$1/2\ O_2 + 2H^+ + 2e^- \rightleftharpoons H_2O \qquad (2)$$

reversible, equilibria are generally not established with other redox systems. The potential, E, and the corresponding pE value for the oxygen/water system are given by equations 3 and 4. At atmospheric oxygen pressure,

$$E = E^0 + \frac{RT}{2F} \ln p(O_2)^{0.5}[H^+] \qquad (3)$$

$$pE = E\left[\frac{RT}{F} \ln 10\right]^{-1} \qquad (4)$$

$p(O_2)$, of 0.21 atm and a pH of 8.1 the potential is 740 mV and pE has a value of 12.5. It is often convenient to use pE, a measure of theoretical oxidative power, in equilibrium calculations [9]. Attempts to measure the oxidation potential of seawater by means of the iodate/iodide redox system [17] gave a pE value of 10.6, which is markedly lower than the calculated value of 12.5. Measurements involving the nitrate/nitrogen system yielded 10.5. Other approaches resulted in much lower values of pE [17].

In areas with strong upwelling the oxygen equilibrium between air and water is so rapidly established, that the concentration of oxygen stays practically constant with depth. Generally, however, the concentration decreases to a minimum at approximately 500 m depth and then increases again in deeper waters, in which oxygen consumption by organic materials is minimal. The concentration at the minimum is often less than one tenth of the surface concentration [19]. Such a reduction of the oxygen concentration has little influence on the oxidative power at equilibrium. The value of pE decreases only to 12.0 even for the drastic reduction of oxygen to one percent of the surface concentration. However, the near depletion of the oxidized component of this redox system severely decreases the redox buffering capacity of the solution. Consequently, further consumption of oxygen might rather easily cause a drastic decrease in the oxidation potential. This has indeed happened in the anoxic zones in the Baltic Sea, the Black Sea, and in the Scandinavian and British-Columbian Fjords. In these waters, impeded circulation and a fairly high rate of oxygen consumption resulted in a complete depletion of oxygen. Under such extreme conditions, the oxidation potential may be determined by the sulfide system (eqn. 5) established by the activity of sulfate-reducing

$$S(s) + 2H^+ + 2e^- \rightleftharpoons H_2S(aq) \qquad (5)$$

bacteria. The standard reduction potential of the sulfide system is 141 mV [18]. In anoxic waters, the sulfide concentration may be 0.5 mM and the pH may drop to 7.5 [20]. Assuming a value of $10^{6.9}$ M^{-1} for the protonation constant of the HS^- ion, pE could reach the very low value of -3.1 corresponding to an oxidation potential of -183 mV.

Factors Determining the Nature of Metal Complexes in Seawater

The chemical forms of a metal ion in seawater are largely determined by the concentrations of anions, other complexing ligands, pH, and the oxidation potential. Temperature and pressure will also influence the equilibria in a manner that is often not well known. Particularly changes induced by the high pressures encountered in the deep parts of the ocean are ill-defined. The ocean floor is on the average at a depth of 3800 m where the pressure is 37,000 kPa (370 atm).

Seawater contains many complexing agents in very low concentrations which vary considerably with time and location. These agents are often of organic origin formed by life- or decay-processes occurring in the ocean or in waters discharged into the ocean. Among the complexing agents arriving with river water are humic and fulvic acids, and man-made chelating agents such as nitrilotriacetic acid. In spite of their low concentrations, some of these agents may be important because of their high affinity for certain metal ions. If the metal ions are present only in trace amounts, they may become - at least locally - completely sequestered. A considerable fraction of certain trace metal ions may be incorporated into various organisms. Metal ions essential for life are especially likely to be taken up from very dilute solutions by highly efficient mechanisms.

Affinities Between Different Classes of Metal Ions and Ligands

Metal ions have preferences for certain ligands. The rules governing these affinities were deduced from a large number of stability constants determined in aqueous solutions. Metal ions are divided into two classes: hard metal ions and soft metal ions [21, 22]. Hard metal ions strongly prefer ligands with F, O, or N as donor atoms. Soft metal ions prefer ligands with heavier donor atoms from the third or a higher period in the periodic chart. The donors preferred by the hard (soft) metal ions are called hard (soft) donors. A hard acceptor has the same affinity sequence toward all groups of donor atoms, whereas the sequences of the soft acceptors differ among the various groups (Table 2). A very important common feature of these sequences are the large affinity differences between donor atoms from the second and third period, though in the opposite sense for hard and soft acceptors.

Table 2. Characteristic Affinity Sequences for Hard and Soft Acceptors toward Ligands with Atoms from the Second and Fifth Period

Donor Atom from Main Group	Oxidation State	Affinity Sequences for	
		Hard Acceptor	Soft Acceptor
7	- I	F>>Cl>Br<I	F<<Cl<Br<I
6	- II	O>>S>Se>Te	O<<S<Se Te
5	- III	N>>P>As>Sb	N<<P>As>Sb

The sequences for hard acceptors are those expected for the formation of complexes via electrostatic interaction. In concordance with this idea, the complexes formed by hard acceptors are invariably stronger the higher the charge and the smaller the radius of the acceptor ion. The most typical hard acceptors are therefore multivalent, small cations. The sequences for soft acceptors are not compatible with mainly electrostatic interactions. The bonds must have considerable covalent character. This conclusion is corroborated by the observation, that complexes of soft acceptors are generally stronger the larger the radius of the acceptor. Strong complexes are also formed by acceptors of low or even zero charge. In most cases, the softness of an element increases with decreasing oxidation state.

Species of Some Important Metals in the Ocean

Main constituents. The alkali metals are certainly almost exclusively present in seawater as the hydrated ions. The alkaline earth ions Ca^{2+} and Mg^{2+} prefer as typical hard acceptors the divalent oxygen donors CO_3^{2-} and SO_4^{2-} to Cl^-. The carbonate complexes are not very stable, and the sulfate complexes have even lower stability. The sulfate complexes predominate, however, because of the higher concentration of sulfate (Table 1) [9]. According to the most reliable estimates, more than 90 percent of magnesium and calcium are present as hydrated cations, eight percent as MSO_4, and less than 0.2 percent as MCO_3 and $MHCO_3^+$ [1, 15]. The affinity of calcium and magnesium for fluoride, a hard anion with a very high charge density, is about the same as their affinity for the dinegative sulfate anion. The concentration of the fluoride complexes, however, is negligible because of the low concentration (0.075 mM) of fluoride in seawater.

The first-row transition elements cobalt, nickel, copper, and zinc. Cobalt, nickel, copper, and zinc are present in standard seawater exclusively as the divalent cations, which are fairly hard acceptors with little affinity for chloride or bromide [9]. Their sulfate complexes are approximately of the same modest stability as those of magnesium and calcium. The carbonate complexes of these transition elements are, however, quite stable. Reliable constants have only been determined for the Cu(II) system [23] at zero ionic strength. Constants pertaining to a seawater medium can be obtained from the zero-ionic-strength-constants. On the assumption that the activity coefficients for divalent ions in seawater are between 0.20 and 0.25, 1.3 log units must be subtracted from the zero-ionic-strength-constants. For the carbonate complexes of cobalt, nickel, and zinc; reasonable constants can be estimated on the basis of the Irving-Williams order [24] of the stability of complexes formed by the first-row transition elements with certain ligands. The stabilities increase from left to right, reach a maximum at copper(II), and then decrease to zinc(II). Electrostatic and covalent bonding forces cooperate to establish this order [25]. The carbonate complexes of Co^{2+},

Inorganic Chemistry of the Ocean

Ni^{2+}, and Zn^{2+} should therefore be less stable than those of Cu^{2+}. Stability constants for these complexes are listed in Table 3.

Table 3. *The Distribution of Cobalt, Nickel, Copper, and Zinc Species in Standard Seawater**

Element	Log Stability Constants**		$\alpha(\%) = \frac{[\text{Species}]}{\Sigma[\text{Species}]} \times 100$				
	$M(OH)^+$	MCO_3	M^{2+}	$M(OH)^+$	MCO_3	MCl^+	MSO_4
Co	3.7	4.0	42	0.4	28	24	5.0
Ni	3.5	4.0	42	0.8	28	24	5.0
Cu	6.0	5.4	5	9.0	77	3	0.5
Zn	4.4	4.0	41	2.0	28	24	5.0

*p[L](-log free ligand concentration) used: OH^- 5.7; CO_3^{2-} 4.17; Cl^- 0.25; Br^- 3.07; SO_4^{2-} 1.90.
**log of stability constants for all MSO_4 = 1.0, for all MCl^+ = 0.
†log stability constant for $Cu(CO_3)_2^{2-}$ = 8.5, α = 6.

The main difficulty in determining the stabilities of these carbonate complexes and those of magnesium and calcium is caused by the formation of slightly soluble carbonate phases. The precipitation reactions of zinc(II) and copper(II) were carefully investigated [23, 26]. Zinc forms $ZnCO_3$, known as the mineral smithsonite. At the pH and the free CO_3^{2-} concentrations in seawater the stable phase is the carbonate hydroxide, $Zn_5(OH)_6(CO_3)_2$. Cooper(II) forms the carbonate hydroxides malachite, $Cu_2(OH)_2CO_3$, and azurite, $Cu_3(OH)_2(CO_3)_2$. The simple carbonate, $CuCO_3$, does not seem to exist. Malachite is the stable phase in equilibrium with seawater, but the rate of formation of malachite is very slow. Several solid carbonate phases exist for nickel(II) and cobalt(II) but their solubility products and rates of formation are not well known. However, the solubilities might be of the same order of magnitude as those of the zinc and copper carbonates. The solubility products for copper and zinc carbonates at ionic strengths of zero and 0.2 were recalculated to the ionic strength of seawater (0.71). The values are listed in Table 4.

The equilibrium constants for hydrolytic reactions pertaining to these extremely dilute solutions are difficult to determine. At metal ion concentrations of 10^{-5} M or higher, at which these reactions were investigated,

Table 4. Ion Products in Seawater and Solubility Products of Slightly Soluble Carbonate Phases of Copper(II) and Zinc(II)

Carbonate	- Log Solubility Product $I=0.2$*	- Log Solubility Product Seawater	- Log Ion Product** in Seawater
$Cu_2(OH)_2CO_3$	31.4	30.8	35.0
$Cu_3(OH)_2(CO_3)_2$	42.1	41.2	48.8
$ZnCO_3$	9.84	9.5	12.9
$Zn_5(OH)_6(CO_3)_2$	70.1	68.7	86.0

*$NaClO_4$ medium
**Calculated using p[OH] 5.7, p[CO_3^{2-}] 4.17, and p[M^{2+}] obtained from the concentrations in Table 1 and $\alpha[M^{2+}]$ in Table 3.

polynuclear complexes predominate [27]. Only very approximate values can be obtained for constants of mononuclear complexes that are expected to be the major species at the very low concentrations present in seawater. The available "mononuclear" constants follow the Irving-Williams order with a very marked maximum at copper(II) (Table 3). The distribution of the metal ions among the various complexes is presented in Table 3. For cobalt, nickel, and zinc the hydrated ions predominate. Because of the high chloride concentration in seawater, approximately 25 percent of cobalt, nickel, and zinc are present as MCl^+ complexes in spite of their low stability. Complexes MCl_2 might also be present, but the available data cannot prove their presence. Although sulfate has a higher affinity than chloride for cobalt, nickel, and zinc, only five percent of the metal ions are present as sulfate complexes because of the low concentration of sulfate. Hydroxy complexes can almost be neglected. Based on the assumed constants the carbonate complexes account for 28 percent of these metal ions. Because the constants are very approximate, this fraction may well be considerably smaller or larger.

Copper(II) shows a distinctly different behavior. The carbonate complex predominates with $Cu(OH)^+$ as the second most abundant species. The hydrated ion, the chloride complex, and the sulfate complex are of minor importance (Table 3).

The concentrations of hydrated (free) metal ions (Table 3) and the known concentrations of hydroxide ion and free CO_3^{2-} in seawater were used to calculate the ion-products for the solid copper and zinc carbonates listed in Table 4. The ion products are all smaller than the solubility products. The ocean is thus not saturated with respect to any of these slightly soluble carbonates. Most probably, the ocean is also unsaturated with respect to similar cobalt and nickel carbonates.

Copper - but not cobalt, nickel, or zinc - is reducible at pE values that could be established in seawater under anoxic conditions. With a standard potential for the Cu(I)/Cu(II) couple of 158.6 mV [9] and the constants for the Cu(I) chloride complexes given in equations 6 and 7 [28, 29], the concentration of Cu(I) is larger than that of

$$Cu^+ + 2Cl^- \rightleftharpoons CuCl_2^- \quad \beta_2 = 10^{5.0} M^{-2} \quad (6)$$

$$CuCl_2^- + Cl^- \rightleftharpoons CuCl_3^{2-} \quad K_3 = 10^{0.35} M^{-1} \quad (7)$$

Cu(II) for pE < 6.5. Copper(I) is so highly stabilized by the high chloride concentration, that a reduction of Cu(I) to elemental copper, for which a standard potential of 518.2 mV is assumed [9], will occur in the absence of any other ligands only for pE < -4.1. However, under typically anoxic conditions sulfide is also present. At pE $\simeq$ 7, at which the copper(II) concentration might still be 10^{-8} M, a sulfide concentration of ~> 10^{-28} must be reached to precipitate CuS ($K_{SO} = 10^{-36.1} M^2$). The highest sulfide concentration attainable at pE $\simeq$ 7 was calculated from the standard potential (141 mV) [18] for the sulfide redox system (eqn. 8) and the constant for the protonation of sulfide (eqn. 9)

$$S(s) + 2H^+ + 2e^- \rightleftharpoons H_2S(aq) \quad (8)$$

$$2H^+ + S^{2-} \rightleftharpoons H_2S \quad K_1K_2 = 10^{21} M^{-2} \quad (9)$$

[30] to be $10^{-30.3}$ M. At this sulfide concentration CuS will not precipitate. To precipitate Cu_2S its solubility product of $10^{-48.9} M^{-3}$ has to be exceeded [18]. With a total concentration of 4 nM for copper(I) the concentration of uncomplexed Cu^+ is 10^{-13} M at the chloride concentration of seawater. A sulfide (S^{2-}) concentration of at least 10^{-23} M is, therefore, required for precipitation to begin. This sulfide concentration can be reached at pE < 3.4. At higher sulfide concentrations, which are possible at lower pE values, soluble sulfide complexes may also be present.

Silver(I) forms such complexes, which have been thoroughly investigated [31]. To what extent copper(I) forms such complexes is presently unknown.

Iron and manganese. At pH and pE values generally found in seawater, iron and manganese predominantly form practically insoluble oxides or hydrous oxides. The extensive hydrolysis of iron(III) at pH 8.1 strongly stabilizes this oxidation state relative to iron(II), which is hardly hydrolyzed at all. Hydrated Fe^{2+} has an acid dissociation constant of $10^{-10.0}$ M at an ionic strength of 0.7, whereas the corresponding constant for Fe^{3+} is $10^{-3.1}$ M. The solid phase of iron(III) stable at the pH of seawater is FeO(OH). The equilibrium constant for FeO(OH) (eqn. 10) was determined for amorphous oxide in equilibrium with a 3 M

$$FeO(OH)(s) + 3H^+ \rightleftharpoons Fe^{3+} + 2H_2O \qquad (10)$$

$$K^*_{SO} = \frac{[Fe^{3+}]}{[H^+]^3} = 10^{3.6} M^{-2}$$

solution of sodium perchlorate [27]. With this solubility constant and the standard potential of 770 mV for the Fe^{3+}/Fe^{2+} couple, the concentrations of hydrated Fe^{3+} and Fe^{2+} in equilibrium with solid FeO(OH) at the pH and pE of seawater was calculated to be $10^{-20.7}$ M for Fe^{3+} and $10^{-20.2}$ M for Fe^{2+}. Because hydrated Fe^{2+} is certainly the most important species of iron(II) in solution, practically no iron(II) is present in seawater at equilibrium. Some iron(II) may be present under non-equilibrium conditions perhaps in organisms. The concentration of hydrated Fe^{3+} is practically also zero. The constants for the hydrolysis reactions of Fe^{3+} (eqn. 11) at the ionic strength of seawater at 25° demand equilibrium concentrations of the

$$Fe^{3+} + j H_2O \rightleftharpoons Fe(OH)_j^{(3-j)} + jH^+ \qquad (j = 1-4) \qquad (11)$$

$$K^*_1 = 10^{3.1} M; \quad K^*_2 = 10^{7.0} M^2; \quad K^*_4 = 10^{22.5} M^4$$

iron(III) hydroxy species of $10^{-10.7}$ M for $Fe(OH)^{2+}$, $10^{-11.5}$ M for $Fe(OH)^{2+}$, and $10^{-10.8}$ M for $Fe(OH)_4^-$ [27]. These concentrations are all much lower than the total iron concentration of 8 nM in seawater. The polynuclear hydroxy complexes are of even less importance at these low concentrations. To what extent uncharged, soluble $Fe(OH)_3$ exists is not known. Most of the iron seems to be present as a finely dispersed hydrous oxide at least partly in colloidal form.

Manganese prefers the Mn(IV) oxidation state. The standard electrode potential for the MnO_2/Mn^{2+} couple (eqn. 12) is 1230 mV [18]. The concentration of hydrated Mn^{2+} in

$$MnO_2(s, \beta\text{-phase}) + 4H^+ + 2e^- \rightleftharpoons Mn^{2+} + 2H_2O \qquad (12)$$

equilibrium in seawater with pyrolusite, the β-phase of solid manganese dioxide, is $10^{-15.8}$ M. The manganese(III) hydroxide is not stable relative to pyrolusite under these conditions. Species with manganese in oxidation states higher than Mn(IV) are reduced to Mn(IV). Once equilibrium has been reached, manganese is present in seawater only in the form of dispersed, solid manganese dioxide.

Eventually, the iron(III) hydrous oxides and the manganese dioxide reach the seabottom sediments. The iron-manganese nodules on the ocean floor were formed by coprecipitation of these oxides perhaps in the vicinity of hot springs.

At lower pE values iron and manganese may exist in lower oxidation states [32]. The data for various solid iron phases are not very precise, but as pE decreases Fe_3O_4 appears to become the stable phase. At still lower pE, iron(II) silicates might become important. Little is known about their compositions and conditions of formation. The iron(II) carbonate, $FeCO_3$, may also be formed. This compound has a fairly large solubility product of $10^{-9.3}$ M² in seawater. As long as the free CO_3^{2-} concentration remains at the level found for aerobic conditions, the equilibrium concentration of hydrated Fe^{2+} could be as high as 10^{-5} M. At low pE values reached in solutions containing sulfide, pyrite, FeS_2, becomes important. The stability regions of the iron(III) and iron(II) species are indicated in Fig. 1, which gives a good idea about the difficulties associated with the equilibrium calculations that - of course - do not provide any information about the kinetic aspects of phase transitions. The quantity log $\{Sp\}$ - log $\{Fe^{2+}\}$ in Fig. 1 denotes the activities of the species "Sp" and Fe^{2+}, respectively. The solid species identified in the figure have an activity of one. The three lines for FeO(OH) and the two lines for Fe_3O_4 correspond to different data sets. FeO(OH) and Fe_3O_4 may exist in equilibrium under the conditions represented by points P_1 and P_2. At P_2, the crossing point of the preferred lines, pE has a value of 3.7.

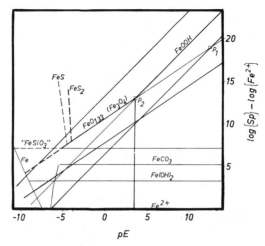

Figure 1. Redox diagram for iron at a temperature of 25° and a pH of 8.1. Reproduced with permission from ref. 32.

Manganese dioxide is seemingly the stable phase over a rather wide range of pE > 7. It is not certain whether MnO(OH) and/or Mn_3O_4 appear as stable phases. At still lower values of pE, $MnCO_3$ is most probably the stable solid phase. Only at pE of -10, too low to ever be reached in the sea, would MnS become stable.

Cadmium. Divalent cadmium is a mildly soft acceptor with properties between those of the fairly hard Zn^{2+} and the very soft Hg^{2+}. Although the chloride complexes are only moderately stable [9], they nevertheless predominate with $CdCl_2$ as the major species (Table 5). The bromide complexes are somewhat more stable than the chloride complexes. Because of the low bromide concentration in seawater, even the most abundant cadmium-bromide complex, the statistically favored CdBrCl, accounts for only 0.2 percent of total cadmium. Sulfate complexes are unimportant because of their low stabilities at the ionic strength of seawater.

Cadmium(II) does not hydrolyze easily (Table 5) [27]. In seawater the formation of chloride complexes prevents hydrolysis completely. Cadmium carbonate is slightly soluble [33] with a solubility product of $10^{-12.0}$ M^2 at zero

Table 5. *Distribution and Stability Constants of Mercury(II), Cadmium(II), and Lead(II) Species in Standard Seawater**

Species	Mercury(II) α(%)	Mercury(II) log β**	Cadmium(II) α(%)	Cadmium(II) log β**	Lead(II) α(%)	Lead(II) log β**
M^{2+}	0	–	3	–	4	–
CH_3M^+	0	–	–	–	–	–
CH_3MCl	60	5.32 †	–	–	–	–
$M(OH)^+$	0	10.1 ††	0	3.6	7	6
MCO_3	0	–	0.2	3	24*†	5
MCl^+	0	6.74	34	1.36	17	0.9
MCl_2	1	13.22	51	1.79	19	1.2
MCl_3^-	5	14.07	12	1.44	27	1.6
MCl_4^{2-}	26	15.07	0	–	0	–
MBr^-	0	9.05	0.1	1.56	0	1.1
$MBrCl$	1	15.9	0.2	2.3	0	–
$MBrCl_2^-$	2	16.4	0	2.1	0	–
$MBrCl_3^{2-}$	5	17.2	0	–	0	–
$MCl(OH)$	0	17.4	0	–	0	–
MSO_4	0	1.3	0.3	1	0.4	1

*p[L](-log free ligand concentration) used: OH^- 5.7; CO_3^{2-} 4.17; Cl^- 0.25; Br^- 3.07; SO_4^{2-} 1.90.

**Stability constants for the cumulative equilibria $M + jL \rightleftharpoons ML_j$.

†Stability constant for $CH_3Hg^+ + Cl^- \rightleftharpoons CH_3HgCl$.

††Calculated from the equilibrium constant for $Hg^{2+} + H_2O \rightleftharpoons Hg(OH)^+ + H^+$: $pK_w = 13.8$.

*†Log β for $Pb(CO_3)_2^{2-}$ is 8; α = 2%.

ionic strength and $10^{-10.7}$ M² in seawater. At the concentrations of Cd^{2+} and CO_3^{2-} in seawater, cadmium carbonate will not precipitate. The stabilities of the soluble carbonate complexes are not known. If their stabilities are somewhat less than those of the copper(II) complexes, cadmium carbonate complexes can be present in seawater only at negligible concentrations.

The influence of nitrilotriacetic acid (NTA), a strong chelating agent used as a component of detergent formulations, on the species distribution of cadmium has been thoroughly investigated [34]. The NTA complexes of cadmium are fairly inert and therefore are reduced polarographically in a separate wave at much more negative potential than hydrated Cd^{2+} and the labile cadmium halide complexes. The two types of cadmium species can therefore by determined by polarography. NTA was added to a 10^{-7} M cadmium(II) solution, a concentration representative of heavily polluted water. The ionic strength of these solutions was adjusted to 0.7, the ionic strength of seawater, by addition of various salts (Fig. 2). The

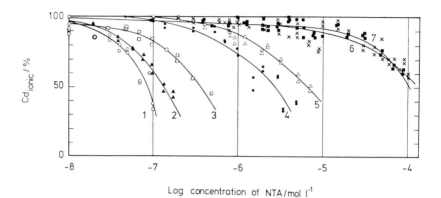

Figure 2. *The dependence of the fraction of ionic and labile Cd in media of ionic strength 0.7 M on the total concentration of nitrilotriacetic acid at a total cadmium concentration of 1×10^{-7} M. The media had pH values between 7.3 and 8.3. Sodium perchlorate was added whenever necessary to achieve an ionic strength of 0.7 M. The media are 1: 0.7 M sodium perchlorate; 2: 0.1 M sodium chloride; 3: 0.59 M sodium chloride; 4: 0.01 M calcium chloride; 5: 0.054 M magnesium chloride; 6: synthetic seawater without trace elements; 7: natural seawater from the Adriatic Sea. Ionic and labile Cd^{2+} refer to the total concentration of hydrated Cd^{2+} and all labile complexes reversibly reduced at the dropping mercury electrode. Reprinted with permission from ref. 34.*

synthetic seawater medium contained all the major constituents of natural seawater but no trace elements. Figure 2, in which the fraction of ionic and labile Cd^{2+} in these solutions is plotted versus the concentration of NTA, reveals that chloride competes with NTA for Cd^{2+}. To have, for instance, 50 percent of the cadmium chelated by NTA, a much higher concentration of NTA is needed in a chloride-containing medium (curve 3) than in a noncomplexing perchlorate solution (curve 1). Calcium (curve 4) and magnesium (curve 5) compete with cadmium for NTA. When the effects of all main constituents of seawater are considered, an NTA concentration of 10^{-4} M is required to sequester half of the 10^{-7} moles/liter of cadmium added to synthetic seawater without trace elements (curve 6) or natural seawater (curve 7). These results demonstrate, that - at least for the Adriatic seawater used in these experiments - the trace elements do not influence the species distribution of cadmium.

Mercury. The species distribution of mercury in seawater is largely determined by the strong softness of mercury(II), the oxidation state stable under the conditions generally prevailing in seawater. The halide complexes [9] and the organometallic methyl compounds of mercury(II) are very stable. Mercury is methylated in an aquatic environment by biological and nonbiological processes that were discovered 15 years ago [35]. The biological significance of these processes and their importance for the cycling of mercury between air, water, and sediments were soon recognized [36]. As a consequence, these processes were extensively studied in later years [37]. Dimethyl mercury, $(CH_3)_2Hg$, and methyl mercury cation, CH_3Hg^+, are both formed in nature. Methyl mercury forms a very stable chloride complex, CH_3HgCl [38]. At the high chloride concentration in seawater, CH_3HgCl ought to be more abundant than dimethyl mercury. The volatile (bp. 96°) and only slightly water-soluble dimethyl mercury should prefer the atmosphere. In the open sea, mercury is exclusively methylated by biological action. In heavily polluted coastal waters chemical agents may also be responsible for methylation [37].

The inorganic mercury halide complexes but not the methyl mercury species are reduced by tin(II) chloride solutions. This reaction is used to differentiate between "reactive" and "nonreactive" mercury [39]. Even in the unpolluted ocean, the "nonreactive" (methylated) mercury accounts for half and often for two thirds of the total mercury. In polluted waters, the "nonreactive" mercury is even more abundant, perhaps because of higher biological activity and the formation of methylated species via nonbiological reactions. The concentration of "reactive" mercury is rather constant at 3 ng/L [39, 40] independent of place and depth. The concentration of "nonreactive" mercury seems to vary much more than the concentration of "reactive" mercury.

The current knowledge about mercury compounds in seawater allows only approximate statements to be made about the species distribution of mercury. The data reported in Table 5 were obtained by assuming that 60 percent of the total mercury in seawater is present as CH_3HgCl. This compound does not dissociate appreciably, nor does it bind additional chloride ions [38]. The concentration of dimethyl mercury is presumed to be negligible. The distribution of mercury among the non-organometallic, labile complexes, which attain thermodynamic equilibrium rapidly, can be calculated from the stability constants characterizing the various equilibria. Generally, values for these constants applicable to seawater are not known, but they can be calculated with sufficient accuracy from experiments performed in media similar to seawater [9]. According to such calculations, almost all of the mercury should be present as CH_3HgCl and $HgCl_4^{2-}$ (Table 5). The complexes $HgCl_3^-$ and $HgBrCl_3^{2-}$ might also be present in seawater. Hydroxy, carbonate, and sulfate complexes are virtually absent.

Mercury - as a typical soft acceptor - has a very strong affinity for sulfur donors. Such donors have not yet been found in seawater at perceptible concentrations, but they are certainly present in organisms. When organisms are abundant, they may locally cause mercury to be completely sequestered. As the organic tissues decay, the sulfur compounds will certainly be oxidized, and the sequestered mercury will be released to form the mercury species listed in Table 5.

Lead. The tetravalent state of lead is favored at high pH values because of the extensive hydrolysis of lead(IV) to PbO_2 and the mixed oxide Pb_3O_4. Divalent lead is not very susceptible to hydrolysis. The acid dissociation constant of hydrated Pb^{2+} is not larger than $10^{-7.8}$ M at the ionic strength of seawater [27]. Calculations using the standard potentials for the reactions 13 and 14 show however, that

$$PbO_2(s) + 4H^+ + 2e^- \rightleftharpoons Pb^{2+} + 2H_2O \qquad E^0 = 1455 \text{ mV} \qquad (13)$$
$$3PbO_2(s) + 2H_2O + 4e^- \rightleftharpoons Pb_3O_4(s) + 4OH^- \qquad E^0 = 295 \text{ mV} \qquad (14)$$

the oxidation of Pb^{2+} to PbO_2 or Pb_3O_4 cannot possibly occur at the pH and the highest conceivable oxidation potential (740 mV) of seawater.

Lead carbonate, $PbCO_3$, should have a solubility product of $10^{-11.8}$ M^2 in seawater [9]. The corresponding constant for $Pb_3(OH)_2(CO_3)_2$ is 10^{-43} M^7. At the pH and free CO_3^{2-} concentration in seawater, the concentration of Pb^{2+} must become approximately $10^{-7.5}$ M before $PbCO_3$ or $Pb_3(OH)_2(CO_3)_2$ begins to precipitate. The total lead concentration in seawater is 10^{-10} M and the free Pb^{2+} concentration because of complex formation is even lower. Precipitation of lead carbonates in seawater, therefore, is unlikely. The stability constant for lead carbonate in a medium of ionic strength 1.0 has been reported [41] to be 10^7 M^{-1}, a value that is not very precise because of the large random errors. The stability constant for $Pb(CO_3)_2{}^{2-}$ of 10^9 M^{-2} found in these experiments [41] is approximately ten times larger than the values of $10^{8.2}$ and $10^{7.9}$ reported by other investigators [42, 43] for media of ionic strength of 1.7. The species distribution for lead (Table 5) was calculated with stability constants of 10^5 M^{-1} and 10^8 M^{-2}. These constants should be regarded as educated guesses.

Lead sulfate has a much larger solubility product at an ionic strength of 1.0 [9] than lead carbonate. Lead sulfate will not precipitate in seawater in spite of its fairly high sulfate concentration. The stabilities of lead sulfate complexes in seawater are not known, but the value of 10 (Table 5) for $PbSO_4$ seems to be a reasonable estimate [44].

The stabilities of lead chloride complexes are fairly well known for solutions of ionic strength similar to seawater [45]. The lead(II) chloride complexes are considerably less stable than the corresponding cadmium(II) complexes (Table 5) but more stable than the chloride complexes of cobalt, nickel, copper, and zinc (Table 3).

The results of the calculation of the species distribution for lead(II) using the stability constants given in Table 5 show, that lead chloride and carbonate complexes are the most abundant lead species in seawater (Table 5). The actual numbers obtained in these calculations depend very much on the unfortunately not very reliable values for the stability constants of lead carbonate complexes.

Uranium. Hexavalent uranium - stable at the normal pE of seawater - is always present as $UO_2{}^{2+}$. The hydrated uranyl ion is a fairly strong acid with a tendency to form polynuclear complexes even at low concentrations [27]. A most remarkable property of $UO_2{}^{2+}$ is the formation of soluble, very stable carbonate complexes. At the pH and free $CO_3{}^{2-}$ concentration in seawater, all uranium is present as the complex $UO_2(CO_3)_3{}^{4-}$ [46]. As a hard acceptor, uranyl ion forms only weak complexes with halide ions. The uranyl sulfate complexes are of only medium strength though much stronger than the sulfate complexes of divalent cations of cobalt, nickel, copper, zinc, cadmium, lead, and mercury.

Because of the formation of the soluble carbonate complex, the concentration of uranium (14 nM) in the sea is relatively high (Table 1) considering the low terrestrial abundance of this element. The concentration of uranium in seawater is so high, that the extraction of uranium from seawater by means of ion exchange has been seriously contemplated as a feasible technical process [47, 48].

Selenium. Selenium is an element that is essential for life and - at the same time - highly toxic even in low doses. Selenium is a good example of an element that exists in seawater largely in a thermodynamically unstable oxidation state. Although selenium is less easily oxidized to the hexavalent state than sulfur, the high pH and pE of seawater nevertheless favor selenium(VI) over selenium(IV) to such an extent, that the ratio Se(VI)/Se(IV) should be 10^{12} at equilibrium. This ratio was calculated with the standard potential of 50 mV for reaction 15 [18] and the constant 10^7 M^{-1} for the protonation of selenite in seawater (eqn. 16).

$$SeO_4^{2-} + 2H_2O + 2e^- \rightleftharpoons SeO_3^{2-} + 2OH^- \qquad (15)$$

$$SeO_3^{2-} + H^+ \rightleftharpoons HSeO_3^- \qquad (16)$$

The corresponding value for this constant in pure water is $10^{8.0}$ [9]. Selenite in seawater would be present predominantly as $HSeO_3^-$ and selenium(VI) as SeO_4^{2-}, because $HSeO_4^-$ is quite a strong acid. To have Se(IV) and Se(VI) present in seawater in equal amounts at equilibrium, pE must be as low as 6.5

The rate of oxidation of Se(IV) to Se(VI) is so slow, that a large fraction of the total selenium in seawater is always found to be in the tetravalent state. In the ocean's surface water, up to 80 percent of the selenium is in the thermodynamically unstable Se(IV) state. In deeper waters selenium seems to be evenly divided between Se(IV) and Se(VI) [49].

Conclusions. The distribution of species reported in this paper is an approximation. Exact answers cannot yet be obtained and perhaps will never be attainable. Variations of temperature and pressure influence the equilibria in largely unknown ways. Many important reactions occur at slow rates that prevent these reactions from reaching equilibrium. Under these conditions equilibrium calculations are of limited usefulness. Biological activities through uptake of trace elements and excretion of chelating agents probably play an important part in determining the chemical nature of trace elements in seawater. These biological influences vary strongly from place to place and from time to time.

The best, currently feasible approach to the determination of species distribution, which combines experimental determinations of the concentrations of inert species with equilibrium calculations for labile complexes, probably gives a rather accurate idea about the species present in seawater. As more data become available these ideas will of course be refined.

References

[1] D. Dyrssen and M. Wedborg, "Equilibrium Calculations of the Speciation of Elements in Seawater," in **"The Sea,"** Vol. 5 of **"Marine Chemistry,"** J. Wiley and Sons: New York, 1974, p. 181.
[2] F. Haber, "Das Gold im Meerwasser," **Z. Angew. Chem., 40** (1927) 303.
[3] B. Schaule and C. C. Patterson, "Lead in the Ocean," in **"Trace Metals in Sea Water,"** C. S. Wong, Ed.; Plenum Press: New York, 1983, p. 487.
[4] L. Brügmann, L.-G. Danielsson, B. Magnusson and S. Westerlund, "Intercomparison of Different Methods for the Determination of Trace Metals in Seawater," **Marine Chem., 13** (1983) 327.
[5] L.-G. Danielsson, "Cadmium, Cobalt, Copper, Iron, Lead, Nickel and Zinc in Indian Ocean Water," **Marine Chem., 8** (1980) 199.
[6] B. Magnusson and S. Westerlund, "The Determination of Cd, Ca, Fe, Ni, Pb and Zn in Baltic Sea Water," **Marine Chem., 8** (1980) 231.
[7] T. Almgren, D. Dyrssen and M. Strandberg, "Determination of pH on the Moles per kg Seawater Scale (M_w)," **Deep-Sea Res., 212** (1975) 635.
[8] R. G. Bates and C. H. Culberson, "Hydrogen Ions and the Thermodynamic State of Marine Systems," in **"The Fate of Fossil Fuel and Carbon Dioxide in the Oceans,"** N. R. Andersen and A. Malahoff, Eds., Plenum Press: New York, 1977, p. 45.
[9] L. G. Sillén and A. E. Martell, "Stability Constants of Metal Ion Complexes," **Chemical Society, London, Special Publications No. 17** (1964) and **25** (1971).
[10] L. G. Sillén, "The Physical Chemistry of Seawater," in **"Oceanography,"** M. Sears, Ed., Amer. Ass. Adv. Science,

Washington, D.C., 1961, p. 549.
[11] F. T. Mackenzie, "Sedimentary Cycling and the Evolution of Seawater," in **"Chemical Oceanography,"** 2nd Ed., J. P. Riley and G. Skirrow, Eds.; Academic Press: London, 1975, p. 309.
[12] J. M. Edmond and K. Von Damm, "Hot Springs on the Ocean Floor," **Scientific American,** 248 (1983) 70.
[13] D. Dyrssen and I. Hansson, "Ionic Medium Effects in Seawater," **Marine Chem.,** 8 (1973) 137.
[14] D. Dyrssen and M. Wedborg, "Major and Minor Elements, Chemical Speciation in Estuarine Waters," in **"Chemistry and Biochemistry of Estuaries,"** E. Clausson and I. Cato, Eds.; Wiley: New York, 1981, p. 72.
[15] S. Ahrland, "Metal Complexes Present in Seawater," in **"The Nature of Seawater,"** E. Goldberg, Ed.; Dahlem Konferenzen: Berlin, 1975, and references cited therein.
[16] R. Parsons, "The Role of Oxygen in Redox Processes in Aqueous Solutions," in **"The Nature of Seawater,"** E. Goldberg, Ed.; Dahlem Konferenzen: Berlin, 1975, and references cited therein.
[17] P. S. Liss, J. R. Hering and E. D. Goldberg, "The Iodide/Iodate System in Seawater as a Possible Measure of Redox Potential," **Nature Phys. Sci.,** 242 (1973) 108.
[18] W. M. Latimer, **"Oxidation Potentials,"** 2nd Ed., Prentice Hall: Englewood Cliffs, N.J., 1972.
[19] **"Marine Chemistry,"** Report of the Marine Chemistry Panel of the Committee on Oceanography, National Academy of Sciences, Washington, D.C., 1971.
[20] T. Almgren, L.-G. Danielsson, D. Dyrssen, T. Johansson and G. Nyquist, "Release of Inorganic Matter from Sediments in a Stagnant Basin," **Thalassia Jugoslavica,** 11 (1975) 19.
[21] S. Ahrland, "Thermodynamics of Complex Formation between Hard and Soft Acceptors and Donors," **Struct. Bonding (Berlin),** 5 (1968) 118.
[22] R. G. Pearson, "Hard and Soft Acids and Bases, HSAB," **J. Chem. Ed.,** 45 (1968) 581, 643.
[23] P. Schindler, M. Reinert and H. Gamsjäger, "Zur Thermodynamik der Metallcarbonate. 2," **Helv. Chim. Acta,** 51 (1968) 1845.
[24] H. M. N. H. Irving ad R. J. P. Williams, "The Stability of Transition-Metal Complexes," **J. Chem. Soc.,** (1953) 3192.
[25] G. Schwarzenbach, "Koordinationsselektivität und die Thermodynamik der Komplexbildung in Lösung," **Chimia,** 27 (1973) 1.
[26] P. Schindler, M. Reinert and H. Gamsjäger, "Zur Thermo-

dynamik der Metallcarbonate. 3," **Helv. Chim. Acta, 52** (1969) 2327.
[27] C. F. Baes, Jr. and R. E. Mesmer, **"The Hydrolysis of Cations,"** Wiley: New York, 1976.
[28] S. Ahrland and B. Tagesson, "Thermodynamics of Metal Complex Formation in Aqueous Solution. XII," **Acta Chem. Scand., A31** (1977) 615.
[29] H. McConnell and N. Davidson, "Optical Interaction between the Chloro-Complexes of Copper(I) and Copper(II) in Solutions of Unit Ionic Strength," **J. Am. Chem. Soc., 72** (1950) 3168.
[30] M. Widmer and G. Schwarzenbach, "Die Acidität des Hydrogen-Sulfidions HS$^-$," **Helv. Chim. Acta, 47** (1964) 266.
[31] G. Schwarzenbach and M. Widmer, "Die Löslichkeit von Metallsulfiden II," **Helv. Chim. Acta, 49** (1966) 111.
[32] L. G. Sillén, "Oxidation State of Earth's Ocean and Atmosphere II.," **Arkiv Kemi, 25** (1966) 159.
[33] H. Gamsjäger, H. U. Stuber and P. Schindler, "Zur Thermodynamik der Metallcarbonate. 1," **Helv. Chim. Acta, 48** (1965) 723.
[34] B. Raspor, P. Valenta, H. W. Nürnberg and M. Branica, "The Chelation of Cadmium with NTA in Seawater as a Model for the Typical Behavior of Trace Metal Chelates in Natural Waters," **Sci. Total Environ., 9** (1978) 87.
[35] G. Westoo, "Determination of Methylmercury Compounds in Food-Stuffs," **Acta Chem. Scand, 20** (1966) 2131.
[36] S. Jensen and A. Jernelöv, "Biological Methylation of Mercury in Aquatic Organisms," **Nature, 223** (1969) 753.
[37] J. O. Nriagu, Ed., **"The Biochemistry of Mercury in the Environment,"** Elsevier: Amsterdam, 1979.
[38] M. Jawald, F. Ingman, D. Hay Liem and T. Wallin, "Solvent Extraction Studies on the Complex Formation between Methylmercury(II) and Bromide, Chloride and Nitrate Ions," **Acta Chem. Scand., A32** (1978) 7.
[39] C. W. Baker, "Mercury in Surface Waters around the United Kingdom," **Nature, 270** (1977) 230.
[40] P. Mukherji and D. R. Kester, "Mercury Distribution in the Gulf Stream," **Science, 204** (1979) 64.
[41] N. N. Baranova, "Investigation of the Carbonato-Complexes of Lead at 25° and 200°C," **Russ. J. Inorg. Chem., 14** (1969) 1717.
[42] J. Faucherre and Y. Bonnaire, "Constitution of Cu and Pb Complex Carbonates," **Compt. Rend. Acad. Sci. Ser. C, 248** (1959) 3705.
[43] F. Fromage and S. Fiorina, "Potassium Carbonatoplumbates(II) and Potassium Hydrogen Carbonatoplumbates(II)," **Compt. Rend. Acad. Sci. Ser. C, 268** (1969)

1764.

[44] A. M. Bond and G. Hefter, "Stability Constant Determination in Precipitating Systems by Rapid Alternating Current Polarography," **J. Electroanal. Chem., 34** (1972) 227.

[45] A. M. Bond and G. Hefter, "Influence of Anion-Induced Adsorption on Half-Wave Potentials and other Polarographic Characteristics," **J. Electroanal. Chem., 42** (1973) 1.

[46] D. Ferri, I. Grenthe and F. Salvatore, "Dioxouranium(VI) Carbonate Complexes in Neutral and Alkaline Solutions," **Acta Chem. Scand., A35** (1981) 165.

[47] N. J. Keen, J. H. Miles and R. Spence, "Extraction of Uranium and other Inorganic Materials from Seawater." Conference on the Technology of the Sea and Sea Bed, A.E.R.E. Harwell, 1967. **Proceedings, AERE-R 5500,** H.M.S.O. 1967, p. 387.

[48] R.V. Davies, J. Kennedy, R. W. McIlroy, R. Spence and K. M. Mill, "Extraction of Uranium from Seawater," **Nature, 203** (1964) 1110.

[49] Y. Sugimara, Y. Suzuki and Y. Miyake, "The Content of Selenium and its Chemical Form in Seawater," **J. Oceanograph. Soc. Japan, 32** (1976) 235.

III

SPECIATION OF METAL COMPLEXES AND METHODS OF PREDICTING THERMODYNAMICS OF METAL-LIGAND REACTIONS

Arthur E. Martell, Ramunas J. Motekaitis and Robert M. Smith

Department of Chemistry, Texas A&M University
College Station, Texas 77843 USA

The problems involved in the determination of the toxicity of metal compounds in the environment are outlined, and the complications resulting from chemical and biological transformations are briefly considered. The importance of knowing the exact compositions and properties of metal compounds in the environment is emphasized, and some examples are given. The methods that have been employed for relating stabilities of metal chelates to their structure and constitution are critically reviewed. Suggestions are made for the use of such correlations for the estimation of the stability constants needed to determine the very large number of metal complexes present in environmental systems. Because accurate calculations from fundamental atomic quantitites and physical principles are not yet possible, the prediction of thermodynamic equilibrium parameters for new complexes based on data for complexes with analogous ligands is the most productive approach. This method requires the development of the largest possible data base of reliable stability constants, associated protonation constants, and hydrolysis constants.

About 50,000 organic compounds are manufactured in the U.S.A., and approximately 1,000 new compounds are being added every year. Toxicity data are available for only one percent of these compounds. Many of these compounds appear to be and probably are rather harmless, but some are very toxic, and many more may be marginally harmful. A significant number of these compounds enter soil and water systems, usually at low concentrations. Many of these compounds persist in the environment for a long time and therefore accumulate as more are added. Some of these toxic materials are concentrated by living organisms. Compounds with appropriate functional groups may form metal complexes and may become less or more toxic depending on the nature of the metal ion and the ligand. Such ligands may also combine with, solubilize, and mobilize toxic metal ions previously present in the environment as insoluble, inaccessible minerals. A large number of organic pollutants may undergo trace metal-catalyzed oxidation by air, or microbial and enzyme-catalyzed bio-oxidation to derivatives that may serve as ligands for metal ions. Thus the hundreds, and potentially thousands, of metal-complexing pollutants can give rise to tens of thousands of complexes of naturally occurring metal ions, and of metal ions introduced with metal-containing pollutants.

It is well known that the cost to test a chemical compound for toxicity is $50,000 to $100,000. It is therefore necessary to select the most important and representative metal complexes for toxicity determinations, because only a very small fraction can be tested thoroughly. To obtain reasonable risk assessments, it will be necessary to first use computational techniques to obtain the best possible estimates of the types of metal complex systems present in the environment. These estimates will help to identify the metal complexes that are the most important in metal-containing environmental systems. Toxicity data are available for a few metal complexes but for most complexes it will be necessary to apply structure-activity relationships (QSAR) to make reasonable estimates of the toxicities of the individual components of complex environmental mixtures.

Dependence of Toxicity on the Chemical Nature of a Complex

The potential interaction of chemicals with metal ions has frequently been overlooked by scientists testing toxicities and carcinogenic effects with experimental animals. Considerable evidence exists in the literature, that the toxicity of an organic compound can be profoundly changed by forming complexes with metal ions [1]. The toxic effects of some metal ions are increased when their complexes are mobile and sufficiently labile to exchange ligands for receptors in biological systems. The formation of a sufficiently stable complex may completely detoxify a metal. The toxicities of organic compounds can be affected similarly. The labile or uncomplexed form of a ligand may be toxic because of its ability to combine with essential metal ions in biological systems, whereas a stable metal-ligand complex may be non-toxic.

Many examples of the application of this principle extend to the aminopolycarboxylate ligands such as ethylenediaminetetraacetic acid (EDTA) and their metal chelates. Intravenous EDTA is lethal to rabbits at 100 mg/kg by hypocalcemic tetany arising from the formation of the physiologically unavailable calcium in calcium EDTA [2]. Similarly, ionic calcium in human plasma is converted to calcium EDTA after intravenous administration of EDTA to humans [3]. The acute, intravenous toxicity of calcium EDTA is only 2 to 4 g/kg [4] because calcium EDTA does not significantly affect calcium levels [5]. This change in toxicity and properties upon conversion of the ligand to the calcium chelate led to the therapeutic use of the calcium chelate for the treatment of lead poisoning [6].

The nature of the metal in the metal chelate has also a significant effect on the biological distribution and toxicity of the ligand. Thus, the low oral absorption of calcium EDTA in humans [7] is in contrast to the considerable absorption of lead EDTA [8]. The administration of calcium diethylenetriaminepentaacetate (DTPA) to pregnant rats led to retardation of fetal growth. Zinc DTPA, however, is reported to be nontoxic [9]. Of direct applicability to the present considerations is the report [10], that the feeding of EDTA to female rats during pregnancy led to malformations in the full-term young. Malformations were not observed, when the EDTA-containing diet included 1000 mg/kg of zinc. This addition of zinc converted the free EDTA ligand to zinc EDTA.

As with other ligands the conversion of NTA to its metal chelates modifies its toxicity. The acute oral LD_{50} for NTA varies from 1000 to 2000 mg/kg in the rhesus monkey, and is in excess of 5000 mg/kg in the dog [11]. Conversion to the calcium chelate decreased the acute oral toxicity, and no deaths or adverse signs were noted with dosages as high as 27,000 mg/kg. The iron chelate of NTA is an effective source of oral iron [12], and the zinc chelate accelerates wound repair by selective delivery of the element to regenerative tissues [13]. Copper NTA serves as a source of oral copper in humans [14].

Changes in the chemical nature of metal species may cause dramatic changes in the toxicities of metals. It is well known, that many forms of chromium and nickel, including simple complexes, are highly toxic and carcinogenic in man when exposure occurs at moderate to high concentrations. Similarly, copper(II) compounds were found to be co-carcinogens at moderate concentrations [15]. At trace levels, however, all three metal ions are essential to life because they are required for the activation of enzymes. Thus, administration of these metal ions at trace concentrations is not harmful, because these ions are complexed by naturally occurring ligands forming metal transport complexes and metal-activated enzymes. When these metals are supplied in excessive amounts, however, the natural metal receptors are overwhelmed, and the excess metals may then have toxic effects on other biomolecules that have coordinating donor groups.

Toxicological effects of NTA salts and complexes. The use of nitrilotriacetic acid (NTA) as a substitute for polyphosphates in detergents stimulated intensive studies of NTA's potential environmental impact and toxicology [16]. Field studies based on anticipated usage established that the NTA concentrations in potable public water supplies are usually less than 10 µg/L with 50 µg/L as an occasional maximum. In Canada, where detergents contain 15 percent NTA and 60 million pounds of NTA are used annually, drinking water contains an average of 5 µg/L of NTA. The concentration of NTA is less than 24 µg/L in 96 percent of all samples. With an intake of two liters of water per day containing 25 µg NTA per liter, a person would receive less than 1.4 g of NTA during a 75-year life.

Of great importance to the toxicity of NTA in the environment is its chemical form, a fact that has been neglected in animal feeding experiments. Nitrilotriacetic acid and its trisodium salt have been almost universally employed in animal tests. These species do not exist in the pH range of 6 to 9 characteristic of natural waters. In this pH range the monoprotonated disodium salt is present. Although the use of the trisodium salt in place of the disodium salt would not make much difference at environmental NTA concentrations of approximately 1 mg/L, significant toxicological differences could exist at the higher concentrations employed in some feeding tests because of the alkalinity of the trisodium salt and the excessive amount of sodium introduced into the diet.

Of equal importance are the reactions of NTA with metals in the environment. To demonstrate the nature and extent of such interactions, model calculations were carried out on a typical natural water system [17-19] using the best available "selected" stability constants of NTA compiled by Martell and Smith [20]. Two concentrations of NTA were selected: 1.0×10^{-5} M, a concentration considerably higher than usually found in the environment and exceeding the trace metal levels, and 1.0×10^{-7} M, a more realistic concentration corresponding to about 20 µg/L and comparable to environmental levels in regions of extensive NTA use. Calculations show that nearly all of the NTA is present over the whole pH range in the form of metal complexes. The calculated distribution of NTA among the various metals present varies with the concentration level of NTA. At low concentrations, a large fraction of the NTA is combined with the heavy trace metals because of the higher stability constants of these complexes. When a considerable excess of NTA is present, most of it is complexed by calcium and magnesium. Thus, when very small amounts of NTA are added to water that contains trace metals NTA becomes coordinated to those metals with the higher stability constants. Metals present in the environment at higher concentrations generally have lower chelate stability constants for NTA and, therefore, are less completely complexed by NTA, until NTA becomes available in higher concentration.

The identification of NTA species in natural waters is complicated by the incomplete knowledge about the nature of the organic complexing agents present in natural waters and the lack of equilibrium constants for many of the organic ligands. Therefore, a model calculation of the distribution of NTA in water was carried out with certain organic and inorganic ligands controlling the metal ion concentrations. The inorganic ligands that may be present are chloride, sulfate, and nitrate. These ligands are relatively weak and may be disregarded in a first approximation calculation. The strongest inorganic complexor that may exist in appreciable concentration in the pH range of interest is the bicarbonate anion. Another complexing anion that may be present in moderate to trace concentrations is monohydrogen phosphate. A review of recent papers on levels of these ligands in natural waters suggests average concentrations of 1.0×10^{-3} M for bicarbonate and 1.0×10^{-5} M for monohydrogen phosphate. Among organic ligands that would complex metals and act as carriers for metal ions are humic acids. The literature reports "humic acid" concentrations for natural waters to be in the range 0.1 to about 100 mg/L. Humic acid has variable composition and molecular weight, and a variety of donor groups distributed almost randomly in the macromolecule. One of the more common combinations of donor groups with considerable metal ion affinity are phenolic hydroxyl groups and carboxyl groups in ortho positions to each other in an aromatic ring. This arrangement of groups was selected as representative of humic acid metal-binding sites. Stability constants of metal chelates of model ligands such as salicylic acid and 4-sulfosalicylic acid were used. Two such chelating sites were conservatively assumed to be present per molecule. If the molecular weight of humic acid is 2000, and its concentration in water is 10 mg/L, the concentration of chelate donor groups will be 1.0×10^{-5} M.

Inclusion of bicarbonate, hydrogen phosphate, and humic acid as carrier ligands for metal ions (Table 1) results in changes of the NTA species distribution. Because of the affinities of these carrier ligands for the heavy metal ions, the transfer of metal ions to NTA is incomplete. NTA is mainly complexed by the alkaline earth ions. These calculations led to the conclusion, that NTA will exist in the environment as calcium and magnesium chelates, unless transition metal ions and Zn(II) are present at a concentration similar to that of NTA.

The results of these calculations clearly show that the nature of the NTA complexes changes drastically with concentration. Such changes are expected to occur in environmental waters and in biological fluids, in which NTA must compete with low but often significant levels of naturally occurring metal ions and ligands. Thus, it is generally not possible to extrapolate toxicity data measured at high concentrations to low environmental or biological levels. This behavior of metal complexes is not found in organic molecules, which do not change composition on dilution. Therefore, extrapolation of toxicity data for organic compounds to low concentrations is sometimes considered valid. However, to predict the toxicity of metal ions in environmental aquatic systems, the metal-ligand concentrations must be determined over the complete range of conditions including concentration, pH, and interaction with competing metals and ligands. With this knowledge the overall toxicities of environmental systems may be estimated using the toxicities of the metal complexes actually present.

Table 1. *Distribution of NTA Species in Tap Water Containing Hydrogen Phosphate, Bicarbonate and Humic Acid* at NTA Concentrations of 1×10^{-5} M and 1×10^{-7} M[†]*

Metal Ion	Total metal concentration, Molarity	Percent NTA bound to Various Metal Ions at pH				
		6.0	6.5	7.0	7.5	8.0
Mg^{2+}	2.0×10^{-4}	1.5	1.7	1.8	1.8	1.9
		<0.1	**<0.1**	**<0.1**	**<0.1**	**0.1**
Ca^{2+}	1.0×10^{-3}	76.0	85.7	89.2	90.5	90.9
		0.4	**0.4**	**0.6**	**1.1**	**2.4**
Sr^{2+}	2.0×10^{-6}	<0.1	<0.1	<0.1	<0.1	<0.1
		<0.1	**<0.1**	**<0.1**	**<0.1**	**<0.1**
Ba^{2+}	3.0×10^{-7}	<0.1	<0.1	<0.1	<0.1	<0.1
		<0.1	**<0.1**	**<0.1**	**<0.1**	**<0.1**
Ni^{2+}	3.0×10^{-7}	3.0	3.0	3.0	3.0	3.0
		8.8	**10.1**	**14.0**	**22.7**	**42.0**
Cu^{2+}	2.0×10^{-7}	2.0	2.0	2.0	2.0	1.9
		82.4	**82.1**	**79.6**	**71.4**	**50.5**
Zn^{2+}	1.0×10^{-7}	1.0	1.0	1.0	1.0	1.0
		0.6	**0.7**	**1.0**	**1.7**	**3.6**
Cd^{2+}	1.0×10^{-8}	0.1	0.1	0.1	0.1	0.1
		<0.1	**<0.1**	**<0.1**	**0.1**	**0.1**
Pb^{2+}	1.0×10^{-7}	1.0	1.0	1.0	1.0	1.0
		7.7	**6.6**	**4.8**	**3.1**	**1.2**
Total Bound NTA	1.0×10^{-5}**	84.7	94.5	98.2	99.4	99.8
	1.0×10^{-7}**	**99.9**	**100.0**	**100.0**	**100.0**	**100.0**

*Total concentrations: HCO_3^- 1.1×10^{-4}; HPO_4^{2-} 1.0×10^{-5} M; humic acid, 10 mg/L.

**Total analytical concentration of NTA.

[†]Representative environmental concentration of NTA.

[††]Percentages in bold refer to the system with 1×10^{-7} M NTA.

Structure-Stability Relationships

Because of the relationship between toxicity and the chemical nature of metal complexes, it is necessary to determine or estimate the stabilities of complexes formed by metal ions with the large number of environmental ligands introduced artificially as chemical pollutants or by natural processes. Knowledge of the stability constants of the metal complexes formed in an environmental system would make it possible to calculate the distribution of the metal ions among the ligands present. Such calculations may now be carried out with high-speed computers for complex, multi-component systems. It is the purpose of this paper to critically examine the relationships that have been developed for the dependence of stability constants on the nature of the metal ion and the structure and constitution of the ligand, in order to extend the very limited current data to the large number of metal ions and complexing agents that may be found in the environment.

Linear free energy relationships. Examples of earlier, representative linear correlations between stability constants and properties of metal ions and ligands are listed in Table 2. The relationships were represented either as algebraic equations or as graphs.

Table 2. *Examples of Early Linear Free Energy Relationships*

Authors	Relationship	Ref.
Larsson (1934)	$\log K_{ML}$ ~ ligand basicity	[21]
Bjerrum (1950)	$\log K_{ML}/\log K_{HL}$ = constant	[22]
Calvin, Wilson (1945)	$\log K_{ML}$ vs. $\log K_{HL}$ (graphical)	[23]
Calvin, Melchior (1948)	$\log K_{HL}$ vs. ionization potential (graphical)	[24]
Irving, Williams (1948)	$\log K_{ML}$ vs. atomic number (graphical)	[25]
Davies (1951)	$\log K_{ML}$ vs. e^2/r	[26]
Martell, Calvin (1952)	$\log K_{ML}$ vs. e^2/r	[27]
Van Uitert, Fernelius, Douglas (1953)	$\ln K_{ML}$ vs. electronegativity (graphical)	[28]
Irving, Rosotti (1956)	$\log K_{ML} = a \log K_{HL} + b$ $\log K_{ML}$ vs. $\log K_{ML'}$ (graphical) $\log K_{ML}$ vs. $\log K_{M'L}$ (graphical)	[29]
Nieboer, McBryde (1970)	$\log K_{ML} = B \log K_{M'L} +$ $+(\log K_{ML'} - B \log K_{M'L'})$ $\log K_{ML} = C \log K_{ML'} +$ $+(\log K_{M'L} - C \log K_{M'L'})$	[30]

Larsson noted in 1934 [21], that the stability constants for the complexes of silver(I) with amines are roughly proportional to the basicity of the ligands. Bjerrum showed in 1950 [22], that the ratio $\log K_{ML}/\log K_{HL}$ is approximately constant for complexes of metal ions with cyanide, ammonia and pyridine. The constancy of this ratio improves and the ratio increases in going from tertiary amines, to secondary amines, to primary amines. The concept of linear free energy relationships was first presented graphically (Fig. 1) by Calvin and Wilson [23] in 1945 with linear plots

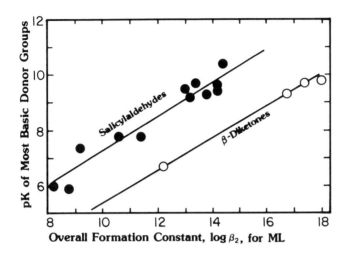

Figure 1. Correlation of chelate stability constants with ligand basicity for β-diketones, salicylaldehyde, and analogous bidentate ligands for ML_2.

of log K_{ML} vs. log K_{HL} for complexes of copper(II) with various salicylaldehydes and acetylacetonates. In 1948 Calvin and Melchior [24] correlated log K of complexes with gaseous ionization potentials, and in 1948 Irving and Williams [25] showed a common dependence of the stability constants of complexes of the first row, divalent transition metal ions with various ligands on the atomic number. Martell and Calvin [27] reviewed in 1952 the linear relationships between log K_{ML} and ligand protonation constants and showed along with Davies [26] the dependence of log K_{ML} for a given ligand on the charge and the radius (e^2/r) of the metal ion. Van Uitert et al. [28] discovered in 1953 an approximate linear correlation of log K values with the electronegativities of the metal ions. Irving and Rossotti [29] used in 1956 a linear equation (eqn. 1) to

$$\log K_{ML} = a \log K_{HL} + b \qquad (1)$$

relate stability constants to the protonation constants. They also showed such relationships to be applicable to a comparison of log K_{ML} with log $K_{M'L}$ for a pair of similar metals with a variety of ligands, and between log K_{ML} and log $K_{M'L}$ for a variety of metal ions with two similar ligands. A typical correlation illustrated by Figure 2 shows that the ratio of stability constants of two closely related metal ions is relatively constant for a wide variety of ligands.

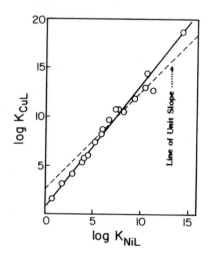

Figure 2. *Log K_{CuL} as a function of log K_{NiL} for complexes of a wide variety of ligands.*

In 1970 Nieboer and McBryde [30] published a pair of equations (Table 2) relating log K_{ML} with the corresponding values of a reference metal M' and a reference ligand L'. Rearrangement of these equations generates the relationships 2 and 3. Equation 2 states that B is the ratio of the

$$B = \frac{\log K_{ML} - \log K_{ML'}}{\log K_{M'L} - \log K_{M'L'}} \quad (2)$$

$$C = \frac{\log K_{ML} - \log K_{M'L}}{\log K_{ML'} - \log K_{M'L'}} \quad (3)$$

relative values of the stability constants of M with L and L' to the relative values of the stability constants of M' with L and L'. According to equation 3, C is similarly the ratio of the relative values of two metal constants with one ligand to the ratio of the same two metal constants with a second ligand. For some combinations of ligands and metal ions B and C were found to be unity.

Hammett equation. The Hammett equation (eqn. 4) was suggested by L. P. Hammett [31] in 1937 to correlate kinetic rates and carboxyl protonation constants of meta- and/or

$$\log(K/K_0) = \sigma\rho \qquad (4)$$

para-substituted benzoic acids. Equation 4 allows the estimation of a protonation constant, K, from a known reference value, K_0, and two adjustable parameters, σ and ρ, that are related to structural features of the compounds. Extensions of this equation to other functional groups, to ortho-subtituted aromatic compounds, and to aliphatic compounds have been extensively used by physical organic chemists. Reviews by Wells [32] in 1963, by Clark and Perrin [33] in 1964, and by Barlin and Perrin [34] in 1966 summarize some of these applications. The equations pertaining to aliphatic compounds are usually referred to as Taft equations.

The Hammett equation was applied by May and Jones [35] to copper(II) complexes of seventeen meta- and para-substituted benzoic acids (Fig. 3). The calculated, new set of sigma

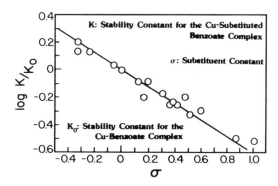

Figure 3. The Hammett relationship applied to Cu(II) complexes of substituted benzoic acids.

values predicted log K_{ML} to within ±0.1 log units. This correlation indicates, that the Hammett relationship applied to stability constants for metal complexes is in principle similar to the linear relationships between metal-ligand stability constants and the protonation constants of the same ligand (Table 2). In 1966 da Silva and Calado [36] correlated the stability constants of copper(II) acetylacetonates, iron(II)-mesoporphyrin complexes containing m- and p-substituted pyridines as ligands, and silver complexes containing m- and p-substituted anilines as ligands with Hammett σ-factors. In 1968, Pettit et al. [37] showed that linear relationships exist between silver complexes of ten m- or p-substituted phenylthioacetic acids and ten phenylselenoacetic acids and original Hammett sigma values. The complexes with o-nitro-, p-nitro-, and p-carboxyphenylchalcogenoacetic acids as ligands did not obey this correlation. Nieboer and McBryde [38] pointed out, that the Hammett equation is a special case of the linear free energy relationship.

Four-Parameter Equations Based on Electrostatic and Covalent Bonding

Edwards equation. In 1954, Edwards [39] suggested a four-parameter equation for the correlation of stability constants of metal ion complexes of unidentate ligands with solubility product constants (eqn. 5). Equation 6, an

$$\log(K/K) = \alpha E_n + \beta H \tag{5}$$

$$\log K_{ML} = \log 55.5 + \alpha(E^0 + 2.60) + \beta(\log K_{HL} + 1.74) \tag{6}$$

E_n: related to the electrode potential of the metal atom
E^0: standard electrode potential of a metal
H: related to the basicity of the ligand
α: correlated with the ionization potential of a metal ion
β: correlated with the charge and ionic radius of a metal ion

alternate form of equation 5, was used for data fitting. More than two-thirds of the calculated stability constants agreed with the experimental values within one log unit. Because the complexes used to derive the equations had a fair amount of covalent bonding, the correlation works best for complexes with considerable covalent character. Later, Edwards [40] suggested that polarizability, P, as measured by the molar refraction should be used instead of the electrode potential (eqn. 7).

$$\log K_{ML} = \log 55.5 + 3.60P + (\beta + 0.0624\alpha)(\log K_{HL} + 1.74) \quad (7)$$

Pearson equation. The concept of hard and soft acids apparently developed from the observation by J. Bjerrum [22] in 1950, that the affinity of metal ions for halide ions establishes two classes of metal ions. Small, highly charged ions such as aluminum(III) and iron(III) have maximal affinity for fluoride, whereas large ions such as mercury(II) have highest affinity for strongly polarizable ions such as iodide ion. In 1956 Schwarzenbach [41] expanded this concept and introduced the terms "A"-metals and "B"-metals. Ahrland, Chatt, and Davies [42] generalized this concept in 1958 and introduced the terms a-acceptors and b-acceptors. In 1963 Pearson [43] extended this classification to bases and coined the now commonly used expressions "hard" and "soft". In 1967, Pearson [44] suggested equation 8 for the estimation of an equilibrium

$$\log K = S_A S_B + \sigma_A \sigma_B \quad (8)$$

K : equilibrium constant
S_A: strength factor for acid
S_B: strength factor for base
A : softness factor for acid
B : softness factor for base

constant of an acid-base reaction. Many quantitative scales of softness have been suggested for metal ions (or acids) [45-54]. There is overall, general agreement in the softness scales. However, many specific points of disagreement exist that point out the difficulties of obtaining a universally acceptable scale.

Hancock and Marsicano [54] also suggested a quantitative scale for unidentate ligands (or bases), and Zhang [55] proposed a quantitative scale for the strength factor of metal ions. It appears, that the necessary scales are being developed for a test of the Pearson and other similar equations. The acceptable scales would be those that can most accurately predict experimental complexation constants.

Drago equation. Drago and Wayland [56, 57, 58] in 1965 suggested a four-parameter equation for correlating enthalpy changes accompanying acid-base reactions in the gas phase or in weakly-solvating solvents (eqn. 9). The E and C parameters are expressed relative to iodine at 1.00. The continuing tabulation of E and C values allows now the prediction of over 2000 reactions, among them some metal ion complexation reactions. Differences between experimental and predicted enthalpies are attributed to steric effects or π back-bonding.

$$-\Delta H = E_A E_B + C_A C_B \qquad (9)$$

E: adjustable parameter related to the electrostatic contribution to bonding
C: adaptable parameter related to the covalent contribution to bonding
A: indicates acceptor (acid)
B: indicates donor (base)

Hancock equation. In 1978 Hancock and Marsicano [54] fitted a four-parameter equation similar to Drago's equation to the stability constants of 27 complexes formed between metal ions, and fluoride, hydroxide, or ammonia in aqueous solution. The calculated log K_{ML} values had a standard deviation of 0.24 log unit.

Six-Parameter Equations Based on Electrostatic and Covalent Bonding

In 1980, Hancock and Marsicano [59] introduced a six-parameter equation with D factors to correct for desolvation and steric effects (eqn. 10). D_A was assumed to be zero for

$$\log K_{ML} = E_A^{aq} E_B^{aq} + C_A^{aq} C_B^{aq} + D_A D_B \qquad (10)$$

silver(I), mercury(II), thallium(III), lead(II), and bismuth(III), whereas D_B was assumed to be zero for fluoride, hydroxide, and ammonia. In general, this equation provided satisfactory predictions of stability constants for the unidentate ligands considered.

The original Drago equation is restricted to reactions in media of low polarity and works best for neutral molecules. To improve the applicability of this equation to ionic systems, Kroeger and Drago [60] introduced in 1981 a six parameter equation (eqn. 11). Tendency of the acid or base

$$-\Delta H = e_A e_B + c_A c_B + t_A t_B \qquad (11)$$

to undergo shifts in electron distribution upon complexation is expressed by the "t" terms. Parameters were adjusted for 364 enthalpies involving 32 acids and 45 bases to provide a more generally applicable equation.

Thermodynamic cycles. In theory, thermodynamic cycles can be used to calculate free energy, enthalpy, and entropy of complexation in solution, when the thermodynamic quantities for the gas phase reaction, the desolvation of the reactants, and the solvation of the products are available. However, thermodynamic data for many of the reactions in these cycles are lacking. Perhaps equations similar to the Drago equation could be developed for solvation and desolvation to give a large data base for calculating complexation reactions in aqueous systems.

Empirical equations with additive adjustments. In 1956 Irving and Rossotti [29] derived equation 12 for the

$$\ln K_{ML} = \ln K_{HL} + \frac{1}{RT}\{\overline{G}^0_M - \overline{G}^0_{ML} - \overline{G}^0_H - \overline{G}^0_{HL} + n_X(t_M - t_{ML} - t_H + t_{HL})\} +$$

reference metal complex   mixed solvent
constant   correction   correction

$$+ \ln(y_M y_{HL}/y_{ML} y_H) \qquad (12)$$

activity correction

$\overline{G}^0$: partial molal free energies
t: free energy of transfer to non-aqueous solvent
n_x: mole fraction of organic component of mixed solvent
y: activity coefficient

calculation of stability constants. In this approach a reference constant is adjusted by corrective terms. A generalized form of this expression is equation 13. A set

$$\ln K_{ML} = \ln K_{HL} + A + B \qquad (13)$$

A: a ligand-complex correction
B: a metal ion correction for a fixed temperature, solvent and ionic strength

of parallel lines are generated when A and B are changed. The distances between the lines are measures of the dissimilarities between the systems. Equation 13 is analogous to the linear free energy relationship (eqn. 1) with unit slope.

Clark and Perrin [33] in 1964 and Barlin and Perrin [34] in 1966 used the additive approach to correlate protonation constants of organic bases and organic acids. Equation 14 was used for substituted acetic acids. Each functional

$$\log K_{HL} = R - \Sigma \Delta \log K_{HL} \qquad (14)$$

R: reference value for parent compound
$\Delta \log K_{HL}$: correction term for various substituents

group has its specific correction term, and each type of compound has its own reference value (Table 3). The correction term for alkyl substituents is assumed to be zero. The correction term for a substituent on the nth carbon atom is $1/2^n$ of the term for this substituent on the first carbon atom.

Table 3. Correction Terms and Reference Values for Equation 14 Correlating Protonation Constants or Enthalpies of Substituted Acetic Acids and Methylamines

X	$\Delta \log K_{HL}$	
	X-CH$_2$COOH*	X-CH$_2$NH$_2$*
NO$_2$	3.08	–
CN	2.29	5.8
C$_6$H$_5$	0.46	1.4
Si(CH$_3$)$_3$	-0.46	-0.4

*Reference value for acetic acids: 4.80; for methylamines: 10.77.

Christensen et al [61] applied in 1969 the additive approach to the enthalpy of protonation of simple aliphatic amines (eqn. 15). The reference value is adjusted by enthalpy increments as carbon branches are added at sequential positions. The expression for primary amines containing the adjusted enthalpy increments is given by equation 16. Methylamine provides the reference value for

$$\Delta H = \Delta H_{ref} + n_\alpha H_\alpha + n_\beta H_\beta + n_\gamma H_\gamma \qquad (15)$$

$$\Delta H = 13.29 + 0.38 n_\alpha + 0.15 n_\beta + 0.11 n_\gamma \qquad (16)$$

n: number of carbon branches at the α, β, or γ positions
H: enthalpy increments

primary amines, dimethylamine for secondary amines, and trimethylamine for tertiary amines. Examples of results obtained with this equation are given in Table 4. This approach could be extended by adding corrective terms for groups other than methyl or methylene. Published data [61] indicate that the corrective terms for $-NH_2$, $-OH$, and $-O-$ groups in equation 15 should be $-1.6n_\alpha$ $-1.0n_\beta$ $-0.4n_\gamma$. The term for $-C=C-$ should be $-0.77n_\beta$ and for $-C_6H_5$ $-1.27n_\beta$.

Table 4. *Experimental Enthalpies of Protonation, and Enthalpies of Protonation Calculated According to Equation 15 for Aliphatic Amines*

Amine	$-\Delta H_{obs}$*	$-\Delta H_{calc}$*	Δ
Methylamine	13.29±0.11	–	–
Ethylamine	13.71±0.05	13.67	-0.04
Propylamine	13.84±0.05	13.82	-0.02
Isopropylamine	13.97±0.05	14.05	+0.08
Butylamine	13.98±0.05	13.93	-0.05
Isobutylamine	13.92±0.08	13.97	+0.05
s-Butylamine	14.03±0.05	14.20	+0.17
t-Butylamine	14.43±0.07	14.43	0.00
Dimethylamine	12.04±0.05	–	–
Diethylamine	12.73±0.06	12.80	+0.07
Trimethylamine	8.80±0.05	–	–
Triethylamine	10.32±0.05	9.94	-0.38

*Kcal/mole

Barbucci, Paoletti, and Vacca [62] extended the enthalpy additivity to include simple, primary diamines (eqn. 17). For ethylenediamine equation 17 produces an enthalpy of protonation of 12.21 kcal/mole (eqn. 18). Other examples of calculated and observed enthalpies are listed in Table 5.

$$\Delta H = \Delta H(NH_2) + \Sigma 0.5^{n-1}\delta(C) + 0.5^{n-1}\delta(NH_2) + 0.5^{n-1}\delta(NH_3^+) \quad (17)$$

$$\Delta H = 12.66 + (1 + 0.5)0.66 + 0.25(-5.78) = 12.21 \quad (18)$$

n: position of group of that type in the molecule relative to the amine
δ: increment due to that type of group

In 1976 Hancock and Marsicano [63] used a similar approach for complexes of copper(II), nickel(II), and cobalt(II) with polyamines containing n amino groups (eqn. 19). The factor 1.15 changes the ammonia reference value to

Table 5. *Experimental Enthalpies of Protonation, and Enthalpies of Protonation Calculated According to Equation 17 for Polymethylenediamines*

Diamine	$-\Delta H_{obs}$*	$-\Delta H_{calc}$*	Δ
Ethylenediamine	12.18	12.21	-0.03
	10.90	10.94	-0.04
Trimethylenediamine	13.09	13.09	0.00
	12.63	12.46	+0.17
Tetramethylenediamine	13.58	13.54	+0.04
	13.16	13.22	-0.06
Pentamethylenediamine	13.86	13.76	+0.10
	13.41	13.60	-0.19
Hexamethylenediamine	13.91	13.87	+0.04
	13.71	13.79	-0.08

*Kcal/mole

that for primary amines. The second term corrects for the decrease of successive constants, and the last term corrects for the chelate effect of 5-membered rings. For polyamino-polycarboxylic acids with m carboxy groups equation 20 was

$$\log K_{ML} = 1.15 \log \beta_n(NH_3) - \sum_{i=1}^{n-1} i(0.50) + (n-1)\log 55.5 \quad (19)$$

$$\log K_{ML} = 1.15 \log \beta_n(NH_3) - \sum_{i=1}^{n-1} i(0.50) + m \log K(\text{acetate}) - \sum_{i=1}^{m} i(0.19) + (n + m - 1)\log 55.5 \quad (20)$$

used. This relation was successful for iron(II), cobalt(II), and nickel(II), but was not applicable to uranyl, copper(II), zinc(II), cadmium(II), and mercury(II) apparently because the complexes departed from an octahedral configuration. More recently a slightly modified form of this equation was applied successfully to calcium(II), scandium(III), yttrium(III), lanthanum(III), lutetium(III), thorium(IV), manganese(II), zinc(II), iron(III), chromium(III), lead(II), bismuth(III), aluminum(III), gallium(III), indium(III), thallium(III), and zirconium(IV) complexes with several polyaminocarboxylic acids. The standard deviation of the calculated values relative to the observed values was 0.17 log units [64]. Hancock and McDougall [65] used equation 19 to obtain an excellent correlation for the polyamine complexes of several metal ions (Figure 4).

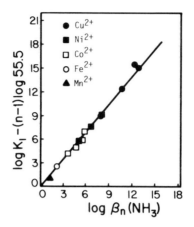

Figure 4. Plot of log K_1 of polyamine complexes (adjusted for symmetry of standard state) vs. overall stability constants, β_n (eqn. 19) of amine complexes.

In 1982 Harris [66] used the additive approach for nickel(II), zinc(II), cadmium(II), and iron(III) complexes of various ligands (eqn. 21). The first term sums the

$$\log K_{ML} = \sum_{i=1}^{n} n_i X_i + [r_5 + \sum_{i=2}^{n_5}(r_5 f_5^{i-1})]_{n_5 \neq 0} + [r_6 + \sum_{i=2}^{n_6}(r_6 f_6^{i-1})]_{n_6 \neq 0} \quad (21)$$

stability increments for the functional groups coordinated to the metal ion, the middle term adds increments for the 5-membered rings formed upon complexation, and the last term adds increments for 6-membered rings formed. The values for the constants employed in equation 21 are given in Table 6. The stability constants were reproduced within 0.8 log units for the ligands used (Fig. 5). Ligands containing strongly electron donating or withdrawing substituents, ligands expected to form 7-membered rings, ligands containing fused rings, ligands having very bulky substituents, and the simplest members of homologous series were not included.

Table 6. Constants Employed in Equation 21

Type of Constant		Ni^{2+}	Zn^{2+}	Cd^{2+}	Fe^{3+}
X_i	Amine	3.0	2.5	2.6	4.7
	Carboxylate	2.1	1.6	1.7	3.4
	Imidazole	2.6	1.7	-	-
	Pyridyl	2.7	1.5	1.5	-
	Thioether	-	-1.6	-0.1	-
	Phenol	-	-	-	9.4
r_5		0.8	0.9	0.09	0.4
f_5		1.0	1.1	2.4	1.0
r_6		-0.3	-0.2	-0.9	-
f_6		0.9	1.7	1.4	-

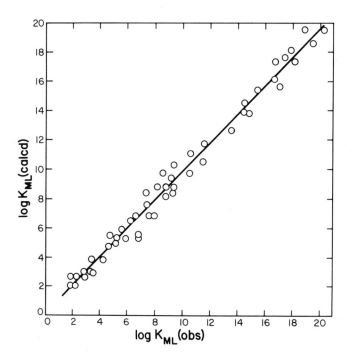

Figure 5. Plot of stability constants of nickel(II) complexes calculated with equation 21 vs. measured stability constants of the same complexes.

Speciation of Metal Complexes

The additive correction approach could be extended to include corrective terms for each effective structural variation such as neighboring electronegative groups and bulky substituents. However, this approach remains primarily empirical with terms added to accomodate factors that contribute to stability constants or enthalpy changes.

The ideal and best method for the calculation of unknown stability constants and enthalpy changes for complexation would use fundamental atomic quantities. Such a method demands a perfect or near-perfect understanding of the forces operating during complexation. Such calculations are not possible at the present time and probably will not become feasible very soon. Munze [67, 68] made a step in this direction with a statistical thermodynamic approach. However, this approach still needs assumptions to obtain approximate values.

Prediction of stability constants by comparison with known values. Thermodynamic complexation values can be predicted by comparing the structure and constitution of a ligand with those of similar ligands. Data such as shown in Table 7

Table 7. Critical Protonation and Formation Constants* [69]

Ligand	Log Protonation Constant		Log Formation Constants, Cu(II)		Log Formation Constants, Ag(I)	
	NR_3	RCO_2	K_1	K_2	K_1	K_2
Primary amine	10.6±0.1		(4)		3.5±0.1	7.4±0.2
Carboxylic acid		4.7±0.1	1.8±0.1	3.2±0.5	0.7	0.6
2-Amino acid	9.6±0.1	2.3±0.1	8.1±0.1	14.9±0.1	3.3±0.3	6.7±0.4
2-Hydroxyamine	9.6±0.1		5.7	9.8	3.1±0.1	6.7±0.0
2-Amidoamine	7.9±0.1		5.0±0.2	5.2±0.2 †		
Dipeptide	8.1±0.1	3.1±0.1	5.5±0.3	4.2±0.7 †		
Secondary amine	11.1±0.2				3.5±0.5	7.0±0.7
Tertiary amine	10.5±0.4				2.8±0.3	4.5±0.7

* 25.0°C, 0.100 M ionic strength.
† Amide protonation constants.

could be extracted from experimental constants for each functional group or set of functional groups. This approach is strictly empirical but has the advantage of quickly producing estimated values. In addition to the primary table, it would be necessary to have tables of correction or equations for temperatures and ionic strengths different from those of the standard table. These tables could probably be developed quickly and stability constants predicted more rapidly than with other methods. The estimates would be accurate enough for most applications to environmental systems. If greater accuracy were needed, a set of linear free energy equations could be tabulated for calculations involving specific systems. This method would serve as a source of useful data until more sophisticated, theoretically-oriented systems are developed.

Examples of how the suggested method might work are shown in Table 8. The predicted stability constants were estimated using constants of analogous complexes listed in Volumes 1-3 of "Critical Stability Constants". The measured values appeared later in Volume 5 [69].

Table 8. Predicted and Measured Stability Constants for Selected Copper(II) Complexes

Ligand	Log Stability Constants		
	Predicted	Measured	Difference
Glycylglycylglycylamide	5.0±0.2	4.77	-0.2
D-Allo-isoleucine	8.1±0.1	8.09	0.0
Glycyl-L-threonine	5.5±0.3	5.57	+0.1

Acknowledgement. This research was supported by Grant ES02971 from the National Institute of Environmental Health Sciences. The authors thank Dr. R. D. Hancock and Dr. W. R. Harris for their helpful comments.

References

[1] A. Albert, **"Selective Toxicity,"** 5th Ed., Chapman and Hall: London, 1973, p. 334.
[2] A. Popovici, C. F. Geschickter, A. Reinovsky and M. Rubin, "Experimental Control of Serum Calcium Levels in-vivo," **Proc. Soc. Exptl. Biol. Med.**, 74 (1950) 415.
[3] A. Saffer and T. Toribora, "Changes in Serum and Spinal Fluid Calcium Effected by Disodium Ethylenediaminetetraacetate," **J. Lab. Clin. Med.**, 58 (1961) 542.
[4] R. O. Bauer, F. R. Rullo, C. Spooner and E. Woodman, "Acute and Subacute Toxicity of Ethylene Diamine Tetraacetic Acid (EDTA) Salts," **Fed. Proc.**, 11 (1952) 321.
[5] H. Spencer, V. Vankinscot, J. Lewin and D. Laszlo, "Removal of Calcium in Man by Ethylenediaminetetra-

acetic Acid. A Metabolic Study," **J. Clin, Invest., 31** (1952) 1023.
[6] M. Rubin, S. Gignac, S. P. Bessman and E. L. Belknap, "Enhancement of Lead Excretion in Humans by Disodium Calcium Ethylenediaminetetraacetate," **Science, 117** (1953) 659.
[7] H. Foreman and T. T. Trujillo, "Metabolism of Carbon-14-Labeled Ethylenediaminetetraacetic Acid in Human Beings," **J. Lab. Clin. Med., 43** (1954) 566.
[8] L. Mosey, cited in J. B. Sidbury, Jr., J. C. Bynum and L. L. Fetz, "Effect of a Chelating Agent on Urinary Lead Excretion. Comparison of Oral and Intravenous Administration," **Proc. Soc. Exptl. Biol. Med., 82** (1953) 226.
[9] J. Bomer, "Metabolismus und Toxizität therapeutischer Chelatbildner," **Strahlentherapie, 142** (1971) 349.
[10] H. Swenerton and L. S. Hurley, "Teratogenic Effects of a Chelating Agent and Their Prevention by Zinc," **Science, 173** (1971) 62.
[11] G. A. Nixon, "Toxicity Evaluation of Trisodium Nitrilotriacetate," **Toxicol. Appl. Pharmacol., 18** (1971) 398.
[12] J. V. Princiotto, E. J. Zapolski, D. H. Bagley, Jr., A. Laskey, R. Morgan and M. Rubin, "Absorption of Oral Chelating Iron," **Biochem. Med., 3** (1970) 289.
[13] G. C. Battistone, F. A. San Filippo, M. I. Rubin, S. Silverman, D. E. Cutright and R. E. Miller, "The Effect of Zinc Injected as Salt or Chelate on Bone Healing in Guinea Pigs," in **"Trace Substances in Environmental Health-IV,"** D. D. Hemphill, Ed., University of Missouri: Columbia, Missouri, 1971, p. 266.
[14] R. I. Henkin and W. D. Grover, "Trichopoliodystrophy (TPD): New Aspects of Pathology and Treatment," in **"Trace Elements Metabolism in Man and Animals-3,"** M. Kirchgessner, Ed., Institute für Ernährungsphysiologie, Technische Universität München, Freising-Weihenstephen, Fed. Rep. Germany, 1978, p. 405.
[15] A. E. Martell, "Chemistry of Carcinogenic Metals," **Envir. Health Perspect., 40** (1981) 207.
[16] M. Rubin and A. E. Martell, "The Implications of Trace Metal-Nitrilotriacetic Acid Speciation on Its Environmental Impact and Toxicology," **Biol. Trace Element Res., 2** (1980) 1.
[17] R. D. Swisher, T. A. Taulli and E. J. Malec, "Biodegradation of NTA Metal Chelates in River Water," in **"Trace Metals and Metal-Organic Interactions in Natural Waters,"** **P. C. Singer, Ed., Ann Arbor Science Publishers:** Ann Arbor, Michigan, 1973, p. 237.

[18] All metal concentrations expect Zn(II) were taken from Table 1-13 in "Major Chemical Constituents in Water," in **"Handbook of Environmental Control," Vol. III, "Water Supply and Treatment,"** R. G. Bond and C. P. Straub, Eds., CRC Press: Cleveland, Ohio, 1973, pp. 15-19.

[19] Zinc concentrations from page 8, Table III, **"Trace Metals in Waters of the United States,"** J. R. Kopp and R. C. Kroner, U.S. Dept. of the Interior, PB215 680, Federal Water Pollution Control Administration, Cincinnati, Ohio (no date). Period covered: 5 years, Oct. 1, 1962 - Sept. 30, 1967.

[20] A. E. Martell and R. M. Smith, **"Critical Stability Constants, Vol. 1, Amino Acids, and Vol. 5, First Supplement,"** Plenum Publishing Corporation: New York, 1974, 1980.

[21] E. Larsson, "The Dissociation Constants of Substituted Ammonium and Silver Diamine Ions and a Relation Between Them," **Z. Phys. Chem., A169** (1934) 208.

[22] J. Bjerrum, "On the Tendency of the Metal Ions Toward Complex Formation," **Chem. Rev., 46** (1950) 381.

[23] M. Calvin and K. W. Wilson, "Stability of Chelate Compounds," **J. Am. Chem. Soc., 67** (1945) 2003.

[24] M. Calvin and N. C. Melchior, "Stability of Chelate Compounds. IV. Effect of Metal Ion," **J. Am. Chem. Soc., 70** (1948) 3270.

[25] H. Irving and R. J. P. Williams, "Order of Stability of Metal Complexes," **Nature (London), 162** (1948) 746; "The Stability of Transition-Metal Complexes," **J. Chem. Soc.,** (1953) 3192.

[26] C. W. Davies, "The Electrolytic Dissociation of Metal Hydroxides," **J. Chem. Soc.,** (1951) 1256.

[27] A. E. Martell and M. Calvin, **"Chemistry of the Metal Chelate Compounds,"** Prentice Hall, Inc.: Englewood Cliffs, N.J., 1952.

[28] L. G. Van Uitert, W. C. Fernelius and B. E. Douglas, "A Comparison of the Chelating Tendencies of β-Diketones Toward Divalent Metals," **J. Am. Chem. Soc., 75** (1953) 2736.

[29] H. Irving and H. Rossotti, "Some Relationships Among the Stabilities of Metal Complexes," **Acta Chem. Scand., 10** (1956) 72.

[30] E. Nieboer and W. A. E. McBryde, "Free-Energy Relationships in Coordination Chemistry. I. Linear Relationships Among Equilibrium Constants," **Can. J. Chem., 48** (1970) 2549; "II. Requirements for Linear Relationships," **Can. J. Chem., 48** (1970) 2565.

[31] L. P. Hammett, "The Effect of Structure upon the

Reactions of Organic Compounds. Benzene Derivatives," **J. Am. Chem. Soc.**, 59 (1937) 96.
[32] P. R. Wells, "Linear Free Energy Relationships," **Chem. Rev.**, 63 (1963) 171.
[33] J. Clark and D. D. Perrin, "Prediction of the Strengths of Organic Bases," **Quart. Rev.**, (1964) 295.
[34] G. B. Barlin and D. D. Perrin, "Prediction of the Strengths of Organic Acids," **Quart. Rev.**, (1966) 75.
[35] W. R. May and M. M. Jones, "The Stability Constants of Several Copper(II) Complexes with Aromatic Carboxylic Acids and Their Correlation with the Hammett Relationship," **J. Inorg. Nucl. Chem.**, 24 (1962) 511.
[36] J. J. R. F. da Silva and J. G. Calado, "Correlation of Stability Constants of Metal Complexes with Hammett's σ-Factor," **J. Inorg. Nucl. Chem.**, 28 (1966) 125.
[37] L. D. Pettit, A. Royston and R. J. Whewell, "Ligands Containing Elements of Group VIB. Part I. The Silver Complexes of Substituted Phenylthioacetic Acids," **J. Chem. Soc. A,** (1968) 2009; L. D. Pettit, C. Sherrington and R. J. Whewell, "Part II. The Silver Complexes of Substituted Phenylselenoacetic Acids," **J. Chem. Soc. A,** (1968) 2204.
[38] E. Nieboer and W. A. E. McBryde, "Linear Free Energy Relationships among Metal Complexes of a Ligand Family," **Inorg. Nucl. Chem. Lett.**, 3 (1967) 569.
[39] J. O. Edwards, "Correlation of Relative Rates and Equilibria with a Double Basicity Scale," **J. Am. Chem. Soc.**, 76 (1954) 1540.
[40] J. O. Edwards, "Polarizability, Basicity and Nucleophilic Character," **J. Am. Chem. Soc.**, 78 (1956) 1819.
[41] G. Schwarzenbach, "Organische Komplexbildner," **Experientia, Suppl. 5,** (1956) 162.
[42] S. Ahrland, J. Chatt and N. R. Davies, "The Relative Affinities of Ligand Atoms for Acceptor Molecules and Ions," **Quart. Rev.**, 12 (1958) 265.
[43] R. G. Pearson, "Hard and Soft Acids and Bases," **J. Am. Chem. Soc.**, 85 (1963) 3533.
[44] R. G. Pearson, "Hard and Soft Acids and Bases," **Chem. Brit.**, 3 (1967) 103.
[45] A. Yingst and D. H. McDaniel, "Use of the Edwards Equation to Determine Hardness of Acids," **Inorg. Chem.**, 6 (1967) 1067.
[46] M. Misono, E. Ochiai and Y. Yoneda, "A New Dual Parameter Scale for the Strength of Lewis Acids and Bases with the Evaluation of their Softness," **J. Inorg. Nucl. Chem.**, 29 (1967) 2685.
[47] G. Klopman, "Chemical Reactivity and the Concept of Charge- and Frontier-Controlled Reactions," **J. Am.**

Chem. Soc., **90** (1968) 223.
[48] R. G. Pearson and R. J. Mawby, "The Nature of Metal-Halogen Bonds," in **"Halogen Chemistry," Vol. 3,** V. Gutmann, Ed., Academic Press: London, 1967, p. 55.
[49] S. Ahrland, "Scales of Softness for Acceptors and Donors," **Chem. Phys. Lett.**, 2 (1968) 303; "Thermodynamics of Complex Formation between Hard and Soft Acceptors and Donors," **Struct. Bonding (Berlin), 5** (1968) 118.
[50] E. Nieboer and W. A. E. McBryde, "Free-Energy Relationships in Coordination Chemistry. III. A Comprehensive Index to Complex Stability," **Can. J. Chem.**, 51 (1973) 2512.
[51] S. Yamada and M. Tanaka, "Softness of Some Metal Ions," **J. Inorg. Nucl. Chem.**, **37** (1975) 587.
[52] C.-T. Liu, "The Bond-Parameter Scale of the Hardness-Softness of Acids and Bases," **Hua Hsueh Tung Pao,** (1976) 346.
[53] A.-P. Tai, "Potential Scale of Softness-Hardness of Acids and Bases, and Preference Strength and Stability of Addition Compounds," **Huaxue Tongbao,** (1978) No. 1, 25.
[54] R. D. Hancock and F. Marsicano, "Parametric Correlation of Formation Constants in Aqueous Solution. 1. Ligands with Small Donor Atoms," **Inorg. Chem.**, **17** (1978) 560.
[55] Y. Zhang, "Electronegativities of Elements in Valence States and Their Applications. 2. A Scale for Strengths of Lewis Acids," **Inorg. Chem.**, 21 (1982) 3889.
[56] R. S. Drago and B. B. Wayland, "A Double-Scale Equation for Correlating Enthalpies of Lewis Acid-Base Interactions," **J. Am. Chem. Soc.**, **87** (1965) 3571.
[57] R. S. Drago, "Quantitative Evaluation and Prediction of Donor-Acceptor Interactions," **Struct. Bonding (Berlin),** 15 (1973) 73.
[58] R. S. Drago, L. B. Parr and C. S. Chamberlain, "Solvent Effects and Their Relationship to the E and C Equation," **J. Am. Chem. Soc.**, 99 (1977) 3203.
[59] R. D. Hancock and F. Marsicano, "Parametric Correlation of Formation Constants in Aqueous Solutions. 2. Ligands with Large Donor Atoms," **Inorg. Chem.**, **19** (1980) 2709.
[60] M. K. Kroeger and R. S. Drago, "Quantitative Prediction and Analysis of Enthalpies for the Interaction of Gas-Phase Ion-Ion, Gas-Phase Ion-Molecule, and Molecule-Molecule Lewis Acid-Base Systems," **J. Am. Chem. Soc.**, **103** (1981) 3250.
[61] J. J. Christensen, R. M. Izatt, D. P. Wrathall and L. D. Hansen, "Thermodynamics of Proton Ionization in Dilute Aqueous Solution. Part XI. pK, ΔH°, and ΔS°

Values for Proton Ionization from Protonated Amines at 25°," **J. Chem. Soc. A,** (1969) 1212.
[62] R. Barbucci, P. Paoletti and A. Vacca, "Predictions of the Enthalpies of Protonation of Amines. Log K, ΔH, and ΔS Values for the Protonation of Ethylenediamine and Tri-, Tetra-, Penta-, and Hexamethylenediamine," **J. Chem. Soc. A,** (1970) 2202.
[63] R. D. Hancock and F. Marsicano, "The Chelate Effect: A Simple Quantitative Approach," **J. Chem. Soc., Dalton,** (1976) 1096.
[64] R. D. Hancock, Department of Chemistry, University of Witwatersrand, Johannesburg, South Africa, personal communication.
[65] R. D. Hancock and G. J. McDougall, "The Affinity of Lead(II) for Nitrogen-Donor Ligands," **J. Coord. Chem.,** 6 (1977) 163.
[66] W. R. Harris, "Structure-Reactivity Relation for the Complexation of Ni, Cd, Zn, and Fe," **J. Coord. Chem.,** 13 (1983) 16.
[67] R. Munze, "Thermodynamische Funktionen der Komplexbildung und Ionenradien 1. Bestimmung der freien Enthalpie, Entropie und Enthalpie von Azetatokomplexen der Lanthaniden und Aktiniden aus zwischenatomaren Donor-Akzeptor Abständen," **J. Inorg. Nucl. Chem.,** 34 (1972) 661. "2. Berechnung der kumulativen freien Bildungsenthalpien und Bildungsentropien der Glykolatokomplexe der Seltenen Erden aus Ionenradien und Polarisierbarkeiten," **J. Inorg. Nucl. Chem.,** 34 (1972) 973.
[68] R. Munze, "Berechnung von Komplexbildungskonstanten für d⁰-Karbonsäure bzw. Aminokarbonsäure-Komplexe aus atomaren und molekularen Parametern," **Z. Phys. Chem.,** 249 (1972) 329; "Zur atomistischen Beschreibung der kumulativen Komplexbildung und des Chelateffekts," **Z. Phys. Chem.,** 252 (1973) 145; "Thermodynamische Funktionen der Komplexbildung und Ionenradien. III. Statistisch-thermodynamische Analyse der Komplexbildungskonstanten in wässrigen Lösungen," **Z. Phys. Chem.,** 256 (1975) 617; "IV. Abhängigkeit der Reaktionsentropie und -enthalpie bei der Komplexbildung harter Kationen von der Struktur der Liganden," **Z. Phys. Chem.,** 256 (1975) 625.
[69] A. E. Martell and R. M. Smith, **"Critical Stability Constants, Volume 1: Amino Acids,"** (1972); **"Volume 2: Amines,"** (1973); **"Volume 3: Other Organic Ligands,"** (1975); **"Volume 5: First Supplement,"** (1982), Plenum Publishing Corporation: New York.

IV

FACTORS INFLUENCING THE COORDINATION CHEMISTRY OF METAL IONS IN AQUEOUS SOLUTIONS

Robert D. Hancock

Department of Chemistry, University of Witwatersrand
Johannesburg, South Africa

Some aspects of the formation of complexes in the gas phase in the absence of a solvent are reviewed. The behavior of the complex-forming reactants in the gas phase is contrasted with their behavior in the aqueous phase. Steric and solvation effects in water may mask the large electronic inductive effects produced by alkyl substitution on amine ligands. In the absence of "masking" solvents, effects such as the "macrocyclic" effect become noticable. The role of steric effects in modifying the Lewis acid behavior of metal ions in solution is discussed. The relatively low affinity of many small metal ions such as Cu(II) for ligands with large donor atoms such as chloride may be the result of steric hindrance between the coordinated chloride and the adjacent coordinated water molecules. The use of a multiparameter equation to predict formation constants of metal ions complexing to unidentate ligands is discussed. Hard and soft acid-base behavior in the gas phase and aqueous phase is examined and shown to be very similar.

The chemistry of metal ions in aqueous solution is of great importance, not only because most chemical reactions in the biosphere occur in an aqueous medium, but also because most of the ideas about factors that govern the formation of metal complexes are based on results obtained in the aqueous phase. To understand the reactions in the aqueous phase leading to metal complexes, the role played by the solvent must be known. This paper discusses some aspects of the formation of complexes in the gas phase in the absence of solvent and contrasts the behavior of the complex-forming reactants in the gas phase with their behavior in the aqueous phase.

Hidden Inductive Effects

The deprotonation constants of protonated ligands determined in aqueous solution have long been regarded as indicators of basicity. Thus, the almost constant value of 10.6 for the negative logarithms of the deprotonation constants, pK_d, for the primary amines with methyl, ethyl, isopropyl, and t-butyl groups is taken to indicate that these amines have the same basicity in aqueous solution. However, the formation constants of the 1:1 Ag(I)-amine complexes increase steadily from methylamine to t-butylamine [1]. The discrepancy between the trends in deprotonation constants and formation constants is inexplicable unless the basicities of the amines in the gas phase are considered [2]. In the gas phase the free energy and the enthalpy of complexation of the proton [2] with these amines increase from methylamine to t-butylamine. Corresponding data for Ag(I)-amine complexes in the gas phase are not available. However, the enthalpy of complexation of Ni(I) with amines increases from methylamine to t-butylamine [3].

Figure 1 shows the dependence of the difference between log K for the formation of a 1:1 amine complex and log K for the formation of the 1:1 methylamine complex in water on the number of methyl groups linked to the carbon atom in the series methylamine through t-butylamine [4]. The effects exerted by the methyl groups are not the same for all metal ions. Whereas the formation constants of the silver complexes increase with increasing number of methyl groups, the constants of the nickel complexes decrease. The most reasonable explanation for these trends is based on the relative increase of inductive and steric effects with increasing methyl substitution. The stability of the complexes and their formation constants are expected to increase in the series from methylamine to t-butylamine, when the inductive effect predominates, and to decrease, when the steric effect prevails. According to the Taft equation [5] the hydrated proton is much more susceptible to steric hindrance than the Ag(I) ion in an aqueous system. The proton basicity of ligands in water has traditionally been regarded as a measure of basicity. However, the proton with its high susceptibility to steric effects is actually a poor choice as a probe for basicity.

Other ligands, for which hidden inductive effects are important, are the C-methyl substituted ethylenediamines [4]. The deprotonation of the ammonium ions determined in aqueous solutions are again independent of the degree of

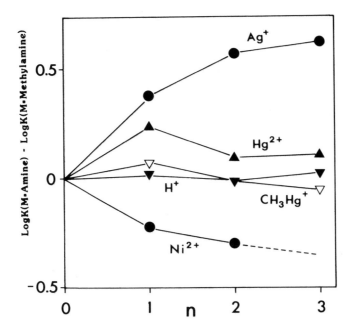

Figure 1. The dependence of [log $K(M \cdot Amine) - $log $K(M \cdot Methyl$-$amine)$] for aqueous systems as a function of the number of methyl groups, n, in $(CH_3)_n CH_{(3-n)} NH_2$. Reproduced with permission from ref. 4.

C-methylation. The response of metal ions to C-methyl substitution illustrates the delicate balance between steric and inductive effects. If C-methylation is symmetrical over the two carbon atoms, inductive effects predominate, and the formation constants increase with the degree of substitution. The inductive effect causes higher ligand field strengths for the Ni(II) and Cu(II) complexes. If methyl substitution is not symmetrical, steric effects predominate, and the stability of the complexes decreases with increasing C-methyl substitution.

The most important example of a hidden inductive effect might be the macrocyclic effect. The N-donor macrocycle cyclam has four secondary nitrogen atoms, whereas its open-chain analog, 1,4,8,11-tetrazaundecane, has two primary and two secondary nitrogen atoms [6]. The deprotonation constants of the primary and secondary ammonium groups are almost equal in aqueous solution indicating that these amino groups have the same basicity. However, Munson [7] demonstrated as early as 1965 that in the gas phase the basicity of amines increases in the sequence $NH_3 < CH_3NH_2 < (CH_3)_2NH < (CH_3)_3N$. In agreement with this trend in gas-phase basicities, the formation constant of the Ni(II) complex of cyclam in aqueous solution is six orders of magnitude larger than the corresponding constant for the complex with 1,4,8,11-tetrazaundecane [8]. The macrocyclic complex has a much higher ligand field strength than the open-chain complex [9].

Why do secondary nitrogen atoms in compounds other than macrocycles not produce similar increases in stability and ligand field strength? The answer is not difficult to find. Complexes of N-alkyl substituted ligands form metal-nitrogen bonds that are much longer than the corresponding bonds in complexes with unsubstituted ligands. Steric effects are responsible for the lengthening of the bonds. For instance, the nickel-secondary nitrogen bonds in [Ni(N,N-dimethylethylenediamine)(NO$_2$)]$^+$ is 6 pm longer than nickel-primary nitrogen bonds (210 pm) [10]. Macrocycles can assume conformations, in which steric hindrance is minimized. Under these conditions, the secondary nitrogen atoms are able to approach the metal ions to "normal" unstrained metal-nitrogen bond lengths, when the effects of their enhanced basicity become apparent.

Correct ligand design is important to minimize steric effects and to take advantage of them. Steric effects are significant in cyclic and in noncyclic ligands. We are using Empirical Force Field calculations to model these effects [11-14]. Several metal ions were found to be so small that they "rattle" in virtually all of their complexes. This situation is reminiscent of the circumstances governed by the radius ratio rules for ionic compounds. A large anion (ligand) cannot approach a small cation sufficiently to form strong bonds. For instance, the small Co(III) cation, which has a strain-free Co-N bond length of 192 pm, forms Co-N bonds with an average bond length of 198 pm with polyamine ligands. The bonds with polyamines are lengthened by Van der Waals repulsion between adjacent donor atoms and their substituents [6, 12]. The stability of complexes of small metal ions such as Co(III), Ni(III), and Ni(IV) appears to depend on how close the donor atoms of the ligand can approach the metal ion. For instance, the ligand 1,4,7-triazacyclononane N,N',N"-triacetate (TACNTA) is able to pack very efficiently around small metal ions allowing the Ni(III)-TACNTA complex to form spontaneously in nitric acid [15]. The Ni-N bond length in the tightly packed TACNTA complex is 192 pm, whereas the corresponding distance in the less stable Ni-hexamethyl-cyclam (5,5,7,12,12,14-hexamethyl-1,4,8-11-tetraazacyclo-tetradecane) complex is 199 pm.

Specific Solvation Effects

Traditionally, bases are divided into protic or hard bases and nonprotic bases [16]. Hard ligands such as fluoride ion and ammonia have a strong affinity for the hard proton, whereas soft ligands such as chloride ion and phosphine have virtually no affinity for the hard proton. The proton is regarded as the typical hard acid, and in the Edwards equation (eqn. 1) [17] the proton basicity is effectively taken as a measure of hardness.

$$\log(K/K_0) = \alpha E_n + \beta H \qquad (1)$$

K : equilibrium constant in nonaqueous medium
K_o : equilibrium constant in aqueous medium
E_n : nucleophilic constant
H : relative basicity of the donor to the proton
α, β: substrate constants

If Klopman's analysis of hard and soft acids and bases [18], which claims hard-hard interactions to involve ionic bonds, is accepted, conceptual difficulties arise. Is not the bond formed between a proton and a carbanion the epitome of covalence? The proton basicity of a carbanion is very high, and the carbanion should be a hard donor. The interactions of the hard proton with the hard carbanion should lead to an ionic bond. Should the cyanide ion be regarded as a hard ligand? The small deprotonation constant of HCN (pK = 9.2) attests to the high proton basicity of the cyanide ion suggesting that this ion is a hard ligand.

The experimental results obtained in the gas phase [2] on acid-base behavior are illuminating. In the gas phase the soft base hydrogen sulfide (dimethyl sulfide) has a higher proton basicity than the hard base water (dimethyl ether). Also, soft trimethylphosphine is a much stronger proton base than hard trimethylamine. Coordination of a single water molecule to each of the protonated bases changes the order of basicity in the gas phase to the order expected in aqueous solution [19]. For example, the reaction between dimethyl sulfide and protonated dimethyl ether (eqn. 2) proceeds to the right in the gas phase with an enthalpy of −8.0 kcal/mol. Coordination of a single water molecule to each of the protonated species causes the reaction (eqn. 3)

$$(CH_3)_2S + (CH_3)_2O-H^+ \rightleftharpoons (CH_3)_2S-H^+ + (CH_3)_2O \qquad (2)$$
$$(CH_3)_2S + (CH_3)_2O-H^+ \cdot H_2O \rightleftharpoons (CH_3)_2S-H^+ \cdot H_2O + (CH_3)_2O \qquad (3)$$

to proceed to the left with an enthalpy of −6.0 kcal/mol [19]. This behavior is in agreement with the basicity expected for dimethyl ether in aqueous solution. The effect caused by coordination of solvent molecules was called "specific solvation" by Taft [2]. In the case of the dimethyl chalcogenides, this effect may be explained by the covalent character of the sulfur-proton bond. The sulfur-proton bond is considerably more covalent than the oxygen-proton bond. Covalent S-H bonding lessens the ability of the proton to form hydrogen bonds with water and prevents stabilization of the cationic species through dispersal of the positive charge.

The importance of the specific solvation effect becomes apparent when the proton basicities of phosphines and amines in water are compared. The deprotonation constants of the trialkylphosphonium and trialkylammonium cations are surprisingly close. For instance, the pK_d of tri(isobutyl)-phosphonium is 8.7 [20] and that for tri(isobutyl)amine is 10.3. In contrast, the pK_d for the ammonium ion is 9.2, whereas the pK_d for the phosphonium ion is −14. The ammonium ion is stabilized by dispersal of the positive charge to the solvent through hydrogen bonding. Such a stabilization cannot be achieved by the phosphonium ion, which is larger and forms more covalent bonds to hydrogen than the ammonium ion [20]. The dramatic increase in aqueous pK_d in the phosphonium series from −14 for the phosphonium ion to +8.7 for the tri(isobutyl)phosphonium ion mirrors the increase of the inductive effect in the gas phase with increasing alkyl substitution. The inductive effect in the corresponding ammonium series is counteracted by charge dispersal through hydrogen bonding, which is specially important for the ammonium ion. Steric effects, which increase dramatically even in the gas phase with alkyl substitution, are also important in the series of ammonium ions [4].

The Balance between Steric Effects and Specific Solvent Effects

In the series methyl-, ethyl-, isopropyl-, and t-butyl-ammonium ion the inductive effects are tempered by steric effects, whereas in the series ammonium, methyl-, dimethyl-, and trimethylammonium ion the inductive effects are largely compensated by specific solvent effects with a contribution from steric effects. It is not yet possible to assess quantitatively the contributions from specific solvent and steric effects. However, the perturbations of the inductive effect in the series of primary ammonium ions may largely be caused by steric effects, because these compounds probably form roughly the same number of hydrogen bonds. The formation constants for metal complexes in aqueous solution - consisting of amines and metal ions such as Ag(I) that are not very susceptible to steric hindrance - decrease with increased methyl substitution in the series ammonia to trimethylamine [21]. This trend suggests a large contribution from the specific solvent effect. The dramatic differences between the proton basicities of first row donors (fluoride ion, O-donors, N-donors) and higher row donors (chloride ion, S-donors, P-donors), and of amines and phosphines as a function of alkyl substitution are probably caused to a large degree by specific solvation effects.

The Formation of Complexes in Aqueous Solution

Drago [22] had considerable success in correlating enthalpies of adduct formation between neutral acids and bases in solvents of low dielectric constant. Equation 4,

$$\log K = E_A^{aq} E_B^{aq} + C_A^{aq} C_B^{aq} - D_A D_B \qquad (4)$$

which is similar to the equation used by Drago but contains E and C parameters different from those employed by Drago, is applicable to formation constants of 1:1 complexes in aqueous solution. The parameter E measures the ability of the acid A and the base B to form ionic bonds in aqueous solution. The parameter C expresses the ability for covalent bond formation. Equation 4 with only the E and C parameters did not predict correctly the formation constants of adducts, when steric hindrance was expected to be important as, for instance, in the case of heavily alkyl-substituted ligands, and when ligands with second or third row donor atoms were present. A two-term expression that considers only ionic and covalent effects in bond formation cannot be successful for aqueous systems, because steric hindrance and specific solvation are not taken into account. Examination of two-term equations such as the Edwards equation (eqn. 1) [17] reveals that observed weak complexes formed, for instance, between Ni(II) - a small Lewis acid - and ligands with large donor atoms cannot be predicted by this approach [23]. At least one additional term is needed to model steric and specific solvation effects satisfactorily. We added the parameters D_A and D_B (eqn. 4) in an attempt to model simultaneously the effects of steric hindrance and specific solvation. Large donor atoms such as phosphorus should cause considerable steric hindrance and - because of the covalent character of their bonds to protons - greatly weaken specific solvation effects. Although these two effects could be separated at a later stage, the use of a single pair of parameters appears to be reasonable. The three-term equation 4 successfully predicts the aqueous-phase chemistry of 31 Lewis acids with sixteen unidentate Lewis bases. In most cases the calculated formation constants agreed with the experimental constants within 0.2 log units [23, 24].

The parameters H (hardness parameter), E (ionic parameter), C (covalent parameter), and D (steric and specific solvation parameter) for acids and bases are listed in Tables 1 and 2, respectively. The E and C parameters may be set arbitrarily [22] for any two acids or bases. This choice then fixes E and C for all other acids and bases. We chose E_B as 1.0 and C_B as 0.0 for fluoride. For hydroxide E_B was chosen as 0.0 and C_B as 14.0 (Table 2). A value of 0.0 was assumed for D_B for both of these ligands. The D_B value of 1.0 was assigned to bromide. A computer program was then used to arrive at a best fit with E and C values for fluoride and hydroxide as fixed points. The hardness parameters, H, obtained as the ratio E/C, allow acids and bases to be arranged in a unique order of hardness even though the E and C parameters are set arbitrarily [24]. Equation 4 with the D parameter predicts the formation constants of 1:1 complexes satisfactorily. The particular values of the parameters E, C, D, and H are also of interest. As expected from the Edwards equation (eqn. 1) [17], a reasonable correlation exists between H and the oxidation potentials of the bases. Values of D correlate well with the Van der Waals radii of the donor atoms. C_A - a measure of the strength of covalent interaction - correlates with NMR metal-hydrogen coupling constants [24], for some 20 complexes of the general formulae CH_3HgL and $(CH_3)_3bipyPt(IV)L$.

Table 1. Values of H_A, E_A^{aq}, C_A^{aq}, and D_A for Lewis Acids*

Lewis Acid**	H_A	E_A^{aq}	C_A^{aq}	D_A
Au^+	−16.	−3.0	0.190	0.0
Ag^+	−10.6	−1.52	0.143	0.0
Cu^+	− 1.30	−0.56	0.430	2.5
Hg^{2+}	1.63	1.35	0.826	0.0
CH_3Hg^+	2.50	1.60	0.640	0.0
Tl^{3+}	2.66	2.55	0.960	0.0
Cu^{2+}	2.68	1.25	0.466	6.0
H^+	3.04	3.07	1.009	20.0
Cd^{2+}	3.31	0.99	0.300	0.6
Ni^{2+}	3.37	1.01	0.300	4.5
Co^{3+}	3.77	3.30	0.875	7.0
Zn^{2+}	4.26	1.33	0.312	4.0
Co^{2+}	4.34	1.20	0.276	3.0
Fe^{2+}	5.94	1.52	0.256	2.0
VO^{2+}	5.97	3.96	0.664	3.5
In^{3+}	6.30	4.49	0.714	0.5
Bi^{3+}	6.39	5.92	0.926	0.0
Pb^{2+}	6.69	2.76	0.413	0.0
Mn^{2+}	7.09	1.58	0.223	1.0
Cr^{3+}	7.14	5.15	0.721	1.5
Fe^{3+}	7.22	6.07	0.841	1.5
Ga^{3+}	7.69	6.06	0.788	1.5
U^{4+}	7.80	7.55	0.968	3.0
Sn^{2+}	8.07	5.65	0.700	0.0
UO_2^{2+}	8.40	4.95	0.589	1.0
Lu^{3+}	10.07	4.57	0.454	0.0
La^{3+}	10.30	3.90	0.379	0.0
Mg^{2+}	10.46	1.86	0.178	1.5
Sc^{3+}	10.49	7.03	0.671	0.0
Al^{3+}	10.50	6.90	0.657	2.0
Y^{3+}	10.64	4.76	0.477	0.0
Ca^{2+}	12.16	0.98	0.081	0.0

*Data from ref. 24

**The metal ions are arranged in order of increasing hardness with Au(I) the softest and Ca(II) the hardest ion.

The dependence of the standard oxidation potentials of Lewis acids on their hardness parameter is shown in Figure 2. D_A does not correlate well with the ionic radii of the metal ions. Metal ions of approximately the same size

Table 2. Values of H_B, E_B^{aq}, C_B^{aq}, and D_B for Lewis Bases*

Lewis Base**	H_B	E_B^{aq}	C_B^{aq}	D_B
F^-	0.0	1.00	0	0.0
CH_3COO^-	0.0	0.0	4.76	0.0
OH^-	0.0	0.0	14.00	0.0
N_3^-	−0.064	−0.067	10.4	0.2
$S=C=N^-$	−0.082	−0.76	9.3	0.2
NH	−0.088	−1.08	12.34	0.0
C_5H_5N	−0.102	−0.74	7.0	0.0
Cl^-	−0.100	−1.04	10.4	0.6
SO_3^{2-}	−0.107	−1.94	18.2	0.4
Br^-	−0.108	−1.54	14.2	1.0
$S_2O_3^{2-}$	−0.119	−3.15	26.5	1.1
I^-	−0.122	−2.43	20.0	1.7
$N=C-S^-$	−0.128	−1.83	14.3	1.0
$(HOCH_2CH_2)_2S$	−0.135	−1.36	10.1	0.6
$PPh_2(4-C_6H_4SO_3^-)$	−0.132	−3.03	23.0	0.7
$As(3-C_6H_4SO_3^-)_3$	−0.135	−1.93	14.3	−
$(NH_2)_2C=S$	−0.135	−2.46	18.2	0.6
$HOCH_2CH_2P(C_2H_5)_2$	−0.141	−4.89	34.7	0.6
CN^-	−0.148	−4.43	30.0	0.38

*Data from ref. 24.
**The Lewis bases are arranged in order of decreasing hardness with fluoride the hardest and cyanide the softest ion.

appear to become less susceptible to steric effects as the ions become more ionic. Under these conditions no simple correlation between H and ionic radii can be expected.

The position of the proton just above cadmium in the hardness order (Table 1) may be satisfying or disturbing depending on one's point of view. If one considers the proton to be the archtypical hard acid, then the position of the proton next to cadmium - implying that the proton is a soft Lewis acid - might be perturbing. If, however, acid-base behavior is analyzed in terms of covalence/ionicity of the bonds formed in an acid-base reaction [18], the classification of the proton as a soft acid is in accord with the proton's ability to form highly covalent bonds. The proton must be placed above cadmium, otherwise the high affinity of the proton in water for ligands such as trimethylphosphine, cyanide, and mercaptans cannot be understood. What makes the proton appear "hard" is its very large D value that greatly reduces the predicted affinity of

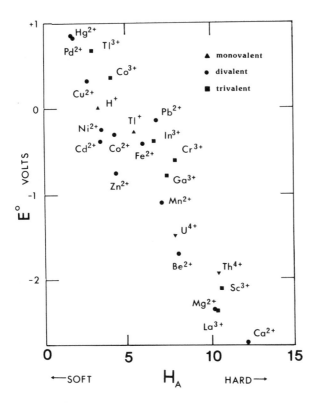

Figure 2. *The relationship between the standard reduction potentials, E^O, of metal ions and the hardness parameter, H_A.*

the proton for ligands with large donor atoms. The proton thus occupies a position in the hardness order close to that of CH_3^+. The proton's position seems reasonable considering the similarity of the covalent character of bonds formed by carbon and by hydrogen.

Linear Free Energy Relation (LFER) Diagrams

When the logarithms of the formation constants of complexes formed from metal ions and unidentate ligands are plotted versus the logarithms of the formation constants of a reference metal ion, the relative hardness of metal ions can be evaluated from the resulting diagram. Such a diagram for Ag(I) versus Hg(II) is shown in Figure 3. The lines

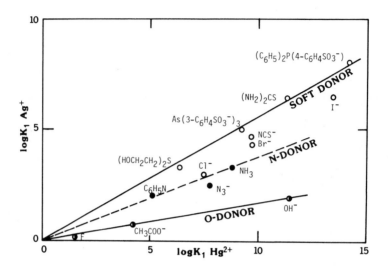

Figure 3. The linear free energy relation diagram of silver ion versus mercuric ion. Reproduced with permission from ref. 23.

connecting groups of related donors have slopes that increase with increasing softness of the donors. The larger the slope of a line the higher is the relative stability of the silver complexes with ligands on this line. Because the line formed by soft ligands has the largest slope in Figure 3, the diagram suggests - in contrast to most other proposed orders of hardness - that Ag(I) is softer than Hg(II). The pattern of the Ag(I)/Hg(II) diagram is reversed in the LFER diagram Bi(III)/Hg(II) (Fig. 4). According to Figure 4, Bi(III) is harder than Hg(II).

If all LFER diagrams resembled Figures 3 and 4, a two-term equation such as the Edwards equation would be quite satisfactory to predict formation constants. However, the diagram proton/Hg(II) (Fig. 5) is different from the Ag(I)/Hg(II) and Bi(III)/Hg(II) diagrams. The ligands with first-row donor atoms produce a pattern similar to those in the Ag and Bi diagrams, whereas the points for ligands with second- and third-row donor atoms are displaced downwards from their expected positions. The largest displacement is associated with iodide, the ligand with the highest D value

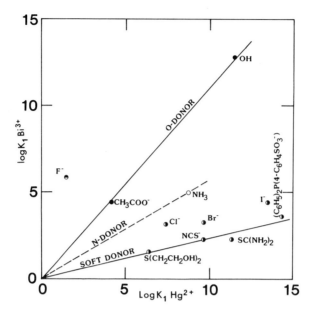

Figure 4. The linear free relation diagram of bismuth(III) ion versus mercuric ion.

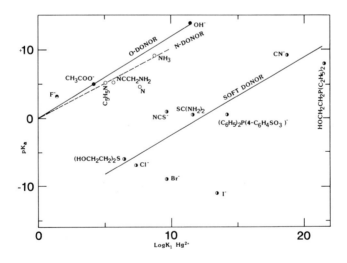

Figure 5. The linear free energy relation diagram of the proton versus mercuric ion.

(Table 2). These displacements of soft ligands, which cannot be modeled with two-term equations, illustrate the need for the D parameter.

Figure 6 is the linear enthalpy relation (LER) diagram for the proton versus Mn(I) in the gas phase [25]. In this Figure the enthalpies for the protonation of the ligands, L, in the gas phase are plotted versus the enthalpies for the reactions, in which methanethiol in the Mn(I)-methanthiol complex is replaced by the ligands, L. This diagram resembles the LFER diagrams Ag(I)/Hg(II) (Fig. 3), Bi(III)/Hg(II) (Fig. 4), and H$^+$/Hg(II) (Fig. 5). The points for ligands with a particular donor atom produce a straight line. The large slope for the ligands with sulfur donor atoms indicates that the proton is softer than Mn(I). Using such diagrams to order Lewis acids in the gas phase, the

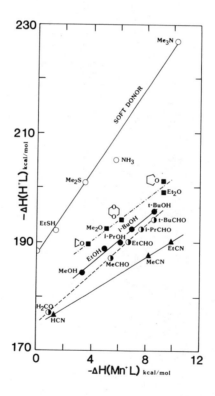

Figure 6. The linear enthalpy relation diagram of the proton versus manganese(I) in the gas phase. Reproduced with permission from ref. 25.

following sequence of increasing softness is obtained [25]: Al(I) < K(I) < Li(I) < Mn(I) < Co(I) < Ni(I)cp < H(I) < Ni(I) < Cu(I). This order is similar to that expected from aqueous-phase chemistry. The proton is again positioned toward the soft end of the series. Virtually all molecules can act as bases in the gas phase. In the absence of the solvent water, which is a strong base, a rich chemistry of weaker bases such as alcohols, aldehydes, and nitriles emerges. Figure 6 suggests that the hardness of the ligands increases in the sequence "sulfur donors ~ trimethylamine < ammonia < ethers < alcohols < aldehydes < nitriles. This sequence is in reasonable agreement with the hardness of representative ligands in solvents of low dielectric constant obtained as the following E/C ratios of Drago et al. [22]: dimethyl sulfide, 0.046; trimethylamine, 0.070; ammonia, 0.393; diethyl ether, 0.296; acetone, 0.424; acetonitrile, 0.662.

We believe that a satisfying picture of complex formation in aqueous solution is emerging. A fairly simple equation such as equation 4 successfully correlates formation constants. The study of complex formation in the gas phase and in solvents other than water will help us to understand the factors governing complex formation in water and to interpret the parameters in equations used for predictive purposes. Water as a solvent masks effectively many of the features of complex formation such as hidden inductive effects.

Acknowledgement. The author thanks his colleagues who helped him in this work, particularly Dr. F. Marsicano, Mr. B. S. Nakani, Dr. N. P. Finkelstein and Professor J. C. A. Boeyens.

References

[1] R. D. Hancock, "Polar and Steric Effects in the Stability of Silver Complexes of Primary Amines," J. **Chem. Soc., Dalton Trans.,** (1980) 416.
[2] R. W. Taft, "Proton Transfer Equilibria in the Gas and Solution Phase," in **"Kinetics of Ion-Molecule Reactions,"** NATO Advanced Study Institute Series B: Physics, Vol. 40, P. Ausloos, Ed., Plenum Press: New York, 1979, p. 271.
[3] M. M. Kappes and R. H. Staley, "Relative Bond Dissociation Energies for Two-Ligand Complexes of Cu^+ with Organic Molecules in the Gas Phase," **J. Am. Chem. Soc.,**

104 (1982) 1813.
[4] R. D. Hancock, B. S. Nakani and F. Marsicano, "Relationship between Lewis Acid-Base Behavior in the Gas Phase and in Aqueous Solution. 1. Role of Inductive, Polarisability, and Steric Effects in Amine Ligands," **Inorg. Chem.**, 22 (1983) 2531.
[5] W. A. Pavelich and R. W. Taft, "The Evaluation of Inductive and Steric Effects on Reactivity. The Methoxide-Ion Catalyzed Rates of Metholysis of 1-Menthyl Esters in Methanol," **J. Am. Chem. Soc.**, 79 (1957) 4935.
[6] R. D. Hancock and G. J. McDougall, "Origin of Macrocyclic Enthalpy," **J. Am. Chem. Soc.**, 102 (1980) 6551.
[7] M. S. B. Munson, "Proton Affinities and the Methyl Inductive Effect," **J. Am. Chem. Soc.**, 87 (1965) 2332.
[8] D. K. Cabbiness and D. W. Margerum, "Macrocyclic Effect on the Stability of Copper(II) Tetramine Complexes," **J. Am. Chem. Soc.**, 91 (1969) 6540.
[9] D. H. Busch, "Distinctive Coordination Chemistry and Biological Significance of Complexes with Macrocyclic Ligands," **Acc. Chem. Res.**, 11 (1978) 392.
[10] R. Birdy, D. M. L. Goodgame, J. C. McCartney and D. Rogers, "Steric Factors in Nitrite Coordination. Part 1. Crystal Structure of Bis(N-N'-dimethylethylenediamine)(nitrito-O,O')nickel(II) tetrafluoroborate," **J. Chem. Soc., Dalton Trans.**, (1977) 1730.
[11] G. J. McDougall, R. D. Hancock and J. C. A. Boeyens, "Empirical Force-Field Calculations of Strain Energy Contributions to the Thermodynamics of Complex-Formation. Part 1. The Difference in Stability between Complexes containing Five- and Six-Membered Chelate Rings," **J. Chem. Soc., Dalton Trans.**, (1978) 1438.
[12] G. J. McDougall and R. D. Hancock, "The N-M-N Bond Angle in the Chelate Ring of Ethylenediamine. The Crystal Structure of Tetra-aqua(ethylenediamine)-nickel(II) nitrate," **J. Chem. Soc., Dalton Trans.**, (1980) 654.
[13] R. D. Hancock, G. J. McDougall and F. Marsicano, "Empirical Force Field Calculations of Strain-Energy Contributions to the Thermodynamics of Complex Formation. 3. Chelate Effect in Complexes of Polyamines," **Inorg. Chem.**, 18 (1979) 2847.
[14] J. C. A. Boeyens, R. D. Hancock and G. J. McDougall, "Empirical Force-Field Calculations of Strain-Energy Contributions to the Thermodynamics of Complex-Formation. Part 2. Further Calculations on Five- and Six-Membered Rings," **S. Afr. J. Chem.**, 32 (1979) 23.
[15] M. J. van der Merwe, J. C. A. Boeyens and R. D.

Hancock, "Optimum Ligand Hole Sizes for Stabilizing Nickel(III). Structure of the Nickel(III) Complex of 1,4,7-Triazacyclononane-N,N',N"-triacetate," **Inorg. Chem.**, 22 (1983) 3489.
[16] R. G. Pearson, "Hard and Soft Acids and Bases," **J. Am. Chem. Soc.**, 85 (1963) 3533.
[17] J. O. Edwards, "Correlation of Relative Rates and Equilibria with a Double Basicity Scale," **J. Am. Chem. Soc.**, 76 (1954) 1540.
[18] G. Klopman, "Chemical Reactivity and the Concept of Charge- and Frontier- Controlled Reactions," **J. Am. Chem. Soc.**, 90 (1968) 223.
[19] Y. K. Lau and P. Kebarle, "Hydrogen Bonding Solvent Effect on the Basicity of Primary Amines CH_3NH_2, $C_2H_5NH_2$, and $CF_3CH_2NH_2$," **Can. J. Chem.**, 59 (1981) 151.
[20] F. M. Jones and E. M. Arnett, "Thermodynamics of Ionization and Solution of Aliphatic Amines in Water," **Prog. Phys. Org. Chem.**, 11 (1974) 263.
[21] R. M. Smith and A. E. Martell, **"Critical Stability Constants, Vol. 2,** Amines," Plenum Press: New York, 1975.
[22] R. S. Drago, G. C. Vogel and T. E. Needham, "A Four-Parameter Equation for Predicting Enthalpies of Adduct Formation," **J. Am. Chem. Soc.**, 93 (1971) 6014.
[23] R. D. Hancock and F. Marsicano, "Parametric Correlation and Formation Constants in Aqueous Solution. 1. Ligands with Small Donor Atoms," **Inorg. Chem.**, 17 (1978) 560.
[24] R. D. Hancock and F. Marsicano, "Parametric Correlation of Formation Constants in Aqueous Solution. 2. Ligands with Large Donor Atoms," **Inorg. Chem.**, 19 (1980) 2709.
[25] J. S. Uppal and R. H. Staley, "Relative Binding Energies of Organic Molecules to Mn^+ in the Gas Phase," **J. Am. Chem. Soc.**, 104 (1982) 1238.

V

ENVIRONMENTAL FACTORS IN THE INORGANIC CHEMISTRY OF NATURAL SYSTEMS: THE ESTUARINE BENTHIC SEDIMENT ENVIRONMENT

Owen P. Bricker

U.S. Geological Survey
Reston, VA 22092 USA

Estuaries receive inputs of metals from natural and anthropogenic sources. Much of this material is strongly associated with particulate matter which is transported by estuarine circulation to various parts of the estuary where it settles to the bottom and accumulates. Bacterially mediated oxidation of organic matter in the sediments creates anoxic conditions in which many of the oxidized detrital phases are not stable. Dissolution of these redox sensitive phases mobilizes metals to the interstitial waters where they may react with products of the decomposition of organic matter (sulfide, carbonate, phosphate) to form new phases that are stable in the anoxic environment, or remain in solution as dissolved species. Physical and biological processes continuously rework the sediment and redistribute both interstitial waters and the particulate bed material. Through this combination of physical, chemical, and biological processes, metals that enter the estuary in particulate form may be mobilized and made available to benthic fauna or to organisms in the water column.

Estuaries form a buffer zone between freshwater rivers and the sea. Historically, their shores have been favored sites for human settlement. Some of the world's highest population densities are found adjacent to major estuarine systems. Parallel industrial development has occurred in these same areas, because estuaries provide an abundant source of water for industrial processes and direct access to marine transport of raw materials and finished products. Estuaries have been used as convenient conduits for the disposal of the broad spectrum of wastes generated by industries and for sewage disposal. As a consequence of these activities, estuaries in industrialized nations commonly are highly stressed environments. Toxic metals represent an obvious threat to the stability of the estuarine ecosystem. Recognition of the role that these substances play in affecting health of an estuary requires a thorough understanding of their chemical, physical, and biological dynamics in the system. In this paper, I would like to address some of the factors that govern the fate of metals in the estuarine benthic environment using the Chesapeake Bay as an example.

The Chesapeake Bay, located in the mid-Atlantic region of the North American Continent, is a typical coastal plain estuary (Fig. 1). It is a geologically young system

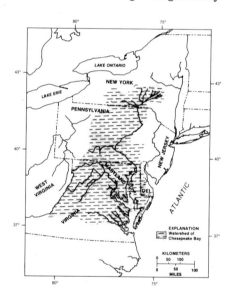

Figure 1. East coast of the U.S. showing location of the Chesapeake Bay and its 98,500 square kilometer watershed.

originating less than 10,000 years ago when the Atlantic ocean, rising in response to meltwaters from receding Pleistocene glaciers, flooded the valley of the Ancestral Susquehanna River. The salinity ranges from that of coastal marine water (26‰) at the mouth of the Bay to fresh Susquehanna River water at its head, three hundred kilometers to the north. Annual cyclic variations in salinity result from seasonal differences in river flow

(Fig. 2). The Bay circulation is basically a two-layer

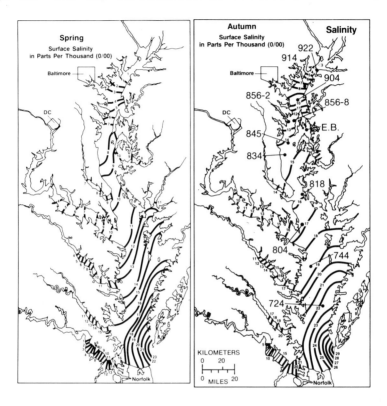

Figure 2. Sampling stations and salinity gradients in the Chesapeake Bay in spring and autumn.

system with the heavier saline waters flowing northward toward the head of the estuary along the bottom and the less dense riverine waters flowing southward toward the sea at the surface [1]. The mixing zone (zone of maximal turbidity) is located near the head of the Bay and varies in position seasonally with discharge of the Susquehanna River [2].

Sources of Chemical Substances

Chemical substances such as nutrients, trace elements, and organic compounds are continuously added to estuaries by inflowing tributary rivers, from shoreline erosion, from the coastal marine environment, from the atmosphere, and from the biosphere (Fig. 3). Much of this material, dissolved

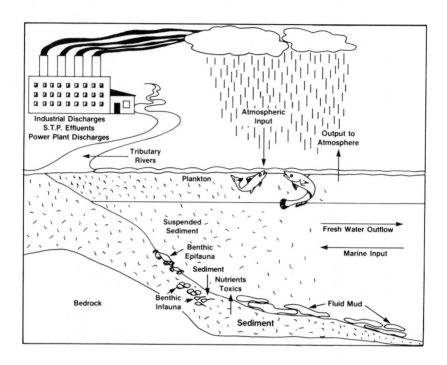

Figure 3. Schematic diagram depicting inputs and sinks for materials in the estuarine system.

and particulate, consists of the natural products of weathering and erosional processes and of biologic activity. In addition, ever increasing amounts of anthropogenic by-products and wastes enter the estuary directly in effluent discharges, in nonpoint source runoff, and via atmospheric deposition. A large proportion of both the natural and the anthropogenic material is intimately associated with sediments, particularly those of fine particle size and large surface area. These sediments are transported to various parts of the estuary where they settle and accumulate on the bottom [3]. In the Chesapeake Bay, the sediments blanketing the bottom constitute the largest reservoir of natural and anthropogenic chemical substances in the estuarine system [4]. This is true for most of the estuaries around the world. The behavior of sediment-associated constituents after deposition and burial will determine their fate in the system. If they remain with the sediment, they may be, for all practical purposes, removed from the system. If, however, they are mobilized by processes active within the sediment, they may become available to the benthic fauna or be returned to the water column. This has important implications in regard to nutrients and toxic materials. To assess the importance of bottom sediments as a source of these substances, it is necessary to understand the chemical, physical, and biological processes that go on in the benthic sediment environment.

Benthic Sediment Environment

Within the bottom sediments the chemical and physical conditions are very different from those in the water above. Sediment particles settle to the bottom and accumulate. Rates of sedimentation in the Chesapeake Bay range from several tenths of a centimeter per year in the southern portion of the Bay to several centimeters per year in the northern portion of the Bay [5]. With the sediments, water is trapped in the interstices between the particles. Initially, while the newly deposited sediment is at or near the sediment-water interface soon after deposition, there is relatively free exchange between the water occupying the interstitial spaces and that in the estuary above. Addition of only a few millimeters of new sediment to the top of the pile, however, strongly inhibits the free exchange of interstitial water within the sediment layer that is now several millimeters beneath the sediment-water interface. With increasing depth of burial, exchange with the water column becomes increasingly difficult. Bacterially mediated decomposition of organic matter in the sediments proceeds utilizing oxygen as an electron acceptor at the sediment-water interface (eqn. 1). Within a very short distance below the interface, oxygen is depleted, and the bacteria switch to sulfate as an electron acceptor (eqn. 2) [6].

$$(CH_2O)_{106}(NH_3)_{16}(PO_4) + 106\ O_2 \longrightarrow 106\ HCO_3^- + 16\ NH_4^+ + HPO_4^{2-} + 89\ H^+ \quad (1)$$

$$(CH_2O)_{106}(NH_3)_{16}(PO_4) + 53\ SO_4^{2-} \longrightarrow 106\ HCO_3^- + 16\ NH_4^+ + HPO_4^{2-} + 53\ HS^- + 36\ H^+ \quad (2)$$

These reactions produce anoxic reducing conditions in the sediment, and the reaction products are released into the interstitial waters where they may reach high concentrations due to the restricted exchange with the open water of the estuary. Figure 4 depicts the conditions of pH and redox potential under which the various products of reactions 1 and 2 are stable. Figure 5 shows the concentrations of some major constituents in the interstitial waters, as a function of depth, for typical Chesapeake Bay sediments. Note the steep gradients in concentration.

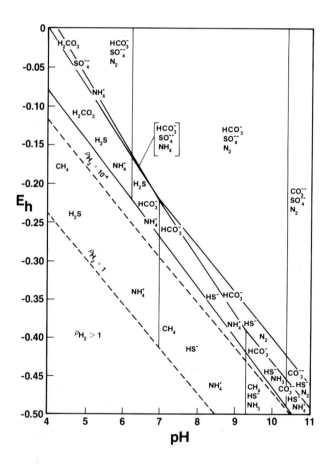

Figure 4. E_h-pH diagram depicting the stability fields of some carbon, nitrogen and sulfur species at 25° and one atmosphere total pressure.

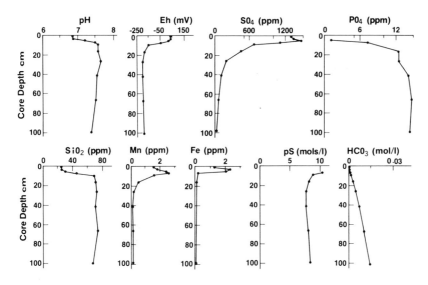

Figure 5. Concentrations of some dissolved constituents in the interstitial waters of a core from the upper Chesapeake Bay as a function of depth beneath the sediment-water interface.

Sediment subjected to weathering, erosional and transport processes, is generally exposed to oxic conditions. Thus, the assemblage of minerals in the sediment entering the estuary is adjusted to an oxidizing environment. Hydrous oxides of manganese and iron commonly occur as coatings on detrital particles transported by rivers. They are excellent scavengers, particularly for trace metals [7, 8]. Redox sensitive minerals transported by riverine systems are oxidized phases, dominantly hydrous oxides, that occur under conditions represented in the upper portion of Figure 4. When this new sediment settles to the bottom and is buried, it is exposed to reducing conditions generated by the reactions described above. Many of the mineral phases are not stable under the new conditions and dissolve. The manganese and iron coatings are reduced, liberating Fe^{2+} and Mn^{2+} to the interstitial water along with the burden of other trace metals scavenged during transport. Concentrations increase until saturation is reached with respect to new mineral phases that are stable in the reducing environment. These are the phases that occur under redox conditions depicted on the lower portion of Figure 4, commonly carbonates, phosphates and sulfides. These phases then control the concentrations of dissolved constituents in the interstitial water (Fig. 5) [9-11]. In the northern part of the Chesapeake Bay, iron is introduced by the Susquehanna River in large amounts as hydrous oxide coatings on detrital particulate matter. In the southern portion of the Bay, each of the major rivers has an estuarine portion that traps detrital material before it reaches the Bay. Conversely, sulfate is present in the saline waters of the southern Bay, but is very low in concentration in the fresher waters of the northern Bay. The relative abundance of iron and sulfur in these environments are reflected in the sediments. Carbonates and phosphates of iron dominate the mineral assemblage of northern Bay sediments, whereas iron sulfides are the major form of iron in southern Bay sediments. This, in turn, is reflected in the chemistry of the interstitial waters from these segments of the Bay (Fig. 7). The northern waters are saturated or supersaturated with respect to siderite and vivianite but are undersaturated with respect to iron sulfide, whereas the reverse is true for interstitial waters from southern Bay sediments (Fig. 6). Generally, the solubilities of the new

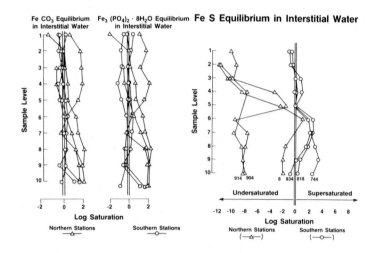

Figure 6. Saturation indices for siderite, $FeCO_3$, vivianite, $Fe_3(PO_4)_2 \cdot 8H_2O$, and mackinawite, FeS, in the interstitial waters of northern and southern Chesapeake Bay sediments. The "sample levels" correspond to the following depth ranges in centimeters beneath the sediment-water interface, 1, 0-2; 2, 2-4; 3, 4-6; 4, 6-8; 5, 8-10; 6, 15-18.5; 7, 20-26; 8, 40-46; 9, 70-76; 10, 94-100.

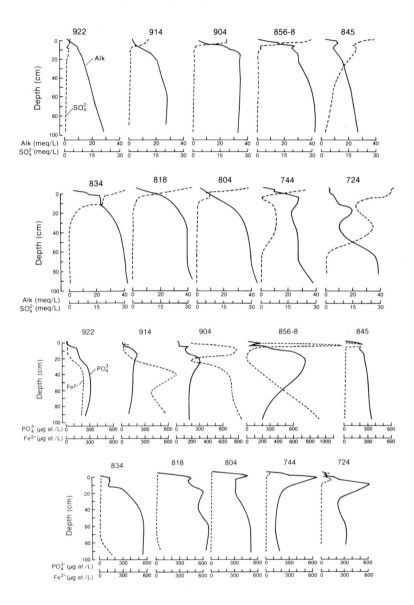

Figure 7. Titration-alkalinities and concentrations of sulfate, phosphate, and iron(II) in the interstitial waters of Chesapeake Bay sediments as a function of depth beneath the sediment-water interface at sites ranging from the head of the Bay (922) to the mouth of the Bay (724). The locations of these sites are identified in Fig. 2.

minerals formed under reducing conditions are considerably larger than the solubilities of the phases stable under oxidizing conditions, so that the concentrations of dissolved species in the interstitial waters are higher than corresponding concentrations in the estuarine water above. Not uncommonly, concentration differences of 100-fold to 1000-fold are observed [4, 12]. The concentrations of dissolved species in the interstitial water are controlled in some cases by equilibrium reactions between the mineral phases and the aqueous medium in which they are bathed.

Figure 8 depicts the phase boundary between siderite and vivianite as a function of phosphate activity versus carbonate activity. Superimposed on this diagram are the corresponding carbonate and phosphate values for interstitial waters from a core in the northern part of the Chesapeake Bay. In certain cases in which the rate of transport of dissolved species through the sediment is larger than the rate of reaction between interstitial waters and sediment minerals, equilibrium will not be achieved.

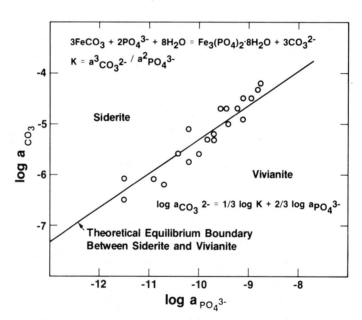

Figure 8. Thermodynamic phase boundary between siderite and vivianite as a function of the activities of carbonate and phosphate. The points represent carbonate and phosphate activities for interstitial waters from a core collected in the northern Chesapeake Bay.

It is thus often possible from a knowledge of the mineralogy of the bottom sediments to predict the composition of the interstitial waters, or conversely, from a knowledge of interstitial water chemistry to predict which reactive phases are present in the sediment. Caution must be exercised, however, because other processes may override the inorganic system. For example, in sulfide-rich anoxic sediments mercury concentrations would be expected to be vanishingly small as a result of the formation of HgS. In natural systems bacterial methylation of mercury occurs in sediments, mobilizing mercury and making it available for uptake by biota [13, 14]. Interstitial waters are the medium through which reactions occur in the sediment, they are the waters in which benthic infauna are constantly bathed, and they are the vehicle for the transport of dissolved substances through the sediment and across the sediment-water interface. The capability of predicting interstitial water composition is thus a powerful tool in understanding the chemistry of the sediment system and assessing its impact on the estuarine environment.

Benthic Processes

Materials in the bottom sediment, both dissolved and particulate, may be reintroduced to the surface environment and water column by a number of processes. Two of the most important processes are physical disturbance of the sediment and the flux of dissolved materials across the sediment-water interface. Physical disturbances are caused by storm-generated waves and currents; biologic activity (bioturbation); gas bubble ebullition; dredging and other engineering projects; propeller wash; and harvesting of bottom organisms such as clams, crabs, and oysters by dredging. The flux of dissolved material across the sediment-water interface is caused by diffusion-driven concentration differences and by life processes of benthic organisms such as irrigation of burrows and benthic feeding.

Physical disturbances are episodic occurrences whereas diffusion is a continuously operating process. Any physical process that disturbs the sediment causes immediate release of the interstitial water and its dissolved constituents to the water column. Exhumation and resuspension of sediment by physical processes can re-expose particulate material that had previously been buried and had been out of direct contact with the surface environment. Gas bubble ebullition creates open, tube-shaped channels in organic-rich muddy sediment [14]. Where these tubes are present, they enhance the flux of dissolved constituents across the sediment-water interface. Apparent bulk diffusivities up to three times greater than molecular diffusivities have been observed in sediments with bubble tubes [15]. In addition to episodic events, there is a continuous and regular migration of dissolved substances across the sediment-water interface as a result of physicochemical processes. In the absence of significant disturbance of the sediment, diffusion gradients develop in response to the concentration differences between the dilute estuarine waters and the more concentrated interstitial waters (Fig. 5). These gradients result in a flux of dissolved substances from the sediment into the estuarine waters above. The magnitude of the flux from the sediments can be estimated using the observed concentration gradient and Fick's first law of diffusion (eqn. 3).

$$J_0 = \Phi D' \frac{dc}{dx}\bigg|_{x=0} \qquad (3)$$

J_0: diffusional flux at sediment water interface in millimoles per square centimeter per year

Φ: porosity at interface

D': effective sediment diffusion coefficient (square centimeter per year)

$\frac{dc}{dx}$: Interstitial water concentration gradient for the dissolved species of interest as a function of depth in the sediment

Estimating fluxes for Chesapeake Bay sediments in this manner, typically yields values within the ranges shown in Table 1. The concentrations of dissolved constituents in

Table 1. *Fluxes of Selected Dissolved Constituents from Chesapeake Bay Sediments*

Constituent	Fluxes in Millimoles/cm²/year at Various Locations		
	Northern Bay	Zone of Turbidity Maximum	Southern Bay
HCO_3^-	770. -1310.	124. -325.	390. -1160.
H_4SiO_4	16.79- 28.47	4.75- 25.92	32.85- 40.15
HPO_4^{--}	0.73- 22.19	1.46- 4.02	5.48- 24.82
NH_4^+	91. - 314.	19. - 95.	68. - 277.
Mn^{2+}	7.30- 34.7	0.37- 1.46	4.01- 6.21
Fe^{2+}	9.49- 45.99	1.83- 19.71	0.37- 6.57

the interstitial waters are higher and the gradients are steeper in the summer months than during the winter, implying larger fluxes from the sediments during the warmer part of the year. Biologic activity is also greater during the warm season and probably accounts for the observed increase in dissolved constituents.

During the cooler months when Bay bottom waters are well oxygenated, a thin brown layer of flocculent sediment is present at the sediment-water interface throughout the Bay. This layer ranges from several millimeters to several centimeters in thickness and strongly inhibits diffusion of iron, manganese, and other redox sensitive species out of the sediment. Diffusion of phosphate and silicate is also inhibited probably due to adsorption onto the hydrous iron oxide in this layer [16-18]. As strong stratification develops in the water column through spring and summer, the bottom waters in deep areas of the Bay go anoxic, the brown oxidized layer disappears, and diffusion of phosphate and silicate takes place freely across the sediment-water interface. The pycnocline, which separates oxic and anoxic waters under the stratified conditions, then becomes the barrier for redox sensitive species.

Flux of the material from the sediment can be directly measured under certain conditions. The most common method currently used is the isolation of a known area of the bottom under a bell jar or dome-type of enclosure. The water contained within the dome is sampled repetitively over a period of time, and changes in the concentrations of dissolved constituents are related to the rate of flux from the bottom sediments. In the Chesapeake Bay the few measurements of benthic fluxes made using the dome technique fall into two categories. Either they correspond well to the fluxes calculated from the interstitial water concentration gradients, or they are larger than the calculated fluxes. It is not uncommon to find measured fluxes exceeding calculated fluxes by factors of two to ten. A number of factors could enhance the rate of transfer across the sediment-water interface. One, in particular, seems to play a major role in augmenting benthic fluxes in the Chesapeake Bay. The benthic infaunal communities have a profound effect on transport of material in the sediment and across the sediment-water interface. Constant reworking of the sediment by these organisms exhumes material from deep within the bottom and deposits it at the surface where it is once again exposed. The burrowing and feeding activities create open tubes and tunnels in the sediment that serve as conduits for the movement of water into and out of the sediment. The bioturbating activity of the organisms generally tends to increase the porosity of the sediment (Figs. 9, 10). Similar effects were reported in

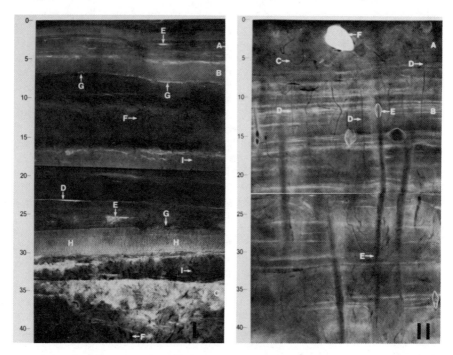

Figure 9. Radiographs of box cores from the northern Chesapeake Bay [23]. Numbers to the left of the radiographs indicate the depths in centimeters beneath the sediment-water interface.
  I. Core from a deep-water area in which the bottom waters become anoxic during the summer. Consequently, little bioturbation has occurred. A: mud bed with silt-clay forset laminae; B: homogeneous mud bed; C: fine sand bed; D: sand lamina; E: sand lenses; F: methane bubbles; G: **Mulinia lateralis** shell lag; H: **Pectinaria gouldii** tube remnants in fine sand bed; I: unidentified polychaete burrows.
  II. Core of a moderately bioturbated sediment from an area in which the bottom waters to not become anoxic during the year. A: homogeneous mud bed; B: interlaminated bed of mud and fine sand; C: active burrow of juvenile **Nereis succinea**; D: traces of **Scolecolepides virides** tubes; E: **Macoma balthica** shell and burrow, methane bubbles in bottom part of burrows; F: **Rangia cuneata** shell.

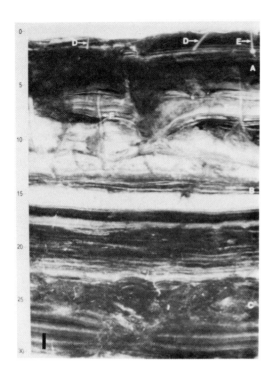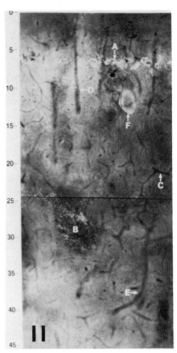

Figure 10. Radiographs of box cores from the southern Chesapeake Bay [23]. Numbers to the left of the radiographs indicate the depths in centimeters beneath the sediment-water interface.
  I. *Core of an area with interbedded mud and sand showing moderate bioturbation. A: bioturbated mud and sand interbeds; B: undisturbed interbedded mud and sand; C: biologically reworked muddy sand layer with* **Mulinia lateralis** *shells; D: active* **Pectinaria gouldii** *tube with feeding trace; F: active* **Loimia medusa** *tubes.*
  II. *Core of a silty clay sediment with extreme bioturbation. A:* **Mulinia lateralis** *shell lag; B:* **Mulinia lateralis** *shell pocket; C: active* **Nereis succinea** *burrow system; D: active* **Macoma balthica** *burrow; E: inactive* **Macoma balthica** *burrow; F: disarticulated* **Macoma balthica** *shell.*

Narragansett Bay [19]. At sites without significant benthic fauna the calculated diffusive flux rates compare well with rates measured in dome experiments. At sites with flourishing benthic infaunal communities, measured flux rates were as much as three times larger than those calculated from pore water concentration gradients. The total effect of the benthic community is to enhance the transfer of material across the sediment-water interface [20].

The massive reservoir of materials contained in the bottom sediments of estuaries have largely been ignored as a potential source of nutrients and trace elements until recent years. On the basis of interstitial water chemistry investigations, it is apparent that there is a very substantial contribution of these substances from the sediment to the water column [4, 18, 21, 22]. Fluxes calculated from concentration gradients in Chesapeake Bay sediments indicate that NH_4^+, HCO_3^-, HPO_4^{2-}, H_4SiO_4, Fe^{2+}, and Mn^{2+} are supplied to estuarine water from the sediment (Table 1). Work is presently in progress to quantify these fluxes and to examine other substances, particularly trace metals.

Summary

The sediment environment is very different from the freely circulating estuarine waters above. Decomposition of organic matter mediated by bacteria creates anoxic conditions within a short distance beneath the sediment-water interface. In this partially restricted environment many trace metals are mobilized and their concentrations are controlled by inorganic reactions such as dissolution and precipitation. Concentration differences between the interstitial waters and the estuarine waters are generated by reactions that occur in the sediments. In response to these concentration differences, diffusional transport of dissolved substances takes place within the sediment and across the sediment-water interface. Benthic fluxes calculated from interstitial water concentration gradients using Fick's law compare well with measured fluxes in areas where the sediment is undisturbed by physical or biological processes. In sediments that have a flourishing benthic infaunal community, calculated flux rates are generally lower than measured flux rates because of the enhancement of material transport due to effects of the physical activities of the benthic organisms on the sediment. Benthic fluxes calculated from interstitial water concentration gradients thus give minimal values for contributions of material to the water column. Based even on these minimal values, however, it is apparent that the bottom sediments are a non-trivial source of material to the estuary. In natural systems, trace metal behavior is strongly dependent upon environmental factors. Adequate models to describe metals in these systems cannot be developed solely from chemical considerations. Account must be taken of the chemical and mineralogical nature of the source material, the sedimentation pattern and rate of sedimentation, the chemical changes that occur in the sediment after burial and the physical and biological processes that continuously work to redistribute the dissolved species in the interstitial water and the particulate bed material.

Acknowledgement. I would like to thank Eli Reinharz for providing radiographs of Chesapeake Bay sediment and Edward Callender and Neil Plummer for reviewing the manuscript.

References

[1] D. W. Pritchard, "Observation of Circulation in Coastal Plain Estuaries," in **"Estuaries,"** G. H. Lauff, Ed., **Am. Assoc. Adv. Sci. Publ. 83,** Washington, D.C., 1967, p. 37.

[2] J. R. Schubel, "Suspended Sediment of the Northern Chesapeake Bay," Chesapeake Bay Institute, The Johns Hopkins University, **Tech. Report No. 35** (1968).

[3] B. N. Troup and O. P. Bricker, "Processes Affecting the Transport of Materials from Continents to Oceans," **ACS Symp. Series, 18** (1975) 133.

[4] O. P. Bricker and B. N. Troup, "Sediment-Water Exchange in Chesapeake Bay," in **"Estuarine Research,"** Vol. 1, L. E. Cronin, Ed., Academic Press, Inc.: New York, 1975, p. 3.

[5] E. D. Goldberg, V. Hodge, M. Koide, J. Griffin, E. Gamble, O. P. Bricker, G. Matisoff, G. R. Holdren and R. Braun, "A Pollution History of Chesapeake Bay," **Geochim. Cosmochim. Acta, 42** (1978) 1413.

[6] A. C. Redfield, B. H. Ketchum and F. A. Richards, "The Influence of Organisms on the Composition of Seawater," in **"The Sea,"** Vol. II, N. M. Hill, Ed., Wiley Interscience: New York, 1966, p. 26.

[7] E. A. Jenne, "Controls on Mn, Fe, Co, Ni, Cu and Zn Concentrations in Soils and Water: The Significant Role of Hydrous Mn and Fe Oxides," in **"Trace Inorganics in Water,"** R. F. Gould, Ed., **Am. Chem. Soc. Advances in Chemistry Series No. 73** (1968) 337.

[8] G. D. Robinson, "Heavy Metal Adsorption by Ferromanganese Coatings on Stream Alluvium: Natural Controls and Implications for Exploration," **Chemical Geology, 38** (1983) 157.

[9] B. N. Troup, "The Interaction of Iron with Phosphate, Carbonate and Sulfide in Chesapeake Bay Interstitial Waters: A Thermodynamic Interpretation," **Ph.D. Dissertation,** The Johns Hopkins University, Baltimore, MD., 1974, 114 pp.

[10] J. T. Bray, "The Behavior of Phosphate in the Interstitial Waters of Chesapeake Bay Sediments," **Ph.D. Dissertation,** The Johns Hopkins University, Baltimore, MD., 1973, 136 pp.

[11] H. Elderfield and A. Hepworth, "Diagenesis, Metals and Pollution in Estuaries," **Mar. Poll. Bull., 6** (1975) 85.

[12] O. P. Bricker, G. Matisoff and G. R. Holdren, "Interstitial Water Chemistry of Chesapeake Bay Sediments," **Maryland Geological Survey, Basic Data Report, No. 9** (1977) 67 pp.

[13] S. Jensen and A. Jernelöv, "Biological Methylation of Mercury in Aquatic Organisms," **Nature, 233** (1969) 753.

[14] B. L. Vallee and D. D. Ulmer, "Biochemical Effects of Mercury, Cadmium, and Lead," **Ann. Rev. Biochem., 41** (1972) 91.

[15] J. V. Klump and C. S. Martens, "Biogeochemical Cycling in an Organic-Rich Coastal Marine Basin: II Nutrient Sediment-Water Exchange Processes," **Geochim. Cosmochim. Acta, 45** (1981) 101.

[16] G. F. Lee, "Factors Affecting the Transfer of Materials between Water and Sediments," **The University of Wisconsin Water Resources Center Literature Review, No. 1,** Madison, WI, 1970, 50 pp.

[17] J. T. Bray, O. P. Bricker and B. N. Troup, "Phosphate in Interstitial Waters of Anoxic Sediments: Oxidation Effects during Sampling Procedure," **Science, 180** (1973) 1362.

[18] E. Callender, "Benthic Phosphorus Regeneration in the Potomac River Estuary," **Hydrobiologia, 92** (1982) 431.

[19] H. Elderfield, N. Luedtke, R. J. McCoffrey and M. Bender, "Benthic Flux Studies in Narragansett Bay," **Am. Jour. Sci., 281** (1981) 768.

[20] R. C. Aller, "The Influence of Macrobenthos on Chemical Diagenesis of Marine Sediments," **Ph.D. Dissertation,** Yale University, 1977, 600 pp.

[21] R. A. Berner, "Kinetics of Nutrient Regeneration of Anoxic Marine Sediments," in **"Origin and Distribution of the Elements,"** L. H. Ahrens, Ed., Pergamon Press: Oxford and New York, 1979, p. 279.

[22] W. R. Boynton, W. M. Kemp and C. G. Osborne, "Nutrient Fluxes Across the Sediment-Water Interface in the Turbid Zone of a Coastal Plain Estuary," in **"Estuarine Perspectives,"** V. S. Kennedy, Ed., Academic Press: New York, 1980, p. 93.

[23] E. Reinharz, K. J. Nilsen, D. F. Boesch, R. Bertelson and A. E. O'Connell, "A Radiographic Examination of Physical and Biogenic Sedimentary Structures in the Chesapeake Bay," **Maryland Geologic Survey, Report of Investigations No. 36,** (1982) 57 pp.

DISCUSSION

J. M. Bellama (University of Maryland): It is interesting that the metals are trapped in river sub-estuaries and are not delivered to the Chesapeake Bay. The situation is different in the Rhine River system. Ane de Groot from Groningen (the Netherlands) reported that metals in the rivers of the Ruhrgebiet are not only carried to the Rhine but also along the Rhine to the Netherlands. Could you comment on the differences in these two systems?

O. P. Bricker: The Susquehanna River and the upper Chesapeake Bay are similar to the Rhine. Material transported by the Susquehanna is carried directly into the upper Chesapeake Bay and is then dispersed within the Bay by estuarine circulation. However, the circulation patterns in the southern Chesapeake Bay are different from those in the Rhine River system. The rivers into the southern Chesapeake Bay have long estuarine segments. The net direction of flow of bottom water in these estuarine segments is upriver, and the net direction of flow of surface water is toward the Chesapeake Bay. The bulk of the particulate material settles to the bottom of the estuarine segments before reaching the main Bay and either stays where it settles out, or is transported toward the head of the estuarine segment by the bottom waters. Thus the bulk of the metals (which are almost all associated with particulate matter) are trapped within the sub-estuaries of the rivers discharging into the southern part of the Chesapeake Bay. A good example of this type of behavior is described by Sinex and Helz [Entrapment of zinc and other trace elements in a rapidly flushed industrialized harbor; Environ. Sci. Technol., $\underline{16}$ (1982) 820] for the Patapsco River sub-estuary.

VI

TRACE METALS IN THE LIGURIAN AND NORTHERN TYRRHENIAN SEAS: RESULTS AND SUGGESTIONS FOR FURTHER STUDIES

R. Capelli and V. Minganti

Institute of General and Inorganic Chemistry
Gruppo Ricerca Oceanologica-Genova
Universita di Genova, Italy

The concentrations of mercury, cadmium, lead, copper, manganese, and zinc in anchovies, striped mullets, shrimp, and mussels, and of methylmercury, total mercury, copper, manganese, and zinc in Atlantic bonitos are summarized in tables. The organisms were collected in the Ligurian and Northern Tyrrhenian Seas and were analyzed by participants in the "Med Pol II" and "Oceanografia e Fondi Marini" projects. The deficiencies in the reported data are noted and recommendations for further research are given.

The examination and the comparison of results obtained by different laboratories in the determination of trace metals in various matrices such as water, sediments, particulate matter, and organisms of a certain area of the sea - even a restricted one - present a number of difficulties that cannot always be overcome. In some cases publications do not provide information about sampling methods, conservation and preparation of samples, size of organisms, method of mineralization, and the choice of dry or wet weight for the calculation of concentrations. Without this information, comparisons are difficult if not impossible.

An attempt to obviate these shortcomings was made during the execution of two projects "MED POL II" by the FAO-UNEP and "Oceanografia e Fondi Marini" of the Italian National Research Council (CNR). Standard methods were adopted for sampling and preparation of marine organisms. Intercalibration tests with participating laboratories using their methodologies made it possible to compare the results. Despite such efforts, it is still difficult to assess the "state of health" of a particular region of the sea from literature data.

In this report we shall consider only marine organisms pertinent to our field of research. We shall summarize our data and data by other authors pertaining to the Ligurian Sea and the Northern Tyrrhenian Sea and suggest topics for future research in this region of the Mediterranean.

Materials and Methods

The marine organisms of the Ligurian and Northern Tyrrhenian Seas (Fig. 1) that were studied during the past few years include mussels, *Mytilus galloprovincialis*; anchovies, *Eugraulis encrasicholus*; striped mullets, *Mullus barbatus*; bluefin tunas, *Thunnus thynnus*; shrimp, *Nephrops norvegicus*; scorpion fish, *Scorpaena porcus*; broadbill swordfish, *Xiphias gladius*; and Atlantic bonitos, *Sarda*

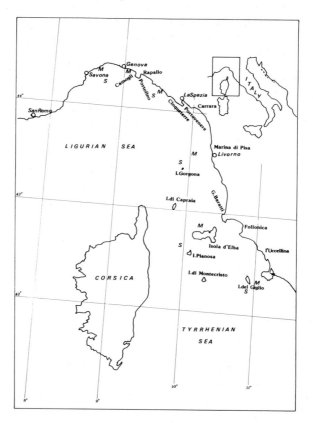

Figure 1. Sampling areas for shrimp (S) and striped mullets (M) in the Ligurian and Northern Tyrrhenian Seas.

Trace Metals in the Ligurian and Tyrrhenian Seas   157

sarda.   However,   only some of these organisms   were investigated   in a coordinated and systematic   manner. Organisms chosen for study by FAO-UNEP and CNR possess the characteristics of bioindicators.   These organisms are easily obtainable for extended periods throughout the year in the entire area being studied,   are representative of the trophic chain,   and have different feeding habits.   Filter-feeder, planktophagi, carnivors in a pelagic environment, omnivores in a benthic environment, and detritus-feeders are represented.   The methods of sampling and sample preparation were in most cases those described in the manual issued   by the   FAO-UNEP   [1].   Muscle tissue   (fillets)   was   most frequently used,   but   other tissues   and organs were   also analyzed [2, 3, 4].

The metals (principally Hg,   Cd, Pb, Mn, Cu, and Zn) were determined in most cases by atomic absorption spectrometry of   the   solutions obtained by wet-ashing of   the   organic matter.   The wet samples were treated with hot,   90-percent nitric acid in a Pyrex or fused-quartz flask equipped with a reflux   condenser   [5].   Previously dried samples   were mineralized with hot,   65-percent nitric acid.   An alternate digestion   method used a mixture of nitric acid and sulfuric acid in an open container equipped with a condenser [4] or hot concentrated nitric acid in a closed Teflon vessel   [6]. Fowler and Oregioni [7] first dried the samples at 60° and then   treated   them with   a mixture of three   volumes concentrated nitric acid and one volume of perchloric acid. Bernhard [1] recommended   the use of nitric acid in a pressurized   container   for the oxidation of   organic materials.

Results and Discussion

The   metal   concentrations found   in anchovies,   striped mullets, shrimp, and Atlantic striped bonitos are summarized in Tables   1 to 4.   If available, concentration ranges and average values are listed.   Data for mussels are more difficult to summarize because the metal concentrations   in these organisms are strongly influenced by local conditions, and samples from a very limited area may have quite different metal concentrations.   Therefore, Tables 5 and 6 present the data for mussels not as averages but as reported by the authors.

The following major shortcomings of literature data made some of the data useless and others difficult to use.

Table 1. Trace Metal Concentrations (mg/kg Wet Weight) in Fillets from Anchovies, *Engraulis encrasicholus*, from the Ligurian, Northern Tyrrhenian and Other Italian Seas

Sampling Area		Length(cm)	Hg	Cd	Pb	Cu	Mn	Zn	Ref.
West of Genoa	min.	12.0	0.02			1.15	0.33	13	[8]
	mean	13.5	0.10	<0.08	<0.3	1.35	0.52	20	
	max.	14.5	0.16			1.65	0.75	28	
East of Genoa	min.	12.0	0.07			0.56	0.40	13	[8]
	mean	13.5	0.12	<0.08	<0.3	1.15	0.75	20	
	max.	14.5	0.22			1.7	1.30	27	
Arcipelago Toscano	min.		0.010			0.12	0.06		[9]
	mean			<0.005	<0.1			1.45	
	max.		0.40						
Arcipelago Toscano	min.	11							[10]
	mean		0.13	<0.005	<0.1	0.65	1.8	18.7	
	max.	15							
Coste di Calabria e Sicilia	min.		0.053	0.011	0.043				[9]
	mean								
	max.		0.230	0.040	0.391				
Alto e Medio Adriatico	min.								[3]
	mean		0.148	0.071	0.695				
	max.								
Adriatico	min.								[11]
	mean		0.18	0.13	0.81		0.93		
	max.								

Table 2. Trace Metal Concentrations (mg/kg Wet Weight) in Fillets from Striped Mullets, *Mullus barbatus*, from the Ligurian, Northern Tyrrhenian and Other Italian Seas

Sampling Area		Length(cm)	Hg	Cd	Pb	Cu	Mn	Zn	Ref.
West of Genoa	min.	12.5	0.040			0.36	0.29	2.4	[8]
	mean	13.6	0.32	<0.08	<0.3	0.52	0.40	4.7	
	max.	15.0	0.97			0.70	0.58	7.0	
East of Genoa	min.	12.5	0.027			0.33	0.14	4.3	[8]
	mean	13.6	0.24	<0.08	<0.3	0.74	0.47	5.5	
	max.	15.0	0.40			1.30	0.96	9.5	
La Spezia-Cinqueterre	min.	13.0	0.051						[12]
	mean	13.9	0.123						
	max.	14.5	0.218						
La Spezia-Portovenere	min.	12.3	0.072						[9]
	mean	13.5	0.221						
	max.	16.0	0.380						
Carrara	min.	11.5	0.041						[12]
18 miles offcoast	mean	12.7	0.069						
	max.	14.0	0.125						
Arcipelago Toscano	min.		0.110						[9]
	mean			<0.005	<0.1	0.214	0.160	3.02	
	max.		2.812						
Costa di Calabria e	min.		0.047	0.005	0.030	0.495	0.769	5.35	[9]
Sicilia	mean								
	max.		0.330	0.051	0.233				
Alto e Medio Adriatico	min.								[3]
	mean		0.074	0.056	0.567				
	max.								

Table 3. *Preliminary Results of Trace Metal Concentrations (mg/kg Wet Weight) in Fillets from Bonito* **Sarda sarda,** *from Camogli (Ligurian Sea) [13]*

Weight (g)	Length (cm)	Methyl Hg	Total Hg	Cu	Mn	Zn
3400	68	1.52	1.72	1.70	0.30	5.0
3280	56	1.10	1.30	1.10	0.15	5.4
3190	57	1.55	1.60	0.60		6.1
2450	55	1.05	1.38	1.70	0.40	6.1
2080	53	0.47	0.60	1.10	0.45	7.9
1300	43		0.29	1.60		4.0
905	38		0.28	1.90	0.25	5.5
676	34		0.41	0.69	0.40	4.2
660	34		0.45	0.71		3.9

- Data are frequently reported as histograms, concentration ranges, and other graphical or tabular forms lacking the required details for purposes of evaluation and comparison. Even when the same data appear in different forms in different publications, lack of detail is still a frequently encountered problem.
- Sometimes weights, concentrations, and similar parameters are reported only as ranges without averages; sometimes only averages are given without the pertinent ranges.
- Often, parameters such as length, weight, sex of the organisms, and sampling dates are missing.
- Sampling locations are often not sufficiently identified. Whereas a general location may be adequate for pelagic organisms, precise locations must be given for sessile organisms.
- Few data are available for trace elements other than Hg, Cd, and Pb.
- Few reports make reference to intercalibration tests or provide results obtained with standard samples under the experimental conditions used.

Despite these shortcomings, some useful conclusions can be drawn from the data listed in the Tables. The data show, that it is difficult, if not impossible, to assess the "state of health" of a body of water using only results from one species. It is very risky to judge the relative "pollution levels" of parts of the ocean by comparing the concentrations of trace metals found in organisms living in those parts. In the case of metals, it is always difficult to distinguish anthropogenic from non-anthropogenic sources.

Table 4. Trace Metal Concentrations (mg/kg Wet Weight) in the Tail Muscles of Shrimp, **Nephrops norvegicus**, from the Ligurian, Northern Tyrrhenian and Other Italian Seas

Sampling Area		Length(cm)	Hg	Cd	Pb	Cu	Mn	Zn	Ref.
West of Genoa	min.	12.5	0.41	0.115	0.400	4.5	0.23	11.0	[8]
	mean	13.5	0.71	0.155	0.850	6.3	1.8	14.0	
	max.	15.0	1.10	0.200	1.70	9.3	2.2	18.0	
East of Genoa	min.	12.5	0.37	0.090		4.0	0.80	7.7	[8]
	mean	13.5	0.66	0.110	0.735	5.9	1.2	13.0	
	max.	15.0	1.05	0.135		8.2	1.6	16.0	
Arcipelago Toscano	min.		0.015	<0.005	<0.1	0.215	0.030	1.35	[9]
	mean								
	max.		2.400						
Coste di Calabria e Sicilia	min.		0.059	0.013	0.145	11.1	3.85	19.3	[9]
	mean								
	max.		0.360	0.111	0.900				
Alto e Medio Adriatico	min.								[3]
	mean		0.333	0.085	0.713				
	max.								

Table 5. Trace Metal Concentrations (mg/kg Wet Weight) in the Soft Parts of Mussels, *Mytilus galloprovincialis*, from the Ligurian Sea

Sampling Area		Length (cm)	Dry wt / Wet wt	Hg	Cd	Pb	Cu	Mn	Zn	Ref.
San Remo	min.	3.0			0.066	1.5	1.3	1.2*	22	[12]
	mean		0.165		0.30	2.5	2.3		41	
	max.	5.0			0.31	3.7	3.5	4.4*	41	
San Remo	min.	3.0						1.4		[5]
	mean		0.233		0.58		4.0			
	max.	4.0						3.7		
Savona	min.	3.0				1.6	1.1	3.7*	25	[12]
	mean		0.165		0.33	2.8	1.6		35	
	max.	5.0				4.9	2.1	9.0*	47	
Savona	min.	3.0								[5]
	mean		0.233			6.5	3.3			
	max.	4.0								
Genova Isola Petroli	min.	4.5		0.017	0.125	0.75	0.80	0.88	16	[8]
	mean		0.187	0.022	0.185	1.6	1.05	1.85	27	
	max.	5.0		0.035	0.066	2.3	1.60	2.6	41	
Genova	min.	3.0			0.30	0.94	1.3	5.0*	21	[12]
	mean		0.165		0.46	2.2	1.9		35	
	max.	5.0			0.16	2.8	2.2	11.5*	43	
Genova-Promontorio di Portofino	min.	4.0		0.010	0.25	1.05	0.50	0.64	19	[8]
	mean	4.5	0.187	0.030	0.37	1.05	0.98	0.95	24	
	max.	5.0		0.065	0.16	1.10	1.50	1.15	31	
Genova-Promontorio di Portofino	min.	3.0		0.002			0.68	1.30	25	[14]
	mean		0.206		0.23		0.86			
	max.	4.0			0.16	5.4	2.28	1.87	29	
Genova Rapallo	min.	3.0		0.031				2.1		[5]
	mean		0.233							
	max.	4.0			0.58	7.5	2.96			
La Spezia	min.	3.0	0.202	0.030	0.40	2.8	1.4	2.4	41	[14]
	mean									
	max.	4.0		0.077	1.37	9.0	6.8	7.6	77	
La Spezia Isola del Tino	min.	4.0		0.006						[12]
	mean			0.026						
	max.	5.0		0.077						[7]

*Data from ref. 7.

Table 6. Trace Metal Concentrations (mg/kg Wet Weight) in the Soft Part of Mussels, **Mytilus galloprovincialis**, from the Northern Tyrrhenian and Other Italian Seas

Sampling Area	Length (cm)	Year	Hg	Cd	Pb	Ref.
ARCIPELAGO TOSCANO:						
Marina di Pisa		1977	0.260	0.11	<0.1-0.	[10]
Marina di Pisa		1978	0.190	0.140	0.300	
Golfo di Baratti		1977	0.130	0.150	1.050	
Golfo di Baratti		1978	0.150	0.180	4.950	
Follonica		1977	0.140	0.130	3.400	
Follonica		1978	0.140	0.170	2.950	
Uccellina (a)		1977	0.780	0.100	<0.1-0.450	
Uccellina (a)		1978	0.370	0.140	0.450	
Uccellina (b)		1977	0.130	0.090	<0.1-0.250	
Uccellina (b)		1978	0.210	0.120	0.460	
OTHER ITALIAN SEAS:						
Siracusa	4.7-5.8	1976	0.044	0.027	0.169	[15]
Siracusa	4.2-5.8	1977	0.037	0.038	0.253	
Siracusa	5.0-5.8	1977	0.047	0.048	0.880	
Siracusa	4.9-5.5	1977	0.093	0.027	0.548	
Siracusa	4.8-6.3	1978	0.105	0.035	0.593	
Siracusa	4.8-7.3	1978	0.133	0.041	0.575	
Alto e Medio Adriatico			0.036	0.284	1.186	[3]

Detailed knowledge of the geochemistry and the hydrodynamics of the pertinent regions of the ocean would be required to make such a distinction.

Mussels are useful as "bioindicators" only for restricted areas in close proximity to each other. The number of samples to be collected and analyzed will depend on the area to be checked [14, 16, 17]. In animals, particularly in large-sized specimens, the red and white muscle fibers may have different metal concentrations (Table 7). For mercury, the metal concentration present in the fillet is often correlated with the size of the organism [10, 19]. Such correlations that are also influenced by factors such as the sex of the organism and the season of the sampling are more evident in organisms collected in the same area, on the same date, and in the same haul. Until now, few values pertaining to metal concentrations in various organs and tissues have been determined. Even less information is available about the various chemical forms in which a metal may be present in an organism.

Table 7. Concentrations of Trace Metals (mg/kg Wet Weight) in Different Types of Muscle Tissue in **Sarda sarda** [18]

Sample	Total Hg	Methyl Hg	Cu	Mn	Zn
White Muscle Tissue	1.20	1.15	0.45	0.20	4.7
Mixed Muscle Tissue	1.35	1.34	1.10	0.15	5.4
Red Muscle Tissue	1.00	0.85	4.10	0.35	7.7

Based on the results from recent studies that are summarized in Tables 1 to 6, the following topics require additional research.

- Analysis of marine organisms for a variety of trace elements such as As, Se, Ag, and Te and investigation of a larger variety of marine organisms.
- Development and use of analytical methods suitable for differentiating the chemical forms in which the elements are present in the various tissues and organs. Standard samples for such analyses are not available.
- Exploration of relationships between the concentrations of metals in various matrices, such as particulate matter and sediments, and the organisms that live in contact with these matrices.
- Repetition of the "Med Pol II" and "Oceanografia e Fondi Marini" studies in 5 to 10 years to obtain information on the change - if any - of the "health condition" of the Ligurian and Northern Tyrrhenian Seas.

This first systematic investigation of marine organisms of the Ligurian and Northern Tyrrhenian Seas carried out during the past six years made possible - in spite of all the drawbacks mentioned - an initial assessment of the condition of this ocean basin. It became clear during this study, that an interdisciplinary approach is necessary for a successful characterization of the environmental condition of the sea. The experience and knowledge gained during this project led to the identification of topics for future research in this field, most of which still wait to be investigated.

References

[1] M. Bernhard, "Manuel des Methodes de la Recherche sur le Environnement Aquatique. Troisieme Partie. Echantillonnage et Analyse du Materiel Biologique (Directives Destinees au Project Commun Coordonne FAO(CGPM)PNEU sur la Pollution en Mediterranee)," **FAO Fish. Tech. Pap., 158** (1976) 132 p.

[2] R. Capelli, A. Cappiello, A. Franchi and G. Zanicchi, "Metaux Lourds Contenus dans Certains Organes de Rougets (Mullus barbatus) et d'Anchois (Engraulis encrasicholus) du Golfe de Genes," **Ves Journees Etud. Pollutions, CIESM, Cagliari,** (1980) 269.

[3] R. Viviani, G. Crisetig, E. Carpene, P. Cortesi, G. Serrazanetti, O. Cattani, L. Mancini and R. Poletti, "Residui di Inquinanti Chimici in Organismi Marini dell'Alto e Medio Adriatico," **Atti del Convegno Scientifico Nazionale CNR, Roma,** 2 (1980) 845.

[4] A. Renzoni, E. Bacci and L. Falciai, "Mercury Concentration in the Water, Sediments and Fauna of an Area of the Tyrrhenian Coast," **Rev. Intern. Oceanogr. Med., 31-32** (1973) 17.

[5] R. Capelli, V. Contardi and G. Zanicchi, "Ecologie et Biologie des Ports de la Mer Ligurienne et Haute Tyrrhenienne," **IIIes Journees Etud. Pollutions, CIESM, Split,** (1976) 83.

[6] A. Renzoni, E. Bacci, S. Focardi and C. Leonzio, "Mercurio e Composti Organoclorurati in Animali Marini del Tirreno." **Atti del Convegno delle Unita Operative Afferenti ai Sottoprogetti: Risorse Biologiche e Inquinamento Marino, Roma,** (1982) 925.

[7] S. W. Fowler and B. Oregioni, "Trace Metals in Mussels from the N.W. Mediterranean," **Mar. Poll. Bull.,** 7 (1976) 26.

[8] R. Capelli, V. Contardi, B. Cosma, V. Minganti and G. Zanicchi, "A Four-Year Study on the Distribution of Some Heavy Metals in Five Marine Organisms of the Ligurian Sea," **Mar. Chem.,** 12 (1983) 281.

[9] UNEP/IG.18/INF 3 "Summary Reports on the Scientific Results of MED POL; Part I" (1980) 202 p.

[10] E. Bacci, S. Focardi, C. Leonzio and A. Renzoni, "Contaminanti in Organismi del Mar Tirreno." **Atti del Convegno Scientifico Nazionale CNR, Roma,** 2 (1979) 885.

[11] R. Viviani, "Aspetti Igienico-Sanitari dei Residui di Inquinanti Chimici nei Prodotti della Pesca." **Atti dei Convegni Lincei, Aspetti Scientifici dell'Inquinamento dei Mari Italiani, Roma,** (1977) 333.

[12] FAO(GFCM) Secretariat, "Heavy Metals and Chlorinated Hydrocarbons in the Mediterranean." **Mid-Term Expert Consultation on the Joint FAO(GFCM)/UNEP Coordinated Project on the Pollution in the Mediterranean, Dubrovnik,** (1977).

[13] R. Capelli, V. Contardi, B. Cosma, V. Minganti and G. Zanicchi, "Resultats Preliminaires d'une Recherche sur la Teneur en Metaux dans les Tissus et les Organes des Pelamydes (*Sarda sarda*) Echantillonnees dans le Golfe de Genes." **VIes Journees Etud. Pollutions, CIESM, Cannes,** (1983) 315.

[14] R. Capelli, V. Contardi, B. Fassone and G. Zanicchi, "Heavy Metals in Mussels (*Mytilus galloprovincialis*) from the Gulf of La Spezia and from the Promontory of Portofino, Italy." **Mar. Chem., 6** (1978) 179.

[15] L. Mojo, S. Martella and C. Martino, "Controllo Stagionale dei Metalli Pesanti (Hg, Cd, Pb) in Alcuni Organismi Marini del Mediterraneo Centrale. Primi Risultati." **Atti del Convegno Scientifico Nazionale CNR, Roma, 2** (1979) 913.

[16] D. J. H. Phillips, "The Common Mussel Mytilus edulis as Indicator of Pollution by Zinc, Cadmium, Lead, Copper. II. Relationship of Metals in the Mussel to those Discharged by Industry." **Mar. Biol., 38** (1976) 71.

[17] J. D. Popham, D. C. Johnson and J. M. D'Anna, "Mussels (*Mytilus edulis*) as 'Point Source' Indicators of Trace Metal Pollution." **Mar. Poll. Bull., 11** (1980) 261.

[18] R. Ferro, R. Capelli, V. Contardi and G. Zanicchi, "Considerazioni sulla Messa a Punto di Metologie Analitiche e sui Risultati Ottenuti durante Cinque Anni di Partecipazione al Sottoprogetto Inquinamento Marino." **Atti del Convegno delle Unita Operative Afferenti ai Sottoprogetti: Risorse Biologiche e Inquinamento Marino, Roma,** (1982) 557.

[19] R. Capelli, V. Contardi, B. Cosma, V. Minganti and G. Zanicchi, "Elements en Traces dans la Chair des Langoustines (*Nephrops norvegicus*) Pechees dans le Golfe de Genes." **VIes Journees Etud. Pollutions, CIESM, Cannes,** (1983) 277.

VII

CHEMISTRY OF METAL OXIDES IN NATURAL WATER: CATALYSIS OF THE OXIDATION OF MANGANESE (II) BY γ-FeOOH AND REDUCTIVE DISSOLUTION OF MANGANESE (III) AND (IV) OXIDES

James J. Morgan, Windsor Sung and Alan Stone

W. M. Keck Laboratories, Environmental Engineering Science
Division of Engineering and Applied Sciences
California Institute of Technology
Pasadena, California 91125 USA

An adequate description of the inorganic chemistry of most trace elements requires consideration of their distributions between aqueous and particulate phases, and their chemical reactions occurring on surfaces in aqueous medium. A surface coordination chemistry framework helps to understand interactions of protons, metals, and ligands at the oxide/water interface and at particle surfaces, and is also useful for the study of surface-catalyzed reactions such as autoxidations and reductive dissolution of minerals. The oxidation of Mn(II) and the reductive dissolution of manganese oxides were investigated. The rate law for the oxidation of Mn(II) in bicarbonate solutions in the pH range 8 to 9 catalyzed by γ-FeOOH was obtained. The dependence of the oxidative removal of Mn(II) from water on pH and available oxide surface is interpreted in terms of a pH-dependent surface coordination of Mn^{2+} at the FeOOH surface and a subsequent oxidation to Mn(III) by bound molecular oxygen. A general rate expression for the oxidation of Mn(II) is proposed to facilitate further work. The reductive dissolution of Mn(III) and Mn(IV) oxides by hydroquinone, dihydroxybenzoic acids, catechols, resorcinol, and other model compounds was studied under simulated natural water conditions in the laboratory. A surface site-binding model was proposed to interpret the observed rate law and the dependence of the dissolution rates on the structures of the different compounds. It is suggested that the dependence of the reaction rates *of a number of redox processes on the properties of particle surfaces and aqueous solutions can be best rationalized and explained with site-binding or coordination models.*

The elucidation of the distribution of elements and chemical species, of their fate and residence times, and of their concentrations on the basis of thermodynamic and kinetic properties is a central problem in aquatic chemistry. A pollutant introduced into the water environment may become distributed among a number of aqueous species and particulate fractions in the water column and the sediments. A description of the inorganic environmental chemistry of most elements in natural water systems must, therefore, include surface chemical reactions of adsorbed or precipitated forms. Among the key reactions are oxidative processes producing insoluble species and reductive processes leading to more soluble forms. The aquatic cycles of iron and manganese exemplify such processes, during which redox species and particulate-solution distributions are significantly altered. Surface reactions, in addition to reactions occurring in homogeneous solution and within biota, can significantly influence the spatial and temporal distributions of elements [1-3]. Adsorption of metals onto mineral and biological particles represents an important scavenging and transport process in fresh and marine waters, influencing residence times of trace metals. Adsorption of phosphate, silicate, and other oxyanions, both inorganic and organic, onto hydrous metal oxide surfaces has been shown to be an important removal mechanism in laboratory and natural systems [4, 5].

The adsorption behavior of metal ions and anions at oxide surfaces in water is becoming better understood in terms of a coordination chemistry framework. This framework is useful in describing scavenging phenomena and fates of elements [6]. It also provides an approach to surface-catalyzed reactions in aqueous systems and the dissolution and precipitation processes of oxides and other minerals [7].

Oxides, in particular those of aluminum, silicon, iron, and manganese, are important components of the earth's surface. In an aqueous medium the surfaces of metal and metalloid oxides become covered with bound water leading to the formation of surface hydroxyl groups [8]. The central metal ion of a hydrous oxide can behave as a Lewis acid, can bind ligands such as sulfate, phosphate, fluoride, and benzoate in exchange for hydroxide, and form inner-sphere surface complexes. The surface OH group can lose a proton or take up a proton (eqns. 1, 2), coordinate to a metal ion from the solution and release a proton (eqns. 3, 4), or be replaced by another anionic ligand (eqns. 5, 6, 7). Equations 1 through 7 describe examples of such reactions occurring at the surface of a hydrous iron oxide [4, 9]. The surface group is represented by >FeOH.

Acid-base reactions

$$>FeOH \rightleftharpoons\, >FeO^- + H^+ \quad (1)$$

$$>FeOH + H^+ \rightleftharpoons\, >FeOH_2^+ \quad (2)$$

Metal-ion complexation

$$>FeOH + Mg^{2+} \rightleftharpoons\, >FeO\text{-}Mg^+ + H^+ \quad (3)$$

$$(>FeOH)_2 + Mg^{2+} \rightleftharpoons\, (>FeO)_2Mg + 2H^+ \quad (4)$$

Ligand exchange

$$>FeOH + SO_4^{2-} \rightleftharpoons\, >FeSO_4^- + OH^- \quad (5)$$

$$(>FeOH)_2 + SO_4^{2-} \rightleftharpoons\, (>Fe)_2SO_4 + 2OH^- \quad (6)$$

$$>FeOH + H_2PO_4^- \rightleftharpoons\, >FeHPO_4^- + H_2O \quad (7)$$

These examples illustrate that a coordination approach to the chemistry at an oxide/water interface can provide a basis for describing the specific influence of pH, metal ions, and ligands on surface species and surface chemical processes. Equilibrium and kinetic properties of aqueous phase reactions can be combined with information about surface reactions to provide models to describe the distribution of inorganic species in complex systems with adsorbing surfaces and precipitating or dissolving phases [10, 11, 12].

Iron and Manganese Cycles and the Role of Oxide Surfaces

Iron and manganese are important in water as essential plant nutrients, as contaminants, and as a source of particulates. Their insoluble hydrous oxides such as FeOOH, MnOOH, MnO_2 coordinate metals and ligands at their surfaces thereby providing regulation and transport mechanisms for other elements in water. Iron and manganese were also reported to catalyze oxidation of reduced sulfur compounds in fog and cloud water [13]. The environmental chemistry of iron and manganese is therefore of considerable interest. This paper strives to advance the idea that the dynamics of iron and manganese cycles in fresh and marine waters are significantly influenced by the coupling of surface coordination processes and redox reactions at oxide and hydroxide surfaces. The surface and redox reactions are, in turn, linked to species in solution and especially pH. The inorganic chemical cycles of iron and manganese are influenced by the organic chemistry of natural waters through the reducing action of organic compounds.

Two examples involving oxides of iron and managanese will be presented. The catalysis of Mn(II) oxidation and removal from aqueous solution by an iron oxide solid, γ-FeOOH, in the presence of dissolved oxygen serves as the first example. This example demonstrates a possible linkage between the iron cycle, in which Fe(II) oxygenation is typically rapid at neutral and slightly alkaline pH, and the manganese cycle, in which Mn(II) oxygenation is very slow unless the pH is elevated or some suitable catalyst such as manganese-oxidizing bacteria is present. The second example describes a laboratory investigation of the reduction and dissolution of Mn(III) and Mn(IV) hydrous oxides by low-molecular weight organic compounds in slightly acidic to slightly alkaline solutions. In natural waters, the manganese cycle is driven by reductive dissolution in deep waters and oxidative precipitation in surface waters. Our laboratory results indicate that the reduction rates are chemically controlled through reactions at manganese oxide surfaces and that adsorption of other species can retard the reductive dissolution.

Catalysis of Mn(II) Oxidation by γ-FeOOH Surface

The product of Fe(II) oxygenation in bicarbonate solutions in the presence of chloride or sulfate at a pH > 6.5 was observed to be poorly crystallized lepidocrocite, γ-FeOOH(s) [14]. Earlier work established that $MnO_2(s)$ catalyzes the oxidation of Mn(II) in alkaline solutions. The mechanism is believed to involve adsorption followed by electron transfer at the $MnO_2(s)$ surface (eqns. 8, 9). It

$$Mn(II) + MnO_2(s) \rightleftharpoons Mn(II) \cdot MnO_2(s) \qquad (8)$$
$$Mn(II) \cdot MnO_2(s) + 1/2 O_2 \xrightarrow{*} 2MnO_2(s) \qquad (9)$$

*Unbalanced with respect to H_2O and H^+

was also observed that γ-FeOOH(s) catalyzes the oxidation of Fe(II) [14]. These observations suggest that hydrous iron oxide surfaces might be suitable catalysts for the oxidative removal of Mn(II).

Experimental. γ-FeOOH was synthesized at pH 7.4 in NaCl - $NaHCO_3$ solutions. The product was identified by X-ray diffraction and IR spectrophotometry. The monolayer capacity was measured by the BET N_2 method. The surface properties of the oxide in water were studied by potentiometric titration, yielding a pH of 6.9 for zero proton charge and approximately eight hydroxyl surface groups per nm^2. The BET surface area was 180 m^2/g [15].

Experiments on the removal of Mn(II) from solution were carried out in 0.7 M NaCl media with a two-phase HCO_3^-/CO_2 buffer system to obtain pH values in the range from 8 to 9. $Mn(ClO_4)_2$ was added to give initially 50 μM solutions. The alkalinity was 5 meq/L, added as $NaHCO_3$. The partial pressure of carbon dioxide was varied in different experiments. Non-oxidative removal was observed by using N_2/CO_2 atmospheres; oxidative removal by changing to O_2/CO_2 atmospheres. Filtration through 0.22 μm Millipore filters was used to separate the solution from the iron oxide particles. Mn(II) in the filtrate was determined spectrophotometrically. All experiments were performed at 25°. A full account of the experimental procedures is available elsewhere [16].

Results. Mn(II) is removed rapidly from solution in the presence of γ-FeOOH and in the absence of oxygen. Equilibrium is reached in less than ten minutes. Figure 1 shows the molar ratio of Mn(II) adsorbed by γ-FeOOH to Mn(II) remaining in solution as a function of pH. The adsorption is strongly influenced by pH, in line with the equilibrium shown in equations 10 and 11. In the

$$\mathord{>}FeOH + Mn^{2+} \rightleftharpoons \mathord{>}FeOMn^+ + H^+ \qquad (10)$$

$$(\mathord{>}FeOH)_2 + Mn^{2+} \rightleftharpoons (\mathord{>}FeO)_2Mn + 2H^+ \qquad (11)$$

$NaCl/NaHCO_3$ solutions (0.7 M NaCl, 5 mM $NaHCO_3$) the important species of manganese are calculated to be Mn^{2+} (~ 20%), $\Sigma MnCl_n$ (~ 75%), and $MnHCO_3^+$ (~ 3%). (Greater adsorption would occur from solutions of lower chloride concentrations.) The γ-FeOOH particles are present at a concentration of 89 mg/L, or 1×10^{-3} M total ferric iron. In the suspension 2.2×10^{-4} moles of hydroxyl surface sites

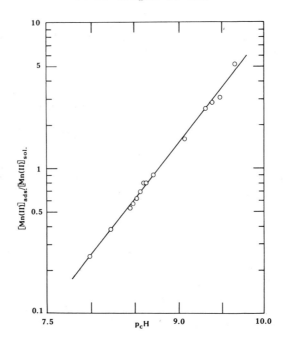

Figure 1. The dependence of the ratio "adsorbed Mn(II)/Mn(II) in solution" on P_cH in the presence of a 1×10^{-3} M γ-FeOOH suspension (89 mg/L γ-FeOOH) in an aqueous 0.7 M NaCl medium at 25°. Reproduced with permission from ref. 16.

are estimated to be present per liter. The surface area of the suspension is 16 m²/L [16].

When dissolved oxygen is present, a rapid initial removal, essentially the same as under nitrogen, is followed by a sustained removal, the rate of which is pseudo-first order in Mn(II) (Fig. 2). The half-time for continuous removal under an oxygen atmosphere is about 40 minutes. The influence of pH on the adsorption equilibrium and the oxidation of Mn(II) in the presence of oxygen is illustrated in Figure 3. The rate of removal is significantly increased as the pH is raised from 8 to 9. At the higher pH values the rate changes probably because of changed surface properties. The half-time for the initial stages of Mn(II) removal with 10^{-3} M γ-FeOOH and oxygen at 0.21 atm ranged from 785 minutes at pH 8.08 to 37 minutes at pH 8.71. The rate appears to be proportional to $1/[H^+]^2$. Decreasing the

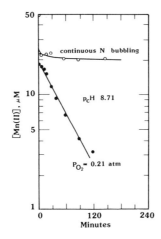

Figure 2. Time dependence of the Mn(II) concentration in a 0.7 M NaCl solution containing 1×10^{-3} M suspended γ-FeOOH under air or nitrogen atmospheres at 25°. Reproduced with permission from ref. 16.

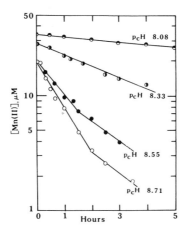

Figure 3. Time and pH dependence of the Mn(II) concentration in a 0.7 M NaCl solution containing 1×10^{-3} M suspended γ-FeOOH at 25°. The nitrogen atmosphere was replaced at time zero by a nitrogen/oxygen atmosphere with 0.21 atm partial pressure of oxygen. Reproduced with permission from ref. 16.

γ-FeOOH concentration from 10^{-3} to 10^{-4} M at a constant pH of 8.55 lengthened the half-time of removal of Mn(II) from 60 to 600 minutes.

Discussion. The experimental results summarized above are representative of high-chloride water, intended to roughly simulate marine conditions. We would expect the removal half-time for Mn(II) at given pH, partial pressure of oxygen, and surface site concentration to vary significantly for different ionic environments, especially ones in which strongly adsorbing metals and ligands might compete effectively for surface sites on the hydrous iron oxide. Thus far it is known, that the rate of oxidative removal of Mn(II) in the presence of γ-FeOOH increases strongly with increasing pH and is proportional to the amount of oxide surface, or surface sites. We can write an empirical rate expression (eqn. 12) to correlate these observations on the

$$-\frac{d[Mn(II)]}{dt} = k[\gamma\text{-FeOOH}]_0[O_2][H^+]^{-2}[Mn(II)] \qquad (12)$$

initial oxidative removal of Mn(II). As the reaction at the FeOOH surface progresses, oxidatively precipitated Mn(II) and Mn(IV) solid phases will continue to accumulate on the surface altering the original catalytic properties toward those of a hydrous manganese oxide surface such as MnOOH or MnO_x. Underlying the empirical equation 12 are the following separate processes:

- adsorption of Mn^{2+} from solution to hydroxyl surface sites, described by equilibria as shown in equation 10 and 11;
- binding of O_2(aq) to $>FeO\text{-}Mn^+$ or $(>FeO)_2Mn$ yielding an intermediate species such as $FeO\text{-}Mn^+\text{---}O_2$;
- an oxidation of Mn(II) to Mn(III) or Mn(IV) in a one- or two-electron transfer process with simultaneous reduction of O_2 to water.

One possible mechanism might involve binding of O_2 to the Mn(II) complex on the iron oxide surface, electron transfer yielding superoxide (O_2^-) bound to a Mn(III) ion, and reaction of an adjacent surface Mn(II) complex with a Mn(III)O_2 intermediate forming a μ-peroxo-bridged surface dimer, -Mn(III)-O-O-Mn(III)-. Further electron transfer and hydrolysis would then yield a mixture of Mn(III)-Mn(IV) solid products. The autoxidations of Fe(II) and Mn(II) in initially homogeneous systems are known to be kinetically favored by hydrolysis of the reduced metal ion [17].

Chemistry of Metal Oxides in Natural Water 175

According to the work of Fallab, electron transfer from Fe^{2+} to O_2 can be accelerated by oxygen ligand atoms bound at appropriate positions in the coordination sphere of Fe(II). It is argued that an additional non-binding electron pair gives π-donor properties and exerts a repulsion on the d electron of Fe(II) [17]. We observed earlier, that complexing of Mn^{2+}(aq) by EDTA, oxalate, pyrophosphate, and tripolyphosphate slows autoxidation of Mn(II), and that sulfate and chloride slow Fe(II) autoxidation [14, 18].

The similar effectiveness of FeOOH(s) and MnO_2(s) surfaces in accelerating the oxygenation of Mn(II) and Fe(II) may indicate that an oxygen-donor surface environment is a common characteristic of such catalysts. It is also known, that $Mn(OH)_2$(s) is rapidly oxidized by O_2. These observations are consistent with the accelerating role of hydrolysis and binding by O-donor ligands in Fe(II) and Mn(II) oxidations in solution. An alternative interpretation of the catalytic behavior of the >Fe(III)(OH) and >Mn(IV)OH surfaces suggests the participation of the central metal, M(Q) in a redox cycle (eqns. 13, 14), in which the oxidation by O_2 is very rapid.

$$M(Q) + Mn(II) \longrightarrow M(Q-1) + Mn(III) \quad (13)$$

$$M(Q-1) + O_2 \longrightarrow M(Q) + H_2O \quad (14)$$

Q: oxidation number

If the expected rate law for surface-catalyzed oxidation is formulated more generally (eqn. 15), then pertinent

$$-\frac{d[Mn(II)]}{dt} = k^S_{ox} *K^S_{Mn} \alpha_{Mn} [>FeOH] K^S_{O_2}[O_2][H^+]^n [Mn(II)] \quad (15)$$

k^S_{ox} : surface oxidation rate coefficient

$*K^S_{Mn}$: adsorption constant for Mn to the >FeOH surface with * emphasizing the proton involvement

α_{Mn} : fraction of Mn(II) in solution available for adsorption

[>FeOH] : concentration of available surface sites; dependent on total sites, pH, ionic strength, nature of the oxide, and composition for sites by other metal ions and ligands

$K^S_{O_2}$: surface binding constant for O_2

$[O_2]$: oxygen concentration

[Mn(II)]: total concentration of Mn(II) in solution

$[H^+]$: hydrogen ion concentration in solution

chemical properties of natural waters with respect to the oxidative part of manganese cycles are evident. The exponent "n" in [H$^+$] reflects the overall dependence of the rate of the Mn(II) oxidation on hydrogen ion concentration incorporating several independent processes. The solution chemistry is represented primarily by the α_{Mn} term, whereas oxide surface properties are embodied in $*K_{Mn}^S$ and [>FeOH]. A goal of surface-chemical research on oxides is to understand the factors that contribute to $*K^S$ and [>FeOH] in complex systems. It should prove valuable to examine the effects of competition for surface sites between various major and minor cations and the effects of adsorbed inorganic and organic ligands on the catalytic power of iron oxides and other metal and metalloid oxides.

Reductive Dissolution of Mn(III) and Mn(IV) Oxides by Organic Compounds

The oxides of Mn(III) and Mn(IV) are of very low solubility in water. For example, the equilibrium solubility product of γ-MnOOH(s) is approximately 10^{-42}. At pH 7 the equilibrium concentration of Mn^{3+} (aq) would be about 10^{-22} M, and that of MnOH^{2+} (aq) about 10^{-15} M. Little is known about other Mn(III) hydrolysis products. The dissolution and mobilization of manganese depends on reduction reactions (eqns. 16, 17). Redox equilibria in

$$MnOOH(s) + 3H^+ + 3e^- \longrightarrow Mn^{2+} + 2H_2O \qquad (16)$$

$$MnO_2(s) + 4H^+ + 2e^- \longrightarrow Mn^{2+} + 2H_2O \qquad (17)$$

aqueous manganese systems were systematically treated by Hem [19], Bricker [20], and others. At neutral to slightly alkaline pH values, reducing potentials below 0.5 to 0.6 volt are required to cause dissolution of MnOOH(s) in micromolar Mn^{2+} solutions.

A number of organic compounds with structures similar to those of natural organics were shown to reduce and dissolve MnO$_x$ compounds [21, 22]. The ability of organic compounds of different structure to reduce MnO$_x$ compounds can be expected to vary widely. A recently completed systematic investigation by Stone [23, 24] examined the physical-chemical factors that influence the rates at which Mn(III) and Mn(IV) oxides are reduced and dissolved in water. Laboratory experiments simulating chemical conditions in slightly acidic to slightly alkaline waters followed the course of the reaction between synthesized colloidal $MnO_x(s)$ particles and model organic compounds. Experimental results for the reduction by hydroquinone lend support to a surface site-binding model, according to which coordination at the hydrous manganese oxide plays an important part in the chemically controlled reductive dissolution process. Competitive adsorption processes can be expected to influence MnO_x dissolution rates in complex systems.

Experimental. A colloidal suspension of oxidized manganese was synthesized by O_2 oxidation of $Mn(OH)_2(s)$ previously precipitated in 0.2 mM NH_4OH under N_2. The suspension slowly coagulated over several months. The particles grew in size from 0.2 μm to 1 μm and had the average composition represented by $MnO_{1.65}$. X-ray diffraction patterns had clear lines for β-MnOOH and fainter lines for γ-MnOOH. The BET surface area was 60 m²/g. Potentiometric titration of two different preparations gave pH values of zero proton charge of 7.4 and 6.9 [23]. Solutions of organic compounds were freshly prepared from reagent grade chemicals and deionized, distilled water. Two-phase CO_2/HCO_3^- buffers were used to adjust the pH in the MnO_x/hydroquinone dissolution experiments. The pH was 7.2 for all survey experiments with other organic compounds. The ionic strength was adjusted with $NaNO_3$. The reactions were carried out under nitrogen in 1-liter jacketed beakers. All systems were stirred at >100 rpm. The MnO_x suspensions flocculated under the typically used ionic strength conditions (5×10^{-2} M). Aliquots of the suspension were filtered through 0.2 μm Nucleopore membrane filters. The filrates were analyzed for dissolved Mn by AAS. UV spectra of the filtrates were obtained to observe changes resulting from the oxidation of individual organic compounds [23].

Results. The reduction of MnO_x by hydroquinone initially yields p-benzoquinone as principal product shown by comparing the UV bands at 288 nm and 244 nm (Fig. 4). Further reduction by p-benzoquinone occurs at a much slower rate than that of the initial step. Figure 4 shows UV spectra of filtered solutions at various times during the course of the reaction of hydroquinone and 2,5-dihydroxybenzoic acid with $MnO_{1.65}$ at pH 7.2. For other reductants such as catechol, the UV spectra showed evidence for polymerization of the organic oxidation products [23].

The order of the reductive dissolution of the manganese oxide by hydroquinone with respect to oxide concentration was examined by varying the initial concentration of the

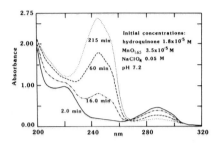
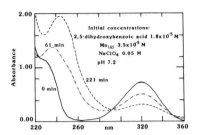

Figure 4. *UV spectra of filtrates of mixtures consisting of a suspension of manganese oxide and hydroquinone or 2,5-dihydroxybenzoic acid. Reproduced with permission from ref. 25.*

manganese oxide at a constant 15-fold excess of reductant. The concentrations of unreacted manganese oxide had a pseudo first-order dependence on time (Fig. 5). When the manganese oxide concentration was kept constant and the hydroquinone concentration varied in the range of 10- to 20-times the manganese oxide concentration, the reaction was first-order in hydroquinone. At hydroquinone concentrations approximately 30-times the manganese oxide concentration, the apparent dissolution order is less than unity suggesting that the reactive sites on the oxide surface become occupied. Variation of pH from 6.5 to 7.7 at fixed initial manganese oxide and hydroquinone concentrations showed the rate to be proportional to the 0.46 power of the hydrogen ion activity. The rate of the reaction decreased approximately 0.4 percent per day as the manganese oxide suspensions aged. The reductive dissolution of manganese oxide by hydroquinone at low hydroquinone concentration, constant temperature, and constant ionic strength under intense stirring is described during the first half of the reaction by equation 18.

$$\frac{d[Mn(II)]_d}{dt} = k_1 \{H^+\}^{0.46} [HQ] \{[MnO_x]_i - [Mn(II)_d]\} \quad (18)$$

$[Mn(II)]_d$: concentration of Mn(II) in solution formed by reduction of MnO_x
$\{H^+\}$: hydrogen ion activity
$[HQ]$: concentration of hydroquinone
$[MnO_x]_i$: initial concentration of suspended manganese oxide

The apparent activation energy for the reduction by hydroquinone is approximately 37 kJ/mol. This value is consistent with a chemically controlled process and with the

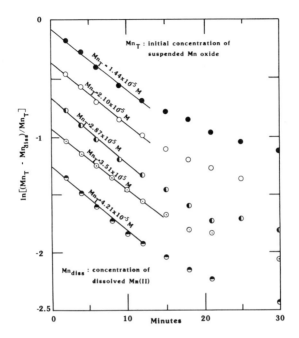

Figure 5. The time dependence of the fraction of undissolved $MnO_{1.65}$ in a 0.05 M aqueous sodium nitrate medium of pH 7.8 at 25° in the presence of a 15-fold excess of hydroquinone. The points corresponding to the four highest manganese oxide concentrations were offset to facilitate comparison. Reproduced with permission from ref. 24.

fact, that a ten-fold change in stirring rates did not affect the rate of the reaction. Variations of ionic strength from 0.01 to 0.5 decreased reaction rates in the hydroquinone and 2,5-dihydroxybenzoic acid systems by approximately 15 to 20 percent.

Surface competition. Two types of experiments were carried out to observe competition for surface sites. First, non-redox species such as phosphate and calcium salts were added to the hydroquinone system. Second, slow-reacting, secondary substrate redox species such as sorbitol at various concentrations relative to hydroquinone were used. Increased levels of phosphate or calcium lower the rate of the manganese oxide-hydroquinone reaction (Fig. 6). Experiments in the absence of hydroquinone showed strong adsorption of phosphate onto the manganese oxide surface at

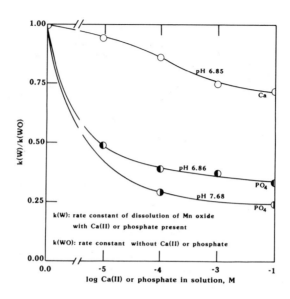

Figure 6. Effects of calcium and phosphate ions on the rate of reductive dissolution of manganese oxide in the presence of hydroquinone at ten-times the concentration of manganese oxide. Reproduced with permission from ref. 24.

neutral pH. Thus, increased adsorption of phosphate to the particle surface is associated with lower rates of oxidative dissolution of manganese oxide. For example, at pH 6.8 a total phosphate concentration of 10 μM lowers the rate of hydroquinone oxidation by 30 percent and at 100 μM total phosphate by 50 percent.

The degree of interference with the reduction of hydroquinone by secondary substrates depended on the structure of the substrate, on relative concentrations, and on pH. Sorbitol and phthalate at 100 times the concentration of hydroquinone reduced the rate of the hydroquinone reaction by 30 percent. Oxalate and salicylate did not interfere with the reduction under the same conditions.

Reactions rates of different organics. The survey experiments at pH 7.2 and fixed ionic strength, temperature, mixing rates, and initial concentrations of oxidant and manganese oxides produced a set of apparent second-order rate constants for the oxidative dissolution of $MnO_{1.65}$. The results are summarized in Figure 7. The second-order rate

Chemistry of Metal Oxides in Natural Water

```
                  Ascorbate ~ Catechol ~ 3-Methoxycatechol ~
                        ~ 3,4-Dihydroxybenzoic Acid
                             4-Nitrocatechol
                              Hydroquinone
                              Thiosalicylate
                       2,5-Dihydroxybenzoic Acid
                              Syringic Acid
                             o-Methoxyphenol
                               Vanillic Acid
                      Orcinol ~ 3,5-Dihydroxybenzoic Acid
                                Resorcinol
                                 Pyruvate
                                 Oxalate

        No Activity:  Formate, Fumarate, Glycerol,
                      Lactate, Malonate, Propanol,
                      Propionaldehyde, Propionate,
                      Salicylate
```

(vertical axis label: INCREASING REACTIVITY)

Figure 7. Relative reactivities of organic reductants in the dissolution of manganese oxide at pH 7.2 and 25°. The second-order rate constants range from 0.005 (resorcinol) to 30 $M^{-1} s^{-1}$ (3-methoxycatechol).

constants cover a range from 5×10^{-3} M^{-1} sec^{-1} for resorcinol to 3×10^{1} for methoxycatechol. The rate constant for hydroquinone is approximately 2 M^{-1} sec^{-1} [25].

Discussion. A synthetic manganese oxide, $MnO_{1.65}$, prepared by oxidation of precipitated $Mn(OH)_2$, was reductively dissolved by twenty-seven model organic compounds with structures similar to those of organics in natural water. The results are consistent with a mechanism that requires the binding of the reductant at the surface of the manganese oxide and involves electron transfer reactions at the surface as rate-limiting steps [24]. That the reductive dissolution is chemically controlled is indicated by the high activation energy of 37 kJ/mol for the oxidation of hydroquinone, the independence of the reaction rates on stirring, the inhibition of reductive dissolution by adsorbed ions such as phosphate and calcium, and the rather large differences in reactivity associated with minor differences in the structure of the reductant. The experimentally derived rate expression for the oxidation of hydroquinone by manganese oxide is consistent with a reaction mechanism in which hydroquinone forms a complex with surface Mn(III) prior to electron transfer. The fractional-order dependence (0.46) of the reaction between manganese oxide and hydroquinone on the hydrogen ion activity may reflect the involvement of Mn-hydroquinone and Mn-deprotonated hydroquinone, or may indicate that protonation of the surface assists in the release of reduced manganese, Mn(II), from the surface. The reaction rates for different reductants such as hydroquinone and catechol show no correlation with reduction potentials. This lack of correlation may be an indication of an inner-sphere surface rather than an outer-sphere mechanism. The overall free energy change of the redox process appears not to be predictive of the rates of reductive dissolution.

It should prove useful to study the relationship between coordination processes at oxide/water interfaces and redox reaction processes in greater detail. Integration of concepts of surface chemical reactions and structure-reactivity relationships appears to provide possibilities for the interpretation and prediction in a number of aquatic enviromental systems.

Conclusions

The examples in this paper illustrate that a coordination chemistry approach to processes at the oxide/water interface can provide an explanation for the specific influences of pH, metals, and ligands on the nature of species at a surface and on chemical reactions at a surface. The surface catalysis by γ-FeOOH of the oxygenation of Mn(II) demonstrates the strong influence of the pH-dependent adsorption on the rate of an important step in the overall manganese cycle in water. Other surfaces may also be effective oxidation catalysts. The coordination reactions of particle surfaces in natural waters should prove to be an important subject for study. The rate of reductive dissolution of Mn(III) and Mn(IV) oxides by organic compounds illustrates another aspect of the manganese cycle. Results for low-molecular weight compounds of different structures indicate that the process at the manganese oxide surface is chemically controlled. The general features of the dependence of the rate on surface and solution properties can be explained in terms of binding at a surface site or in terms of a coordination model. Extension of these laboratory methodologies and models to more complex natural systems should help to increase understanding of the manganese and iron cycles.

Acknowledgements. The work on γ-FeOOH catalysis was supported in part by grants to J. J. Morgan and W. Sung from Dupont and Union Oil. The work on reductive dissolution of manganese oxides was assisted by a fellowship to A. Stone from the Noyes Foundation and to J. J. Morgan from Union Oil.

References

[1] W. Stumm and J. J. Morgan, **"Aquatic Chemistry,"** 2nd Ed., Wiley-Interscience: New York, 1981, p. 736.
[2] A. W. Morris and A. J. Bale, "Effect of Rapid Precipitation of dissolved Mn in River Water on Estuarine Mn Distributions," **Nature, 279** (1979) 318.
[3] E. R. Sholkovitz and D. Copland, "The Chemistry of Suspended Matter in Esthwaite Water, a Biologically Productive Lake with Seasonally Anoxic Hypolimnion," **Geochim. Cosmochim. Acta, 46** (1982) 393.
[4] L. Sigg and W. Stumm, "Interaction of Anions and Weak Acids with Hydrous Geothite (α-FeOOH) Surface," **Colloids and Surfaces, 2** (1981) 101.
[5] J. A. Davis, "Adsorption of Natural Dissolved Organic Matter at the Oxide/Water Interface," **Geochim. Cosmochim. Acta, 46** (1982) 2381.
[6] L. Balistrieri, P. G. Brewer and J. W. Murray, "Scavenging Residence Times of Trace Metals and Surface Chemistry of Sinking Particles in the Deep Ocean," **Deep-Sea Res., 28A** (1981) 101.
[7] W. Stumm, G. Furrer and B. Kunz, "The Role of Surface Coordination in Precipitation and Dissolution of Mineral Phases," **Croat. Chem. Acta, 56** (1984) 593.
[8] P. W. Schindler, "Surface Complexes at Oxide-Water Interfaces," in **"Adsorption of Inorganics at Solid-Liquid Interfaces,"** M. A. Anderson and A. J. Rubin, Eds., Ann Arbor Science: Ann Arbor, Michigan, 1981, p. 2.
[9] H. Hohl, L. Sigg and W. Stumm, "Characterization of Surface Chemical Properties of Oxides in Natural Waters: The Role of Specific Adsorption in Determining the Surface Charge," **Adv. Chem. Ser., 189** (1980) 1.
[10] J. Vuceta and J. J. Morgan, "Chemical Modeling of Trace Metals in Fresh Waters: Role of Complexation and Adsorption," **Environ. Sci. Tech., 12** (1978) 1302.
[11] J. Westall, "Chemical Equilibrium Including Adsorption on Charged Surfaces," **Adv. Chem. Ser., 189** (1980) 33.
[12] F. M. M. Morel, J. G. Yeasted and J. C. Westall, "Adsorption Models: A Mathematical Analysis in the

Framework of General Equilibrium Calculation," in **"Adsorption of Inorganics at Solid-Liquid Interfaces,"** M. A. Anderson and A. J. Rubin, Eds., Ann Arbor Science Publishers: Ann Arbor, Michigan, 1981, p. 263.

[13] M. R. Hoffmann and S. D. Boyce, "Catalytic Autoxidation of Aqueous Sulfur Dioxide in Relationship to Atmospheric Systems," in **"Trace Atmospheric Constituents,"** S. E. Schwartz, Ed., Wiley-Interscience: New York, 1983, p. 148.

[14] W. Sung and J. J. Morgan, "Kinetics and Product of Ferrous Iron Oxygenation in Aqueous Systems," **Environ. Sci. Tech.**, 14 (1980) 561.

[15] W. Sung, "Catalytic Effects of the γ-FeOOH Surface on the Oxygenation Removal Kinetics of Fe(II) and Mn(II)," **Thesis**, California Institute of Technology, 1980.

[16] W. Sung and J. J. Morgan, "Oxidation Removal of Mn(II) from Solution Catalyzed by the γ-FeOOH (Lepidocrocite) Surface," **Geochim. Cosmochim. Acta**, 45 (1981) 2377.

[17] S. Fallab, "Reactions with Molecular Oxygen," **Angew. Chem. Internat. Ed. Engl.**, 6 (1967) 496.

[18] H. Bilinski and J. J. Morgan, "Complex Formation and Oxygenation of Mn(II) in Pyrophosphate Solutions," Div. Water, Air and Waste Chem., Amer. Chem. Soc., Minneapolis, April 1969.

[19] J. D. Hem, "Chemical Equilibria and Rates of Manganese Oxidations," **U.S. Geol. Surv. Water Supply. Pap., 1667-A,** 64 pp.

[20] O. P. Bricker, "Some Stability Relationships in the System $Mn-O_2-H_2O$ at 25°C and 1 Atmosphere Total Pressure," **Am. Mineral.**, 50 (1965) 1296.

[21] W. E. Baker, "The Role of Humic Acids from Tasmanian Podzolic Soils in Mineral Degradation and Metal Mobilization," **Geochim. Cosmochim. Acta,** 37 (1973) 269.

[22] J. D. Hem, "Reduction and Complexing of Manganese by Gallic Acids," **U.S. Geol. Surv. Water Supply Pap., 1667-D** (1965) 28 pp.

[23] A. T. Stone, "The Reduction and Dissolution of Mn(II) and Mn(IV) Oxides by Organics," **Thesis,** California Institute of Technology, 1983.

[24] A. T. Stone and J. J. Morgan, "Reduction and Dissolution of Manganese(III) and Manganese(IV) Oxides by Organics. 1. Reaction with Hydroquinone," **Environ. Sci. Technol.,** 18 (1984) 450.

[25] A. T. Stone and J. J. Morgan, "The Reduction and Dissolution of Mn(III) and Mn(IV) Oxides by Organics. 2. Survey of Reactivity of Organics," **Environ. Sci. Technol.,** 18 (1984) 617.

VIII

SECONDARY MINERALS: NATURAL METAL ION BUFFERS

Peter A. Williams

Department of Chemistry, University College
P. O. Box 78, Cardiff CF1 1XL, Wales, U.K.

Secondary minerals can exert a significant control on the concentrations of metal ions in natural aqueous solution. Evidence is presented to support this with particular emphasis on lead(II) and copper(II) species containing chloride ions. The role of secondary minerals in limiting metal concentrations may be more important than has been previously thought, although many physical and chemical phenomena do also play important roles.

Studies of the effects and causes of pollution by transition and other heavy metals in the natural environment sometimes obscure the fact that the presence of such metals is entirely natural. Some kind of distinction might be made on the basis of the origins of the metals. Increased inputs as the result of human activities can give rise to problems that have been enumerated elsewhere in these proceedings. However, unusually high concentrations of elements such as copper, lead, mercury, and zinc are not necessarily a reflection of extraordinary processes. Indeed, the identification of such patterns of metal distribution forms the basis of a branch of geochemistry applied to the search for ore bodies [1].

Whereas the dispersion of metallic and other elements in the aqueous environment is controlled by many factors, simple naturally occurring inorganic species can play a significant role. Some minerals, especially those that have classically been described as secondary, may be viewed as natural buffers of metal ions between comparatively insoluble primary species such as sulfides (Table 1) and dissolved inorganic ions and complexes that may be transported over large distances.

The relevance of such secondary minerals to environmental inorganic chemistry is obvious and an understanding of them embraces the fields of chemistry, geology, and mineralogy

Table 1. Typical Primary and Secondary Minerals of Copper, Lead, and Zinc*

Primary Minerals		Secondary Minerals	
$CuFeS_2$	Chalcopyrite	$Cu_2CO_3(OH)_2$	Malachite
Cu_5FeS_4	Bornite	$Cu_3(CO_3)_2(OH)_2$	Azurite
$CuFe_2S_3$	Cubanite	$Cu_4SO_4(OH)_6$	Brochantite
PbS	Galena	$PbSO_4$	Anglesite
		$PbCO_3$	Cerussite
		$Pb_3(CO_3)_2(OH)_2$	Hydrocerussite
ZnS	Blende,	$ZnCO_3$	Smithsonite
	Wurtzite**	$Zn_5(CO_3)_2(OH)_6$	Hydrozincite

*Names are those given in reference [2] and those approved by the I.M.A. Commission on New Minerals.
**A number of polymorphs, being slight variants on the wurtzite structure, are known [2].

among others. It is the aim of this article to highlight a few of the relationships between secondary minerals, some of them rather complex, and the dispersion of transition and other heavy metal elements from ore bodies and other sources. Because some attention at this meeting has been directed toward the marine environment, it seems appropriate to particularly consider chloride-containing phases.

Secondary Minerals of Copper and Lead as Metal Ion Buffers

It is certainly beyond the scope of this paper to explore the entire chemistry of the naturally occurring chlorides, let alone all the halides. This discussion is limited to lead and copper species in the secondary environment. In Tables 2 and 3 are listed the known chloride minerals of these metals, and most of the other species containing necessary anions other than oxide and hydroxide.

The choice to examine in detail lead and copper minerals is not entirely arbitrary. The two metals are of interest from a nutritional-biochemical and a pollution viewpoint. Secondly, a good deal of thermodynamic data are available for compounds of copper and lead and for many of the minerals listed in Tables 2 and 3. Therefore, these compounds and minerals are suitable for incorporation into equilibrium models. Some of the minerals in Tables 2 and 3, including pseudocotunnite, eriochalcite, calumetite, and anthonyite, are too soluble to be considered to control the

Table 2. Chloride-Containing Minerals of Lead and Copper

Formula	Name	Ref.*
$PbCl_2$	Cotunnite**	
$PbFCl$	Matlockite	
$PbClOH$	Laurionite, Paralaurionite	
Pb_2Cl_3OH	Penfieldite	
$Pb_3Cl_2O_2$	Mendipite	
$Pb_3Cl_4(OH)_2$	Fiedlerite	[3]
K_2PbCl_4 or $3KPbCl_3 \cdot H_2O$	Pseudocotunnite	
$Pb_2Cl(O,OH)_{2-x}$ (x~0.3)	Blixite	
$PbBiO_2Cl$	Perite	
$Pb_2AgCl_3(F,OH)_2$	Bideauxite	[4]
$Pb_4Fe_4O_9(OH,Cl)_2$	Hematophanite	
$CuCl$	Nantockite**	
$CuCl_2 \cdot 2H_2O$	Eriochalcite	
$Cu(OH,Cl)_2 \cdot 2H_2O$	Calumetite	
$Cu(OH,Cl)_2 \cdot 3H_2O$	Anthonyite	
$Cu_2Cl(OH)_3$***	Atacamite,** Paratacamite,** Botallackite**	
$CuClOH(?)$	Melanothallite	
$CuClOH \cdot H_2O(?)$	Hydromelanothallite	
$K_2CuCl_4 \cdot 2H_2O$	Mitscherlichite	
$Cu_4Cl(OH)_7 \cdot nH_2O$ (n~½)	Claringbullite	[5]
$Pb_2CuCl_2(OH)_4$	Diaboleite	[6]
$Pb_5Cu_4Cl_{10}(OH)_8 \cdot 2H_2O$	Pseudoboleite**	[7]
$Pb_3CuCl_2O_2(OH)_2$	Chloroxiphite	[8]
$Pb_{19}Cu_{24}Cl_{42}(OH)_{44}$	Cumengeite	[7]
$Pb_{26}Cu_{24}Ag_9Cl_{62}(OH)_{47} \cdot H_2O$	Boleite**	[9]
$Pb_6Cl_6Cr(O,OH,H_2O)_8$	Yedlinite	[10]

*In general ref. [2]. Other useful references to unusual species are also given.

**These minerals were reported as corrosion products on archaeological artifacts [11-15].

***A zincian variety of paratacamite sometimes called anarkite [2] was reported from corrosion crusts on ancient brasses [15].

concentrations of copper(II) and lead(II) in aqueous solution except under the most unusual circumstances. A few of the compounds such as wherryite [16] and chloroxiphite [18] are known only from one locality, and many others are extremely rare. However, some of the minerals, even those with the most exotic stoichiometries, occur widely. An example is the remarkable species boleite. Whereas this mineral most often occurs in the oxidized zones of base-

Table 3. *Further Lead and Copper Minerals Other than Oxides and Hydroxides with Essential Chloride Anions**

Formula	Name	Ref.**
$Pb_2CO_3Cl_2$	Phosgenite***	
$Pb_4CuCO_3(SO_4)_2(OH,Cl)_2O$	Wherryite	[16]
$CuBO_2Cl \cdot 2H_2O$	Bandylite	
$Cu_{37}(SO_4)_2Cl_8(OH)_{62} \cdot 8H_2O$	Connellite	[17]
$Cu_{37}(NO_3)_4Cl_8(OH)_{62} \cdot 8H_2O$	Buttgenbachite	[17]
$K_2CuSO_4Cl_2$	Chlorothionite	
$Cu_6AlSO_4(OH)_{12}Cl \cdot 3H_2O$	Spangolite	
$Na_3Pb_2(SO_4)_3Cl$	Caracolite	
$Cu_4Pb_2So_4Cl_6(OH)_4 \cdot 2H_2O$	Arzrunite	
$Pb_5IO_3Cl_3O_3$	Schwartzembergite	
$NaCaCu_5(PO_4)_4Cl \cdot 5H_2O$	Sampleite	
$Pb_5(PO_4)_3Cl$	Pyromorphite	
$Pb_3AsO_4Cl_3$	Georgiadesite	
$Pb_{14}(AsO_4)_2O_9Cl_4$	Sahlinite	
$Pb_5(AsO_4)_3Cl$	Mimetite	
$Pb_5(VO_4)_3Cl$	Vanadinite	
$Pb_3IO_3Cl_3O$	Seeligerite	
$(Ca,Na)_2Cu_5(AsO_4)_4Cl \cdot 4-5H_2O$	Lavendulan	

*Several chloro-arsenites and antimonites are also known [2].
**Generally ref. [2] but other useful references are also listed.
***A common corrosion product on lead artifacts [11-15].

metal ore deposits in arid environments together with other members of the Pb(II)-Cu(II) oxychloride group [3, 7, 9-22], it has also been found in the oxidized portion of a mineral vein outcropping in the Sea in Cornwall, England [23]. Boleite has also been reported among corrosion products on objects from the "Batavia" shipwreck [14], and in oxidized lead slags from Laurium, Greece. These slags had been dumped in the sea in ancient times [24]. This latter deposit is especially noteworthy, because phosgenite, laurionite, paralaurionite, fiedlerite, matlockite, georgiadesite, pseudoboleite, cummengeite, diaboleite, and several of the secondary minerals listed in Table 1 were also identified.

The occurrence of many of these halide minerals and most of the secondary species in Table 1 in marine corrosion products is of particular interest. It is clear that these minerals are functioning as intermediates in the dispersion of copper and lead from metal objects under oxidizing conditions in salt water. Thus, we might view these minerals as phases controlling the release of the metals to the solution and call them metal ion buffers. Of course, these examples are rather exceptional, but all of the minerals do occur naturally and are not only the result of human intervention.

The geochemical significance of this fact is obvious. Under some conditions these secondary minerals do control the dispersion of transition and related metals in natural surface and groundwaters. A general approach to the understanding of the kinds of conditions giving rise to common species in the vicinity of oxidizing sulfide ore bodies was outlined by Garrels [25], who later developed his equilibrium thermodynamic model with Christ in a now classic work [26].

In recent experiments Giblin [27] showed, that during the oxidation of base-metal sulfides copper, lead, and iron show varying mobilities in surrounding soils, but zinc is relatively mobile. The complex sulfide posnjakite, $Cu_4SO_4(OH)_6 \cdot H_2O$, was formed in these experiments as a discrete phase that in part limits the concentrations of Cu(II) ions in soil pore waters. The general spatial relationships between some of the common secondary minerals of copper, lead, and zinc are shown in Figure 1 and are now well-understood. Other workers have also shown that simple inorganic species can control the solubility of the metals in groundwater near ores being oxidized [28, 29]. The main minerals implicated include atacamite, malachite, antlerite, $Cu_3SO_4(OH)_4$, and brochantite, $Cu_4SO_4(OH)_6$, for copper(II), and for lead(II) and zinc(II) anglesite, cerussite, and smithsonite. The ubiquitous presence of these minerals in oxide zones is well-known [3]. Mann and Deutscher [29] also pointed out the probable importance of the compounds $Zn(OH)_2$, $Pb_2Cl(OH)_3$ and $Zn_4SO_4(OH)_6 \cdot nH_2O$, which were unknown as minerals at the time of their study. These species assumed some interest because calculations on the basis of thermodynamic relationships predicted that they should occur naturally. It was suggested that these phases might be "fine-grained, well dispersed, and difficult to detect." Subsequently, however, the coordination compound $Zn_4SO_4(OH)_6 \cdot H_2O$ was described as a new mineral [30].

Ample evidence is available, therefore, for the important role that secondary minerals can play in the control of metal ion concentrations in aqueous solution in some natural environments. This control by secondary minerals does, of

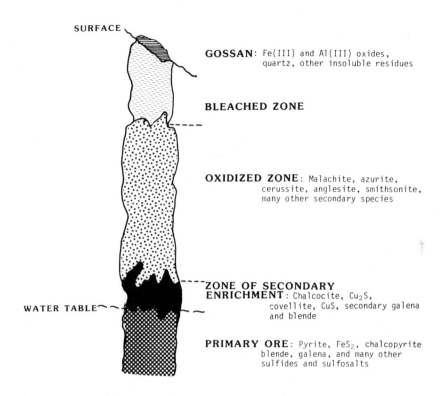

Figure 1. Diagrammatic representation of main zones in an oxidizing base-metal ore body and typical common minerals.

course, not discount the importance of metal sulfide deposition under reducing conditions. We may also conclude that a simple thermodynamic approach to secondary mineral formation can provide a framework within which the aqueous chemistry of such compounds can be rationalized to a significant extent.

Such considerations have been made with respect to a number of the halide minerals listed in the Tables. Conditions for the formation of boleite, pseudoboleite, cumengeite, diaboleite, chloroxiphite, and mendipite were recently described [31, 32]. These equilibrium studies led to the prediction of paragenetic sequences for several kinds of mineral assemblages, which were seen to be followed in the deposits of the Mendip Hills [18], Boleo [21], and in the famous oxidized ore zone at the Mammoth-St. Anthony mine, Tiger, Arizona [33, 34]. Cumengeite was also recently found in an oxidized mineral vein containing galena, bournonite, $PbCuSbS_3$, and tetrahedrite, $(Cu,Ag,Fe)_{12}Sb_4S_{13}$, outcropping in the sea [35]. It was pointed out that gross seawater chemistry, in terms of chloride ion activity and pH, would explain the mineral's occurrence in such an environment. Many other mineralogical and chemical studies of a similar kind have been reported [26, 36].

The role of secondary minerals as metal ion buffers has received comparatively little attention except in the special environments discussed above. Part of the reason for this neglect is the need to take into account many species, for which thermodynamic properties are not available. One element for which this problem is particularly acute is copper. In addition to the species listed in the Tables, scores of sulfate, arsenate, and phosphate minerals were reported from numerous localities [2, 3]. Many synthetic complexes exist that have not yet been found in nature. Clearly much more work is required in this field. Indeed, secondary minerals may be of greater importance than hitherto suspected in relation to aqueous processes. A few examples from the literature are worth recalling. Hem, in a study of the chemistry of cadmium and zinc in surface and groundwaters [37], suggested that willemite, Zn_2SiO_4, might significantly control zinc ion concentrations. Another study of the chemistry of copper in natural aqueous systems [38] pointed out that malachite can be important in limiting copper ion levels, especially in freshwater environments.

A full explanation and understanding of the chemistry of secondary mineral formation will not give a comprehensive picture of the behavior and concentration of metal ions in natural aqueous solutions. Controls on the solution geochemistry of metals are complex and indeed far more complicated than can be understood by simple equilibrium calculations involving well-defined solid phases. Metal ion activities in natural solutions are known to be sometimes controlled by adsorption and coprecipitation phenomena, interaction with organic species, and other factors. Nevertheless, to quote previous workers [28], "compound formation .... is of prime importance to an understanding of the solution geochemistry of trace elements and is fundamental to studies which might seek to consider these other factors."

References

[1] F. R. Siegel, **"Applied Geochemistry,"** Wiley: New York, 1974.

[2] M. H. Hey, **"An Index of Mineral Species and Varieties,"** 2nd Ed., British Museum of Natural History: London, 1975.

[3] C. Palache, H. Berman and C. Frondel, **"The System of Mineralogy,"** Vol. II, 7th Ed., Wiley: New York, 1951.

[4] S. A. Williams, "Bideauxite, a New Arizona Mineral," **Mineral. Mag.**, **37** (1970) 637.

[5] E. E. Fejer, A. M. Clark, A. G. Couper and C. J. Elliott, "Claringbullite, a New Hydrated Copper Chloride," **Mineral. Mag.**, **41** (1977) 433.

[6] R. C. Rouse, "The Crystal Chemistry of Diaboleite," **Z. Kristallogr.**, **134** (1971) 69.

[7] R. E. Winchell and R. C. Rouse, "The Mineralogy of the Boleite Group," **Mineral. Rec.**, **5** (1974) 280.

[8] J. J. Finney, E. J. Graeber, A. Rosenweig and R. D. Hamilton, "The Structure of Chloroxiphite, $Pb_3CuO_2(OH)_2Cl_2$," **Mineral. Mag.**, **41** (1977) 357.

[9] R. C. Rouse, "The Crystal Structure of Boleite − a Mineral Containing Silver Atom Clusters," **J. Solid State Chem.**, **6** (1973) 86.

[10] M. M. Wood, W. J. McLean and R. B. Laughon, "The Crystal Structure and Composition of Yedlinite," **Amer. Mineral.**, **59** (1974) 1160.

[11] A. Lacroix, "Sur quelques Mineraux Formes par l'Action de l'Eau de Mer sur des Objets Metalliques Romains Trouves en Mer au Large de Mahdia (Tunisie)," **Compt. Rend.**, **151** (1910) 276.

[12] A. Russell, "On the Occurrence of Cotunnite, Anglesite, Leadhillite and Galena on Fused Lead from the Wreck of the Fire-ship 'Firebrand' in Falmouth Harbour, Cornwall," **Mineral. Mag.**, **19** (1920) 64.

[13] R. M. Organ, "The Current Status of the Treatment of Corroded Metal Artifacts," in **"Corrosion and Metal Artifacts,"** B. F. Brown, H. C. Burnett, W. T. Chase, M. Goodway, J. Kruger and M. Pourbaix, Eds., **National Bureau of Standards Special Publication, 479**, Washington, D.C., 1977, p. 107.

[14] R. C. Gorman, **"Report of the Government Chemical Laboratories of Western Australia for the Year 1975,"** W. Australian Govt.: Perth, 1976.

[15] I. D. MacLeod, "Formation of Marine Concretions on Copper and its Alloys," **Int. J. Naut. Arch. Underwater Explor.**, **11** (1982) 267.

[16] W. J. McLean, "Confirmation of the Mineral Species

Wherryite," **Amer. Mineral.**, **55** (1970) 505.
[17] W. J. McLean and J. W. Anthony, "The Disordered, 'Zeolite-like' Structure of Connellite," **Amer. Mineral.**, **57** (1972) 426.
[18] R. F. Symes and P. G. Embrey, "Mendipite and Other Rare Oxychloride Minerals from the Mendip Hills, Somerset, England," **Mineral. Rec.**, **8** (1977) 298.
[19] A. Liversidge, "Boleite, Nantockite, Kerargyrite and Cuprite from Broken Hill, New South Wales," **J. Proc. Roy. Soc. N.S.W.**, **28** (1894) 94.
[20] L. Fletcher, "On Crystals of Percylite, Caracolite, and an Oxychloride of Lead (Daviesite) from Mina Beatriz, Sierra Gorda, Atacama, South America," **Mineral. Mag.**, **8** (1889) 171.
[21] I. F. Wilson and V. S. Rocha, "Geology and Mineral Deposits of the Boleo Copper District Baja California, Mexico," **U.S. Geol. Surv. Prof. Paper**, **273** (1955).
[22] P. Bariand, "Contributions a la Mineralogie de l'Iran," **Bull. Soc. Franc. Miner. Crist.**, **86** (1963) 17.
[23] A. C. Dean, University College, Cardiff, personal communication.
[24] W. Kohlberger, "Minerals of the Laurium Mines, Attica, Greece," **Mineral. Rec.**, **7** (1976) 114.
[25] R. M. Garrels, "Mineral Species as Functions of pH and Oxidation Reduction Potentials, with Special Reference to the Zone of Oxidation and Secondary Enrichment of Sulfide Ore Deposits," **Geochim. Cosmochim. Acta**, **5** (1965) 153.
[26] R. M. Garrels and C. L. Christ, **"Solutions, Minerals and Equilibria,"** Harper and Row: New York, 1965.
[27] A. M. Giblin, "Experiments to Demonstrate Mobility of Metals in Waters near Base-Metal Sulfides," **Chem. Geol.**, **23** (1978) 215.
[28] A. W. Mann and R. L. Deutscher, "Solution Geochemistry of Copper in Water Containing Carbonate, Sulphate and Chloride Ions," **Chem. Geol.**, **19** (1977) 253.
[29] A. W. Mann and R. L. Deutscher, "Solution Geochemistry of Lead and Zinc in Water Containing Carbonate, Sulphate and Chloride Ions," **Chem. Geol.**, **29** (1980) 293.
[30] R. E. Bevins, S. Turgoose and P. A. Williams, "Namuwite, $(Zn,Cu)_4SO_4(OH)_7 \cdot 4H_2O$, a New Mineral from Wales," **Mineral. Mag.**, **46** (1982) 51.
[31] D. A. Humphreys, J. H. Thomas, P. A. Williams and R. F. Symes, "The Chemical Stability of Mendipite, Diaboleite, Chloroxiphite and Cumengeite, and their Relationships to Other Secondary Lead(II) Minerals," **Mineral. Mag.**, **43** (1980) 901.

[32] F. A. Abdul-Samad, D. A. Humphreys, J. H. Thomas and P. A. Williams, "Chemical Studies on the Stabilities of Boleite and Pseudobleite," **Mineral. Mag., 44** (1981) 101.
[33] R. A. Bideaux, "Famous Mineral Localities: Tiger Arizona," **Mineral. Rec., 11** (1980) 155.
[34] F. Abdul-Samad, J. H. Thomas, P. A. Williams, R. A. Bideaux and R. F. Symes, "Mode of Formation of Some Rare Copper(II) and Lead(II) Minerals from Aqueous Solution, with Particular Reference to Deposits at Tiger, Arizona," **Transition Met. Chem., 7** (1982) 32.
[35] A. C. Dean, R. F. Symes, J. H. Thomas and P. A. Williams, "Cumengeite from Cornwall," **Mineral. Mag., 47** (1983) 235.
[36] K. W. Bladh, "The Formation of Goethite, Jarosite and Alunite During the Weathering of Sulfide-Bearing Felsic Rocks," **Economic Geology, 77** (1982) 176, and references therein.
[37] J. D. Hem, "Chemistry and Occurrence of Cadmium and Zinc in Surface Water and Groundwater," **Water Resources Research, 8** (1972) 661.
[38] D. T. Rickard, "The Chemistry of Copper in Natural Aqueous Solutions," **Stockholm Contrib. Geol., 23** (1970) 1.

IX

ENVIRONMENTAL INORGANIC CHEMISTRY OF MAIN GROUP ELEMENTS WITH SPECIAL EMPHASIS ON THEIR OCCURRENCE AS METHYL DERIVATIVES

F. E. Brinckman

Chemical and Biodegradation Processes Group
Center for Materials Science
National Bureau of Standards
Washington, D.C. 20234 USA

Dedicated to the memory of Professor Frederick Challenger (1888-1983), who more than fifty years ago began our present task.

This review covers several special topics of environmental inorganic chemistry including biotic and abiotic methylation of main group elements and several transition elements, the transfer of methyl groups from organometallic compounds such as trimethylchlorostannane to inorganic substrates, the role of photomethylation in environmental chemistry, reactions of enviromental importance occurring on surfaces, reactions aided by inorganic or organic templates, and correlations between molecular geometry and biological activity. The importance of kinetics for an understanding of environmental inorganic chemistry and the necessity to analyze environmental samples for trace element compounds and not only for total trace element concentrations is stressed.

To understand inorganic processes in the environment, chemists must cope with unfamiliar molecular sites and events occurring on both macro- and microscales. For example, the sequestration or release of metals and metalloids from mineral phases via bacterial metabolism occurs both anaerobically and aerobically with the cells usually attached to the inorganic substrate [1]. Biogenic "concentration cells" may form from precipitated minerals either as elastic or hard-shelled structures, called nodules or tubercles, that effectively enclose the biochemical system. As a consequence, extreme concentration gradients develop over micron to centimeter dimensions, far surpassing gradients generally encountered in the laboratory. Living cells sustain large chemical concentration gradients also across their protective membranes, resulting sometimes in remarkable "bioaccumulation factors" of both essential and toxic elements over many orders of magnitude [2]. On the macroscale in our planet's atmosphere quite a different situation confronts the environmental chemist. Very large, but not necessarily homogeneous, dispersions of inorganic molecules occur in highly diluted forms as gases, dissolved in aerosols, or adsorbed on particulates [3, 4].

At interfaces between condensed, concentrated microscale-systems and dispersed macroscale-systems, slow and very fast transport phenomena take place about which little is yet quantitatively known on a molecular level [5]. To increase our understanding of these phenomena, the following topics need to be investigated:

- Metal-ion-specific transport and accumulation of metal ions by single-celled organisms from dilute aqueous solutions.
- Interactions between aquatic surface microlayers [6, 7] with very high concentrations of metals and metalloids with the atmosphere and underlying bulk waters.
- Mediation of the global transformations and transport of metals and metalloids by organic and inorganic particulates in the atmosphere and in water [3-8].

Progress in these areas is mainly limited by lack of methods for molecule-specific detection of inorganic compounds in complex environmental matrices and in model systems. The study of relevant model systems is expected to provide results, which will assist in predicting the behavior of inorganic substances in the much more complex "natural" systems.

This review emphasizes the environmental chemistry of main-group metalloids and metals. Many important parallels and contrasts exist between main-group metals and transition metals in terms of biogeochemical cycles [9] and their underlying chemistry. In several instances, useful but limited comparisons will be drawn between these two classes of elements.

The inorganic chemistry of main-group elements in the environment deals with their sequestration, mobilization, transport, and transformation mediated by and affecting geological or biological processes on local or global scales. Free (hydrated) ions in aqueous solutions are not the major concern of this chemistry. Environmental inorganic chemistry must explore the complexation of metal ions and metalloid ions with a variety of common and unusual ligands under homogeneous and heterogeneous conditions over enormous concentration ranges.

Growing concerns for essentiality, nutritional value, and toxicity are reflections of the great capacity of these elements to interact with and control diverse biological functions as catalysts or as stoichiometric partners. Equally great interest in the technological and economic uses and value of these elements depends on our ability to comprehend their environmental pathways and, in turn, devise practical means for their recovery, use, and safe disposal.

In what forms are we likely to find main group elements in the environment? Will these molecular forms be similar to those chemists deal with in the laboratory? Not many environmental processes achieve thermodynamic equilibrium, but rather proceed in a series of "steady state" conditions dependent on local compartments, which confound the purely laboratory-based chemist [5, 6, 9]. Frequently, anticipated chemical equilibria, demanding certain forms and concentrations of inorganic molecular species, do not correspond to the actual conditions observed in the environment. In such cases, biological activity is often responsible. Thus, the observed "non-thermodynamic" predominance of arsenite over arsenate in aerobic seawater is caused by bacterial reduction of arsenate [10] and by the unexpected long half-life of arsenite in aerobic waters [11]. Similar, unresolved questions remain concerning the oxidation states of other bioactive elements such as Sn, Sb, and Pb in anaerobic zones [5, 12]. Much future progress in the environmental chemistry of main group metals and metalloids will depend on our quickness in discerning forms and reactivities of unexpected molecules not predicted to be stable by present-day, laboratory-based chemistry.

Several sections of this paper address the chemistry of methyl derivatives of selected elements. These methylated compounds mediate interesting environmental processes. In particular, these derivatives provide extensions of recent laboratory practice for searching out and estimating the "chemodynamics" [13] of transient species affected by both familiar thermodynamic and unknown kinetic environmental factors. Figure 1 illustrates our approach to assessing

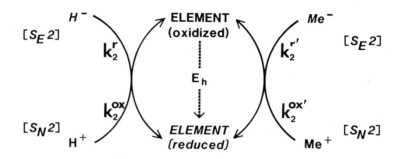

Figure 1. A schematic two-electron redox cycle applicable to environmental reactions.

stability domains based on thermodynamic and kinetic data. In this simplified two electron cycle, which can be easily adapted to radical reactions, the oxidized and reduced forms of an element – connected by the thermodynamic redox potential E_h [14] – can be converted into each other by the biogenic, two-electron carriers "methyl" or "hydrogen". Hydride or methyl-carbanion may reduce an oxidized form of an element at rates k_2^r or $k_2^{r'}$, whereas the proton or a methyl-carbonium ion may oxidize a reduced form of an element at rates k_2^{ox} or $k_2^{ox'}$. The oxidation state of an element can usually not be predicted on the basis of thermodynamic data, when unspecified or unknown kinetic events occurring on short time scales influence expected redox equilibria. Many environmental reactions, catalyzed by enzymes or non-enzymic materials, may involve one- or two-electron redox steps caused by small, membrane-permeable "carrier" moieties such as hydrogen and methyl [15].

Evidence for "hydridase" and "methylase" enzyme factors operating as endo-cellular agents, and for biogenic methylcobalamin [16, 17] and trimethyltin [18] acting as exo-cellular methyl carbanion donors, is available. An important example is the bacterial production of gaseous methylarsines [19] and methylstannanes [20]. The metalloid or metal center traps in this example both methyl and H "carriers" and forms stable molecules.

Photomethylation and reactions on surfaces are also important environmental processes. Discussion of photomethylation permits consideration of prevalent abiotic influences of sunlight on metal or metalloid chemistry in certain environmental compartments. Surface reactions are responsible for some remarkable heterogeneous interactions between dissolved biogenic ligands and suspended particulates. These interactions cause the solubilization of bulk elements or their mineral forms. Turekian contends that marine particulates mediate most metal transport from land to sea and to the ocean floor [21]. This fascinating idea can be experimentally checked and offers the possibility to gauge from sediment records the Earth's transition from reduced anoxic environments to our current aerobic conditions [22].

The last parts of the paper focus on reactions between bioactive inorganic and organometallic molecules; on special biological or man-made macromolecular compounds that permit template reactions to take place or facilitate the transport of molecules or ions across cell membranes that otherwise selectively reject or impede molecular mobility; on recent advances in biological structure-activity predictions based on molecular topology [23]; and on the experimental methods for the verification of such predictions.

The recognition that organisms transform metals and metalloids into characteristic molecular forms and accumulate these molecules gave rise to the idea [24], that elements such as arsenic in fossil substrates may contain a biogenic "organometallic record". Organometallic compounds may, therefore, join the familiar organic derivatives which have proven so valuable in unraveling our distant past [25]. It is noteworthy that flora now extant in many regions on our planet's surface can be used as a practical biogeochemical indicator of precious or harmful elements underlying soils [26] because plants selectively take up elements and transform them into novel compounds that can be identified and determined.

Transmethylation Reactions

Major recent advances in environmental inorganic chemistry were achieved in understanding the formation and fate of sigma-bonded methylmetal and methylmetalloid molecules in aquatic media and biota [27]. Many of the stimuli for these advances were provided by the disastrous "mercury" episodes in Japan during the late 1950's. Inorganic mercury from industrial effluents reached humans as methylmercury via a diet consisting mainly of fish. Similar but less severe health problems with environmental methylmercury were rapidly uncovered throughout the world.

Biogenic methylmercury was always involved in these episodes [28]. Methylmercury cation, CH_3Hg^+, has long been recognized [29] as a water-stable species, but the extent of its occurrence in the environment is still not known with certainty. Wood's pioneering work [30] with anaerobic *Methanobacterium*, isolated from polluted Dutch canal sediments, set the stage for much of the research performed during the past fifteen years. The basic questions for biologists and chemists were:

- Is "environmental methylmercury" of biogenic or anthropogenic origin?
- Which laboratory experiments will provide reasonable models for predicting the formation and fate of environmental methylmercury?

Chemists recognized that methyl derivatives of many main group metals and metalloids form relatively stable solvates similar to those of mercury(II) in aqueous solutions [31]. A vigorous search for other biogenic organometals ensued. Table 1 summarizes the most notable examples reported to date. With exception of a few unconfirmed, scattered reports, only biomethylation of metals and metalloids seems to occur [27]. Unfortunately, organometallic chemistry in aquatic media is not a common topic in the literature. Consequently, initial efforts in this area were directed toward gaining a qualitative picture through surveys of pertinent reactions and preliminary rate studies [27, 31].

Ionic reactions of organometals. Kinetic studies have shown that transmethylation between metal centers in aqueous solution usually are bimolecular reactions (eqn. 1). Many

$$CH_3-E + E' \longrightarrow E + CH_3-E' \qquad (1)$$

$$\text{Rate} = k_2[CH_3E][E']$$

E,E' : metal centers
CH_3-E: methyl donor
E' : methyl acceptor
k_2 : bimolecular rate constant

of the metal and metalloid reactants are charged. Hence the ionic strength of the solution, specific ion (ligand) effects, and the effective charge numbers, Z, are vitally important for the prediction of reaction rates [18].

Activity rate theory applied to ionic reactions [59] allows predictions to be made about the influence of the

Environmental Inorganic Chemistry: Main Groups

Table 1. Methylmetal and Methylmetalloids in the Environment

Compound	Properties	Occurrence* and Formation	Ref.
$(CH_3)_n HgX_{2-n}$	good STW;** volatile (n=2)	A, W, S, B; biomethylation	[32-35]
$(CH_3)_n CdX_{2-n}$	poor STW, stabil. ligand-dependent; volatile (n=2)	A, W, S, B; biomethylation(?)	[36-37]
$(CH_3)_n AsX_{3-n}$	C-As bond stable in water; As-X may hydrolyze	A, W, S; biomethylation	[38]
$(CH_3)_n AsO(OH)_{3-n}$	stable in water	W, S, B; biomethylation + oxidation	[38]
$(CH_3)_n SbO(OH)_{3-n}$	poor to fair STW;	n=1,2 in anoxic S: biomethylation?	[38-40]
$(CH_3)_n SiX_{4-n}$	C-Si good STW; Si-X may hydrolyze; methylator	W, A(?); man-made	[41-42]
$(CH_3)_n GeX_{4-n}$	C-Ge fair, pH-dependent STW	water(?) biomethylation(?)	[43,44]
$(CH_3)_n SnX_{4-n}$	C-Sn good STW; methylator	seawater, freshwater, A, S, B; biomethylation	[45-48]
$(CH_3)_n PbX_{4-n}$	C-Pb poor, ligand-dependent STW (n=1,2); methylator	A, W, S, B; biomethylation(?)	[49-51]
$(CH_3)_2 Se_n$	good STW; may also exist as R_2SeX or R_2SeO	A, W, S, B; biomethylation	[52-53]
$(CH_3)_2 Te$	fair STW	A, W(?), S, B;	[54,55]
$(CH_3)_2 TlX_{3-n}$	poor STW for n=1,2; unstable in saline water for n=1	occurrence uncertain; biomethylation(?)	[56-58]

*A = Air; W = Water; S = Soil or Sediment; B = Biota
**STW: Stability toward water

primary kinetic salt effect on the rates of transmethylation reactions (eqn. 2). It is assumed that neither specific ion (ligand) nor "large ion" (charge delocalization) effects are

$$\ln(k_2/k_2^0) = (Z_E Z_{E'})\alpha\sqrt{\mu} \qquad (2)$$

k_2^0: rate constant at zero ionic strength
α: units constant
μ: ionic strength

operative. Equation 2 predicts the following relations between the rate constant and the ionic strength:

- an increase of k with μ, if Z_E and $Z_{E'}$ have the same sign;
- a decrease of k with μ, if Z_E and $Z_{E'}$ have opposite sign;
- k to be independent of μ, if Z_E or $Z_{E'}$ is zero.

At environmental concentrations, S_E2 transmethylation reactions might be very slow, because the half-lives for such reactions are proportional to $k_2[E]$. Thus for k_2 in the typical range of 10^{-2} to 10^{-4} $M^{-1}s^{-1}$ [27, 31] with reactants at 100 nM concentrations, half-lives would be in the range of 32 to 3170 years. If, however, chemists have available techniques for the determination of trace element compounds sensitive to ng/L levels (as element), then the products of these reactions could be detected after about 0.1 percent completion within perhaps 1 to 100 days after initiation of a reaction. This consideration forms the basis and prospects for much new quantitative environmental chemistry.

Reactions of inorganic species in environmental media occur in the presence of a multitude of competing electrophiles and ligands. In familiar media, such as seawater with high ionic strengths and chloridity, which are also characteristic of cellular fluids, the metal centers favoring complex formation with chloride as ligands will tend to produce at equilibrium increased concentrations of molecules with reduced charge and diminished electrophilicity [27, 60]. Mercuric ion in seawater [61] is a classic example (eqn. 3). Other elements such as Tl, Sn,

$$Hg^{2+} + nCl^- + mOH^- \rightleftharpoons HgCl_n(OH)_m^{(-n-m+2)} \qquad (3)$$

Al, Ga, As, Sb, Cd, Pb, and many transition metals form analogous chlorohydroxy species, whose compositions depend on pH and pCl [62].

To quantitate likely and principal reactants for mechanistic studies, it is necessary to recognize the large range of lifetimes of species. In dilute aqueous solutions, scission of Hg-Cl bonds is very rapid (eqn. 4), and within

$$HgCl_2 \longrightarrow HgCl^+ + Cl^- \quad k_1 = 10^{9.9} \, M^{-1} s^{-1} \quad (4)$$

an order of magnitude of the exchange rates of other divalent metal ions with chloride or hydroxide [63]. For the case of methylmetals or methylmetalloids at neutral pH, usually associated with environmental solutions, the lifetimes of metal-carbon bonds are much longer [64] (eqn. 5) than those of metal-halogen bonds. The exchange of anions on the methylmetal moiety is as rapid [65] (eqn. 6) as on inorganic centers.

$$(CH_3)_n M + H^+ \longrightarrow CH_4 + (CH_3)_{n-1} M^+ \quad k_2 << 10^{-6} M^{-1} s^{-1} \quad (5)$$

$$CH_3HgCN + Cl^- \longrightarrow CH_3HgCl + CN^- \quad k_2 = 10^{+9.9} \, M^{-1} s^{-1} \quad (6)$$

Certain biological donors, particularly those with sulfhydryl sites, react in-vivo with methylmetal cations to produce complexes of remarkable longevity [9, 27, 31]. As a result, the increased mobility, volatility, and lipophilicity imparted to the metal center by covalent carbon-metal sigma-bonding, combine with facile anion exchange to make it possible for these polar agents to migrate across ionic gradients of otherwise protective membranes to crucial sites in living organisms [15, 27]. Prominent examples are methylmercury and trimethyltin cations, both of which are neurotoxic and have long residence times in mammals [28, 66, 67].

Transmethylation is a relatively slow process in water, but one that imparts special biological characteristics to solvated metal and metalloid ions. We showed, using proton NMR, that concurrent reactions [60] were responsible for the methylation of mercury by trimethyltin moieties in saline solutions (eqn. 7).

$$(CH_3)_3 Sn^+ + Hg^{2+} \rightleftharpoons (CH_3)_2 Sn^{2+} + CH_3Hg^+ \quad (7)$$

Bimolecular rate constants were determined under controlled pH, pCl, and ionic strength for this reaction employing available stability constants to estimate the starting concentrations of chlorohydroxy reactants [18]. Chloride ions retarded methyl transfer, whereas perchlorate had little influence (Fig. 2). The specific ligand effect

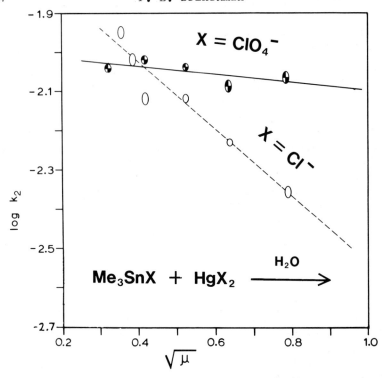

Figure 2. *The dependence of the bimolecular rate constant for the S_E2 methylation of mercury(II) chloride by trimethyltin species on the nature of the anion and the ionic strength of the medium. Adapted with permission from ref. 18.*

exerted by chloride indicates, that anionic species are involved in some of the activated complexes. The transmethylation reaction was found to proceed via six concurrent pathways (eqn. 8).

$$\text{Rate} = k_2^a[R_3Sn^+][HgCl_2^0] + k_2^b[R_3Sn^+][HgCl_3^-] +$$
$$+ k_2^c[R_3Sn^+][HgCl_3^-] + k_2^d[R_3SnCl^0][HgCl_2^0] +$$
$$+ k_2^e[R_3SnCl^0][HgCl_3^-] + k^f[R_3SnCl^0][HgCl_4^{2-}] \quad (8)$$

The main reaction pathways are associated with the rate constants $k_2^a = 0.0174$, $k_2^d = 0.116$, and $k_2^e = 0.139$, which are two- to fifteen-fold larger than the constants k_2^b, k_2^c and k_2^f. The rate constant k_2^f is smaller than 10^{-5} $M^{-1}s^{-1}$; the reaction between trimethylchlorostannane and tetrachloromercurate(II) is, therefore, of no importance. At least one species in each principal pair of reactants is neutral. Therefore, coulombic repulsion is reduced and the reactions have low overall activation energies of 59 kJ comparable to the activation energies of 41-63 kJ for anologous reactions in organic solvents [67a, 68].

Variable results were obtained for reactions between other isoelectronic methylmetal(loid)s and the electrophile mercury dichloride. In a series of reactions conducted under nearly the same conditions [69], the trimethyllead moiety methylated mercury dichloride approximately 10,000-times faster than trimethyltin, whereas dimethylthallium did not react even at higher temperatures. However, dimethylthallium became a methyl donor in the presence of acetate [70]. Trimethylarsenic(V) and trimethylantimony(V) species did not react with mercury dichloride [71]. A preliminary study [72] of the aqueous chemistry of trimethylarsenic(V) species revealed an alternate reaction of potential environmental importance, which involves an irreversible, rapid, pH-dependent redox pathway between two well-established bacterial metabolites, trimethylarsine and elemental mercury [19, 27, 71] (eqn. 9).

$$(CH_3)_3As + HgCl_2 \longrightarrow (CH_3)_3AsCl_2 + Hg^0 \quad (9)$$

Unstable methylation products. In most reactions so far reported, the initial bimolecular transmethylation step is followed by a unimolecular reductive elimination of a substituted methane [27, 60] from the methylmetal(loid) compound [31, 68, 69]. The gaseous products of a number of such S_E2 transmethylation reactions using trimethyltin chloride as the methyl donor (Fig. 3) were examined in our laboratories by NMR and mass spectrometry. No reaction was observed with cadmium chloride and indium chloride. Mercury chloride formed the very stable methylmercury cation (eqn. 7). All other halides of main group and transition elements shown in Figure 3 produced unstable methylated intermediates that subsequently lost their methyl groups (eqn. 10).

$$(CH_3)_3SnCl + MCl_2 \longrightarrow (CH_3)_2SnCl_2 + CH_3MCl_{n-1} \quad (10)$$
$$CH_3MCl_{n-1} \longrightarrow CH_3Cl + MCl_{n-2}$$

The Au, Pd, and Pt intermediates evolved methyl chloride and ethane, and deposited colloidal metal. NMR signals

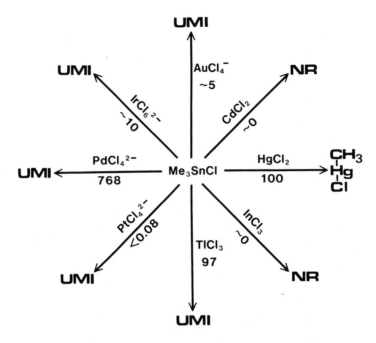

Figure 3. Reactions of trimethylchlorostannane with main-group and transition element halides. The numbers are rates ($M^{-1}s^{-1}$) of the S_E2 transmethylation reactions. "NR" indicates "no reaction". "UMI" means "Unstable Methylated Intermediate".

observed for the Pd and Pt intermediates indicated that cis- and trans-isomers of dimethyl compounds were present, which subsequently decomposed unimolecularly to form methyl chloride and ethane [68, 69]. The Ir and Tl intermediates produced in addition to methyl chloride and ethane soluble ionic metal species with an oxidation state two units lower than the starting materials. These observations have interesting implications for in-vivo intracellular and extracellular methylation of metals. Not only have pathways been shown to be available for irreversible ejection of "active methyl" as gases, but it seems also possible that very finely divided, active metal particles can be deposited within or near biologically active methylation sites with presently unknown consequences. The extent and rates of such deposition would depend on the specific methylated intermediates and on local ligands available for the stabilization of these intermediates. Some metals and metalloids are regarded as toxic to biota in a finely divided, elemental form. Concerns about the possible carcinogenicity of elemental Al, Au, Hg, Se, Ag, Sn, and Zn have been expressed [73, 74]. However, certain metallophilic microorganisms are known to extract dissolved metal ions from their growth media or from particulates and to concentrate these metal ions as pure elements or compounds in form of crystallites or films within the cell or on the outside of the cell in the process of satisfying their energy requirements [1, 2, 75, 76].

Biomethylation of lead is still controversial [27, 77-80]. Any aqueous, saline medium supporting the expected stepwise methylation of lead by abiotic or biotic mechanisms would produce mono- and dimethyllead(IV) species. These species are known to decompose rapidly (eqn. 10) [81]. For lead to be methylated in the environment very selective sites and special ligands must be available that inhibit the cleavage of Pb-C bonds and counteract the strong tendency for methyllead(IV) compounds to revert to lead(II). The nucleophilic S_N2 methylation of Pb(II), which may actually be the more prevalent route for the formation of the observed environmental tetramethyllead, can proceed by extracellular reactions [82, 83] using widely occurring metabolites such as methyl iodide [84].

The questions associated with the biomethylation of cadmium and other electropositive metals have not yet been answered definitively. The gaseous products generated by a bacterial culture growing in a cadmium-containing medium were passed into a sterile aqueous solution of mercury(II) [85]. Methylcadmium is expected to transfer methyl groups to mercury forming methylmercury. The ratio of methylmercury/cadmium in the aqueous solution far exceeded the ratio expected even if all the cadmium had been transported into the mercury solution as dimethylcadmium and all the "cadmium methyls" had been transferred to mercury. These experiments indicated that volatile methylcadmium metabolites might have formed but were not the main carriers of "active methyl groups" in these experiments. The stability of dimethylcadmium, generally believed to be easily decomposed by water, was examined in aqueous solution. Dimethylcadmium was found to survive protolysis for periods sufficient to transfer most of its methyl groups to mercury dichloride [86]. Figure 4 shows the proton nmr spectra of a solution of dimethylcadmium in chilled water. Although dimethylcadmium is only slightly soluble in water, it hydrolyzes rather quickly to a gelatinous precipitate of polymeric $[CH_3Cd(OH)]_x$, which has a half-life of more than

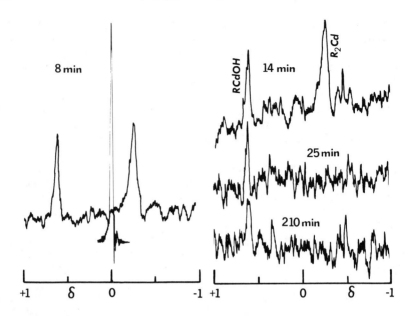

Figure 4. Proton NMR spectra of a solution of dimethylcadmium and methylcadmium hydroxide in chilled water at 8°.

three hours [87] and is capable of methylating aquated metal electrophiles such as mercury(II). The dimethylcadmium signal (upfield from TMS) had disappeared 25 minutes after preparation of the solution. The signal from $CH_3Cd(OH)$ was still discernible after more than three hours. The pathways for the formation of methylcadmium compounds in the environment are still not known with certainty. Biomethylation was claimed for cadmium in a recent report [88]. Experiments to demonstrate biomethylation of cadmium and other electropositive metals by cell-free extracts or the isolated methylase enzyme under carefully controlled conditions could resolve this issue. This approach was successful in identifying and characterizing the enzyme responsible for the reduction of mercury(II) to elemental mercury in *Thiobacillus ferrooxidans* [89].

Photomethylation

Environmental view. Sunlight sustains life on this planet and directly or indirectly influences most crustal chemical reactions. Environmental inorganic chemists should, therefore, examine photo-assisted events that mediate

- homogeneous and heterogeneous reactions in the atmosphere, where - depending on elevation, solar angle, and local smog or dust concentrations - photons of a great range of energies prevail [90];
- reactions occurring in thin, but globally distributed, highly concentrated biofilms such as aquatic microlayers [91, 92] or macrofloral derma [93, 94];
- reactions in bulk homogeneous solutions of diluted inorganic molecules in fresh or saline surface and ground waters [95];
- interactions between biogenic or abiotic ligands and bulk or particulate metal and mineral phases in shallow and benthic zones exposed to sunlight [96].

Solar photoactivation of inorganic metal and metalloid species will not produce large populations of reactive intermediates with long half-lives, hence detectable reactions must occur within microvolumes with adequate transparency to sunlight and with substantial substrate concentrations. Moreover, ubiquitous photochemical formation of destructive oxidants in surface waters, such as peroxide [97], can locally regulate the lifetimes of methylmetals or compete widely with methyl carriers in their formation.

In photosynthesis, highly structured cells provide a necessary, photoactive metal sited in a stabilizing ligand within a transparent medium protected by a membrane that assures the availability of reaction substrates (gaseous carbon dioxide) and selectively excludes competing molecules. Porphyrins are now suggested to be metallo-complexes capable of supporting prebiotic "photosyntheses" on Earth [98], though these satisfy only a few requirements that seem to be necessary even in an archaic anoxic soup [22].

Care must be taken in assessing the likely courses taken by large-scale photocatalyzed environmental reactions at inorganic centers, especially with respect to aerobic versus anoxic conditions [5]. An intriguing modern example of a photocatalyzed reaction in an oxic medium is provided by recent evidence for the global photoreduction of atmospheric nitrogen to ammonia occurring on TiO_2 surfaces in terrestrial sands. This reaction is estimated to contribute approximately 10 million tons of ammonia to the atmosphere annually [99, 100]. In contrast, major global fluxes of carbon or sulfur depend on non-photoassisted reductions of oxycarbonaceous substrates and sulfate by anaerobic bacteria in which both metal and metalloid centers play a role [1].

In certain environmental compartments where organic and inorganic precursors are available, strong diurnal photochemical cycling can arise. Akagi showed [101, 102] that methylmercury species were produced in sunlight by the reaction between acetate and mercury(II) (eqn. 11). This

$$Hg^{2+} + CH_3CO_2^- \longrightarrow CH_3Hg^+ + CO_2 \qquad (11)$$

reaction is photosensitized by elemental sulfur and sulfide. The acetate and sulfur could arise from either anthropogenic or natural sources. These substances are common bacterial metabolites. Such photomethylation reactions may produce dimethylmercury (eqn. 12), which subsequently can be cleaved to ethane and elemental mercury (eqn. 13) [103]. The net

$$CH_3HgOOCCH_3 \xrightarrow{h\nu} (CH_3)_2Hg + CO_2 \qquad (12)$$

$$(CH_3)_2Hg \xrightarrow{h\nu} C_2H_6 + Hg^0 \qquad (13)$$

effect in an intertidal flat polluted with mercury(II) and bacterial nutrients (not an uncommon situation) during daylight hours is the production of methylmercurials, stable gaseous elemental Hg, and ethane, which escape into the atmosphere [104]. A similar photochemical reaction sequence may explain the presence of tetramethyllead in the air near tidal flats in England [105]. Tetramethyllead was shown not to come from automotive fuel. Subsequent studies [106] revealed a relatively long half-life of 1.7 hours for tetramethyllead in daylight.

Table 1 lists a number of other methyl derivatives of main group elements that appear in the atmosphere, generally at the very low concentrations of less than 0.1 nM m^{-3}. Residence times or photolytic half-lives in sunlight have been reported only for dimethylmercury [3, 107] and for tetramethyllead [106]. Because methyl derivatives of Se, Sn, and As are now regarded also as significant for the atmospheric transport of these elements in global fluxes [3, 5], the lifetimes and fates of these and related molecules subject to solar photodissociation and the modes of their abiotic and/or biological formation must be determined.

The importance of photo-assisted processes is indicated by the formation of methylmercury species upon irradiation of an alpha-amino acid complex of inorganic Hg(II) [108]. The photomethylation was enhanced by the presence of copper(II).

Photomethylation studies in the laboratory. We studied reactions related to Akagi's photomethylation of Hg(II). Earlier, Janzen and coworkers [109] performed spin-trapping ESR measurments on a series of heavy methylmetal acetates with the aim of identifying organoxy radical intermediates, such as the elusive acetoxy radical. They found that photomethylation definitely occurred with mercury, was possible with lead, but could not be confirmed with Sn. We have examined Tl(I) and Sn(II) acetates in aqueous solutions exposed to 254 nm light. Jewett and coworkers [69, 103] employed quartz tubes to follow the rate of the thallium(I) reaction (eqn. 14) in heavy water by proton NMR. Production

$$TlOOCCH_3 \xrightarrow[D_2O]{h\nu} CH_3D + CO_2 + TlOD \qquad (14)$$

of monodeuteromethane and carbon dioxide was confirmed by mass spectrometry. These results, along with the concomitant formation of TlOD, imply that a transient methylthallium(I) species underwent rapid deuterolysis. Photolysis of 10^{-5} M solutions of tin(II) acetate was monitored [110] at much lower concentrations by examination of headspace gases by GC-MS. Small aliquots of the reaction mixture were also treated with borohydride. The products of this reduction were analyzed with a gas chromatograph fitted with a tin-selective flame photometric detector [20]. No tetramethyltin was detected in the head space after several hours of irradiation with 254 nm light. Traces of less-methylated tin species were seen in solution. Much of the borohydride-reducible tin(II) or tin(IV) was lost from the solution by unknown means. Further work on these reactions is underway at NBS to determine the rates of pertinent reactions of tin and other metals and metalloids in the presence of biogenic ligands and photo-assisting metal ions.

Environmental Surface Chemistry

Direct reactions. In their early classic experiments, Frankland [111] and Paneth [112] demonstrated solubilization and volatilization of refractory bulk metals by small organic molecules. Chemists since have actively sought to understand in detail and exploit these powerful heterogeneous, often stereoselective processes occurring on inorganic substrates. Indeed, Rochow [113] created an industrial organosilicon chemistry based on the simple "direct reaction" between haloalkanes and silicon, and later, other metalloids. The literature is replete with related condensation or coupling reactions, such as Wurtz or Grignard procedures [114], which are usually successful only in anhydrous organic solvents. The synthetic value of such direct reactions forces us to explore their potential scope and occurrence in environmental media. Moreover, as our needs mount for cost-effective catalysts based on localized or highly tailored aggregates of metal atoms [115] in gases, in cryogenic solutions, and on diverse inorganic substrates, a reconsideration of the relation between laboratory and environmental "direct" processes is very appropriate.

Burdette [116] points out the dichotomy that exists between our favorable theoretical stance on the relationship between structure and reactivity for isolated molecules based upon symmetry and isolobality principles [117] and the limits of theory applicable to even simple solids. Rules for local geometries meeting mutual stereospecific requirements of a reactant molecule and a solid must be available, before likely reaction sites can be predicted. We must also seek to identify stoichiometric aggregates in the solid that are most capable of participating in the low-energy translocations characteristic of direct syntheses. As inorganic chemists devote increased attention to environmental processes dependent on such phenomena, a picture relating reactivity of "free" molecules to sites on both micro- and macrosurface scales is likely to be unraveled. Such progress will aid biogeochemistry and especially the chemistry of those remarkable microbial metabolic pathways that largely depend on such "direct" mobilization of metals and metalloids from bulk surfaces under extreme conditions of temperature and pressure [75, 76, 118, 119] to provide the energy for sustenance of life.

Solubilization of minerals and sediments. Extraction, assimilation, and translocation of essential metalloids and metals from inorganic mineral substrates were identified as major contributors to crustal cycling of elements by microorganisms [75, 76]. Some bacteria, for example, obtain iron necessary for their growth by synthesizing and excreting powerful chelators called siderophores that react with ferric ions in insoluble minerals to form soluble iron(III) complexes. These complexes are then assimilitated [120]. Some bacteria are even capable of utilizing these ligands in several Fe "turnovers", demonstrating the facility with which relatively small molecules can pass membranes as metal carriers. In seemingly less complex fashion, other microbial communities efficiently extract metalloids and metals from refractory sulfide minerals by metabolic oxidation of sulfide to sulfate [121] (eqn. 15). The very low pH generated by the cells in this reaction probably acts to dissolve the remaining metal compounds. Iron(II) that is dissolved in this manner may be converted to iron(III) by bacterial oxidation (eqn. 16) The iron(III)

$$MS(s) + 2O_2 \xrightarrow[H_2O]{bacteria} M^{2+}(aq) + SO_4^{2-} \qquad (15)$$

$$MS(s) + 2Fe^{3+} \xrightarrow{bacteria} M^{2+}(aq) + 2Fe^{2+} + S^0 \qquad (16)$$

thus formed is known to dissolve metal sulfides (eqn. 16). The combined action of bacteria and iron(III) causes the dissolution of metal sulfides to proceed at rates much faster than attainable with either process alone [122].

Still another mineral solubilization process involving exocellular metabolites was recently discovered. Thayer [123, 124] found that dilute aqueous solutions of methylcobalamin, a powerful biogenic methylator [9, 27, 29], dissolved common metal and metalloid oxides in particulate form apparently by a two-step mechanism that may involve direct formation of intermediate methylelement species. The well-known presence of corrinoids in bioactive sediments and activated sludges [125] suggests that extracellular resolubilization of heavy metals commonly present in these materials is possible, especially under the anaerobic conditions usually prevailing.

In our laboratory we surveyed similar solubilizing metabolites of micro- and macrobiota occurring as widely dispersed exocellular molecules. One of these metabolites is methyl iodide, which is produced by seaweeds in amounts exceeding 40 million tons per year [5, 126]. Methyl iodide is a well-known nucleophilic methylator in protic media [127], providing facile oxidative methylation of suitable donor elements (Fig. 1). The methylation of tin(II) and lead(II) in the environment may significantly depend upon this oxidative pathway rather than on direct, cell-mediated methylcarbanion transmethylation [9, 27, 80]. Present knowledge of oxidative methylation in aqueous solution is limited to divalent tin and lead but other candidates for methylation are likely to be present as solids in anaerobic sulfidic sediments [12, 27, 62]. Our initial experiments, therefore, involved natural and metal-spiked, high-sulfide sediments collected from anoxic sites in the Chesapeake Bay.

Direct treatment of sediments with methyl iodide released within hours metals such as Pb, Mn, and Cu into supernatant water [128]. Concomitant volatilization of dimethyl sulfide and dimethyl disulfide was detected consistent with the behavior of authentic metal and metalloid sulfides under similar conditions (eqn. 17). Some methyltin(IV) but no

$$MS_n(s) + 2CH_3I \xrightarrow{n = 1,2} M^{2+}(a) + 2I^- + (CH_3)_2S_n \quad (17)$$

methyllead products were isolated [129]. Experiments under strictly anoxic conditions using sediments amended with pure compounds of metals and metalloids at their lowest oxidation states are underway to evaluate the possibilities of in-situ redox reactions of methyl iodide-solubilized metals and to determine whether known sulfide-catalyzed redistribution reactions of any methylelement products affect the species distributions of tin or lead [130].

Environmental Template Reactions

Hetero-templating reactions. Weiss made the intriguing proposal [131] that layer silicates such as clays might function as non-biological templates for chemical replication and evolution. Ubiquitous montmorillonite and similar silicates consist of stacked silicate sheets separated by 0.8-1.0 nm. This distance is controlled by the ionic strength of the medium and by specific ions in penetrant waters. The introduction of metal, metalloid or organic species into the space between the sheets is important for the environmental distribution of these species and for the nucleation of new "daughter" generations of clay macromolecules from the kinetically "isolated" parent fragments of seed crystals. X-ray analysis shows that many successive generations of these macromolecular hetero-template hosts can form with a high degree of structural fidelity [131], which is a basic requirement for a self-multiplying chemical information carrier. During intercalation clay sheets are not only elastically deformed to increase or decrease Van der Waals interlamellar distances, but they also experience extensive substitution of backbone silicon by aluminum, magnesium or other locally available, common ions including transition metal ions. These chromatographic-like events produce charge-deficient and charge-excess sites with Lewis-acid and Lewis-base properties. These sites have highly specific catalytic activities for producing monomeric and polymeric guest products [132]. For example, interlamellar sites in Bentonite clay exchanged with copper(II), on which water of hydration was replaced with methanol, strongly catalyzed the conversion of 1-alkenes to di(2-alkyl) ethers, including sterically crowded t-butyl ethers [133]. Other reports suggest, that clays have similar capacities for catalyzing organometallic reactions and stereoselective metal transformations.

Three-dimensional inorganic materials may also serve as host templates for environmental reactions. These materials, however, are not well suited to mimick DNA transcription. Major classes of such naturally occurring materials are zeolites and fibrous clay minerals that display selective catalytic properties for useful organic reactions [134]. Of related interest are reverse bonded-phases on suitable materials serving as "immobilizing carriers" for reactive organometallic moieties. An example of "carrier supported" organometallic reaction sites is shown in formula 1. The

1

cationic mercuriphenyl site is bonded to the silica substrate through organosiloxy groups. Tailored hydrophobic organosilane-bonded surfaces on small (<5 μm) silica particles for high-performance liquid chromatography (HPLC) are now commercially prepared with a high level of reproducibility [135]. "Carrier-supported" organotin hydrides on silica and alumina provided, for instance, improved and regenerable hydridic reducing beds [136]. Organomercurials were immobilized via organosilane coupling agents to high surface-active HPLC columns for selective separations of polar organic ligands [137]. It is reasonable to expect that similar carrier-supported active sites will be synthesized in the future in the more restrictive, "two-dimensional" space of a layer host to emulate efficient environmental materials.

The broad and useful field of intercalation chemistry does not, however, rely solely on naturally occurring, two- or three-dimensional host matrices. Extensive research has shown that many layered-sheet or globular cryptate structures can be synthesized in the laboratory [138]. Among those that readily and selectively accommodate bulky aromatic or organometallic guest molecules are transition metal dichalcogenides such as titanium disulfide, and the long-known intermetallic Hofman-type complex hosts such as $Cd[NH(CH_3)_2]_2Ni(CN)_4$ shown in cross-section in structure 2

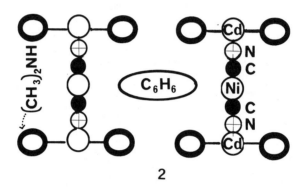

2

[139]. This compound acts as an efficient host for benzene and its derivatives with bulky substituents. The nature of the amines coordinated to cadmium determine the kind of aromatic guest molecule that can be incorporated into the complex. The potential for cooperative chemical effects such as directed delocalization by two or more dissimilar metal centers in the same "ligand cavity" is possible. Recently, our understanding of polynuclear metallo-bioligands and the role main group and transition metals play in them were discussed [140, 141].

Homo-templating reactions. To model reaction sites in a complex solid-phase host, synthetic compounds mimicking such hosts in solution would be invaluable. Recently, the well-known application [142, 143] of transition metal ions as template centers for the cis-closure of extended bifunctional heteroorganic ligands into planar macrocyclic ligands was extended to magnesium, an environmentally important main group metal [144]. Magnesium is of crucial importance to living systems for the biosynthesis of porphyrins and the activation of enzymes. The successful use of magnesium as a template suggests that main-group s-elements may have isosteric and isoelectronic properties similar to those of bioactive transition metals [31, 60, 145].

Porphyrins, a class of biologically important planar macrocyclic ligands, are not only viewed as primordial photosynthetic hosts for redox reactions of organic molecules in the presence of colloidal platinum [98], but also as environmentally important ligands capable of forming unusual inclusion complexes with main group elements [146]. With increasing frequency, methylmetal derivatives, particularly those considered to be involved in biological cycles [9, 31], are shown in the laboratory to form stable complexes with porphyrins or macrocyclic crown compounds. Recent examples include dimethyltin(IV) [147] and dimethylthallium(III) [148] complexes with 18-crown-6 ethers. The x-ray structure of the thallium complex reveals the linear $CH_3-Tl-CH_3$ moiety to be "threaded" perpendicularly through the central cavity restricting the Tl atom to the plane of the cyclic ether. A related arrangement may be present in complexes of $(CH_3)_2Tl^+$ with the "N_6" expanded porphyrin-like macrocycle. Tetradentate N_4 porphyrins cannot accept even the CH_3Tl^{2+} ion within their cavity [149]. Such models may provide useful information about intermolecular rates and paths controlled by the cavity exclusion volume in polar solvents common under environmental conditions [150]. In addition, the peculiar "shielding" of otherwise highly reactive guest molecules such as the methylmetal compounds appears to provide an unusual pathway for crossing protective membrane barriers that otherwise exclude the "free" organometal species. An important example is the selective transport of mono- and divalent main group cations including Ag, Na, K, Pb, Cs, Ca, Br, Ba, Pb, and Tl across artificial membranes with the assistance of macrocyclic ether derivatives of appropriate cavity sizes [151]. It is not known whether metal and organometal complexes can be similarly transported across membranes of living cells. The presence of many metal porphyrins in the environment, including even a gallium derivative found in coal [152], suggests that such pathways exist.

Diebler, Eigen, and coworkers [153] surveyed in 1969 a number of reactions of main group metal ions with biological carriers and concluded that the most effective carrier

- must involve a ligand with sufficient conformational flexibility to encrypt the metal stepwise by the lowest energy pathway available during removal of coordinated

solvent molecules;
- must replace as many inner-sphere coordinated solvent molecules as possible with coordination sites from the carrier;
- must be able to form a cavity adapted to the size of the guest metal ion;
- must orient the ligand's electrophilic groups inward toward the crypt and its lipophilic groups outward to produce a globular solvophobic "surface" for the complete complex.

It will be instructive to see whether these qualitative requirements apply also to macrocyclic ligands capable of transporting organometals across membranes. These simple topological rules do indeed describe a large class of biogenic metal carriers including antibiotics [153] and siderophores [120].

Molecular Geometry and Biological Response

Quantitative structure-activity relationships. Chemists have long sought reliable and independent predictors of physical and biological properties of molecules based on their structure. Most notable are the "quantitative structure-activity relationships", QSAR, developed on semi-empirical grounds by Hansch and coworkers [154]. QSARs are mainly derived as molecular fragment- or substituent-dependent correlations for chemical systems at or near equilibrium. In consequence, many substituent parameters familar to inorganic chemists, such as Taft-Hammett sigma values or water-octanol partition coefficients (related to Hansch's π values), bear a linear free-energy dependence on the experimental parameter as shown in equation 18.

$$\Pi_x = \log P_x - \log P_p \qquad (18)$$

P_x: partition coefficient of derivative
P_p: partition coefficient of parent compound

Thus, π is proportional to the free energy of phase transfer of substituent X. Of great importance, π is an additive-constitutive property accounting well for hydrophobic properties of many classes of large and small organic molecules [155]. Mackay pointed out [156] that attempts to correlate biological dose-response relationships with independently determined QSARs will fail, if the toxicant is short-lived and achieves only steady-state between lipid uptake, depuration, and abiotic degradation. However, increasing numbers of reports have appeared that demonstrate that the effects of subacute exposures of many types of test organisms to a host of organic molecules correlate highly with various substituents. This general approach is the basis of a vigorous commercial approach to the design of chemotherapeutics [157].

The reader is now alerted to two unanswered questions. First, why not develop a more comprehensive scheme for relating structural-electronic features of molecules in-toto to dose-response? Secondly, cannot suitable molecules bearing metals or metalloids be subjected to similar predictive tests? Both questions intrigued us. While many environmentalists expressed concern over fate and effects of metal compounds in the biosphere, no rigorous attempts were made to apply the data base carefully assembled by Hansch and subsequent workers to metal and metalloid compounds.

Molecular topology approaches. In general, QSARs best predict solution properties of sparingly soluble molecules, hence hydrocarbons and heteroorganics are widely studied classes of compounds. In 1968 Hermann provided a theoretical treatment [158] of water solubility in terms of the surface area of a solvent cavity [150] and showed that best correlations were obtained with computed "total surface areas" (TSA) of solutes without a monolayer of water molecules on even the most polar species. Since this work a number of approaches to generating representations of the "chemical coding" inherent in individual molecular shapes and sizes has emerged. Principal methods in current use are summarized in Table 2 along with statements about their theoretical bases and approximate physical significance [162]. Generally, all methods are in good agreement with correlation coefficients greater than 0.92. Inorganic chemists will find centrist "bond connectivity" correlations appealing because of their dependence on central, atom-oriented symmetry arguments [159]. In our ongoing work we employ correlations of several properties with TSA because of its conceptual and physical simplicity, especially in dealing with conformational questions.

Many of the metalloids and metals discussed in this paper, including coordinatively-saturated, homoleptic alkyls and metallo-organic complexes enclosed by hydrophobic, macrocyclic ligands, are suitable candidates for study. In most instances, requisite bond distances and angles can be estimated to a reasonable degree of precision, if they have

Table 2. *Approaches to Molecular Topology*

Method	Theoretical Base	Physical Relevance
Molecular Connectivity Index [159]	Bond centered summation of electronegativity and electron density	H positions implicit, related to molecular volume and valence states
Branching Index [160]	Perturbational analysis of central and adjacent atoms	Related to molecular refraction and volume
Total Surface Area [161]	Sums all Van der Waals radii for specified atom positions (bond lengths, angles) in any conformations with occluded surfaces accounted for	Molecular volume and "cybotactic volume"; highly sensitive to conformational changes

not already been measured. Van der Waals radii needed for estimating minimum contact (minimum energy) conformations in TSA computations are also available [163, 164]. We selected a group of well-known permethyl derivatives of main group elements, $(CH_3)_nM$, for which aqueous solubility data were available from the literature (M = Br, I, C, Hg, Sb) or our laboratory (Sn). Using standard bond parameters from handbooks and compilations [163, 164] we computed TSAs for each molecule and performed a linear regression analysis on their solubility values (Fig. 5). The correlation coefficient is excellent in view of the experimental difficulties in measuring solubilities of toxic and even pyrophoric molecules.

We examined alternative methods for estimating the total surface areas and finding correlations for solution-partition properties of organometallic molecules of environmental interest. The HPLC retention indices, or capacity factors, were shown in theory and practice to correlate directly with solvophobic properties encoded in TSA and the other molecular shape-dependent topologies summarized in Table 2 [159]. Using carefully equilibrated reverse bonded-phase columns, we measured [165] a large number of capacity factors, k', (Fig. 6) for various neutral organometals including the tetraorganotins listed in Figure

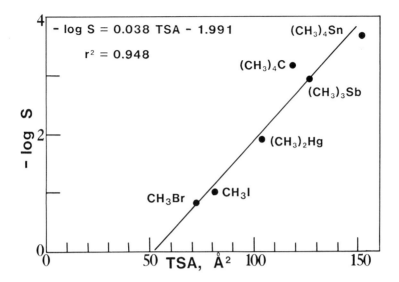

Figure 5. *The linear correlation between the experimental solubility (S, moles/kg) and the total surface area (TSA) calculated from bond distances, bond angles and Van der Waals radii of biogenic permethyl compounds.*

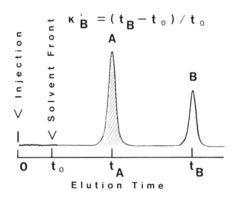

Figure 6. *The definition of the capacity factor k'. Adapted with permission from ref. 165.*

7 [166]. The capacity factors are directly proportional to the free energy of partition of the species interacting with the column if the chromatographic system is at equilibrium. The capacity factors correlate well with TSAs and related properties. Total surface areas can, therefore, be

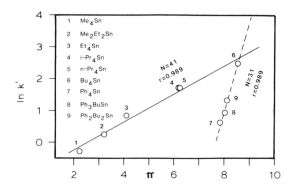

Figure 7. The correlation between the capacity factor k' and Hansch's π parameter for tetraalkyltin compounds.

estimated from experimentally easily accessible capacity factors. For example, tetrabutyltin has a solvophobic response to the lipophilic organosilane-bonded surface different from the response of butyltriphenyltin in spite of similar TSA and π values. We draw the conclusion that both electronic and steric or conformational differences are discernible under the chromatographic conditions applied. Evidence has been presented elsewhere that conformers can be distinguished by HPLC [167].

Recently we calculated [168] the TSAs for triorganotin moieties dissolved in seawater. Stability constants suggest that either the neutral, approximately tetrahedral tributylchlorostannane or its trigonal bipyramidal monohydrate is prevalent. Replicate groups of crab larvae, *Rhithropanopeus harrisii*, were subjected to subacute doses of a series of eight triorganotins representing TSAs from 1.55 to 4.08 nm^2 for a period of one week to insure equilibration. The survival of the larvae was quantitated as LC_{50}. The remarkable relation between survival and TSA is shown in Figure 8. Similar experiments are underway in our laboratories with diorganotin species and other test organisms to further compare conventional π (water-octanol partition coefficients) and TSA correlations with toxicity [169].

The applications of TSA or other topological methods for predicting solution, phase-transport, and biological properties of more complex metal-containing molecules are in a primitive stage. Basic questions, for example, can be

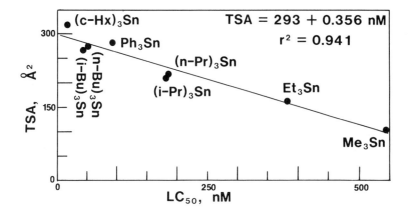

Figure 8. The correlation between the total surface areas (TSA) of trialkyltin compounds and their LC_{50} values (nanomoles) obtained with common crab-larvae, **Rhithropanopeus harrisi**.

raised concerning the solvophobic effects of different metals complexed by the **same** macrocyclic ligand. Will such molecules, nominally of the same TSA, display quite different solution properties or toxicities toward different organisms? Moreover, can we extend topological methods to predictions of preferred guest cavities or stereoselective channels for given molecules in the plethora of synthetic or naturally occurring macrocyclic ligands that increasingly claim our attention?

Acknowledgements

At the outset I feel it is important to give special recognition to a colleague and friend, no longer among us yet assuredly with us in every turn we make, present and future, in the field of environmental inorganic chemistry. I speak of the late Professor Frederick Challenger, who pioneered for more than fifty years that subtle and pervasive relationship between inorganic chemistry and living cells. He showed us just how beautifully our emerging renaissance of organometallic chemistry was very much a part of nature's biogeochemical design. In his last years, I became his friend, and this gave me pleasure and new inspiration to press on in his wake. I am honored to dedicate this paper to him and his works.

I wish to also thank others, too many to cite here, who shared their insights and valued criticisms with me many times. Some colleagues merit special acknowledgement for the present work: Professors A. E. Martell and K. J. Irgolic were instrumental in my participating in this San Miniato Workshop, and I thank them for their help and guidance. Dr. Oren Williams also convinced me that the time was right to finally inaugurate a series of such workshops, seeking to attract the international cooperation and facilities prerequisite to seeding truly effective programs in this infant field of environmental inorganic chemistry. To all of the organizers, both in Italy and the United States, I am deeply grateful for the opportunity to join with such a dedicated and talented community to explore the exciting future of environmental inorganic chemstry.

In addition to the many excellent and stimulating literature reports I have freely employed in this review, I have cited much published and unpublished material from my own laboratory. To my coworkers, past and present, particularly Mssrs. W. R. Blair, R. H. Fish, W. P. Iverson, K. L. Jewett, R. B. Johannesen, G. J. Olson, G. E. Parris, J. S. Thayer, and C. S. Weiss, I extend my heartfelt thanks. The sustained support and guidance of the Office of Naval Research, National Science Foundation, Deutsche Forschungsgemeinschaft, Enviromental Protection Agency, Department of Energy, and the National Bureau of Standards over many years made the work possible.

References

[1] T. Fenchel and T. H. Blackburn, **"Bacteria and Mineral Cycling,"** Academic Press: London, 1797.

[2] S. E. Schumate, G. W. Stradberg and J. R. Parrott, "Biological Removal of Metal Ions from Aqueous Process Streams," in **"Biotechnol. Bioeng. Symp., 8,"** C. D. Scott, Ed., Wiley: New York, 1979, p. 13.

[3] R. J. Lantzy and F. T. MacKenzie, "Atmospheric Trace Metals: Global Cycles and Assessment of Man's Impact," **Geochim. Cosmochim. Acta, 43** (1980) 511.

[4] E. Murad, W. Swider and S. W. Benson, "Possible Roles for Metals in Stratospheric Chlorine Chemistry," **Nature, 289** (1981) 273.

[5] F. E. Brinckman, G. J. Olson and W. P. Iverson, "The Production and Fate of Volatile Molecular Species in the Environments: Metals and Metalloids," in **"Atmospheric Chemistry,"** E. D. Goldberg, Ed., Springer-

Verlag: Berlin, 1982, pp. 231-249.
[6] L. W. Lion and J. O. Leckie, "The Biogeochemistry of the Air-Sea Interface," **Ann. Rev. Earth Plan. Sci., 9** (1981) 449.
[7] N. J. Pattenden, R. S. Cambray and K. Playford, "Trace and Major Elements in the Sea-Surface Microlayer," **Geochem. Cosmochim. Acta, 45** (1981) 93.
[8] S. Horstan, W. Barkley, E. Larson and E. Bingham, "Aerosols of Nickel and Cadmium," **Arch. Environ. Health, 26** (1973) 75.
[9] P. J. Craig, "Metal Cycles and Biological Methylation," in **"The Handbook of Environmental Chemistry,"** Vol. 1, Part A, O. H. Hutzinger, Ed., Springer-Verlag: Berlin, 1980, p. 169.
[10] D. L. Johnson, "Bacterial Reduction of Arsenate in Sea Water," **Nature, 240** (1972) 44.
[11] D. E. Tallman and A. U. Shaikh, "Redox Stability of Inorganic Arsenic(III) and Arsenic(V) in Aqueous Solution," **Anal. Chem., 52** (1980) 196.
[12] F. E. Brinckman, J. A. Jackson, W. R. Blair, G. J. Olson and W. P. Iverson, "Ultratrace Speciation and Biogenesis of Methyltin Transport Species in Estuarine Waters," in **"Trace Metals in Sea Water,"** C. S. Wong, E. Boyle, K. W. Bruland, J. D. Burton and E. D. Goldberg, Eds., Plenum Press: New York, 1982, p. 39.
[13] L. J. Thibodeaux, **"Chemodynamics. Enviromental Movement of Chemicals in Air, Water, and Soil,"** Wiley: New York, 1979.
[14] M. Pourbaix, **"Atlas of Electrochemical Equilibria,"** Pergamon Press: New York, 1966.
[15] R. J. P. Williams, "On First Looking Into Nature's Chemistry. Part I. The Role of Small Molecules and Ions: The Transport of Elements," **Chem. Soc. Rev., 9** (1980) 281.
[16] J. M. Pratt, **"Inorganic Chemistry of Vitamin B ,"** Academic Press: New York, 1972, p. 150.
[17] G. Agnes, S. Bendle, H. O. A. Hill, F. R. Williams and R. J. P. Williams, "Methylation by Methyl Vitamin B ," **Chem Comm.,** (1971) 850.
[18] K. J. Jewett, F. E. Brinckman and J. M. Bellama, "Influence of Environmental Parameters on Transmethylation between Aquated Metal Ions," **ACS Symp. Ser., 82** (1978) 158.
[19] C. N. Cheng and D. D. Focht, "Production of Arsine and Methylarsines in Soil and in Culture," **Appl. Environ. Microbiol., 38** (1979) 494.
[20] J. A. Jackson, W. R. Blair, F. E. Brinckman and W. P. Iverson, "Gas-Chromatographic Speciation of Methyl-

stannanes in the Chesapeake Bay Using Purge and Trap Sampling with a Tin-Selective Detector," **Environ. Sci. Technol., 16** (1982) 110.

[21] K. K. Turekian, "The Fate of Metals in the Oceans," **Geochim. Cosmochim. Acta, 41** (1977) 1139.

[22] L. Margulis and J. E. Lovelock, "The Biota as Ancient and Modern Modulator of the Earth's Atmosphere," in **"Influence of the Biosphere on the Atmosphere,"** H. U. Dütsch, Ed., Birkhauser Verlag: Basel, 1978, p. 239.

[23] J. Josephsen, "Is Predictive Toxicology Coming?" **Environ. Sci. Technol., 15** (1981) 379.

[24] R. H. Fish, R. S. Tannous, W. Walker, C. S. Weiss and F. E. Brinckman, "Organometallic Geochemistry. Isolation and Identification of Organoarsenic Compounds from Green River Formation Oil Shale," **J. Chem. Soc. Chem. Comm.,** (1983) 490.

[25] G. Eglington, S. Hajibrahim, K. Saleh, J. R. Maxwell and J. M. E. Quirke, "Petroporphyrins: Structural Elucidation and Application of HPLC Fingerprinting to Geochemical Problems," **Phys. Chem. Earth, 12** (1980) 193.

[26] A. L. Kovalevskii, **"Biogeochemical Exploration for Mineral Deposits,"** Amerind Publ. Co. Pvt. Ltd: New Delhi, 1979.

[27] J. S. Thayer and F. E. Brinckman, "The Biological Methylation of Metals and Metalloids," **Adv. Organometal. Chem., 20** (1982) 313.

[28] P. D'Itri and F. M. D'Itri, **"Mercury Contamination: A Human Tragedy,"** Wiley-Interscience: New York, 1977.

[29] D. L. Rabenstein, "The Aqueous Solution Chemistry of Methylmercury and Its Complexes," **Acc. Chem. Res., 11** (1978) 101.

[30] J. M. Wood, F. S. Kennedy and C. G. Rosen, "Synthesis of Methylmercury Compounds by Extracts of a Methanogenic Bacterium," **Nature, 220** (1967) 173.

[31] F. E. Brinckman and J. M. Bellama, Eds., "Organometals and Organometalloids: Occurrence and Fate in the Environment," **ACS Symp. Series, 82,** 1978.

[32] S. G. Berk and R. R. Colwell, "Transfer of Mercury through a Marine Microbial Food Web," **J. Exp. Mar. Biol. Ecol., 52** (1981) 157.

[33] J. Gavis and S. F. Ferguson, "The Cycling of Mercury through the Environment," **Water Res., 6** (1972) 909.

[34] T. Fägerstrom and A. Jernelöv, "Formation of Methyl Mercury from Pure Mercuric Sulfide in Aerobic Organic Sediment," **Water Res., 5** (1971) 121.

[35] W. J. Spangler, J. L. Spigarolli, J. M. Rose, R. S.

Flippin and H. H. Miller, "Degradation of Methylmercury by Bacteria Isolated from Environmental Samples," **Appl. Microbiol.**, 25 (1973) 488.

[36] H. Babich and G. Stotzky, "Reduction in the Toxicity of Cadmium to Microorganisms by Clay Minerals," **Appl. Environ. Microbiol.**, 33 (1977) 696.

[37] M. Webb, Ed., **"The Chemistry, Biochemistry, and Biology of Cadmium,"** Elsevier: Amsterdam, 1979.

[38] G. E. Parris and F. E. Brinckman, "Reactions Which Relate to Environmental Mobility of Arsenic and Antimony. II. Oxidation of Trimethylarsine and Trimethylstibine," **Environ. Sci. Technol.**, 10 (1976) 1128.

[39] K. K. Bertine and D. S. Lee, "Antimony Content and Speciation in the Water Column and Interstitial Waters of Saanich Inlet," in **"Trace Metals in Sea Water,"** C. S. Wong et al., Eds., Plenum Press: New York, 1982, pp. 21-38.

[40] M. O. Andreae, J.-F. Asmode', P. Foster and L. Van't Dack, "Determination of Antimony(III), Antimony(V), and Methylantimony Species in Natural Waters by Atomic Absorption Spectrometry with Hydride Generation," **Anal. Chem.**, 53 (1981) 1766.

[41] R. Pellenbarg, "Environmental Poly(organosiloxanes) (Silicones)," **Environ. Sci. Technol.**, 13 (1979) 565.

[42] R. E. DeSimone, "Organosilanes as Aquatic Alkylators of Metal Ions," **ACS Symp. Ser.**, 82 (1978) 149.

[43] M. O. Andreae and P. N. Froelich, "Determination of Germanium in Natural Waters by Graphite Furnace Atomic Absorption Spectrometry with Hydride Generation," **Anal. Chem.**, 53 (1981) 287.

[44] R. S. Braman and M. A. Tompkins, "Atomic Emission Spectrometric Determination of Antimony, Germanium, and Methylgermanium Compounds in the Environment." **Anal. Chem.**, 50 (1978) 1088.

[45] R. S. Braman and R. A. Tompkins, "Separation and Determination of Nanogram Amounts of Inorganic Tin and Methyltin Compounds in the Environment," **Anal. Chem.**, 51 (1979) 12.

[46] V. F. Hodge, S. L. Seidel and E. D. Goldberg, "Determination of Tin(IV) and Organotin Compounds in Natural Waters, Coastal Sediments, and Macro Algae by Atomic Absorption Spectrometry," **Anal. Chem.**, 51 (1979) 1256.

[47] T. Ishii, "Tin in Algae," **Bull. Jap. Soc. Sci. Fish**, 48 (1982) 1609.

[48] G. J. Olson, F. E. Brinckman and J. A. Jackson, "Purge and Trap Flame Photometric Gas Chromatography Technique for the Speciation of Trace Organotin and

Organosulfur Compounds in a Human Urine Standard Reference Material," **Intern. J. Environ. Anal. Chem.,** **15** (1983) 249.

[49] W. R. A. De Jonghe and F. C. Adams, "Measurements of Organic Lead in Air - A Review," **Talanta, 29** (1982) 1057.

[50] P. T. S. Wong, Y. K. Chau, O. Kramar and G. A. Bengert, "Accumulation and Depuration of Tetramethyllead by Rainbow Trout," **Water Res., 15** (1981) 621.

[51] P. Grandjean and T. Nielson, "Organolead Compounds: Environmental Health Aspects," **Residue Rev., 72** (1979) 97.

[52] D. C. Reamer and W. H. Zoller, "Selenium Biomethylation Products from Soil and Sewage Sludge," **Science, 208** (1980) 500.

[53] S. Jiang, H. Robberecht and F. Adams, "Identification and Determination of Alkylselenide Compounds in Environmental Air," **Atmos. Environ., 17** (1983) 111.

[54] R. W. Fleming and M. Alexander, "Dimethylselenide and Dimethyltelluride Formation by a Strain of *Penicillium*," **Appl. Micobiol., 24** (1972) 424.

[55] M. L. Bird and F. Challenger, "The Formation of Organo-metalloidal and Similar Compounds by Microorganisms. Part VII. Dimethyl Telluride," **J. Chem. Soc.,** (1939) 163.

[56] F. Huber, U. Schmidt and H. Kirchmann, "Aqueous Chemistry of Organolead and Organothallium in the Presence of Microorganisms," **ACS Symp. Ser., 82** (1978) 65.

[57] G. S. Schneiderman, T. R. Garland, R. E. Wildurg and H. Druker, "Growth and Thallium Transport of Microorganisms Isolated from Thallium Enriched Soils," **Abstracts, Ann. Mtg. Amer. Soc. Microbiol.,** (1974) 2.

[58] P. Norris, W. K. Man, M. N. Hughes and D. P. Kelly, "Toxicity and Accumulation of Thallium in Bacteria and Yeast," **Arch. Microbiol., 110** (1976) 279.

[59] C. W. Davis, "Salt Effects in Solution Kinetics," in **"Progress in Reaction Kinetics, Vol. 1,"** G. Porter, Ed., Pergamon Press: New York, 1961, p. 163.

[60] M. D. Johnson, "Reactions of Electrophiles with Sigma-Bonded Organotransition-Metal Complexes," **Acc. Chem. Res., 11** (1978) 57.

[61] T. Anfält, D. Dyrssen, E. Ivanova and D. Jägner, "The State of Divalent Mercury in Seawater," **Svensk Kem. Tidskraft, 80** (1968) 340.

[62] A. J. Rubin, Ed., **"Aqueous-Environmental Chemistry of Metals,"** Ann Arbor Science Publ.: Michigan, 1974.

[63] M. Eigen and E. M. Eyring, "Second Wien Effect in

Aqueous Mercuric Chloride Solution," **Inorg. Chem.,** 2 (1963) 636.
[64] N. L. Wolfe, R. G. Zepp, J. A. Gordon and G. L. Baughman, "Chemistry of Methylmercurials in Aqueous Solution," **Chemosphere,** 2 (1973) 147.
[65] R. B. Simpson, "Kinetics of Anion Exchange between Methylmercuric Complexes," **J. Chem. Phys.,** 46 (1967) 4775.
[66] A. W. Brown, W. N. Aldridge, B. W. Sheet and R. D. Verschoylo, "The Behavioral and Neuropathological Sequelae of Intoxification by Trimethyltin Compounds in the Rat," **Amer. J. Pathol.,** 97 (1979) 59.
[67] R. S. Dyer, T. J. Walsh, W. E. Wonderlin and M. Bercegeay, "The Trimethyltin Syndrome in Rats," **Neurobehav. Tox. Terat.,** 4 (1982) 127.
[67a] J. E. Lockhart, **Redistribution Reactions,** Academic Press: New York, 1970.
[68] F. E. Brinckman, "Environmental Organotin Chemistry Today: Experiences in the Field and Laboratory," **J. Organometal. Chem. Libr.,** 12 (1981) 343.
[69] K. L. Jewett, Ph.D. Dissertation, University of Maryland, 1978; **Diss. Abstr.,** 40 (1979) 741.
[70] H. Kurosawa and R. Okawara, "The Existence of Monomethylthallium(III) Species," **Inorg. Nucl. Chem. Lett.,** 3 (1967) 21.
[71] F. E. Brinckman, G. E. Parris, W. R. Blair, K. L. Jewett, W. P. Iverson and J. M. Bellama, "Questions Concerning Environmental Mobility of Arsenic: Needs for a Chemical Data Base and Means for Speciation of Trace Organoarsenicals," **Environ. Health Perspect.,** 19 (1977) 11.
[72] G. E. Parris and F. E. Brinckman, unpublished results.
[73] D. R. Williams, "Metals, Ligands, and Cancer," **Chem. Rev.,** 72 (1972) 203.
[74] P. J. Sadler, "Book Review: Carcinogenicity and Metal Ions," **J. Inorg. Biochem.,** 15 (1981) 185.
[75] R. C. Charley and A. T. Bull, "Bioaccumulation of Silver by a Multispecies Community of Bacteria," **Arch. Microbiol.,** 123 (1979) 239.
[76] H. L. Ehrlich, "Inorganic Energy Sources for Chemolithotrophic and Mixotrophic Bacteria," **Geomicrobiol. J.,** 1 (1978) 65.
[77] K. Reisinger, M. Stoeppler and H. W. Nürnberg, "Evidence for the Absence of Biological Methylation of Lead in the Environment," **Nature,** 291 (1981) 228.
[78] P. T. S. Wong, Y. K. Chau and P. L. Luxon, "Methylation of Lead in the Environment," **Nature,** 253 (1975) 263.

[79] U. Schmidt and F. Huber, "Methylation of Organolead and Lead(II) Compounds to $(CH_3)_4Pb$ by Microorganisms," **Nature, 259** (1976) 157.
[80] P. J. Craig, "Methylation of Trimethyllead Species in the Environment; An Abiotic Process?" **Environ. Technol. Lett.**, 1 (1980) 17.
[81] H. J. Haupt, F. Huber and J. Gmehling, "Zersetzungsreaktion von Dimethylbleichlorid in Lösungen mit und ohne Fremdsalzzusatz," **Z. Anorg. Allg. Chem., 390** (1972) 31.
[82] I. Ahmad, Y. K. Chau, P. T. S. Wong, A. J. Carty and L. Taylor, "Chemical Alkylation of Lead(II) Salts to Tetraalkyllead(IV) in Aqueous Solution," **Nature, 287** (1980) 716.
[83] A. W. P. Jarvie, R. N. Markall and H. R. Potter, "Chemical Alkylation of Lead," **Nature, 255** (1975) 217.
[84] J. E. Lovelock, "Natural Hydrocarbons in the Air and in the Sea," **Nature, 256** (1975) 193.
[85] C. W. Huey, F. E. Brinckman, W. P. Iverson and S. O. Grim, Bacterial Volatilization of Cadmium," **Proc. Int. Conf. Heavy Met. Environ.**, Natl. Res. Council, Ottawa, 1975, p. 27.
[86] C. W. Huey, Ph.D. Dissertation, University of Maryland, 1976; **Diss. Abstr., 37** (1976) 2823-B.
[87] K. Cavanaugh and D. F. Evans, "Raman and Infrared Spectra of Methylcadmium Halides and Related Compounds," **J. Chem. Soc. (A)**, (1969) 2890.
[88] J. W. Robinson and E. L. Kiesel, "Methylation of Cadmium with Vitamin B : A Possible Method of Detoxification", **J. Environ. Sci. Health, A16** (1981) 341.
[89] G. J. Olson, F. D. Porter, J. Rubinstein and S. Silver, "Mercuric Reductase Enzyme from a Mercury-Volatilizing Strain of *Thiobacillus ferrooxidans*," **J. Bacteriol., 151** (1982) 1239.
[90] M. D. Carabine, "Interactions in the Atmosphere of Droplets and Gases," **Chem. Soc. Rev.**, 1 (1972) 411.
[91] L. W. Lion and J. O. Leckie, "Accumulation and Transport of Cd, Cu, and Pb in an Estuarine Salt Marsh Surface Microlayer," **Limnol. Oceanogr., 27** (1981) 111.
[92] L. W. Lion and J. O. Leckie, "Chemical Speciation of Trace Metals at the Air-Sea Interface: The Application of an Equilibrium Model," **Environ. Geol., 3** (1981) 293.
[93] W. Beauford, J. Barber and A. R. Barringer, "Heavy Metal Release from Plants into the Atmosphere," **Nature, 256** (1975) 35.
[94] W. Beauford, J. Barber and A. R. Barringer, "Release

[95] R. G. Zepp, N. L. Wolf and J. A. Jordon, "Photodecomposition of Phenylmercuric Compounds in Sunlight," **Chemosphere,** 2 (1973) 93.
[96] K. S. Jackson, I. R. Jonasson and G. B. Skippen, "The Nature of Metals-Sediment-Water Interactions in Freshwater Bodies with Emphasis on the Role of Organic Matter," **Earth-Science Rev.,** 14 (1978) 97.
[97] W. J. Cooper and R. G. Zika, "Photochemical Formation of Hydrogen Peroxide in Surface and Groundwaters Exposed to Sunlight," **Science,** 220 (1983) 711.
[98] J. A. Mercer-Smith and D. G. Mauzerall, "Photochemistry of Porphyrins: Models for the Origin of Photosynthesis," **Abstr. Papers, 194th Natl ACS Mtg.,** Seattle, Washington, March, 1983, PHYS 32.
[99] A. Henderson-Sellers and A. W. Schwartz, "Chemical Evolution and Ammonia in the Early Earth's Atmosphere," **Nature,** 287 (1980) 526.
[100] G. N. Schrauzer and T. D. Guth, "Photolysis of Water and Photoreduction of Nitrogen by Titanium Dioxide," **J. Am. Chem. Soc.,** 99 (1977) 7189.
[101] H. Akagi and E. Takabatake, "Photochemical Formation of Methylmercuric Compounds from Mercuric Acetate," **Chemosphere,** 2 (1973) 131.
[102] H. Akagi, Y. Fujita and E. Takabatake, "Photochemical Methylation of Inorganic Mercury in the Presence of Mercuric Sulfide," **Chem. Lett.,** (1975) 171.
[103] K. L. Jewett, F. E. Brinckman and J. M. Bellama, "Chemical Factors Influencing Metal Akylation in Water," in **"Marine Chemistry in the Coastal Environment,"** T. M. Church, Ed., Amer. Chem. Soc.: Washington, D.C., 1975, p. 304.
[104] M. Inoko, "Studies on the Photochemical Decomposition of Organomercurials - Methylmercury(II) Chloride," **Environ. Pollut. (Ser. B),** 2 (1981) 3.
[105] R. M. Harrison and D. P. H. Laxen, "Natural Source of Tetraalkyllead in Air," **Nature,** 275 (1978) 738.
[106] R. M. Harrison and D. P. H. Laxen, "Sink Processes for Tetraalkyllead Compounds in the Atmosphere," **Environ. Sci. Technol.,** 12 (1978) 1384.
[107] D. L. Johnson and R. S. Braman, "Distribution of Atmospheric Mercury Species Near Ground," **Environ. Sci. Technol.,** 8 (1974) 1003.
[108] K. Hayashi, S. Kawai, T. Ohno and Y. Maki, "Photomethylation of Inorganic Mercury by Aliphatic α-Amino Acids," **J. Chem. Soc. Chem. Comm.,** (1977) 158.
[109] E. G. Janzen and B. J. Blackburn, "Detection and

Identification of Short-Lived Free Radicals by Electron Spin Resonance Trapping Techniques (Spin Trapping). Photolysis of Organolead, -tin, and -mercury Compounds," **J. Am. Chem. Soc., 91** (1969) 4481.

[110] G. J. Olson, W. R. Blair and F. E. Brinckman, National Bureau of Standards, Washington, D.C., unpublished results.

[111] E. Frankland, "On the Isolation of Organic Radicals," **J. Chem. Soc., 2** (1849) 263.

[112] F. Paneth and W. Hofeditz, "Über die Darstellung von freiem Methyl," **Chem. Ber., 62** (1929) 1335.

[113] E. G. Rochow, "The Direct Synthesis of Organometallic Compounds," **J. Chem. Ed., 43** (1966) 58.

[114] G. B. Sergeev, V. V. Zogorsky and F. Z. Badaev, "Mechanism of the Solid-Phase Reaction of Magnesium with Organic Halides," **J. Organometal. Chem., 243** (1983) 123.

[115] S. C. Davis and K. J. Klabunde, "Unsupported Small Metal Particles: Preparation, Reactivity, and Characterization," **Chem. Rev., 82** (1982) 153.

[116] J. K. Burdette, "New Ways to Look at Solids," **Acc. Chem. Res., 15** (1982) 34.

[117] R. Hoffmann, "Theoretical Organometallic Chemistry," **Science, 211** (1981) 995.

[118] E. G. Ruby, C. O. Wilson and H. W. Jannasch, "Chemolithotrophic Sulfur-Oxidizing Bacteria from the Galapagos Riff Hydrothermal Vents," **Appl. Environ. Microbiol., 42** (1981) 317.

[119] F. Fischer, W. Zillig, K. O. Stetter and G. Schreiber, "Chemolithoautotrophic Metabolism of Anaerobic Extremely Thermophilic Archaebacteria," **Nature, 301** (1983) 511.

[120] K. N. Raymond and C. J. Carrano, "Coordination Chemistry and Microbial Iron Transport," **Acc. Chem. Res., 12** (1979) 183.

[121] P. R. Norris and D. P. Kelly, "The Use of Mixed Microbial Cultures in Metal Recovery," in **"Microbial Interactions and Communities,"** A. T. Bull and J. H. Slater, Eds., Academic Press: London, 1982, p. 443.

[122] R. E. Cripps, "The Recovery of Metals by Microbial Leaching," **Biotechnol. Lett., 2** (1980) 225.

[123] J. S. Thayer, "The Reaction Between Lead Dioxide and Methylcobalamin," **J. Environ. Sci. Health, A18** (1983) 471.

[124] J. S. Thayer, "Demethylation of Methylcobalamin: Some Comparative Rate Studies," **ACS Symp. Ser., 82** (1978) 189.

[125] R. A. Beck and J. J. Brink, "Production of Cobalamins during Activated Sludge Treatment," **Environ. Sci. Technol., 12** (1978) 435.

[126] J. E. Lovelock, R. J. Maggs and R. J. Wade, "Halogenated Hydrocarbons in and over the Atlantic," **Nature, 241** (1973) 194.

[127] J. Koivurinta, A. Kyllonen, L. Leinonen, K. Valaste and J. Koskikallio, "Nucleophilic Reactivity. Part VIII. Kinetics of Reactions of Methyl Iodide with Nucleophiles in Water," **Finn. Chem. Lett.,** (1974) 239.

[128] J. S. Thayer, G. J. Olson and F. E. Brinckman, "Iodomethane as a Potential Metal Mobilizing Agent in Nature," **Environ. Sci. Technol., 18** (1984) 726.

[129] W. F. Manders, G. J. Olson, F. E. Brinckman and J. M. Bellama, "A Novel Synthesis of Methyltin Triiodide with Environmental Implications," **J. Chem. Soc., Chem. Comm.,** (1984) 538.

[130] P. J. Craig and S. Rapsomanikis, "A New Route to Tris-(dimethyltin sulfide) with Tetramethyltin as Co-Product; the Wider Implications of This and Some Other Reactions leading to Tetramethyltin and -lead from Iodomethane," **J. Chem. Soc. Chem. Comm.,** (1982) 114.

[131] A. Weiss, "Replication and Evolution in Inorganic Systems," **Angew. Chem. Int. Ed. Engl., 20** (1981) 850.

[132] J. A. Ballantine, M. Daires, H. Purnell, M. Rayanakoru, J. M. Thomas and K. J. Williams, "Chemical Conversions using Sheet Silicates: Novel Intermolecular Dehydrations of Alcohols to Ethers and Polymers," **J. Chem. Soc. Chem. Comm.,** (1981) 427.

[133] A. Bylina, J. M. Adams, S. H. Graham and J. M. Thomas, "Chemical Conversions Using Sheet Silicates: A Simple Method for Producing Methyl t-Butyl Ether," **J. Chem. Soc. Chem. Comm.,** (1980) 1003.

[134] Y. Kitayama and A. Michishita, "Catalytic Activity of Fibrous Clay Mineral Sepiolite for Butadiene Formation from Ethanol," **J. Chem. Soc. Chem. Comm.,** (1981) 401.

[135] J. C. Liao and C. R. Vogt, "Bonded Reverse Phase Ion Exchange Columns for the Liquid Chromatographic Separations of Neutral and Ionic Organic Compounds," **J. Chromatogr. Sci., 17** (1979) 237.

[136] H. Schumann and B. Pachaly, "Heterogeneous Hydrogenation of Organic Halogen Compounds by Carrier-Supported Organotin Hydrides," **Angew. Chem. Int. Ed. Engl., 20** (1981) 1043.

[137] J. Chmielowiec, "Organomercuric Bonded Phase for High Performance Liquid Chromatographic Separations of pi-Electron-, Sulfur-, Oxygen-, and Nitrogen-Containing Compounds," **J. Chromatogr. Sci., 19** (1981) 296.

[138] M. S. Whittingham and M. B. Dines, "Intercalation Chemistry," in **"Surv. Progr. Chem."**, **9**, A. F. Scott, Ed., Academic Press: New York, 1980; p. 55.

[139] S. Nishikiori and T. Iwamoto, "Inclusion of Aromatic Guest Molecules with Bulky Substituents in Layered Metal Complex Host trans-Bis[Dimethylamine Cadmium(II)] Tetra-catena-μ-Cyano Nickelate(II)," **Chem. Lett.**, (1982) 1035.

[140] D. E. Fenton and P. A. Vigato, "The Challenge of Polynuclear Inorganic Compounds," **Inorg. Chim. Acta, 62** (1982) 1.

[141] A. Scozzafava, "Cooperative Phenomena in Polynuclear Metalloproteins," **Inorg. Chim. Acta, 62** (1982) 15.

[142] D. H. Busch, "The Significance of Complexes of Macrocyclic Ligands and Their Synthesis by Ligand Reactions," **Record Chem. Progr.**, 25 (1964) 107.

[143] G. A. Melson, Ed., **"Coordination Chemistry of Macrocyclic Compounds,"** Plenum Press: New York, 1979.

[144] M. G. B. Drew, A. Hamid bin Othman, S. G. McFall and S. M. Nelson, "The Mg^{2+} Ion as a Template for the Synthesis of Planar Nitrogen-Donor Macrocyclic Ligands: Pentagonal Bipyramidal Mg^{2+} Complexes," **J. Chem. Soc. Chem. Comm.**, (1975) 818.

[145] P. Schramel, Ed., **"Trace Element Analytical Chemistry in Medicine and Biology,"** W. De Gruyter: New York, 1980.

[146] P. Sayer, M. Gouterman and C. R. Connell, "Metalloid Porphyrins and Phthalocyanines," **Acc. Chem. Res.**, 15 (1982) 73.

[147] P. J. Smith and B. N. Patel, "Synthesis and Spectroscopic Studies of Organotin Complexes of Crown Ethers," **J. Organometal. Chem.**, 243 (1983) C73.

[148] K. Henrick, R. W. Matthews, B. L. Podejma and P. A. Tasker, "Restriction of the Environment of the Dimethylthallium(III) ion by 'Threading' through a Crown Ether: X-Ray Crystal Structure of Dimethyl(dibenzo-(b,k)(1,4,7,10,13,16)hexaoxacyclo-octadecinthallium-(III) 2,4,6-trinitrophenolate," **J. Chem. Soc. Chem. Comm.**, (1982) 118.

[149] Y. Kawasaki and N. Okuda, "Preparation of Diorganothallium(III) complexes with an 'N_6' Macrocyclic Ligand and Their H- and 13-C NMR Spectra," **Chem. Lett.**, (1982) 1161.

[150] F. M. Menger, "Reactivity of Organic Molecules at Phase Boundaries," **Chem. Soc. Rev.**, 1 (1972) 229.

[151] R. M. Izatt, J. D. Lamb, R. T. Hawkins, P. R. Brown, S. R. Izatt and J. J. Christensen, "Selective M^+-H^+ Coupled Transport of Cations through a Liquid Membrane

by Macrocyclic Calixarene Ligands," **J. Am. Chem. Soc.**, 105 (1983) 1782.
[152] R. Bonnett and F. Czechowski, "Gallium Porphyrins in Bituminous Coal," **Nature,** 283 (1980) 465.
[153] H. Diebler, M. Eigen, G. Ilgenfritz, G. Maas and R. Winkler, "Kinetics and Mechanism of Reactions of Main Group Metal Ions with Biological Carriers," **Pure Appl. Chem.**, 20 (1969) 93.
[154] C. Hansch, "A Quantitative Approach to Biochemical Structure-Activity Relationships," **Acc. Chem. Res.**, 2 (1979) 232.
[155] C. Hansch, J. E. Quinlan and G. L. Lawrence, "The Linear Free-Energy Relationship between Partition Coefficients and Aqueous Solubility of Organic Liquids," **J. Org. Chem.**, 33 (1968) 347.
[156] D. Mackay, "Correlation of Bioaccumulation Factors," **Environ. Sci. Technol.**, 16 (1982) 274.
[157] A. Albert, "The Long Search for Valid Structure-Action Relationships in Drugs," **J. Med. Chem.**, 25 (1982) 1.
[158] R. B. Hermann, "Theory of Hydrophobic Bonding. II. The Correlation of Hydrocarbon Solubility in Water with Solvent Cavity Surface Area," **J. Phys. Chem.**, 76 (1968) 2754.
[159] L. B. Kier and L. H. Hall, "Derivation and Significance of Valence Molecular Connectivity," **J. Pharm. Sci.**, 70 (1981) 583.
[160] A. Cammarata, "Molecular Topology and Aqueous Solubility of Aliphatic Alcohols," **J. Pharm. Sci.**, 68 (1979) 839.
[161] S. C. Valvani, S. H. Yalkowsky and G. L. Amidon, "Solubility of Nonelectrolytes in Polar Solvents. VI. Refinements in Molecular Surface Area Computations," **J. Phys. Chem.**, 80 (1976) 829.
[162] L. B. Kier and L. H. Hall, "Molecular Connectivity Analyses of Structure Influencing Chromatographic Retention Indixes," **J. Pharm. Sci.**, 69 (1979) 120.
[163] H. J. M. Bowen, Ed., "Tables of Interatomic Distances and Configuration in Molecules and Ions," **Special Publication No. 11,** The Chemical Society: London, 1958, and supplemental editions.
[164] P. A. Cusack, P. J. Smith, J. D. Donaldson and S. M. Grimes, "A Bibliography of X-Ray Crystal Structures of Tin Compounds," **International Tin Research Institute, Publication No. 583,** London, 160 pp.
[165] K. L. Jewett and F. E. Brinckman, "The Use of Element-Specific Detectors Coupled with High Performance Liquid Chromatographs," in **"Liquid Chromatography Detectors,"** T. M. Vickrey, Ed., Marcell Dekker: New

York, 1983, p. 205.
[166] C. S. Weiss, K. L. Jewett, F. E. Brinckman and R. H. Fish, "Application of Molecular Substituent Parameters for the Speciation of Trace Organometals in Energy-Related Process Fluids by Element-Selective HPLC," in **"Environmental Speciation and Monitoring Needs for Trace Metal-Containing Substances from Energy-Related Processes,"** F. E. Brinckman and R. H. Fish, Eds., **NBS Spec. Publ. 618,** Washington, D.C., 1981, p. 197.
[167] L. R. Snyder and J. T. Kirkland, **" Introduction to Modern Liquid Chromatography,"** 2nd Ed., Wiley Inter.-Science: New York, 1979.
[168] R. B. Laughlin, W. French, R. B. Johannesen, H. E. Guard and F. E. Brinckman, "Predictive Toxicology Using Computed Molecular Topologies: The Example of Triorganotin Compounds," **Chemosphere, 13** (1984) 575.
[169] R. B. Laughlin, R. B. Johannesen, W. French, H. E. Guard and F. E. Brinckman, "Stucture-Activity Relationships for Organotin Compounds," **Environ. Toxicol. Chem.,** in press.

DISCUSSION

J. M. Bellama (University of Maryland): With respect to the transmethylation reactions, I'd like to have your comments on the relationship between kinetics and environmental impact. For example, the reaction between the trimethyltin species and a mercury(II) moiety is rapid enough to be followed easily on the NMR time scale. Some of our organosilicon species will methylate mercury(II) ten million times faster than trimethyltin. Conversely, consider a reaction that occurs too slowly to be seen on the NMR time scale. How reliably can one draw conclusions about likely environmental impact based on such kinetic studies?

F. E. Brinckman: The transmethylation reaction I described in my review involve S_E2 mechanisms with the electrophilicity of the carbanion acceptor playing a major role in activation energies and concomitant bimolecular rates. In the real world, in which large numbers and kinds of ligands and many competitive electrophilic centers on other metal species exist, there would be a tendency for a broadened range of transmethylation rates, say, between Sn and Hg. I estimated such rates [reference 68] for freshwater, estuarine water, and seawater, media with greatly different salinity (chloridity) and ionic strength. Based upon our laboratory kinetic data [references 18, 103], we estimated a probable upper and lower transmethylation rate for the Sn-Hg pair in the absence of other competitors. Presumably, with patience, a great deal of experimental work, and complicated kinetic analyses, one could extend such rate/lifetime studies to additional metal centers, a few more competing ligands, and environmental parameters such as ionic strength, salinity, pH, and E_h. Our best current estimates for transmethylation rates of chloromercury(II) compounds involving the half-dozen main group and transition metals so far studied cover a range of 10^8 L mol^{-1}s^{-1} [reference 27].

J. M. Bellama (University of Maryland): In a number of these transmethylation reactions, the presence of chloride ion retards the reaction. Would you please comment further on the mechanistic role played by the chloride ion in such cases?

F. E. Brinckman: Chloride ion specifically retards bimolecular transmethylation by reducing the "electrophilicity" [reference 60] of the methyl carbanion in the manner I just described for Hg(II). This seems to be the case for all the methylmetal transfers to Tl(III) or Hg(II) acceptors thus far reported. We found that less polarizable or "harder" gegenions, such as perchlorate or nitrate do not appreciably affect the second-order transmethylation rate, except as increased concentrations retard the rates by the ionic strength dependence [see Fig. 2]. Other polarizable, or bridging, ligands such as cyanide, bromide, iodide, as well as bidentate ligands such as acetate will also retard transmethylation in a fashion similar to chloride, mainly by reducing the electrophilic character of the methyl acceptor. In cellular media, for example, both high chloridity, high ionic strength, and a plethora of ligating sites would also serve to inhibit transmethylation, it seems to me, mainly by attenuating strong electrophiles by the means I described.

J. M. Wood (Gray Freshwater Biological Institute, University of Minnesota): How do you reconcile the increased toxicity of methylmercury versus Hg(II), when the methylmercuric ion is much less reactive with nucleophiles than inorganic mercuric ion?

F. E. Brinckman: Just as dative-bond or covalent coordination of neutral donors of "soft" anions will serve to diminish the electrophilicity of metals in cellular media, formation of comparatively long-lived covalent sigma methylmercury bonds serve to reduce the acidic character of CH_3Hg^+. With true organometals, however, several unique features prevail that sharply distinguish their toxic behavior from parent "inorganic" metal ions, including: a) hydrophobic or lipophilic properties; b) steric or conformation possibilities for inserting into biologically essential sites in the manner discussed by Diebler et al. [reference 153]; c) ion-exchange capabilities at rapid rates; and, finally, d) a considerable degree of toxicity "information" encoded into the organometal ion by the kind and number of organic groups sigma-bonded to the metal center. Methylmercury possesses all of these features in varying degrees, depending on the membrane traversed and the final tissue site reached. Organotins are a better example. We can now confidently say that the basic topology of the kind and number of organic ligands bound to tin(IV) principally dictates subacute lethality [reference 168]. Thus, electrophilicity or acidity must be considered together with rather selective aspects of membrane transport and intracellular siting. So, methylmercury and other organometals apparently get the job done more effectively than the "bare" metal ion.

X

RATES OF METHYLATION OF MERCURY (II) SPECIES IN WATER BY ORGANOTIN AND ORGANOSILICON COMPOUNDS

Jon M. Bellama, K. L. Jewett and J. D. Nies

Department of Chemistry, University of Maryland
College Park, Maryland 20742 USA

Metals can not only be methylated in the environment by microorganisms but also by other methylmetal compounds. Water-soluble and water-stable main group metals and metalloids were surveyed for the identification of species that can either donate or accept methyl groups. The methylation of mercury(II) by trimethyltin cation was found to be a bimolecular reaction, the reaction rate of which decreased with increasing chloride concentration. Kinetic investigations using NMR techniques showed that the most important pairs of reactants were $(CH_3)_3Sn^+$ + $HgCl_2$, $(CH_3)_3SnCl$ + $HgCl_2$, and $(CH_3)_3SnCl$ + $HgCl_3^-$. Trimethyltin and thallium(III) produced an unstable methylthallium(III) compound. Sodium 2,2,3,3-tetradeutero-3-(trimethylsilyl)propionate and sodium 2,2-dimethyl-2-silapentane-5-sulfonate methylated mercury(II). The chloride ions released in these reactions probably form Si-Cl bonds. Organylsilatranes were found to transfer their organic groups readily to mercury(II) to produce organic mercury compounds. Other metal salts [$SnCl_4$, $Pb(NO_3)_2$, $CdCl_2$, $ZnCl_2$, $FeCl_3$] were not alkylated or arylated by silatranes under these conditions.

Microorganisms are known to methylate heavy metals in the aquatic environment. Organometallic species with methyl groups in their molecules can also serve as methylators. The methylation of heavy metals in aqueous solution by organometallic species is often considered to be not important, because most organometallic species are judged to be unstable in water. The vigorous decomposition of dimethyl cadmium by water is frequently cited to illustrate the extreme reactivity of organometallic compounds toward water.

Environmental processes are a complex mixture of biological and abiotic reactions. To completely understand environmental processes, the interactions among the multitude of substances in various environmental compartments must be known. A first step toward acquiring such knowledge is the study of the interactions between appropriate pairs of substances in a controlled laboratory setting. The results of these studies and the conclusions drawn from them will add to the data base required to unravel complex environmental processes. To understand in detail the transformations of organometallic species in the environment, the rates at which these transformations occur must be known. Nuclear magnetic resonance spectroscopy (NMR) is a useful technique for the identification of species and the determination of reaction rates. Integrals of appropriate NMR signals provide information about kinetic and thermodynamic parameters of transmethylation reactions. In the course of transmethylation reactions a variety of intermediates may be formed. Ligands bound to metals may facilitate or retard the transfer of methyl groups. The ionic strength of the solution also may influence reaction rates. This paper summarizes the results of NMR studies of the methylation of heavy metals by trimethyltin and methylsilicon compounds.

Survey of Methylation Reactions

Water-soluble and water-stable main-group metal and metalloid compounds were surveyed to identify the species that can either donate methyl groups to other metals in aqueous solution or accept methyl groups with formation of carbon-metal sigma bonds. Under appropriate conditions (pH, ligand, ionic strength), Hg(II) and Tl(III) accepted methyl groups most readily from $(CH_3)_2Tl^+$, $(CH_3)_3Sn^+$, and $(CH_3)_3Pb^+$. Cadmium(II) and indium(III) did not readily accept methyl groups. Compounds such as dimethylarsinic acid and trimethylsulfonium chloride showed no sign of methylating Hg(II).

Methylation of Hg(II) by Trimethyltin Species

The trimethyltin cation, a water-soluble and water-stable species with labile methyl groups, is a good choice as a methyl donor for a detailed study of the methylation of heavy metals. Trimethyltin cation reacts with mercuric ion to form methylmercury cation and dimethyltin dication. Kinetic data were obtained for this reaction over a wide range of conditions. A bimolecular rate constant of

0.0104 ± 0.0002 L mol^{-1}s^{-1} was obtained, when the experimentally determined concentrations of trimethyltin cation and methylmercury cation were used. The significantly different constant 0.0091 ± 0.0002 L mol^{-1}s^{-1} resulted, when data for trimethyltin cation and dimethyltin dication were employed [1, 2].

Chloride ion drastically influences the methylation of Hg(II) by trimethyltin cation. At an ionic strength of approximately 0.1 M (adjusted with sodium perchlorate) the bimolecular rate constant increased linearly from 0.0036 to 0.009 L mol^{-1}s^{-1} when the chloride concentration changed from 0.024 to 0.076 M [2, 3]. The rate was not affected by the replacement of sodium perchlorate by sodium nitrate. Perchlorate [4] and nitrate [5] coordinate only weakly – if at all – with Hg(II) and thus do not change the concentrations of mercuric chloride complexes. In the absence of chloride, the rate was too low to be measured.

When the chloride ion concentration was held constant, the bimolecular rate constant decreased only very slightly with increasing perchlorate concentration [2, 3]. According to the activity rate theory [6], the major pathways of this transmethylation reaction cannot involve pairs of oppositely charged species or pairs of species with the same sign. Such combinations would cause a decrease or an increase of the rate constant with increasing ionic strength. The experimental results indicate that the reacting species are either without charge or that only one species of a reacting pair is charged.

When the chloride and the perchlorate concentrations were increased, the rate constant decreased substantially [2, 3]. This decrease could be caused by the involvement of oppositely charged species, e.g., positively charged trimethyltin cation and a negatively charged chloromercuric complex [6]. However, the formation of chloride complexes of mercury(II) [7] is the more likely reason for the decrease of the rate constant. Dodd et al. [8] demonstrated that the demethylation rates of a variety of methyl donors decreased drastically when $HgCl_2$ was replaced by $HgCl_3^-$ or $HgCl_4^{2-}$ as methyl acceptor. Formation constants from the literature [2] were used to calculate [9] the concentrations of mercury(II) and tin species during the reaction between trimethyltin and mercury(II) [1, 2]. As the ratio chloride concentration/mercury concentration increased from 3 to 23, $HgCl_4^{2-}$ became the dominant species. At a ratio of 3, $HgCl_2$ (80%), $HgCl_3^-$ (20%), and $(CH_3)_3Sn^+$ (100%) were the predominant species in solution. Even at a ratio of 23, $(CH_3)_3SnCl$ did not account for more than ten percent of the tin species. Methylmercury, the product of the reaction, was present almost completely as CH_3HgCl. More than 90 percent of the dimethyltin species remained as $(CH_3)_2Sn^{2+}$; $(CH_3)_2SnCl^+$ accounted for five to nine percent of all dimethyltin species. No neutral $(CH_3)_2SnCl_2$ was found. A series of calculations were performed to determine the contributions of various pairs of tin/mercury reactants to the overall transmethylation reaction. In a medium corresponding to seawater ($\mu = 0.6$, $[Cl^-] = 0.5$ M) [10] the most important pairs were $(CH_3)_3Sn^+ + HgCl_2$, $(CH_3)_3SnCl + HgCl_2$, and $(CH_3)_3SnCl + HgCl_3^-$. The specific rate constant for the reaction between $(CH_3)_3SnCl$ and $HgCl_2^{2-}$ is almost zero.

Methylation of Thallium(III) by Trimethyltin Species

Because Tl(III) and Hg(II) are isoelectronic, they are expected and found to react in a similar manner. Each ion accepts only one methyl group from the trimethyltin cation in a reaction that is initially of second order. The methylthallium dication, expected to be the initial product, was never detected by NMR. Instead, methyl chloride and an insoluble precipitate consisting only of thallium and chlorine were identified. Most methylthallium(III) compounds are unstable [11] and decompose to a Tl(I) compound and a methyl derivative. Only methylthallium(III) diacetate was reported to have some stability in aqueous solution [12, 13].

Methylation of Mercury(II) by Methylsilicon Compounds

Sodium 2,2,3,3-tetradeutero-3-(trimethylsilyl)propionate (TSP) and sodium 2,2,-dimethyl-2-silapentane-5-sulfonate (DSS), water-soluble nuclear magnetic resonance standards, methylated mercury(II) in a second-order reaction. In a D_2O solution of 0.1 M ionic strength ($NaClO_4$) and 0.025 M concentrations of TSP and $HgCl_2$ the rate constant increased with temperature from 0.00148 at 30° to 0.0084 L mol^{-1}s^{-1} at 45°. The lnk/T versus temperature plot is linear over the temperature range studied. The rate of methylation of mercury(II) by DSS is strongly influenced by the chloride concentration. The second order rate constant increased from 0.0000025 at a Cl/Hg ratio of 2 to 0.2 at a ratio of 0.5. At a ratio of 0.75 the rate constant was 0.095

L mol^{-1}s^{-1}. These experiments were carried out in D$_2$O at 25° at 0.1 M ionic strength (NaClO$_4$) with DSS and Hg(II) concentrations of 0.025 M.

The reactions of TSP and DSS with Hg(II) may proceed via a four-center transition state that would provide electrophilic assistance for the methide ion, which is a very poor leaving group. With a good leaving group a five-coordinate trigonal bipyramidal transition state is expected. If the transmethylation proceeds via the four-center pathway, the rate of this reaction is expected to increase with increasing electrophilicity of mercury(II). In the series of HgCl$_n^{2-n}$ compounds (n = 0-4) the electrophilicity is highest for Hg^{2+} (n = 0) and lowest for HgCl$_4^{2-}$ (n = 4). Therefore, the rate of transmethylation is expected to increase with decreasing Cl/Hg ratios that imply decreasing concentrations of chloromercury complexes of low electrophilicity. Experiments with TSP and DSS as methyl donors and Hg(II) as methyl acceptors confirmed these expectations.

During the transmethylation reaction, one chloride ion must be either released into solution, become bonded to the silicon atom or attached to methylmercury for each methyl group transferred. If the chloride ions are released into the solution, the concentration of chloride will increase as the reaction progresses. Transmethylation reactions have lower rate constants in high chloride media than in low chloride media. If the rate constant changes during a reaction, the reciprocal of the concentration of a reacting substance is no longer a linear function of time.

The observation of the linearity of the reciprocal of the DSS concentration in the reaction with HgCl$_{0.75}$(ClO$_4$)$_{1.25}$ with time suggests that the chloride ion is removed from solution. Species that could bind chloride are CH$_3$Hg$^+$, Hg^{2+}, HgCl$^+$, and the demethylated silicon compound. The logarithm of the formation constant for CH$_3$HgCl is 5.25 [14], for HgCl$^+$ 6.72, and for HgCl$_2$ (from HgCl$^+$ and Cl$^-$) 6.51 [15]. When the methylation reaction is 63 percent complete, the total concentration of Hg(II) is 0.00925 M, of CH$_3$Hg(II) 0.0158 M, and of chloride 0.0188 M. The calculated concentration of HgCl$_2$ is 0.0083 M accounting for most of the unreacted Hg(II) concentration of 0.00925 M. At such a concentration of HgCl$_2$ the rate constant should have decreased [16] several orders of magnitude to a value (0.000003) approaching that observed for the reaction between DSS and HgCl$_2$. Because such a change was not observed, a silicon-chloride bond that does not hydrolyze during the reaction is probably formed. It should be noted that the hydrolysis of the Si-F bond in trimethylfluorosilane was reported not to be thermodynamically favored at 25° [17].

Transfer of Aliphatic and Aromatic Groups from Organylsilatranes to Mercury(II) Species

Organylsilatranes 1 react readily with mercury(II) to produce organylmercury species (eqn. 1) despite the presence of three deactivating oxygen substituents on silicon.

$$R-\overset{\frown}{Si(OCH_2CH_2)_3N} + HgX_2 \rightarrow RHgX + X-\overset{\frown}{Si(OCH_2CH_2)_3N} \qquad (1)$$

However, organyltrialkoxysilanes, which do not have a nitrogen atom in the molecule, did not react with mercury salts [18]. Silatranes are pentacoordinate organic silicon compounds, in which the nitrogen atom donates electron density to the silicon atom [19, 20]. The apical Si-C bond is very susceptible to direct electrophilic attack by Hg(II). The relative reaction rates of organylsilatranes toward mercury(II) and other metal salts are summarized in Table 1. The order of reactivities are in accord with the

Table 1. *Relative Rates of Reactions of Organylsilatranes with Mercury(II) in Acetone-d_6*

$R-\overset{\frown}{Si(OCH_2CH_2)_3N}$ R	Electrophile	Relative Rate of Reaction
CH_3	$HgCl_2$	1.00*
CH_3	$HgCl_2$	0.06**
CH_3	$HgCl_2$	0.5
CH_3	HgI_2	0.001
$ClCH_2$	$HgCl_2$	<0.000015
Cl_2CH	$HgCl_2$	<0.000015
C_2H_5	$HgCl_2$	0.0002
C_3H_7	$HgCl_2$	0.0002
$c-C_6H_{11}$	$HgCl_2$	<0.000015
$CH_2=CH$	$HgCl_2$	1.5
C_6H_5	$HgCl_2$	1.5
C_6H_5	HgI_2	0.08
$4-ClC_6H_4$	$HgCl_2$	1.5

*In D_2O as solvent.
**In CD_3OD at 35°; apparent bimolecular rate constant: 0.004 L mol^{-1} s^{-1}.

relative rates observed for cobalt-carbon bond cleavage in reactions of organylbis(dimethylglyoximato)cobalt(III) with mercury(II) [21].

The rates of reactions of organylsilatranes with mercury(II) are the results of electronic and steric effects. The nature of the mercury(II) electrophile and the counter ions present in solution can have a pronounced effect on the rate of the metathetical reaction. For instance, mercury(II) iodide is substantially less reactive than mercury(II) chloride. No reaction was observed with CH_3Hg^+ and $C_6H_5Hg^+$ [22, 23]. Other metal electrophiles ($SnCl_4$, $Pb(NO_3)_2$, $CdCl_2$, $ZnCl_2$, $FeCl_3$) were also screened for their reactivity toward methylsilatrane. The reaction rates in acetone-d_6 were less than 0.000015 L mol^{-1}s^{-1}. No stable methylmetal species were detected by proton NMR for Zn(II), Cd(II), Pd(II), Fe(III), or Sn(IV). The solvolysis of the silatranes to triethanolamine in protic media was observed to be more rapid in the presence than the absence of these electrophiles.

Acknowledgements. We wish to thank the National Science Foundation and the University of Maryland Sea Grant Program for financial support and the Gillette Research Corporation for the fellowship for J. D. Nies. Research conducted by K. L. Jewett was carried out at the National Bureau of Standards. We thank Dr. F. E. Brinckman for extensive discussions.

References

[1] K. L. Jewett, "Chemical Factors Influencing the Transmethylation of Heavy Metal Species in Polar Media," **Ph.D. Dissertation,** Department of Chemistry, Unversity of Maryland, 1978.
[2] K. L. Jewett, F. E. Brinckman and J. M. Bellama, "Influence of Environmental Parameters on Transmethylation between Aquatic Metal Ions," **ACS Symp. Ser.,** 82 (1978) 158.
[3] F. E. Brinckman, "Environmental Inorganic Chemistry of Main Group Elements with Special Emphasis on Their Occurrence as Methyl Derivatives," in **"Environmental Inorganic Chemistry,"** San Miniato, Italy, June 2-7, 1983, VCH Publishers: Deerfield Beach, Florida, 1985, p. 195.
[4] L. Johansson, "The Role of Perchlorate Ion as Ligand in Solution," **Coord. Chem. Rev.,** 12 (1974) 241, and

references therein.

[5] R. E. Hester and R. A. Plane, "A Raman Spectrophotometric Comparison of Interionic Association in Aqueous Solution of Metal Nitrates, Sulfates, and Perchlorates," **Inorg. Chem.**, **3** (1964) 769.

[6] C. W. Davis, "Salt Effects in Solution Kinetics," **Progress in Reaction Kinetics," Vol. 1,** G. Porter, Ed., Pergamon Press: New York, 1961, p. 163.

[7] M. Sandström, "An X-Ray Diffraction and Raman Study of Mercury(II) Chloride Complexes in Aqueous Solution. Evidence for the Formation of Polynuclear Complexes," **Acta. Chem. Scand.**, **A31** (1977) 141.

[8] D. Dodd, M. D. Johnson and D. Vamplew, "Mechanism of Electrophilic Substitution at a Saturated Carbon Atom. Part XVI. Rates of Reaction of Metallic Electrophiles with the 2-, 3-, and 4-Pyridiniomethylchromium(III) Ions," **J. Chem. Soc. B,** (1971) 1841.

[9] D. Dryssen, D. Jagner and F. Wengeln, **"Computer Calculations of Ionic Equilibria and Titration Procedures,"** Almqvist and Wiskell: Stockholm, 1968.

[10] W. Stumm and J. J. Morgan, **"Aquatic Chemistry,"** 2nd Ed., Wiley-Interscience: New York, 1981.

[11] C. R. Hart and C. K. Ingold, "Mechanism of Electrophilic Substitution of a Saturated Carbon Atom. Part VIII. Kinetics and Stereochemistry of a Thallium-for-Mercury Substitution. The Unimolecular Mechanism in Thallium-for-Mercury, Mercury-for-Thallium, and Thallium-for-Thallium Substitutions. Appendix: the Dubious Existence of Alkylthallic Salts," **J. Chem. Soc.,** (1964) 4372.

[12] H. Kurosawa and R. Okawara, "The Preparation of Methylthallium(III) Diacetate," **Inorg. Nucl. Chem. Lett.,** **3** (1967) 93.

[13] H. Kurosawa and R. Okawara, "Mono- and Dialkylthallium Chemistry," **Trans. N.Y. Acad. Sci., Ser. 2, 30** (1968) 962.

[14] G. Geier and I. W. Erni, "Kinetik und Mechanismus von Methylquecksilber-Komplexbildungen," **Chimia, 27** (1973) 635.

[15] I. L. Ciavatta and M. Grimaldi, "Equilibrium Constants of Mercury(II) Chloride Complexes," **J. Inorg. Nucl. Chem.,** **30** (1968) 197.

[16] M. Eigen and E. M. Eyring, "Second Wien Effect in Aqueous Mercuric Chloride Solution," **Inorg. Chem.,** **2** (1963) 636.

[17] J. A. Gibson and A. F. Janzen, "N.M.R. Study of Base-Catalyzed Hydrolysis of Trimethylfluorosilane," **Can. J. Chem.,** **50** (1972) 3087.

[18] M. G. Voronkov, N. F. Chernov and T. A. Dekina, "Reaction of Mercury Salts with Si-Substituted Phenylsilane, $C_6H_5SiX_3$," **Dokl. Akad. Nauk SSSR, 230** (1976) 853.
[19] C. L. Frye, G. A. Vincent and W. A. Finzel, "Pentacoordinate Silicon Chemistry. V. Novel Silatrane Chemistry," **J. Am. Chem. Soc., 93** (1971) 6805.
[20] L. Parkanyi, K. Simon and J. Nagy, "Crystal and Molecular Structure of β-1-Phenylsilatrane, $C_{12}H_{17}O_3NSi$," **Acta Crystallogr., Sect. B, 30** (1974) 2328.
[21] P. Abley, E. R. Dockal and J. J. Halpern, "Alkylation and Arylation of Mercury(II) and Thallium(III) by Organocobalt Compounds," **J. Am. Chem. Soc., 95** (1973) 3166.
[22] H. Saffer and T. DeVries, "Dipole Moments of Aromatic Derivatives of Trimethylsilane," **J. Am. Chem. Soc., 73** (1951) 5817.
[23] K. L. Jewett, F. E. Brinckman and J. M. Bellama, "Chemical Factors Influencing Metal Alkylation in Water," **ACS Symp. Ser., 18** (1975) 304.

DISCUSSION

J. M. Wood (Gray Freshwater Biological Institute, University of Minnesota): The reaction of silanes with mercury(II) was discovered by R. E. DeSimone in 1972. He showed that NMR internal standards reacted with mercury(II).

XI

THE ENVIRONMENTAL CHEMISTRY OF METALS WITH EXAMPLES FROM STUDIES OF THE SPECIATION OF CADMIUM

Peter J. Sadler, Denise P. Higham and Jeremy K. Nicholson*

*Department of Chemistry, Birkbeck College
University of London, Malet Street, London WC1E 7HX, U.K.*

**Toxicology Unit, School of Pharmacy
University of London, Brunswick Square, London WC1, U.K.*

The factors that influence the speciation of metals in the environment such as complexation, redox potential, pH, and metal transformation by microorganisms are discussed. The species of a metal present in the environment is often more important than its concentration, because the chemical properties of a species affect its availability to organisms. The dependence of bioavailability on species-specific properties is illustrated by studies on the speciation of cadmium including the binding of cadmium to metallothionein, the binding of cadmium to glutathione in red cells, and the uptake of cadmium by cadmium-resistant bacteria. Multinuclear resonance spectroscopy has been particularly useful in these studies of intact systems, isolated proteins, and small molecules. Attention is also drawn to our lack of understanding of the biological reactivity of polymeric metal complexes including minerals and their surfaces. We need to further our knowledge of the biological chemistry of newly discovered essential metals such as V, Cr, Mn, and Ni, and of the reactions through which Cr, Ni, Pt, and other metals exert potent immunological effects.

All forms of life utilize inorganic chemistry. The unravelling of the interrelationships between the inorganic chemistry of the atmosphere, hydrosphere, lithosphere, and living organisms such as animals, plants, and microorganisms is a major and exciting challenge for the inorganic chemist today. The metals, the topic of this paper, are involved in many of these interrelationships. The chemistry of the metals of the first three groups (Ia, IIa, IIIa) of the periodic chart (Fig. 4) in aqueous media is best described by an ionic equilibrium model. A one-electron redox chemistry is characteristic of transition metals. The covalency of element-transition metal bonds increases within a period with increasing atomic number of the metal. The metal ions of groups Ib, IIb, and IIIb are strong Lewis acids that achieve equilibrium quickly but show little redox activity [1]. These and other metal ions usually establish equilibria rapidly unless the metal is in a high oxidation state. Metal ions in high oxidation states acquire oxide or hydroxide ions as ligands in aqueous media. These complexes are rather inert toward substitution. The non-metals form strong covalent bonds and their reactions are kinetically controlled.

The covalent and ionic properties of metal ions in water are illustrated by their affinities for ligands [2]. The division of metals by Ahrland, Chatt, and Davies [3] into "class a" and "class b" and by Pearson [4] into the equivalent "soft" and "hard" classes is based on the affinities of metal ions toward ligands. Nieboer and Richardson [5] related these concepts to the formal charges, radii, and electronegativities of metal ions (Fig. 1.). Examples of ligands preferred by these classes of metal ions are for "class a" F^-, O^{2-}, OH^-, H_2O, CO_3^{2-}, SO_4^{2-}, HPO_4^{2-}, RCO_2^-, $ROPO_3^{2-}$; for "class b" H^-, I^-, CN^-, S^{2-}, RS^-, R_2S, RSH, heterocyclic N. This classification holds for aqueous media but may not be true for other solvents as illustrated by silver(I) halide complexes (Fig. 2) [6]. In water, silver(I) behaves as a typical "class b" metal ion, with iodide bound much stronger than bromide, and bromide bound stronger than chloride. However, in dimethyl sulfoxide and acetonitrile, silver(I) binds all three halides with approximately equal strength. In the silver-halide system the major difference between the aqueous and organic media is the weaker solvation of the halide ions in the organic media. Therefore, solvents may drastically influence metal-ligand affinities. Solvents need to be carefully chosen, when model systems are constructed for the study of environmental problems associated with metals.

Availability and Analysis: Concentration of the Elements in the Environment

The concentration and availability of metals in the environment will vary geographically and chronologically. The chemical forms of metals may be highly dependent on

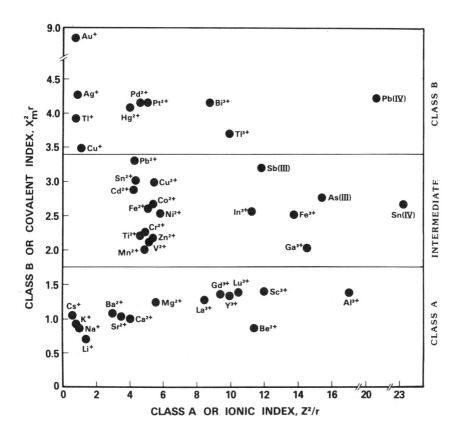

Figure 1. The division of metal ions into "class a" (hard), "class b" (soft), and an intermediate class on the basis of electronegativity (X_m), formal charge (Z), and ionic radius (r). Reprinted with permission from ref. 5.

factors such as pH, redox potential, ligand availability, and transformation by microorganisms. Microorganisms are known to influence the forms of a variety of metals in the environment through chelation, reduction, oxidation, and alkylation. Marine bacilli reduce Fe(III) in Fe_2O_3. Several bacteria and fungi convert Mn(IV) to a lower oxidation state of manganese. Certain pseudomonads transform Hg(II) to metallic mercury. Other microorganisms oxidize metal sulfides such as FeS_2 to iron(II) sulfate, a process important in acid mine drainage, and convert Mn(II) to MnO_2, a reaction leading to manganese nodules in marine

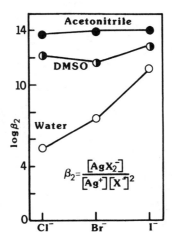

Figure 2. The trends of the formation constants for dihaloargentate(I) complexes in acetonitrile, dimethyl sulfoxide, and water [6].

environments [7]. Mercury(II) can also be methylated by methyl donors such as vitamin B_{12} and choline or transalkylated to produce, for example, phenyl mercuric compounds [8]. Other organisms can dealkylate mercury compounds to produce volatile elemental mercury and benzene or methane. Tellurium, selenium, and arsenic are also methylated to toxic, volatile dimethyl and trimethyl derivatives. Methylation of lead and tin has also been reported [7].

The average concentrations of some of the elements in man are plotted in Figure 3 versus the average concentration of the elements in the earth's crust [9]. If man had the same composition as the earth's crust, all elements would be arranged along the line of equal concentration or abundance. Elements below this line are more abundant and elements above the line are less abundant in man than in the earth's crust. To some degree man's composition - largely that of an aqueous electrolyte solution containing protons, calcium, magnesium, sodium, potassium and chloride ions, and organic compounds consisting of C, O, H, S, P, and N atoms - generally reflects the composition of his environment [10, 11]. With respect to the environment Si, Na, Al, Li, Fe, Cu, Ba, Mg, and K are depleted and Br, P, S, Cl, N, C, and Cd are enriched in man. Eleven bulk elements and 14 trace

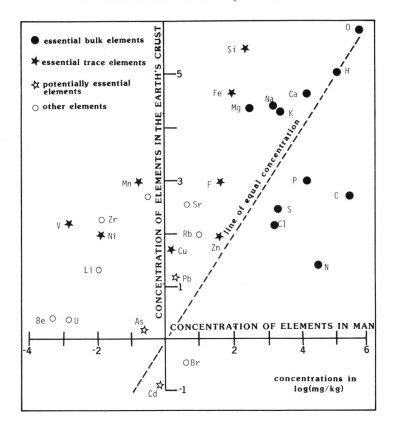

Figure 3. A comparison of the concentrations of elements in man and the environment showing accumulation and exclusion of certain elements by man. Data from ref. 9.

elements are currently thought to be essential to warm-blooded animals (Fig. 4). Whereas all elements of the first transition series except Sc and Ti are essential, only Mo of the second and third series is apparently required [12]. Arsenic, cadmium, and lead are potentially essential. These three elements at the low levels found in normal diets enhanced the growth of experimental animals. Perhaps other elements will also be proven to be essential. Normal human blood usually contains more than 55 elements.

The results of many of the older determinations of elements in man and animals are unreliable. Considerable gaps still exist in our knowledge of the concentrations of elements and especially of trace metals in biological

							GROUP									
Ia	IIa	IIIa	IVa	Va	VIa	VIIa	VIII			Ib	IIb	IIIb	IVb	Vb	VIb	VIIb	0
H				essential bulk elements								essential trace elements				H	He
Li	Be			potentially essential elements								B	C	N	O	F	Ne
Na	Mg											Al	Si	P	S	Cl	Ar
K	Ca	Sc	Ti	V	Cr	Mn	Fe	Co	Ni	Cu	Zn	Ga	Ge	As	Se	Br	Kr
Rb	Sr	Y	Zr	Nb	Mo	Tc	Ru	Rh	Pd	Ag	Cd	In	Sn	Sb	Te	I	Xe
Cs	Ba	La	Hf	Ta	W	Re	Os	Ir	Pt	Au	Hg	Tl	Pb	Bi	Po	At	Rn
Fr	Ra	Ac															

Figure 4. Periodic table identifying the elements currently thought to be essential for warm blooded animals.

systems. However, the situation is improving with the availability of standard reference materials and multi-element instruments. The results of measurements of element concentrations would be much more useful for solving biological problems, if information about the type of compounds, in which these elements occur, were also available and were correlated with biological functions.

We should avoid classifying metals as "toxic" and "non-toxic". The Bertrand Diagram (Fig. 5) illustrates the relationship between the dose of a metal compound and the physiological response. At sufficiently high doses all compounds will be toxic, and the distinction between "toxic" and "non-toxic" elements ceases to exist. For any particular metal the Bertrand Diagram consists of a family of curves. Each curve is characteristic for a particular compound, a particular route for its administration, and the prevailing biochemical status of the host [13].

Redox Potentials in External and Internal Environments

Metal transport. Whereas the internal compartments of biological systems operate under mildly reducing conditions at redox potentials of approximately -0.2 to 0.0 V, the external environments may generate greatly fluctuating potentials. Ahrland [14] discussed the metal species

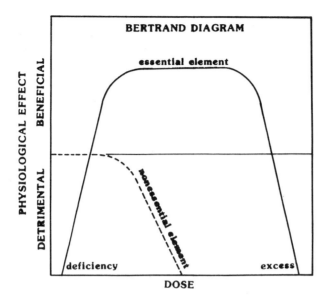

Figure 5. A Bertrand diagram illustrating the continuous relationship between the physiological effect of a compound (element) and the dose administered.

present in seawater. Under aerobic conditions in surface waters the metal ions are likely to be in high oxidation states and extensively hydrolyzed. Examples of metal species stable under these conditions are HVO_4^{2-}, CrO_4^{2-}, MnO_2, $FeO(OH)$, Fe_3O_4, UO_2^{2+}. Chromium(III) is not thermodynamically stable in seawater. Thus, in seawater, animals meet anionic species such as chromates and vanadates. These anions may interfere with natural transport pathways for other anions such as phosphate. The deprotonation constants (pK) of phosphoric acid (1.7, 6.5, 12.1) are close to those of vanadate (3.5, 7.8, 12.5).

Erythrocytes can accumulate orthovanadate, which is known to enter the cells via phosphate channels. Inside the cell vanadate is reduced to V(IV) and then binds to haemoglobin [15]. Curiously, the biochemistry of vanadium seems to be the same whether vanadium is administered to animals as V(III), V(IV), or V(V) probably because all vanadium species are rapidly converted to the vanadyl ion, VO^{2+}. The vanadyl ion is then transported to the tissues by the Fe(III) transport protein, transferrin. Like Fe(III) hydroxide, vanadium(IV) hydroxide, $VO(OH)_2$ is highly insoluble.

Vanadate and vanadyl ions are potent inhibitors of phosphatases, ATPases, phosphortransferases, nucleases, and kinases. Recently, vanadium was suggested to be a natural regulator of ATPase activity. ATP isolated from muscle can contain one molecule of vanadium per 3000 molecules of ATP [16]. Marine tunicates accumulate vanadium as V(III) in vanadocytes. $H_2VO_4^+$ appears to enter this animal also through anion channels. In the cell the vanadate is rapidly reduced [17]. The intracellular pH is now thought to be 7.2 and not, as originally suggested, highly acidic. Because V(III) hydrolyzes, dimerizes, and precipitates at pH > 2.2 it must be complexed inside the cells perhaps with tunichrome.

Metals and the immune response. The immune system of the body provides protection against foreign substances and pathogenic organisms. The possible roles of metal ions in mediating the immune response are largely unexplored. Chromium(VI) as chromate or dichromate is of widespread environmental concern. Chromate ion readily penetrates cell membranes, a property routinely exploited to radio-label cells. Once inside a cell chromate is reduced. Exposure of the skin to chromate commonly leads to an allergenic reaction and dermatitis. Such immune responses – observed for a number of metal compounds (Table 1) – need to be studied further.

Table 1. Examples of Metal Compounds that Provoke Immune Responses [18, 19]

Metal	Compound	Effects
Be	BeF_2, BeO	Dermatitis
		Chronic beryllium disease
Al	$Al(OH)_3$	Intradermal granulomas
Cr	$K_2Cr_2O_7$	Dermatitis
Ni	$NiSO_4$	Dermatitis
Pt	Na_2PtCl_6, K_2PtCl_4	Allergenic reaction
Hg	R-Hg-SR, R-Hg$^+$	Skin sensitization

Contact dermatitis from exposure to nickel and nickel salts, the "nickel itch" [20], is very common. People can become sensitized to metal allergies, and minute amounts of metals can then elicit a response. For instance, sufficient nickel is often present in stainless steel to cause such a response. A dramatic example is provided by platinum. As little as 10^{-15} g of platinum can cause an allergenic reaction in a sensitized person [21] in a skin prick test. Whether the true allergen is a Pt-protein complex is unknown. The most active complexes such as $PtCl_4^{2-}$ and $PtCl_6^{2-}$ have reactive Pt-Cl bonds. These commonly used ions can be a serious hazard in the laboratory. Platinum compounds provide good examples for the dependence of physiological effects on the type of species. Tetrachloroplatinate(II), $PtCl_4^{2-}$, and hexachloroplatinate(IV), $PtCl_6^{2-}$, are potent allergens. Cis-diaminodichloroplatinum(II), cis-$Pt(NH_3)_2Cl_2$, is an anti-cancer agent but the trans isomer is not. Tetrakis(ethylenediamine)platinum(IV), $Pt(en)_3^{2+}$, is a neuromuscular blocking agent.

Studies on the Speciation of Cadmium

Environmental studies on cadmium are usually concerned with its toxicity. However, Schwarz has shown that the growth of experimental animals is impaired when trace levels of cadmium are removed from the diet [22]. Therefore, cadmium must be considered as a "potentially essential element". A major hazard to animals is the body burden of cadmium that increases with age and may cause renal damage. Most natural deposits of cadmium consist of relatively insoluble cadmium sulfide, CdS. This sulfide is rendered soluble by weathering processes. About five years ago we decided to investigate whether bacteria from natural environments could biotransform cadmium. Such bacterial action might produce new and potentially more toxic cadmium compounds, which then might be released by the bacteria into the environment.

Cadmium accumulation by bacteria. The *P. putida* bacteria, isolated from sewage sludge, readily adapt to and grow in a chemically-defined medium containing 3 mM cadmium. Growth in the presence of the Cd(II) is strongly influenced by the amounts of Zn(II) and Ca(II) in the medium. The chemistry of cadmium closely resembles that of zinc and calcium. However, cadmium has a high affinity for sulfur of thiolate ligands whereas calcium does not. Bacteria exposed to cadmium accumulate cadmium inside the cells and exhibit gross morphological changes [23].

Proton nuclear magnetic resonance spectroscopy (NMR) is a useful method for the simultaneous estimation of the concentrations of diamagnetic metal ions that form chelates with ethylenediaminetetraacetic acid (EDTA). Over the pH range of 6 to 10 free EDTA is exchanging slowly on the NMR timescale with EDTA in metal complexes [24]. The signal for the four equivalent NCH_2CH_2N protons of complexed EDTA can be used to estimate the concentration of the metal. Spectra for seven metal-EDTA complexes are shown in Figure 6. The

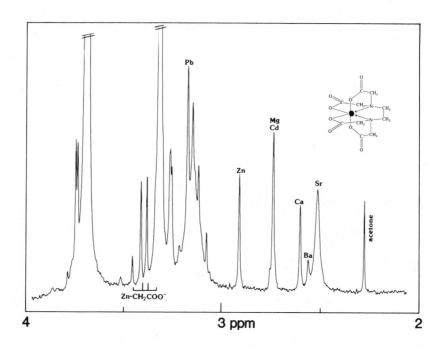

Figure 6. *400 MHz proton NMR spectrum of a D_2O solution at pH 7 containing seven diamagnetic metal ions and excess EDTA.*

signals for the ethylenic protons in the metal complexes are singlets, whereas those of the acetate protons are complex multiplets (AB patterns). The magnesium and cadmium signals overlap, but they can be resolved. The cadmium satellites are caused by Cd-113 and Cd-111. The strontium signal is broadened by exchange. Metal concentrations as low as 50 μM can be estimated. We used this NMR method to determine Ca, Mg, Zn, and Cd concentrations in animal plasmas, lysed red cells [25], and intact bacteria, and to study the zinc and cadmium affinity constants of metallothionein.

Proton NMR spectra of intact bacteria are shown in Figure 7. After EDTA had been added, signals for EDTA complexes of

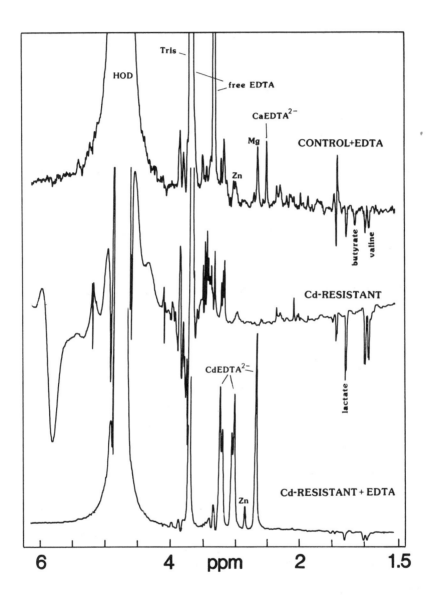

Figure 7. *400 MHz spin-echo proton NMR spectra of normal and cadmium-resistant, intact P. **putida** bacteria in the absence and presence of EDTA.*

Zn, Mg, and Ca were present in the spectrum of the control cells. The cadmium-resistant cells had signals for Zn-EDTA and Cd-EDTA in their spectrum. Addition of EDTA to the cells leads to the extraction of Zn, Mg, and Ca from the control cells in a ratio of approximately 0.1:1:1. In cadmium-resistant cells the ratio of extracted Zn to extracted Mg+Cd is 0.1:1. The cadmium concentration inside the resistant cells is so high that little free EDTA is present even at a total EDTA concentration of 25 mM. Atomic absorption measurements gave an intracellular Mg to Ca ratio of 400:1 for control cells. Therefore, EDTA probably extracts these metals from the cell walls, in which the Mg to Ca ratio is known to be 1:1. No calcium was extracted from the Cd-resistant bacteria. It is possible that Cd replaced some of the Ca in the cell walls.

We are also using NMR to monitor metabolic differences between control cells and resistant cells. We found that Cd-resistant cells utilize almost all the alanine and lysine in the medium as well as glucose, serine, and glycine. Control cells use less glucose and only trace amounts of the amino acids. Cadmium also affects the storage products accumulated by the cells and the waste products that are excreted. Perhaps our most significant discovery are novel cadmium proteins present in cadmium-resistant cells. Three proteins of 4,000 to 7,000 molecular weight containing up to seven metal ions (Cd, Zn, Cu) were isolated from the cells during different phases of their growth. The proteins are rich in cysteine and appear to be related to metallothioneins [23]. One of these proteins was released into the growth medium during cell lysis in the phase of declining growth.

Preferential accumulation of cadmium in the kidney was observed after administration of cadmium metallothionein to mice [26]. Binding of cadmium to metallothionein is probably responsible for the long term build up of cadmium levels in animal kidneys that leads in some cases to renal failure. It is therefore of importance to learn more about the nature of cadmium binding to metallothionein, in the hope that this knowledge will aid the design of chemotherapeutic agents to remove cadmium from the body.

Cadmium-binding sites in metallothionein. Mammalian metallothioneins are low molecular weight proteins (~ 7,000) that are usually devoid of aromatic amino acids and histidine but are rich in cysteine [27]. Twenty of the 61 amino acids in this protein are cysteine. Metallothioneins are thought to be involved with the uptake, storage, and regulation of the essential metals zinc and copper, with their insertion into apoenzymes, and with the detoxification of other metals [28]. The synthesis of metallothionein is induced at the transcriptional level [29]. Metals with a high affinity for sulfur (cysteine) ligands are known to bind to metallothionein. For in-vivo systems such metals are Zn(II), Cd(II), Hg(II), Cu(I), Ag(I), Au(I), and Bi(III). These same metals and Co(II) and Ni(II) bind to metallothionein in in-vitro systems. Using potentiometric titration data Vasak and Kagi [30] estimated, that the affinity constants of the protein at pH 7 are approximately 10^{16} for Cd(II) and 10^{12} for Zn(II). We established by Cd-113 NMR [31], that a natural Cd-Zn rat liver metallothionein with seven metal ions per mole of protein contains more than 20 slightly different cadmium sites (Fig. 8). The cadmium and zinc ions probably are tetrahedrally coordinated to four cystein sulfur atoms that can also serve as bridges between metal ions. Bridging between two Cd-113 ions produces the multiplet splitting shown in the uppermost spectrum in Figure 8. These sites complicate attempts to understand affinity constants. Are there high and low affinity sites? Are some bound metal ions more reactive than others?

We gained insight into this problem using proton NMR spectroscopy. The rate of extraction of Cd(II) and Zn(II) from plaice metallothionein containing seven bound metal ions with a Zn:Cd ratio of 1:10 [32] was monitored simultaneously. The rate of removal was determined by following the intensity changes of the NCH_2CH_2N singlets with time. The calcium-EDTA signal is probably ascribable to slight contamination of the protein (Fig. 9). Over a period of a few hours EDTA removed most of the Zn but little cadmium (Fig. 10). Only one seventh of the bound metal was accessible to EDTA. Structural changes in the protein that accompany metal removal can be studied at the same time. Further work along these lines is in progress.

The interplay between thermodynamics and kinetics deserves further attention in work relating to the environmental chemistry of metals. A striking example of a reaction with apparently high energy barriers despite strong thermodynamic driving forces is the removal of iron from the transport protein transferrin. Although the affinity of desferrioxamine for ferric ions is two orders of magnitude greater than that of transferrin, transferrin lost only 30 percent of the iron to desferrioxamine after 30 hours [33]. This reaction is important for the treatment of iron

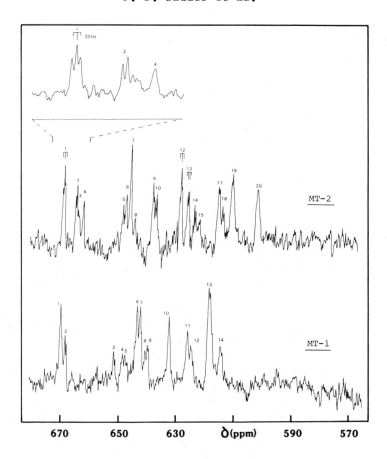

Figure 8. Cadmium-113 proton-decoupled, 88 MHz NMR spectra of native rat liver metallothioneins illustrating the power of NMR for speciation studies. Reprinted with permission from The Biochemical Society, London, from ref. 31.

overload diseases. The transfer is catalysed by citrate and nitrilotriacetate, two ligands that do not bind appreciable amounts of iron at any instant.

Cadmium entry into red cells. Red cells appear to be involved in the metabolism of cadmium in a variety of animal species [34]. The intracellular concentration of the tripeptide glutathione, γ-Glu-Cys-Gly, is about 2 mM. In a model solution at pH 7, cadmium binds to the sulfur atom of cysteine causing a downfield shift of the cysteine methylene

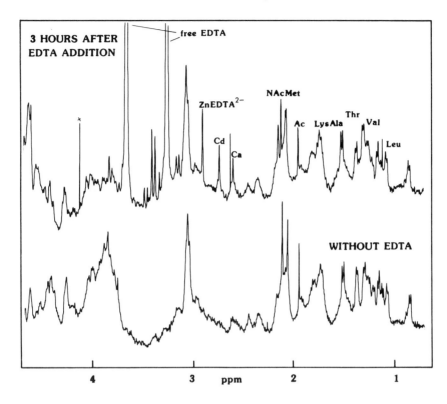

Figure 9. *400 MHz proton NMR spectra of Zn/Cd metallothionein from plaice liver before and three hours after addition of excess EDTA.*

protons. These protons become non-equivalent in the metal complexes (Fig. 11). The exchange of the glutathion-SH group between free and bound forms is sufficiently rapid to produce only one signal. A remarkable similarity in the binding and exchange rates of Cd(II) and Zn(II) was observed. Polarography appears to be another promising method for the study of this system. The peak at -0.73 V, which disappeared below pH 5, is probably caused by a Cd-S species. The peak at -0.59 V, characteristic of aquated Cd(II), increased in intensity below pH 6.2 (Fig. 12). At lower pH sulfur is readily diplaced from cadmium.

Cadmium(II) added to suspension of red blood cells also binds to the sulfur of intracellular glutathione (Fig. 13). We used the spin-echo technique to remove the broad signals

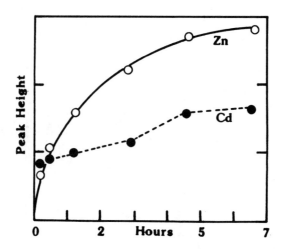

Figure 10. *The time dependence of the NCH_2CH_2N proton NMR singlets of Zn-EDTA and Cd-EDTA in a system containing Zn/Cd metallothionein from plaice liver and EDTA.*

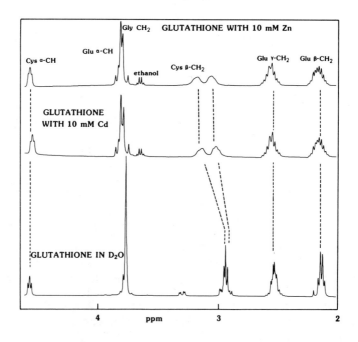

Figure 11. *400 MHz H-1 NMR spectra of a 20 mM D_2O solution of glutathione at pH 7 in the presence of 10 mM Cd or Zn.*

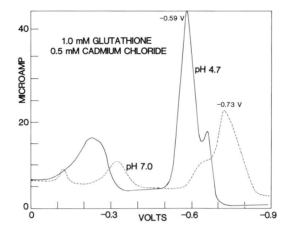

Figure 12. *Differential pulse polarograms of solutions containing per liter 1 millimole glutathione, 0.5 millimoles cadmium chloride, and 0.1 mole sodium nitrate at pH 7 and 4.7.*

of macromolecules from the spectrum. Only signals from low molecular weight metabolites remain [34]. The time-dependent binding of cadmium to glutathione inside the cell (Fig. 14) was monitored by following the decrease in intensity of the glycine (glutathione) signal relative to free glycine. The time dependence may be caused by a slow passage of cadmium through the cell membrane or a slow diffusion inside the cell. Calcium is expected to compete for cadmium binding sites on the membrane. Addition of calcium did delay the binding of cadmium to glutathione (Fig. 14). Rabenstein et al. have carried out similar spin-echo NMR experiments on the binding of other metals such as zinc and mercury to glutathione in red blood cells [35-37].

Solids, Minerals and Surfaces

Mineralization in biology is carefully controlled precipitation. Examples of minerals that are precipitated in biological systems are calcium carbonates, phosphates, oxalates, sulfates, and silicates in bones and teeth [38]. Magnetite, Fe_3O_4, is used as a magnetic device by some bacteria. We need to understand how biological control of mineralization is achieved. Evidently the occurrence of minerals in the wrong places can lead to serious disorders.

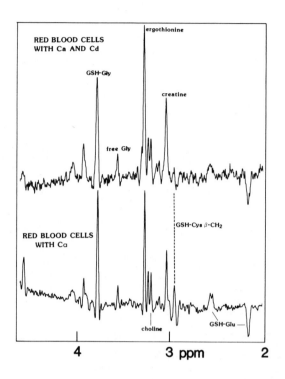

Figure 13. *400 MHz spin-echo proton NMR spectra of red blood cells showing binding of cadmium to intracellular glutathione.*

Calcium pyrophosphate crystals may give rise to a pseudo-gout syndrome, and calcium hydroxyapatite crystals may cause joint inflammation. Chrysotiles (asbestos) are haemolytic and toxic to macrophages. These effects are probably attributable to the surface properties of the crystals [39]. This area needs further consideration. Crystal surfaces can profoundly influence the course of biological reactions. Rhodes and Klug [40] showed, that DNA and other polydeoxynucleotides can be immobilized on the surfaces of mica and calcium phosphate microcrystals. Their cleavage by enzymes then occurs selectively every 10.6 ± 1 base pairs. Nickel sulfide, Ni_3S_2, a highly insoluble compound, is a potent carcinogen when administered as dust to experimental animals [41]. The importance of surface adhesion to cells and coating by immunoglobulins to carcinogenicity and the prevention of cancer are not known. Curiously, manganese dust exerts a protective effect [42]. The injection of

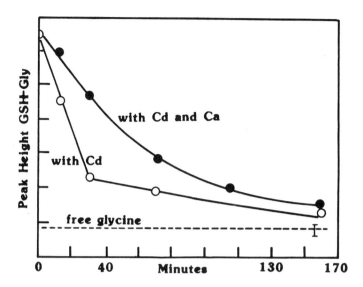

Figure 14. *The time dependence of the intracellular glycine (glutathione) signal relative to free glycine in a system consisting of red blood cells and cadmium (0.5 mM), or red blood cells, cadmium (0.5 mM) and calcium (0.5 mM).*

chromium, nickel, and cadmium oxides can also lead to sarcomas [41]. Defects that are well known to occur in the structures of transition metal oxides could have a large effect on biological activity.

Our general understanding of how biological systems control the uptake and release of metals from polynuclear stores of metal ions is poor. In addition to metallothionein, a potential storage protein for zinc and cadmium, and ferritin, a depot protein for iron, many other biological storage sites (Table 2) exist that require

Table 2. Examples of Polynuclear Metal Inclusion Bodies

Metal	Condition	Deposition site
Fe	Overload Hemmorrhage	"sidersomes" (lysosomes)
Au	Drug therapy	"aurosomes" (lysosomes)
Pb	Exposure	intranuclear (kidney, liver)
Bi	Drug therapy	intranuclear (kidney)

further investigation. For example, aggregates of lead ions bound to acidic proteins in inclusion bodies are found within the cell nucleus [26].

Experimental

NMR spectra were recorded on a Bruker WH400 spectrometer at 400 MHz, 298° K, in 5 mm tubes. A pulse sequence of 90°-τ-180°-τ-collect, was used for spin-echo experiments, with τ = 60 ms. Red cells were obtained by centrifugation and were washed three times with isotonic saline to exchange H_2O for D_2O.

The *P. putida* bacteria were originally isolated from sewage sludge in Southern England and were selected because of their ability to grow in the presence of low levels of metal ions. These bacteria are referred to as control cells. Subsequently, the cells adapted their growth to high levels of cadmium (3 mM) under chemically-defined conditions. These bacteria are referred to as resistant cells. For NMR work, the bacteria were harvested in the exponential growth phase, washed once with 0.9 percent saline containing 50 mM ammonium bicarbonate at pH 8 and finally resuspended (10^{10} cells per mL) in the same deuterated buffer.

Differential pulse polarography was carried out on nitrogen-purged solutions with a PAR model 174A analyser and a model 303 dropping mercury electrode system.

Acknowledgements. We are grateful to the SERC, MRC, Rio Tinto Zinc Services, The Royal Society, and University of London Intercollegiate Research Service for their support of this work. We also benefitted from collaborations and discussions with a large number of colleagues.

References

[1] C. S. G. Phillips and R. J. P. Williams, **"Inorganic Chemistry,"** Oxford University Press, 1965.
[2] M. N. Hughes, **"Inorganic Chemistry of Biological Processes,"** 2nd Ed., John Wiley & Sons: New York, 1981.
[3] S. Ahrland, J. Chatt and N. R. Davies, "Relative Affinities of Ligand Atoms for Acceptor Molecules and Ions," **Quart. Rev. (London), 12** (1958) 265.
[4] R. G. Pearson, "Hard and Soft Acids and Bases," J. Am. **Chem. Soc., 85** (1963) 3533.

[5] E. Nieboer and D. H. S. Richardson, "The Replacement of the Nondescript Term 'Heavy Metals' by a Biologically and Chemically Significant Classification of Metal Ions," **Environ. Pollut., Ser. B: Chem. Phys., 1** (1980) 3.

[6] J. Bjerrum, "Complex Formation in Aqueous Solution with Special Regard to Metal Phosphine Complex Formation," **Kemiai Kozlemenyek., 41** (1974) 67; **Chem. Abstr., 81** 96804.

[7] H. L. Ehrlich, "How Microbes Cope with Heavy Metals: Arsenic and Antimony in the Environment," in **"Microbial Life in Extreme Environments,"** D. J. Kushner, Ed., Academic Press: New York, 1978, p. 381.

[8] S. Silver, "Bacterial Transformations of and Resistance to Heavy Metals," in **"Changing Metal Cycles and Human Health,"** Report of the Dahlem Workshop, March 20-25, 1983, Berlin, J. O. Nriagu, Ed., Springer Verlag: Berlin, 1984, p. 199.

[9] H. J. M. Bowen, **"Environmental Chemistry of the Elements,"** Academic Press: London, 1979.

[10] R. J. P. Williams, "Natural Selection of the Chemical Elements," **Proc. Roy. Soc. London, B213** (1981) 361.

[11] J. J. R. F. da Silva and R. J. P. Williams, "The Uptake of Elements by Biological Systems," **Struct. Bonding (Berlin), 29** (1976) 67.

[12] N. J. Birch and P. J. Sadler, "Metal Ions in Biology and Medicine," in **"Specialist Periodical Reports: Inorganic Biochemistry,"** Vols. 1-3, H. A. O. Hill, Ed., The Chemical Society: London, 1979, 1980, 1982.

[13] P. J. Sadler, "Inorganic Pharmacology," **Chem. Brit., 18** (1982) 12.

[14] S. Ahrland, "Metal Complexes Present in Sea Water," in **"The Nature of Seawater,"** Report of the Dahlem Workshop, March 10-15, 1975, E. D. Goldberg, Ed., Dahlem Konferenzen: Berlin, 1975, p. 219.

[15] N. D. Chasteen, "The Biochemistry of Vanadium," **Struct. Bonding (Berlin), 53** (1983) 105.

[16] N. J. Birch and P. J. Sadler, "Inorganic Elements in Biology and Medicine," in **"Specialist Periodical Reports: Inorganic Biochemistry,"** Vol. 2, H. A. O. Hill, Ed., The Chemical Society: London, 1981, p. 315.

[17] K. Kurstin, G. C. McLeod, T. R. Gilbert and L. B. R. Briggs, "Vanadium and Other Metal Ions in the Physiological Ecology of Marine Organisms," **Struct. Bonding (Berlin), 53** (1983) 140.

[18] G. Kazantsis, "The Role of Hypersensitivity and the Immune Response in Influencing Susceptibility to Metal Toxicity," **Environ. Health Perspect., 25** (1978) 111.

[19] J. L. Turk and D. Parker, "Granuloma Formation in Normal Guinea Pigs Injected Intradermally with Aluminum and Zirconium Compounds," **J. Invest. Dermatol.**, **68** (1977) 336.

[20] M. Blomberg, E. Hellsten, A. Hendrikson-Enflo, M. Sundbom and H. Vokel, "Nickel," **Univ. Stockholm Inst. Phys. Rep.**, **77-07** (1977).

[21] M. J. Cleare, E. G. Hughes, B. Jacoby and J. Pepys, "Immediate (Type I) Allergic Responses to Platinum Compounds," **Clin. Allergy**, **6** (1976) 183.

[22] K. Schwarz and J. Spallholm, "Growth Effects of Small Cadmium Supplements in Rats Maintained under Trace Element Controlled Conditions," **Fed. Proc.**, **35** (1976) 114.

[23] D. P. Higham, M. D. Scawen and P. J. Sadler, "Bacterial Cadmium-Binding Proteins," **Inorg. Chim. Acta**, **79** (1983) 140.

[24] R. J. Kula, D. T. Sawyer, S. I. Chin and C. M. Finley, "Nuclear Magnetic Resonance Studies of Metal Ethylenediaminetetraacetic Acid Complexes," **J. Am. Chem. Soc.**, **85** (1963) 2930.

[25] J. K. Nicholson, M. J. Buckingham and P. J. Sadler, "High Resolution 1-H NMR Studies of Vertebrate Blood and Plasma," **Biochem. J.**, **211** (1983) 605.

[26] R. A. Goyer and M. G. Cherian, "Tissue and Cellular Toxicology of Metals," Proceedings, International Symposium **"Clinical Chemistry and Chemical Toxicology of Metals,"** S. S. Brown, Ed., Elsevier: Amsterdam, 1977, p. 89.

[27] J. H. R. Kägi and M. Nordberg, Eds., **"Metallothionein,"** Elsevier: Amsterdam, 1979.

[28] M. Webb, Ed., **"The Chemistry, Biochemistry and Biology of Cadmium,"** Elsevier: Amsterdam, 1979.

[29] N. Glanville, D. Durnam and R. D. Palmiter, "Structure of Mouse Metallothionein I Gene and its mRNA," **Nature**, **292** (1981) 267.

[30] M. Vasak and J. H. R. Kägi, "Spectroscopic Properties of Metallothionein," in **"Metal Ions in Biological Systems,"** Vol. 15, H. Sigel, Ed., Marcel Dekker: New York, N.Y., 1983, p. 213.

[31] J. K. Nicholson, P. J. Sadler, K. Cain, D. E. Holt, M. Webb and G. E. Hawkes, "88-MHz Cd-113 NMR Studies of Native Rat Liver Metallothioneins," **Biochem. J.**, **211** (1983) 251.

[32] J. K. Nicholson, P. J. Sadler and J. Overnall, "Proton NMR Studies of Plaice Liver Metallothionein: Removal of Metal by EDTA," in **"Metals in Animals,"** Proc. ITE Symp. No. 12, Inst. Terrest. Ecology, Cambridge, 1984, p. 72.

[33] P. Aisen, "Some Physiochemical Aspects of Iron Metabolism," **Ciba Symposium, 51** (1977) 1.
[34] A. Daniels, R. J. P. Williams and P. E. Wright, "Nuclear Magnetic Resonance Studies of the Adrenal Gland and Some Other Organs," **Nature, 261** (1976) 321.
[35] B. J. Fuhr and D. L. Rabenstein, "NMR Studies of the Solution Chemistry of Metal Complexes. IX. The Binding of Cd, Zn, Pb and Hg to Glutathione," **J. Am. Chem. Soc., 95** (1973) 6944.
[36] D. L. Rabenstein and A. A. Isab, "Complexation of Zinc in Intact Human Erythrocytes Studied by H-1 Spin Echo NMR," **FEBS Lett., 121** (1980) 61.
[37] D. L. Rabenstein and A. A. Isab, "A Proton Nuclear Magnetic Resonance Study of the Interaction of Mercury with Intact Human Erythrocytes," **Biochem. Biophys. Acta, 721** (1982) 374.
[38] F. G. E. Pautard and R. J. P. Williams, "Biological Minerals," **Chem. Brit., 18** (1982) 14.
[39] P. J. Sadler, "Multimetal Centres: Powerful Devices in Biology and Pharmacology," **Inorg. Perspect. Biol. Med., 1** (1978) 233.
[40] D. Rhodes and A. Klug, "Sequence-Dependent Helical Periodicity of DNA," **Nature, 292** (1981) 387.
[41] F. W. Sunderman, Jr., "Carcinogenic Effects of Metals," **Fed. Proc., 37** (1978) 1790.
[42] F. W. Sunderman, Jr., K. S. Kasperzak, T. J. Lau, P. P. Minghetti, R. M. Maenza, N. Becker, C. Onkelinx and P. J. Goldblatt, "Effects of Manganese on Carcinogenicity and Metabolism on Nickel Subsulfide," **Cancer Res., 36** (1976) 1790.

XII

DESIGN OF METAL-SPECIFIC LIGANDS

D. E. Fenton*, U. Casellato and P. A. Vigato

*Department of Chemistry, The University, Sheffield S3 7HF, U.K.
Istituto di Chimica e Tecnologia dei Radioelementi del CNR
Corso Stati Uniti, I-35100 Padova, Italy

Ligands selective for specific metals will find many applications. The extraction of copper by selective ligands in form of macrocyclic, square-planar copper complexes is presented as an example. Macrocyclic ligands provide to the metal ions coordination cavities, the size of which can be adjusted by use of appropriate reagents and templating metal ions for the synthesis of ligands. The results of such synthetic efforts employing 2,5-diacetylpyridine and similar diketones, α,ω-diamines, α,ω-diaminoethers and α,ω-diaminothioethers, and several metal ions as templating agents are reviewed.

Ligands that are selective for specific metals are useful in a wide variety of areas such as hydrometallurgy, chelation therapy, and pollution control. Copper was chosen to illustrate the utility of metal-selective ligands.

Selective Extraction of Copper

Copper ores were originally deposited as sulfides. Weathering converted the sulfides in the surface layers of the ore body to oxides. The classic approach to the extraction of copper has been deep shaft mining followed by crushing and flotation of the sulfidic ore. Because oxides, phosphates, and silicates do not float, the sulfide ore can be selectively removed and then treated pyrometallurgically. This process produces sulfur dioxide, which upon release into the air damages the environment severely. The oxide ore cannot be treated with pyrometallurgical techniques. It is processed by leaching with sulfuric acid and solvent extraction of the leachate. Almost all high-grade copper ores have been consumed. Low grade ores must now be used, which together with the gangue from flotation can be leached economically with sulfuric acid. The resulting solutions can be processed by solvent extraction processes.

Several 2-hydroxyaryloximes 1 have been effectively used for the solvent extraction of copper(II) [1, 2] which is described by equation 1. Strict control of pH ensures the

$$CuSO_4(a) + 2HL(o) \underset{strip}{\overset{extract}{\rightleftharpoons}} CuL_2(o) + H_2SO_4(a) \qquad (1)$$

selective extraction of copper. The copper(II) complex 2 is pseudo-macrocyclic, square-planar. The complex is stabilized by intramolecular hydrogen bonds. The hydrophobic groups on the surface of the complex facilitate the

extraction of the copper complex from an aqueous to a kerosene medium [1]. Recently, the natural product cardanol, a compound in the liquid pressed from Cashew nut shells, was converted to the oxime of 4-pentadecylsalicylaldehyde 3, which proved to be a reagent for the efficient extraction of copper(II) [3]. Because Cashew nuts grow in India, East Africa, and South America, extensive use of compound 3 for the recovery of copper could bring financial benefits to Third World Countries.

The cyclic polyethers known as "crowns" or "coronands" are discriminatory reagents for alkali metals [4] and have been shown to selectively extract copper acting as second-sphere ligands [5]. When 18-crown-6 was added to a solution containing equimolar amounts of $[Co(NH_3)_6](PF_6)_3$ and $[Cu(NH_3)_4(H_2O)](PF_6)_2$ in aqueous ammonia, the 1:1 crown ether-tetraminecopper complex precipitated leaving cobalt in solution. When the filtered precipitate was treated with aqueous HCl, an aqueous solution of copper(II) and the insoluble ammonium crown-ether complex, which can be extracted with methylene chloride, was obtained. This cyclic process (eqn. 2) was found to separate in one pass more than 90 percent of the copper from cobalt [4].

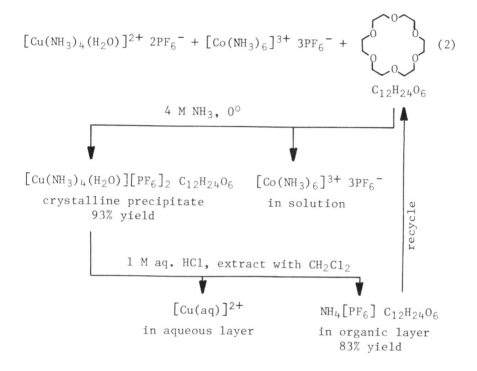

Macrocyclic Ligands as Reagents for the Selective Extraction of Metals

Several factors influence the design and choice of ligands for the selective extraction of metals. Macrocyclic ligands are specially promising as reagents for hydrometallurgical processes. The classification of metal ions into hard (class a) and soft (class b) acceptors [6, 7] and the corresponding classification of donors or ligand groups provide useful criteria for the design of ligands. The properties of a metal ion dictate to some extent the nature of the preferred donor. For instance, the "hard" alkali and alkaline metal ions prefer "hard", oxygen-based donors, and the "soft" ions of lead, cadmium, and mercury choose "soft" sulfur-based ligands. The features desired in a ligand and required for selectivity can be built into macrocycles formed through Schiff-base reactions (eqn. 3) using template techniques [8-10]. In general, α,ω-diaminoethers react with 2,6-diacetylpyridine to form polymers in the absence of metal ions. In the presence of metal ions the nature of the product is determined by factors such as the charge/radius ratio of the metal ion, the compatibility of the size of the ion with the size of the cavity in the macrocycle, the site geometry, and the nature of the donor atom. The influence of the charge/radius ratio is evidenced by the fact that alkaline earth cations help to produce Schiff-base macrocycles whereas alkali metal cations are ineffective. The nature of the ligand also influences the formation of macrocycles. α,ω-Diaminothioethers and diketones form macrocycles in the presence of lead ions but not in the presence of the "harder" alkaline earth ions, whereas diaminoethers produce macrocycles with both, lead and alkaline earth ions.

The dependence of the size of the ligand cavity on the ionic radius of the metal ion is shown by the results summarized in equation 3. Magnesium with an ionic radius of

65 pm guides the reagents to form the "1+1" macrocycle **6** (one molecule diketone **4** + one molecule α,ω-diaminoether **5**). With larger cations such as calcium (99 pm), strontium (113 pm), barium (135 pm) or lead (121 pm) the "2+2" macrocycle **7** with a larger and more flexible cavity is formed. A "1+1" macrocycle **8** is produced in the presence of lead, calcium, strontium or barium only if the α,ω-diaminoether **9** with three oxygen atoms in the chain is used. Direct formation of such macrocycles using transition metal ions as templates is rarely successful. These ions do not have the required ionic radii and – in addition – promote the hydrolysis of imine bonds in the macrocycles. Transition metals – by virtue of the orthogonality of their d-orbitals – impose also a more rigid geometry than do the spherically symmetric alkaline earth cations and thus introduce an additional constraint to the template control of syntheses. The macrocycle **7** forms a dinuclear lead complex that is kinetically labile. The lead ions can be exchanged for transition metal ions. The reaction of calcium-containing macrocycles with lead ions produces insoluble lead complexes. The stability of complexes of metal ions with macrocyclic ligands is strikingly dependent on the ratio ionic radius/radius of cavity. This dependence is clearly shown by the stability constants for the complexes of alkali metal ions with cryptand ligands [11, 12] (Table 1). An ion with a radius matching closely the radius of the ligand cavity will have greater stability than an ion with a smaller or larger radius. These differences in stability can be exploited to selectively extract a metal ion that forms a very stable complex with a cryptand from other metal ions that complex inefficiently. Such selectivity is observed, for example, with the naturally occurring antibiotics valinomycin and the actins, which perferentially complex with alkali metal ions [13].

If a ligand possesses flexibility, it may adjust its cavity to fit a metal ion. When lanthanide ions with large radii were used as templating agents in the preparation of macrocycles from 2,6-diformylpyridine and ethylenediamine, the rigid "2+2" macrocycle **10** (eqn. 4) was obtained. Lanthanide ions with small radii produced the flexible ligand **11** formed by addition of water to one of the imine groups of the rigid compound **10** (eqn. 4). The flexible macrocycle **11** can somewhat decrease its cavity size to better fit the small ion [14]. The use of barium ion as a templating agent in the reaction between 2,6-diacetyl-pyridine and 1,3-diamino-2-propanol (eqn. 5) gave the expected "2+2" macrocycle **12**. The smaller lead cation

Table 1. Stability Constants for Complexes of Alkali Metal Ions Formed with Cryptands of Various Cavity Size in Aqueous Medium[†]

				Log Stability Constant				
x	y	z	Radius of Cavity, pm	Li 86*	Na 112*	K 144*	Pb 158*	Cs 184*
0	0	1	80	**4.30**	2.80	2.0	2.0	2.0
1	1	0	115	2.50	**5.40**	3.95	2.55	2.0
1	1	1	140	2.0	3.90	**5.40**	4.35	2.0
1	1	2	180	2.0	2.0	2.2	2.05	**2.20**
2	2	1	210	2.0	2.0	2.0	0.7	2.0
3	3	3	240	2.0	2.0	2.0	0.5	2.0

[†]Compiled from ref. 12. The bold numbers are stability constants for the complexes with a close match between cavity size and radius of the cation.
*Ionic radii in picometers.

(4)

produced the contracted macrocycle **13**, which was formed by addition of propanol-hydroxy groups to the imine bonds [15]. Functional groups in macrocyclic compounds in addition to those forming the cycle might increase its selectivity for metal ions.

The geometry of the coordination site is particularly important for the selectivity of ligands for transition metal ions as illustrated by bovine erythrocyte superoxide dismutase [16]. The bimetallic site **14** of this enzyme provides a square-planar environment for copper(II) and a

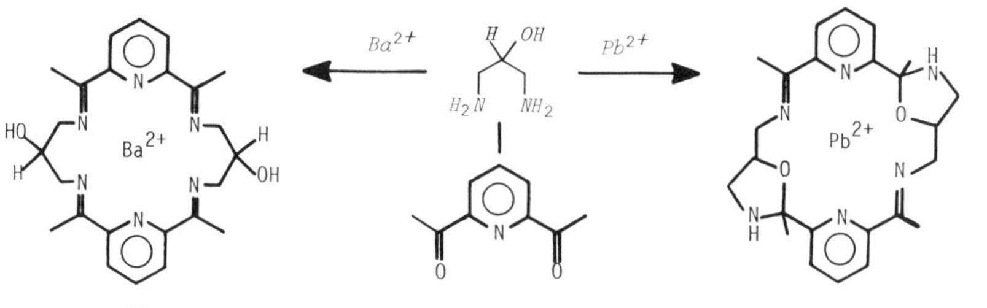

Design of Metal-Specific Ligands   279

(5)

12   **13**

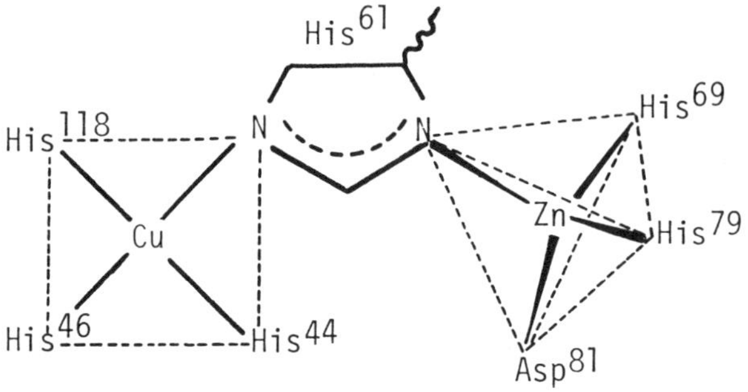

14

tetrahedral environment for zinc(II). The planar and tetrahedral coordination sites are bridged by an imidazolate unit. Copper(II) prefers square-planar environments as evidenced by its reaction with compartmental Schiff-base ligands **15, 16** and **17** derived from 1-(2-pyrryl)butane-1,3-dione [17, 18]. These compartmental ligands provide

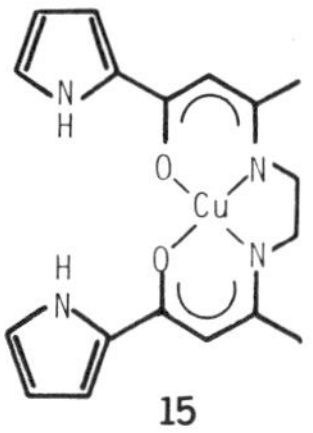

15 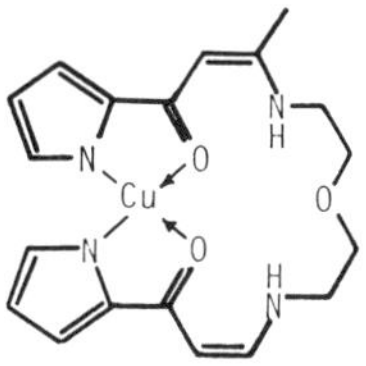   **16** 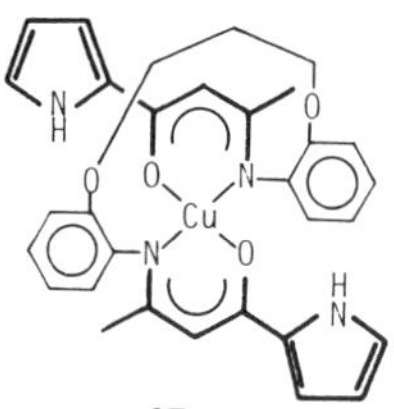   **17**

adjacent, dissimilar sets of donors, and a choice of environment and site geometry to metal ions. When the bridging unit in a simple Schiff-base derived from a diketone becomes longer, tetrahedral complexes are formed in preference to square planar complexes. Binuclear complexes

18 may also be produced but with retention of square-planar geometry at the metal. Copper(II) changes sites within compartmental ligands to retain planarity. In ligands **15** with short bridges such as ethylene units, copper occupies the inner site. In a ligand **16** with an extended bridge, copper prefers the outer, square-planar site to the tetrahedral inner environment. In ligands **17** with long bridges the metal returns to the inner site, which now has like donor atoms in trans-positions and a "strapping" bridge similar to the bridge present in a "strapped" porphyrin [19]. This arrangement avoids ligand-ligand interactions and provides copper with the preferred planar arrangement of the donor atoms. Binuclear, square-planar complexes of ligands with extended bridges are known [20].

We investigated homobinuclear copper(II) and dioxouranium(IV) complexes of compartmental Schiff-base ligands as reagents for the selective extraction of these metals. It might be possible to remove two metals simultaneously with one ligand from a solution [21]. The two metals might then be separated by selectively stripping one of the metals under appropriate conditions from the extract. Such a process is attractive because of its potential cost effectiveness. The complexes thus far investigated are very insoluble. The ligands must be modified to increase the solubility of the complexes. Tasker and Lindoy [22-25] synthesized a series of macrocycles **19**, in which the number

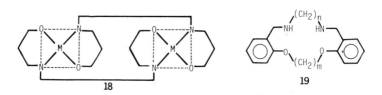

of methylene groups joining the nitrogen and oxygen atoms determine the size of the cavity. The stability constants for the nickel(II), copper(II), and zinc(II) complexes and the selectivity of the ligands for these metal ions change with the size of the cavity (Table 2). The stabilities of the nickel complexes are clearly related to the cavity sizes, because the nickel ion is forced to reside in the cavity in these pseudo-octahedral complexes. The irregular five-coordinated copper complexes and the pseudo-tetrahedral zinc complexes have their metal ions outside the cavity.

Table 2. *Log K Values for Complexes of Nickel(II), Copper(II) and Zinc(II) with Macrocycles, 19 at 25° in 95 Percent Methanol [23, 24]*

Ligand 19		Ring Size	Log K values		
m	n		Ni	Cu	Zn
2	2	14	3.7	8.2	3.0
3	2	15	5.4	7.2	4.1
3	3	16	5.8	7.7	4.3
3	4	17	3.5	7.2	4.1

Therefore, the stabilities of the copper and zinc complexes are not strongly influenced by the dimension of the cavity. On this basis ligands can be designed that discriminate against nickel(II). When the ligand provides only a poor fit for nickel, the nickel complex will be less stable than the copper complex. The stability differences will be larger than estimated from the Irving-Williams order.

The first steps appear to have been taken toward the design and preparation of macrocyclic ligands that have selectivity toward specific transition metal cations. Selective extraction procedures for metal ions are of great importance to industry and the environment.

References

[1] R. Price and J. A. Tumilty, "An Interpretation of Some Aspects of Solvent Extraction as Related to the Extraction of Copper using o-Hydroxyaryl Oximes," in **"Hydrometallurgy,"** G. A. Davis and J. B. Scuffham, Eds., **Institute of Chemical Engineers Symp. Series (London),** 42 (1975) 18.1.

[2] A. J. van der Zeeuw, "Shell Metal Extractant 529- A New Selective Extractant for Copper and Nickel," in **"Extractive Metallurgy of Copper,"** Vol. 2, J. C. Yannopoulos and J. C. Agarwal, Eds., **Metall. Soc. A.I.M.E.,** New York, (1976) 1039.

[3] J. H. P. Tyman, "m-Alkylphenol Derivatives and Their Use in Copper Extraction," U.K. Patent Application GB 2,104,516A (1982), **Chem. Abstr.,** 99 (1983) 5349.

[4] C. J. Pedersen, "Cyclic Polyethers and Their Complexes with Metal Salts," **J. Am. Chem. Soc.,** 89 (1967) 7017.

[5] H. M. Colquhoun, D. F. Lewis, J. F. Stoddart and D. J.

Williams, "Crown Ethers as Second-Sphere Ligands. The Interactions of Transition-Metal Ammines with 18-Crown-6 and Dibenzo-18-Crown-6." **J. Chem. Soc., Dalton Trans.**, (1983) 607.

[6] R. G. Pearson, "Hard and Soft Acids and Bases," **J. Am. Chem. Soc.**, 85 (1963) 3533.

[7] S. Ahrland, J. Chatt and N. R. Davies, "The Reactive Affinities of Ligand Atoms for Acceptor Molecules and Ions," **Quart. Rev. Chem. Soc.**, 12 (1958) 265.

[8] D. H. Cook and D. E. Fenton, "Alkaline-Earth Metal Complexes of Macrocyclic Schiff-Base Derived from Furan-2,5-dicarbaldehyde," **J. Chem. Soc., Dalton Trans.**, (1979) 810.

[9] D. H. Cook, D. E. Fenton, M. G. B. Drew, A. Rodgers, M. McCann and S. M. Nelson, "Mononuclear and Dinuclear Lead(II) Complexes of Macrocyclic Schiff Bases," **J. Chem. Soc., Dalton Trans.**, (1979) 414.

[10] N. A. Bailey, M. M. Eddy, D. E. Fenton, G. Jones, S. Moss and A. Mukhopadhyay, "The Non-Template Synthesis of Dinucleating Macrocyclic Schiff Bases Derived from Thiophen, and their Di-Silver Complexes; X-Ray Crystal Structures of two of the Products," **J. Chem. Soc., Chem. Commun.**, (1981) 628.

[11] J. M. Lehn, "Cryptates: The Chemistry of Macropolycyclic Inclusion Complexes," **Accts. Chem. Res.**, 11 (1978) 49.

[12] J. M. Lehn and J. P. Sauvage, **J. Chem. Soc., Chem. Commun.**, (1971) 446.

[13] D. E. Fenton, "Across the Living Barrier," **Chem. Soc. Rev.**, 6 (1977) 325.

[14] K. K. Abid and D. E. Fenton, "Lanthanide Complexes of Some Macrocyclic Schiff Bases Derived from Pyridine-2,6-dicarboxaldehyde and α,ω-Primary Diamines," **Inorg. Chim. Acta**, 95 (1984) 119.

[15] N. A. Bailey, D. E. Fenton, I. T. Jackson, R. Moody and C. Rodriguez de Barbarin, "Metal Ion Controlled Ring Contraction to Produce an Oxazolidine-Containing Schiff Base Macrocycle and the X-Ray Structure of the Pb(CNS)$_2$ Complex," **J. Chem. Soc., Chem. Commun.**, (1983) 1463.

[16] J. S. Richardson, K. A. Thomas, B. H. Rubin and D. C. Richardson, "Crystal Structure of Bovine Cu, Zn Superoxide Dismutase at 3Å Resolution: Chain Tracing and Metal Ligands," **Proc. Natl. Acad. Sci. (USA)**, 72 (1975) 1349.

[17] H. Adams, N. A. Bailey, D. E. Fenton and M. S. Leal Gonzales, "Compartmental Ligands. Part 5. Copper(II) Complexes of Schiff Bases Derived from 1-(Pyrrol-2-yl)-butane-1,3-dione: Crystal and Molecular Structure of

[3,11-Dimethyl-1,13-bis(pyroll-2-yl)-7-oxa-4,10-diaza-trideca-2,11-diene-1,13-dionato(2-)-N,N',O,O']copper-(II)," **J. Chem. Soc., Dalton Trans.**, (1983) 1345.

[18] N. A. Bailey, A. Barrass, D. E. Fenton, M. S. Leal Gonzales, R. Moody and C. O. Rodriguez de Barbarin, "Compartmental Ligands. Part 10. Synthesis and Crystal Structure of the Copper(II) Complexes of Schiff Bases Derived from 1-(Pyrrol-2-yl)butane-1,3-dione and 1,3-Bis(2-aminophenoxy)propane, and from Pyrrole-2-carbaldehyde and 3,6-Dioxaoctane-1,8-diamine," **J. Chem. Soc., Dalton Trans.**, (1984), in press.

[19] J. E. Baldwin and P. Perlmutter, "Bridged, Capped, and Fenced Porphyrins," **Topics in Curr. Chem.**, 121 (1984) 181.

[20] D. L. Barber, S. J. Loeb, J. W. L. Martin, N. C. Payne and C. J. Willis, "Fluorinated Alkoxides. 16. Structure of a Dinuclear Imino Alkoxy Complex of Copper(II)," **Inorg. Chem.**, 20 (1981) 272.

[21] R. C. Coombes, D. E. Fenton, P. A. Vigato, U. Casellato and M. Vidali, "Mono- and Homobinuclear Dioxouranium(VI) Complexes of Various Compartmental Ligands," **Inorg. Chim. Acta.**, 54 (1981) L155.

[22] K. R. Adam, L. F. Lindoy, R. J. Smith, G. Anderegg, K. Henrick, M. McPartlin and P. A. Tasker, "Macrocyclic Ligand Control of the Thermodynamic Stabilities of Nickel Complexes Containing 14- to 17-Membered Macrocyclic Rings. X-Ray Crystal Structure of the Trans-Nickel(II) Bromide Complexes of the 16-Membered Ring." **J. Chem. Soc. Chem. Commun.**, (1979) 812.

[23] A. Ekstrom, L. F. Lindoy and R. J. Smith, "Cyclic Ligand Control of Kinetic Lability. Kinetics of Dissociation of Nickel(II) Complexes of a Series of O_2N_2-Donor Macrocycles in Acid," **Inorg. Chem.**, 19 (1980) 724.

[24] K. R. Adam, G. Anderegg, L. F. Lindoy, H. C. Lip, M. McPartlin, J. H. Rea, R. J. Smith and P. A. Tasker, "Metal-Ion Recognition by Macrocyclic Ligands. Synthetic, Thermodynamic, Kinetic, and Structural Aspects of the Interaction of Copper(II) with 14- to 17-Membered Cyclic Ligands Containing an O_2N_2-Donor Set." **Inorg. Chem.**, 19 (1980) 2956.

[25] L. F. Lindoy, H. C. Lip, J. H. Rea, R. J. Smith, K. Henrick, M. McPartlin and P. A. Tasker, "Metal-Ion Recognition. Interaction of O_2N_2-Donor Macrocycles with Cobalt(II), Zinc(II), and Cadmium(II) and Structure of the Zinc Complex of One Such 15-Membered Macrocycle." **Inorg. Chem.**, 19 (1980) 3360.

DISCUSSION

K. N. Raymond (University of California, Berkeley): Cavity size is important for the design of metal-specific ligands. However, cavity size alone does not determine metal-specificity of a ligand. This statement by Dr. Fenton deserves amplification. Metal specificity is also influenced by the free energy of complex formation. As the charge/radius ratio of the cation increases, the enthalpy of complex formation becomes dominant in establishing the free energy. The enthalpy is largely due to charge neutralization.

XIII

MIXED-METAL COMPLEXES OF BIOFUNCTIONAL LIGANDS: FORMATION, STABILITY AND POSSIBLE ROLE IN MODEL SYSTEMS WITH METAL ION OVERLOADS

Paola Amico*, Giuseppe Arena†, Pier Giuseppe Daniele*, Giorgio Ostacoli*, Enrico Rizzarelli† and Silvio Sammartano§

*Istituto di Analisi Chimica Strumentale
Universita di Torino, I-10125 Torino, Italy
†Istituto Dipartimentale di Chimica e Chimica Industriale
Universita di Catania, I-95125 Catania, Italy
§Istituto di Chimica Analitica
Universita di Messina, I-98100 Messina, Italy

Mixed-metal complexes of copper(II), cadmium(II), zinc(II), and cobalt(II) with some biological ligands [L-histidine, histamine, L-carnosine, 3-(3,4-dihyroxyphenyl)-L-alanine, L-aspartyl-L-alanyl-L-histidyl-N-methylamide, DL-isoserine and citric acid] are discussed. The protonation constants of the ligands and the formation constants of their binary and mixed-metal complexes were determined potentiometrically. The mixed metal complexes are generally more stable than statistically predicted with respect to their parent binary complexes. This work shows that homobinuclear species need not be present in order for heterobinuclear complexes to form. These results are not consistent with previous hypothesis that require ligands possessing donor sites of different softness, a "hard" metal ion, and a "soft" metal ion for the formation of mixed-metal complexes. Because in-vitro studies and computer simulations show that mixed-metal complexes form in the presence of high concentrations of metal ions, such complexes may play a role in systems with metal ion overloads.

In 1959 Booman and Holbrock [1] described the first, characterized heterobinuclear aluminum-uranium-citrate complex. The change of the properties of uranium citrate solutions on addition of aluminum(III) served as the basis for a controlled-potential coulometric method for the specific determination of uranium(VI). Studies related to the separation of nuclear fission products provided evidence for heteronuclear chromium(III)-cerium(III) hydrocarboxylates [2]. The possibility, that heteronuclear rare earth complexes interfer with the separation of individual rare earths, was explored by Russian chemists [3-5]. The formation of heteronuclear complexes has been shown to be rather widespread. The awareness of their presence and their roles in analytical and bioinorganic chemistry is growing as the number of reported species increases. However, the reasons for the higher stability of heteronuclear complexes in comparison with homobinuclear complexes are still not fully understood.

It has been proposed that mixed metal complexes may play a role in in-vivo metal-metal stimulation phenomena. Dissimilar metal ions may stimulate through their attachment to different sites of a fairly large ligand and thus change the pH dependent distribution of the species expected for the parent binary complexes [6]. To form such heteronuclear complexes, a ligand with two complexing sites of different softness, a "hard" metal ion, and a "soft" metal ion are needed [7, 8]. Heteronuclear complexes of a particular type named "wishbone" complexes may serve as useful models for the study of natural systems, in which concerted effects of two metal ions are involved. In our laboratories we have been studying a number of mixed ligand complexes for more than a decade with a focus on ligands that occur naturally or may serve as models for natural systems [9, 10]. These studies illuminated the role of mixed ligand complexes in mimicking systems of biological interest [11] and their influence on the distribution of metal species in biofluids [12, 13].

The paucity of quantitative data for heteronuclear complexes caused us to investigate these complexes to obtain a better understanding of the forces leading to their formation, to explore their influence on species distribution, and to identify the influences metal ions exert on each other. Metal ions in natural systems do influence each other synergistically [14] or antagonistically. For instance, the uptake and transport of cadmium in plants is reduced at increased zinc or potassium concentrations [15]. We believe, that heteronuclear complexes may serve as useful models for compartments [16] and other natural systems, in which large or, as it is often the case, excessive concentrations of metal ions produce undesired effects.

Mixed-Metal Complexes

We report in this paper our results for mixed metal complexes with L-histidine, histamine, L-carnosine, 3-(3,4-dihydroxyphenyl)-L-alanine, L-aspartyl-L-alanyl-L-histidyl-N-methylamide, DL-isoserine, and citrate as ligands. Work with malic acid, glycyl-L-hystidine, and ATP is in progress. Preliminary results indicate the existence of heterobinuclear complexes. The ligands were chosen because they promised to provide information about the type of donor atoms required for the formation of heterobinuclear complexes. In addition, these ligands were expected to produce heterobinuclear complexes, even if homobinuclear species did not exist. All the metal ions studied were transition metal ions. All systems were investigated potentiometrically at 25° and an ionic strength of 0.1 mol/L. The experiments with carnosine were carried out at 37° and 0.15 mol/L ionic strength. Histidine (his) and histamine (hm) systems were also investigated calorimetrically [17].

The results for the histidine systems are summarized in Table 1. Four types of heterobinuclear copper-metal

Table 1. Formation Contants of L-Histidine (his)* Mixed-Metal Complexes at 25° and 0.1 mol/L Ionic Strength Adjusted with Potassium Nitrate [18]

Reaction	log β
$Cu^{2+} + Ni^{2+} + H^+ + 2\ his^- \rightleftharpoons [CuNi(his^0)(his^-)]^{3+}$	25.56
$Cu^{2+} + Zn^{2+} + H^+ + 2\ his^- \rightleftharpoons [CuZn(his^0)(his^-)]^{3+}$	25.69
$Cu^{2+} + Cd^{2+} + H^+ + 2\ his^- \rightleftharpoons [CuCd(his^0)(his^-)]^{3+}$	25.45
$Cu^{2+} + Ni^{2+} + 2\ his^- \rightleftharpoons [CuNi(his^-)_2]^{2+}$	21.20
$Cu^{2+} + Zn^{2+} + 2\ his^- \rightleftharpoons [CuZn(his^-)_2]^{2+}$	20.78
$Cu^{2+} + Cd^{2+} + 2\ his^- \rightleftharpoons [CuCd(his^-)_2]^{2+}$	20.73
$Cu^{2+} + Ni^{2+} + 2\ his^- \rightleftharpoons [CuNi(his^-)(his^{--})]^+ + H^+$	14.00
$Cu^{2+} + Zn^{2+} + 2\ his^- \rightleftharpoons [CuZn(his^-)(his^{--})]^+ + H^+$	13.02
$Cu^{2+} + Cd^{2+} + 2\ his^- \rightleftharpoons [CuCd(his^-)(his^{--})]^+ + H^+$	12.65
$Cu^{2+} + Ni^{2+} + 2\ his^- \rightleftharpoons [CuNi(his^{--})_2]^0 + 2H^+$	5.45
$Cu^{2+} + Zn^{2+} + 2\ his^- \rightleftharpoons [CuZn(his^{--})_2]^0 + 2H^+$	5.48
$Cu^{2+} + Cd^{2+} + 2\ his^- \rightleftharpoons [CuCd(his^{--})_2]^0 + 2H^+$	<3.9

*his^0: neutral histidine; his^-: histidine that lost one proton; his^{--}: histidine that lost two protons.

complexes are formed: complexes with one neutral and one mononegative histidine ligand, complexes with two mononegative histidine ligands, complexes with one mononegative and one dinegative histidine ligand, and complexes with two dinegative histidine ligands. The main species are [CuM(his$^-$)$_2$] and [CuM(his$^-$)(his^{--})] [18]. The protonated species [CuM(his^0)(his$^-$)] reach their maximal concentrations at pH 4 and the deprotonated species [CuM(his$^-$)(his^{--})] above pH 7. The species [CuM(his^{--})$_2$] begin to form at pH >7. These systems cannot be investigated above pH 8 because of precipitation [17]. Table 2 shows the resuls for histamine. Neither protonated species nor copper-nickel heteronuclear complexes were detected.

Table 2. Formation Constants of Histamine (hm)* Mixed-Metal Complexes at 25° and 0.1 mol/L Ionic Strength Adjusted with Potassium Nitrate [17]

Reaction	log β
Cu^{2+} + Zn^{2+} + 2 hm^0 ⇌ [CuZn(hm^0)$_2$]$^{4+}$	18.00
Cu^{2+} + Cd^{2+} + 2 hm^0 ⇌ [CuCd(hm^0)$_2$]$^{4+}$	18.36
Cu^{2+} + Zn^{2+} + 2 hm^0 ⇌ [CuZn(hm^0)(hm$^-$)]$^{3+}$ + H$^+$	10.6
Cu^{2+} + Cd^{2+} + 2 hm^0 ⇌ [CuCd(hm^0)(hm$^-$)]$^{3+}$ + H$^+$	10.64

*hm^0: neutral histamine; hm$^-$: deprotonated histamine.

Copper(II)-zinc(II) and copper(II)-cadmium(II) heterobinuclear complexes are formed with L-carnosine (car) in suitable concentrations and ratios [19]. Significant differences exist between the experimental titration curves and the curves calculated with the assumption that only homonuclear binary complexes form (Table 3). These differences disappear when the two heterobinuclear species [CuM(car^{--})]$^{2+}$ and [CuM(car^{---})]$^+$ with appropriate stability constants (Table 4) are included in the calculation. The homobinuclear species [Cu$_2$(car^{--})]$^{2+}$ but not the corresponding zinc or cadmium complex was detected. These results suggest, that the presence of several donor groups within a ligand favors the binding of another metal ion, when the two coordination centers are sufficiently apart, and indicate that an "encouragement" factor is promoting the formation of heterobinuclear complexes.

Heteronuclear species were also reported for 3-(3,4-dihydroxyphenyl)-L-alanine (L-DOPA) [20]. The complexes [CuM(L^{--})]$^{2+}$ and [CuM(L^{---})]$^+$ were identified as the main

Table 3. *Experimental and Calculated Hydroxide Ion Concentrations for the Titration of Copper-Zinc-L-Carnosine and Copper-Cadmium-L-Carnosine Systems* [19]*

pH	[OH]calc	[OH]exp	Δ%**	[OH]calc	[OH]exp	Δ%**
4.0	2.95	2.96	0.33	2.95	2.95	–
4.5	3.71	3.75	1.06	3.71	3.77	1.59
5.0	5.77	6.14	6.02	5.77	6.12	5.71
5.5	8.77	9.32	5.95	8.77	9.30	5.69
6.0	10.94	11.19	2.23	10.94	11.26	2.84
6.5	11.70	11.84	1.18	11.70	12.15	3.70
7.0	11.91	12.09	1.48	11.91	13.01	8.45
7.5	11.98	12.42	3.54			

*Initial carnosine concentration: 9.00 mM.

** $\Delta\% = \dfrac{[OH]exp - [OH]calcd}{[OH]exp} \times 100$

†[OH]calc is the hydroxide concentration calculated with neglect of binuclear species.

species (Table 4) by comparing appropriately calculated with experimental titration curves [20]. Should the catechol group and the amino acid moiety in L-DOPA, and perhaps the metal ions act independently, then the formation constants

Table 4. *Formation Constants for Cu-M (M = Zn, Cd) Mixed-Metal Complexes with L-Carnosine [19]† or 3-(3,4-Dihydroxyphenyl)-L-Alanine (L-DOPA) [20] as Ligands*

Reaction	log β Zn²⁺	log β Cd²⁺
$Cu^{2+} + M^{2+} + Car^- \rightleftharpoons [CuM(Car^{--})]^{2+} + H^+$	4.42*	4.47*
$Cu^{2+} + M^{2+} + Car^- \rightleftharpoons [CuM(Car^{---})]^+ + 2H^+$	-2.46*	-3.43*
$Cu^{2+} + M^{2+} + L^- \rightleftharpoons [CuM(L^{--})]^{2+} + H^+$	7.98**	8.15**
$Cu^{2+} + M^{2+} + L^- \rightleftharpoons [CuM(L^{---})]^+ + 2H^+$	1.13**	0.40**

†Car⁻: carnosine that lost one proton; Car⁻⁻: carnosine that lost two protons. Car⁻⁻⁻: carnosine that lost three protons; L⁻: L-DOPA that lost one proton; L⁻⁻: L-DOPA that lost two protons; L⁻⁻⁻: L-DOPA that lost three protons.

*Ionic strength 0.15 M potassium nitrate at 37°.
**Ionic strength 0.10 M potassium nitrate at 25°.

for the heterobinuclear Cu-Zn and Cu-Cd L-DOPA complexes are expected to be equal to the sum of the constants for the corresponding complexes with catechol and alanine (Table 5). However, the formation constants for the heterobinuclear L-DOPA complexes were found to be 3 to 4 log units larger (Table 4) than the values calculated under the assumption of "independent action" (Table 5).

Table 5. Formation Constants for Catechol and Alanine Complexes of Copper, Zinc, and Cadmium

M	log Formation Constant $(\log\beta)$*		$\log\beta[Cu(c^-)]^+$ $+\log\beta[M(a^-)]^+$	$\log\beta[Cu(a^-)]^+$ $+\log\beta[M(c^-)]^+$
	$[M(c^-)]^+$	$[M(a^-)]^+$		
Cu(II)	0.96	8.13	9.09	9.09
Zn(II)	-3.10	4.58	5.44	5.03
Cd(II)	-4.80	3.96	4.92	3.33

*c^-: catechol (1,2-dihydroxybenzene) that lost one proton; a^-: alanine that lost the carboxyl proton.

Heterobinuclear species of Cu(II) with Zn(II) or Cd(II) were found for the tripeptide L-aspartyl-L-alanyl-L-histidyl-N-methylamide [21] (eqn. 1). Such a complex had

$$Cu^{2+} + M^{2+} + L^- \rightleftharpoons [CuM(L^{---})]^+ + 2H^+ \quad (1)$$

$\log\beta$ at 25^0 and 0.1 M ionic strength (KNO$_3$):
Cu-Zn 0.39; Cu-Cd 0.46 [21]

L^-: tripeptide that lost one proton
L^{---}: tripeptide that lost three protons

never been found before, probably because the experiments were carried with metal/ligand ratios <1 [11, 22, 23]. The zinc or cadmium are probably bound to the carboxylate group of aspartyl. However, the participation of the carboxylate group must be confirmed by methods other than potentiometry. Definite evidence for carboxylate bonding would prove that the carboxylate group serves as a donor site not only in the rather rigid "wishbone" ligand [1,4-bis(2,5,5-tris(carboxymethyl)-2,5-diazapentyl)benzene] [8] but also in flexible tripeptide ligands, in which the coordination sites are sufficiently apart.

Heterobinuclear complexes of the type [MM'(IS^{---})]$^+$ with DL-isoserine were also detected. Homobinuclear complexes of this ligand with Zn(II), Cu(II) [24], Cd(II), and Co(II) [25] are known. The formation constants for these complexes are listed in Table 6. If equation 2 (L = isoserine) is

Table 6. Formation Constants of Homonuclear and Heterobinuclear Complexes of Cadmium, Cobalt, and Zinc with DL-Isoserine* at 25° and 0.1 M Ionic Strength Adjusted with Potassium Nitrate [25]

Reaction	log β
$Cd^{2+} + I^- + H^+ \rightleftharpoons [Cd(IS^0)]^{2+}$	11.14
$2Cd^{2+} + 2I^- \rightleftharpoons [Cd_2(IS^{--})_2]^0 + 2H^+$	-6.62
$Cd^{2+} + 2I^- \rightleftharpoons [Cd(IS^{--})_2]^{2-} + 2H^+$	-11.75
$Zn^{2+} + I^- + H^+ \rightleftharpoons [Zn(IS^0)]^{2+}$	11.03
$2Zn^{2+} + 2I^- \rightleftharpoons [Zn_2(IS^{--})_2]^0 + 2H^+$	-3.50
$Zn^{2+} + 2I^- \rightleftharpoons [Zn(IS^-)_2]^{2-} + 2H^+$	-9.1
$Co^{2+} + I^- + H^+ \rightleftharpoons [Co(IS^0)]^{2+}$	10.4
$2Co^{2+} + 2I^- \rightleftharpoons [Co_2(IS^{--})_2]^0 + 2H^+$	-5.61
$Zn^{2+} + Cd^{2+} + 2IS^- \rightleftharpoons [ZnCd(IS^{--})_2]^0 + 2H^+$	-4.33
$Co^{2+} + Cd^{2+} + 2IS^- \rightleftharpoons [CoCd(IS^{--})_2]^0 + 2H^+$	-4.97
$Co^{2+} + Zn^{2+} + 2IS^- \rightleftharpoons [CoZn(IS^{--})_2]^0 + 2H^+$	-3.74

*IS^0: neutral isoserine; IS$^-$: isoserine that lost one proton; IS^{--}: isoserine that lost two protons.

$$\Delta \log \beta = 2 \log \beta_{[MM'(L^{--})_2]} - \log \beta_{[M_2(L^{--})_2]} - \log \beta_{[M'_2(L^{--})_2]} \quad (2)$$

used to evaluate the stability of the heterobinuclear complexes with respect to the homobinuclear complexes, positive values (1.46 for Zn-Cd, 2.29 for Co-Cd, 1.63 for Co-Zn) are obtained for Δ log β. These positive values show, that the heterobinuclear complexes are more stable than the homobinuclear complexes.

The experimental alkalimetric titration curve of the Cu(II)-Ni(II)-citrate system [26] (Fig. 1) does not coincide with the curve calculated under the assumption, that mixed-metal complexes are not present. However, the curves are superimposible when heterobinuclear complexes are included in the calculation. Mixed-metal complexes do not exist in

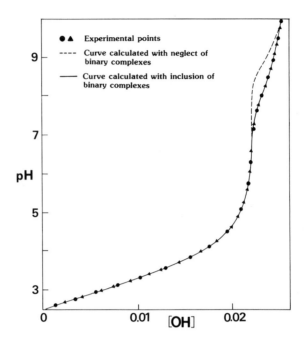

Figure 1. Alkalimetric titration curve for Cu(II)-Ni(II)-citrate system. Reproduced with permission from Ref. 26.

an acidic medium. They begin to form above pH 7 (Fig. 1). The formation constants for heterobinuclear citrate complexes [27] are listed in Table 7, and the constants for homonuclear citrate complexes [26] are reported in Table 8.

Table 7. Formation Constants of Mixed-Metal Citrate* Complexes at 25° and 0.1 M Ionic Strength Adjusted with Potassium Nitrate

| Metal | | Log Formation Constant | |
M	M'	$[MM'L_2^{4-}]^{4-}$	$[MM'L^{3-}L^{4-}]^{3-}$
Cu	Ni	1.55 [26]	–
Cu	Zn	1.51 [26]	–
Ni	Zn	−2.94 [26]	–
Cd	Mn	−5.75	–
Cd	Ni	−4.22	3.65
Cd	Zn	−3.30	4.12

*L^{3-}: citric acid that lost three protons; L^{4-}: citric acid that lost four protons.

Table 8. *Formation Constants of Binary Citrate* Complexes at 25° and 0.1 M Ionic Strength Adjusted with Potassium Nitrate*

Metal M	$\left[\begin{array}{c}M\\L^{3-}\end{array}\right]^{-}$	$\left[\begin{array}{c}M\\L^{2-}\end{array}\right]^{0}$	$\left[\begin{array}{c}M_2\\L_2^{3-}\end{array}\right]^{2-}$	$\left[\begin{array}{c}M_2\\L^{3-}\\L^{4-}\end{array}\right]^{3-}$	$\left[\begin{array}{c}M_2\\L_2^{4-}\end{array}\right]^{4-}$	$\left[\begin{array}{c}M\\L^{4-}\end{array}\right]^{2-}$
Cu(II)	–	9.47	14.60	10.75	6.00	–
Ni(II)	5.30	8.84	–	–	-4.71	–
Zn(II)	4.83	8.43	–	–	-2.94	–
Cd(II)	3.65	7.80	–	–	–	-3.81
Mn(II)	3.81	8.15	–	–	-6.28	–

*L^{2-}: citric acid that lost two protons; L^{3-}: citric acid that lost three protons; L^{4-}: citric acid that lost four protons.

Based on equation 2 and the appropriate formation constants from Tables 7 and 8 the heterobinuclear complexes $[MM'(L^{4-})_2]^{4-}$ are generally more stable than the homobinuclear complexes ($\Delta \log \beta$ for Cu-Ni 1.81, for Cu-Zn -0.04, for Ni-Zn 1.77). The importance of the heterobinuclear complexes is well expressed by the distribution curves for citrate (Fig. 2). At pH 8 approximately 50 percent of citrate is part of the heterobinuclear complex in the Cu(II)-Ni(II) system.

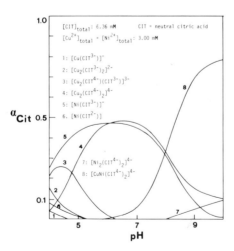

Figure 2. *The distribution of citrate among metal complexes in the Cu(II)-Ni(II)-citrate system. Reproduced with permission from Ref. 26.*

Conclusions

Lack of sufficient data prevents a detailed discussion of the conditions, under which heterobinuclear complexes form. However, a few generalizations are possible. The presence of homobinuclear species does not seem to be mandatory for the formation of heterobinuclear complexes. Heteronuclear complexes were found in systems [17, 19-21], in which only one of the two metal ions formed homobinuclear species. For heterobinuclear complexes to form, a ligand does not need to have a "hard" and a "soft" coordination site [6]. Heterobinuclear complexes were detected in systems that had multidentate ligands with only one type of donor atom. Such a ligand with only oxygen as donor atom is citrate [26]. In the systems thus far investigated the heterobinuclear complexes are more stable than their parent homobinuclear complexes. The available data [7, 8, 21] indicate, that the two coordinating centers in a ligand do not influence each other, when their separation is sufficiently large. However, when flexible ligands have coordination sites that are not too distant from each other they act synergistically in favor of the heterobinuclear complex.

In the systems thus far investigated heterobinuclear complexes begin to form at pH $\geq$ 6. They are the predominant species in the pH range 6 to 9. Neglect of these mixed-metal complexes may produce an "unrealistic" species distribution particularly in fluids that hyperaccumulated metal ions for physiological or pathological reasons. "Normal" aqueous media in nature often have a pH between 7 and 8. At these pH values and at high concentrations of metal ions, mixed-metal species might be important. Therefore, such species must be considered when the transport and adsorption of metal ions in "overloaded" systems are investigated. A model of a multi-metal ion system that considers only homonuclear species is inadequate. An overloaded citrate system with Pb(II) (9.0 mM), Zn(II) (5.0 mM), Cu(II) (2.0 mM), Cd(II) (0.04 mM) and citrate (2.5 mM) at pH 8 was simulated. The calculations indicated that as much as fifteen percent of the total citrate may be associated with heterobinuclear species. Concentrations of metal ions similar to those used in these calculations may be found in regions polluted by agricultural activities and waste disposal. Localized, significant concentrations of metal ions may be achieved through partitioning of metal ions among different chelating agents.

Metal ions were observed to compete [28, 29] for uptake by plants, but also to synergistically enhance their uptake [15]. A model of such a system neglecting heterobinuclear species can account only for competition, whereas a model with heterobinuclear species can also explain synergistic effects. Chemical modeling is important for environmental sciences and will certainly help to understand the role of metal ions in the environment.

The area of mixed-metal complexes is only a small part of the growing field of environmental chemistry that attempts to identify and quantitate metal species as a necessary sequel to the determination of "total" metal concentrations. The "total" concentration of a metal is not a measure of its bioavailability. Bioavailability determines to a large degree the effects that a metal may exert. The chemical and biological interactions in systems with a metal overload should be studied with consideration of mixed-metal complexes. Such investigations will help to understand the generally very complex environmental processes.

References

[1] G. L. Booman and W. B. Holbrook, "An Extraction, Controlled-Potential Coulometric Method Specific for Uranium," **Anal. Chem.**, **31** (1959) 10.

[2] W. W. Schulz, J. E. Mendel and J. F. Phillips, Jr., "Evidence for a Chromium(III)-Cerium(III)-Citrate Complex," **J. Inorg. Nucl. Chem.**, 28 (1966) 2399.

[3] A. V. Stepanov, V. P. Shvedov and A. P. Rozhnov, "Complex Formation of Cerium(III) and Europium(III) with Tartaric Acid," **Russ. J. Inorg. Chem.**, **10** (1965) 750.

[4] N. S. Polyektov and V. T. Mishchenko, "Mixed Sulfosalicylate Complexes of the Rare Earth Elements," **Russ. J. Inorg. Chem.**, **9** (1964) 986.

[5] R. S. Lauer, V. T. Mishchenko and N. S. Polyektov, "Mixed Polynuclear Citrate Complexes of Neodymium and Some Other Metals," **Russ. J. Inorg. Chem.**, **13** (1968) 1246.

[6] M. L. D. Touche and D. R. Williams, "Thermodynamic Considerations in Co-ordination. Part XXV. Formation of Ternary Complexes Containing two Dissimilar Metal Ions and the Implication for Metal-Metal Stimulation Phenomena in-vivo," **J. Chem. Soc. Dalton Trans.** (1976) 1355.

[7] C. Y. Ng, R. Motekaitis and A. E. Martell, "New Multi-

dentate Ligands. 18. Synthsis of 1,4-Bis[bis(2-aminoethyl)aminomethyl]benzene: Binuclear Chelating Tendencies and Mixed-Ligand Binuclear Chelate Formation," **Inorg. Chem., 18** (1979) 2982.

[8] C. Y. Ng, A. E. Martell and R. Motekaitis, "1,4-Bis[2,5,5-tris(carboxymethyl)-2,5-diazapentyl]-benzene (PXED3A): Synthesis, Binuclear Chelating Tendencies, and Iron(III) μ-Oxo Bridge Formation," **Inorg. Chem., 22** (1983) 721.

[9] G. Ostacoli, "Mixed Complexes," in **"Bioenergetics and Thermodynamics: Model Systems,"** A. Braibanti, Ed., Proceedings, NATO Advanced Study Institute, Tabiano, Palermo, May 21-June 1, 1979, D. Reidel Publishing: Dordrecht, 1980, p. 181; and references therein.

[10] E. Rizzarelli, "Thermodynamics of Mixed Complexes in Aqueous Solution," **Gazz. Chim. Ital., 112** (1982) 139; and references therein.

[11] G. Arena, E. Rizzarelli and B. Sarkar, "Thermodynamic Studies of Copper(II) Transport Site of Human Serum Albumin," **Inorg. Chim. Acta, 37** (1979) L 555.

[12] G. Arena, S. Musumeci, E. Rizzarelli, S. Sammartano and D. R. Williams, "Formation and Stability of Ternary Complexes of Copper(II) with EDTA and Some Aminoacids," **Ann. Chim. (Rome), 68** (1978) 535.

[13] G. Arena, C. Rigano and S. Sammartano, "Computer Analysis of Multicomponent Systems," **Ann. Chim. (Rome), 68** (1978) 693.

[14] M. K. John, "Interrelations between Plant Cadmium and Uptake of Some Other Elements from Culture Solutions by Oats and Lettuce," **Environ. Poll., 11** (1976) 85.

[15] M. A. Turner, "Effect of Cadmium Treatment on Cadmium and Zinc Uptake by Selected Vegetable Species," **J. Environ. Qual., 2** (1973) 118.

[16] R. J. P. Williams, "The Symbiosis of Metal Ion and Protein Chemistry," **Pure Appl. Chem., 55** (1983) 35.

[17] P. Amico, G. Arena, P. G. Daniele, G. Ostacoli, E. Rizzarelli and S. Sammartano, "Mixed-Metal Complexes in Solution. Part III. Thermodynamic Study of Heterobinuclear Copper(II)-L-Histidine or -Histamine Complexes in Aqueous Solution," **Inorg. Chem., 20** (1981) 772.

[18] P. Amico, G. Arena, P. G. Daniele, G. Ostacoli, E. Rizzarelli and S. Sammartano, "Mixed-Metal Complexes in Solution. Part I. Potentiometric Study of Heterobinuclear Copper(II)-L-Histidinate Complexes with Nickel(II), Zinc(II) and Cadmium(II) Ions in Aqueous Solutions," **Inorg. Chim. Acta, 35** (1979) L 383.

[19] P. G. Daniele, P. Amico and G. Ostacoli, "Heterobinuclear Cu(II)-L-Carnosine Complexes with Cd(II) or

Zn(II) in Aqueous Solution," **Inorg. Chim. Acta, 66** (1982) 65.
[20] P. G. Daniele, P. Amico, G. Ostacoli and V. Zelano, "Heterobinuclear Cu(II)-L-DOPA Complexes with Zn(II) or Cd(II) in Aqueous Solution," **Ann. Chim. (Rome), 73** (1983) 199.
[21] P. G. Daniele, P. Amico, G. Ostacoli and M. Marzona, "Mixed Metal Complexes of L-Aspartyl-L-Alanyl-L-Histidine-N-Methylamide with Cu(II), Cd(II) and Cu(II), Zn(II) in Aqueous Solution," **Ann. Chim. (Rome), 73** (1983) 299.
[22] K. S. Iyer, S. J. Lau, S. H. Laurie and B. Sarkar, "Synthesis of the Native Copper(II) Transport Site of Human Serum Albumin and its Copper(II)-Binding Properties," **Biochem. J., 169** (1978) 61.
[23] B. Sarkar and Y. Wigfield, "Evidence for Albumin-Cu(II)-Amino Acid Ternary Complex," **Can. J. Biochem., 46** (1968) 601.
[24] A. Braibanti and G. Mori, "Protonation Equilibria of DL-3-Amino-2-Hydroxypropanoic Acid and its Complexes with Cobalt(II), Nickel(II), Copper(II), and Zinc(II)," **J. Chem. Soc. Dalton Trans.**, (1976) 826.
[25] P. Amico, G. Arena, V. Cucinotta, P. G. Daniele, G. Ostacoli, E. Rizzarelli and S. Sammartano, "Mixed Metal Complexes in Solution. Equilibrium and Spectroscopic Studies of Metal Complexes of Hydroxyl Containing Aminoacids," **Abstracts of Papers, 29th IUPAC Congress, Cologne,** Germany, June 5-10, 1983, p. 39.
[26] P. Amico, P. G. Daniele, G. Ostacoli, G. Arena, E. Rizzarelli and S. Sammartano, "Mixed Metal Complexes in Solution. Part II. Potentiometric Study of Heterobinuclear Metal(II)-Citrate Complexes in Aqueous Solution," **Inorg. Chim. Acta, 44** (1980) L 219.
[27] P. Amico, P. G. Daniele, G. Ostacoli, G. Arena, E. Rizzarelli and S. Sammartano, "Mixed Metal Complexes in Solution. Part IV. Formation and Stability of Heterobinuclear Complexes of Cadmium(II)-Citrate with Some Bivalent Metal Ions in Aqueous Solutions," **Transition Met. Chem., 9** (1984) in press.
[28] J. V. Lagerwerff and G. T. Biersdorf, "Interaction of Zinc with Uptake and Translocation of Cadmium in Radish," in **"Proceedings, 5th Conference on Trace Substances in Environmental Health,"** D. D. Hemphill, Ed., June 29 - July 1, 1971, University of Missouri-Columbia: Columbia, Missouri, 1972, p. 515.
[29] L. R. Hawf and W. E. Schmid, "Uptake and Translocation of Zinc by Intact Plants," **Plant Soil, 27** (1967) 249.

XIV

BEHAVIOR OF RADIONUCLIDES IN THE MARINE ENVIRONMENT: PRESENT STATE OF KNOWLEDGE AND FUTURE NEEDS

P. Scoppa* and C. Myttenaere†

Contribution No. 2022 of the Radiation Protection Programme Commission of the European Communities—DG XII/F/1
*ENEA, Centro Ricerche Energia Ambiente, I-19100 La Spezia, Italy
†200, Rue de la Loi, Bruxelles, Belgium

Information about the behavior of radionuclides in the sea and their pathways from the sea to man is a prerequisite for the assessment of the radiological impact of human activities involving releases of artificial radionuclides and redistribution of natural radioisotopes. The relative importance of the different radionuclides for man and its environment must be defined, and their sources, their dispersion and reconcentration processes, their biogeochemical behavior, and their migration through food chains must be investigated. Present knowledge is adequate for conservative evaluations of the radiological impact of effluents released by nuclear industry in normal operations, but much additional work is required for the application of dose optimization procedures. However, the existing equilibrium models for radionuclides are not applicable for unplanned releases. Therefore, dynamic models must be developed. Such models require more information than presently available about the kinetic aspects of biogeochemical processes responsible for the transfer of radionuclides within the different marine ecosystems. A much better knowledge of the long-term behavior of longer-lived radionuclides is very important in relation to the disposal of high-level radioactive wastes.

Radiation ecology has received varying attention during the last several decades. Prior to the early 1960s the environmental behavior of fission products generated in atmospheric nuclear tests and of radionuclides formed in power reactors was studied. Concern about non-nuclear pollution subsequently overshadowed nuclear issues. Later, the problems associated with the development of nuclear energy became of increasing interest with special emphasis on transuranic nuclides that present special disposal problems and other radioisotopes such as tritium and krypton that are difficult to confine. Within the past several years, radio-ecologists broadened their activities to include radiological and nonradiological problems associated with all forms of energy production.

The highly complex natural processes that determine the distribution of radionuclides in aquatic and terrestrial ecosystems operate also on other chemical substances such as heavy metals and nutrients. Therefore, the basic knowledge acquired from radio-ecological investigations can be used not only to provide protection from radiation, but also to control other pollutants, and to explore normal ecosystems.

Present State of Knowledge about Radionuclides in the Marine Environment

Research on the environmental behavior of radionuclides and their movement within different ecosystems is of basic importance for radiation protection. The results of numerous investigations [1, 2] are currently used to evaluate the impact of radioactive effluents released during normal operations of nuclear facilities [3]. The marine environment has been contaminated with radioactive materials primarily from testing of nuclear weapons. Reactor effluents, nuclear fuel reprocessing plants, and miscellaneous sources of radionuclides are responsible for localized contamination. Tritium and carbon-14 were shown to be very mobile and to be important for the global dispersion of radioactive contamination. The data defining the behavior of fission and activation products in seawater, in sediments, and in biota were used to develop mathematical models describing the equilibrium distribution of radionuclides in the marine environment.

The physical and chemical properties of radionuclides determine their environmental behavior. The diversity of radioactive sources and the large number of chemical substances present in seawater make it very difficult to determine the physicochemical state of most radionuclides in the marine environment. Cesium and strontium exist probably in ionic form, whereas Ru, Cr, Zr, Y, Nb, and Fe tend to be present as suspended particles and colloids. Very little is known about the kinetic aspects of the reactions of radionuclides in seawater. Some of these reactions might be very slow. Ionic Zn-65, for instance, requires more than one year to become equilibrated with non-radioactive zinc naturally present in seawater. Nuclides such as Co-60, Mn-54, and Fe-59 originally present in ionic form may precipitate or coprecipitate with other ions. Physical, chemical, and biological processes occurring in the different marine ecosystems found in the open oceans, in abyssal depths, on continental shelves, in coastal waters, in tidal ponds, and in estuaries influence the movement and the distribution of radionuclides. Radionuclides released into the marine environment are primarily dispersed through the movement of water masses. The movement is dependent on physical and meteorological factors such as currents, winds, and the temperature, density, and salinity of waters.

Chemical processes strongly influence the distribution of radionuclides among a marine ecosystem's different compartments such as seawater, sediments, and biota. Although organisms serving as human food may transfer radionuclides from the environment to man, living organisms have otherwise a minor influence on the environmental behavior of radionuclides. Organisms may accumulate radionuclides by adsorption on their surfaces, by absorption through membranes exposed to the aqueous phase, and by ingestion. Plankton and higher-order organisms may accumulate significant amounts of radionuclides from seawater. The fate of these radionuclides, if not rapidly excreted, is determined by the fate of the organisms. Populations of small organisms undergoing rapid turnover carry upon their death measurable quantities of radionuclides to deeper layers of the water column and to the bottom sediments. The bioavailability of radionuclides appears to be strictly dependent on their physicochemical properties. For instance, nuclides found preferentially as particles or colloids are readily available to organisms in food chains that include filter feeders. In general, the present state of knowledge about radionuclides in the environment can be considered to be adequate for overall control purposes. However, greater, future use of fission energy demands further efforts to improve our understanding of the environmental processes affecting radionuclides in the marine environment.

Future Needs

In spite of more than 30 years of radio-ecological studies, detailed knowledge of the processes leading to the observed distributions of radionuclides in different marine ecosystems is still lacking. Further work is needed to ensure that assessments of radiological impacts are based on a solid scientific background. Such a background is especially needed for the evaluation of the significance of longer-lived radionuclides in relation to human exposure over long periods.

The evolution of basic criteria of radiation protection [4-6] dictates a reorientation of radioecological research, because fundamental information is needed to assess the collective dose commitment, the infinite time integral of the product "size of a specified population times the per capita dose rate for that population," over wider geographic areas. According to the recommendations of the International Commission on Radiological Protection, exposures should be kept "as low as reasonably achievable". For dose optimization [6], conservative values should not be used anymore in dose-assessment procedures. The various pathways of radionuclides from their source to man must be described as realistically as possible by actual transfer rates.

Radio-ecological studies, including investigations of the dispersion and reconcentration of radionuclides by inorganic and organic materials and by living organisms, are not only very useful for radiation protection, but also make a significant contribution to oceanography and marine ecology in areas such as the circulation and mixing of water masses, the chemistry and physical chemistry of seawater and sediments, and the metabolism of marine organisms.

Disposal of long-lived radionuclides (transuranic elements, technetium-99) is of particular concern. Too little is now known about the long-term distribution of these man-made elements in terrestrial and aquatic ecosystems. Therefore, accurate assessments of collective dose commitments, needed to make decisions based on recent criteria of radiation protection, are not possible. The quantity and quality of data on most fission and activation products must be improved before optimization procedures can be successfully implemented and the radiological consequences of accidental releases evaluated. Exposure of man to radionuclides will in most cases come from coastal areas rather than from the open ocean or the deep sea.

Therefore, shallow waters should be examined first with respect to their potential for release of radioactive and nonradioactive wastes. Nevertheless, more attention than in the past should be given to the deep sea as a receiving environment.

Small scale (1-10 km) and medium scale (10-100 km) distribution patterns are now in general sufficiently understood for predicting radionuclide dispersion and for establishing control procedures. However, more data are required for large-scale circulation patterns (100-1000 km) in order to predict long-range effects. Knowledge about the dispersion processes in the open sea is incomplete and particularly scanty for the deep sea. The deep sea, remote from human activities and from fishing grounds, is expected to be more suitable than coastal waters for receiving wastes. However, before radioactive and especially high level radioactive wastes are disposed in the deep sea, the vertical and horizontal dispersion processes and the interactions between waste components and sediments must be better understood.

Estuaries and adjacent shallow sea areas are unique. Changes in salinity and suspended sediment load occur at the freshwater-saltwater interface. The distribution of radioactive materials under these conditions can be approximately described in terms of salinity, temperature, and suspended matter, but the processes involved and their mechanisms are not sufficiently understood.

Physical oceanography can make an important contribution to the understanding of the dispersion of radionuclides in different marine ecosystems. A thorough understanding of the long-term distribution of radionuclides requires a much better marine geochemical background than available now, and more detailed information about water-sediment interactions and the physicochemical behavior of radionuclides in marine sediments.

Extrapolations of experimental data obtained with systems containing millimolar concentrations of radionuclides to the much lower concentrations in the environment are extremely risky. The few comparable data so far available indicate that at very low concentrations the kinetic behavior may be considerably different from that observed at much higher concentrations. Predictions are further complicated by the complex chemical composition of seawater, sediments, and the waste solutions. Matrix effects, particle size distribution, adsorption-desorption on surfaces, competition between ions, isotopic exchange, and complex formation must be investigated at extremely low concentrations. The elucidation of the environmental distribution of radionuclides may be facilitated by the determination of the behavior of stable isotopes of the same or chemically similar elements in seawater, sediments, and organisms.

Recently, the dependence of the behavior of radionuclides on physicochemical parameters such as valence, solubility, colloidal nature, and particle size has become appreciated. However, the poor state of knowledge of marine trace element chemistry and particularly of those elements that are directly relevant to radioactive contamination does not allow reliable predictions to be made. When waste solutions mix with seawater, the various components interact, and very often change their physicochemical states. Liquid effluents often have freshwater character and may contain large quantities of complexing agents. The interactions between freshwater and seawater modifying the environmental behavior of radionuclides must be understood before long-term predictions on their distribution in estuaries and adjacent areas can be made.

Ultimately, some radioisotopes will be adsorbed on marine sediments. Fixation by sediments is important for the short-term cycling of most radionuclides. The present knowledge of the processes involved in this fixation and of the sorption capacities of the various types of marine sediments is limited and needs to be developed.

The biological aspects of marine radiocontamination have received much attention during the past years because marine food has been recognized to be a source of radionuclides for man. Fishermen and their families are considered as a "critical group" of the population because of their high consumption of marine food. However, whenever the optimization concept is applied, it is necessary to assess the collective dose commitment. Living organisms may disperse or concentrate radionuclides, modify the chemical form of radioisotopes, and transfer a fraction of them to man via the food chains. Information should be acquired about the distribution of organisms in space and time, about their capacity to disperse or concentrate radionuclides and stable elements, and about the pathways of radionuclides through food chains.

The distribution of marine biota in space and time has long been a subject of study, but even today quantitative investigations are hampered by sampling techniques inadequate for the deep ocean and for many levels of the food chains. The assessment of radiological impact sometimes requires knowledge about the distribution and migration of marine organisms in relation to seasonal changes and various stages of their life cycles. Special attention should be paid to species that are not directly consumed by man but constitute important links in the food chain of commercial species. The proper use of these species as pollution indicators should be investigated. Experiments on the uptake, accumulation and loss of radionuclides have been hindered by a lack of suitable methods for culturing and breeding marine organisms under laboratory conditions. Much progress has been made during the last few years, but whether the experimental systems are truly representative of natural environments often remains doubtful. More investigations should be performed under controlled conditions at least for certain elements, for which more information is needed.

Although the importance of food chains as the major pathways for many radionuclides to man has received much attention, more experimentation on specific topics such as the bioavailability of different chemical forms influencing, for instance, the metabolism of transuranic and other elements is needed. These biochemical studies will provide information about uptake, turnover, and excretion of complex mixtures containing radionuclides and synthetic or natural chemicals. These investigations should be supplemented by observations in disposal areas and by semi-controlled experiments in natural environments.

An effort must also be made to improve instrumentation for physical oceanography, techniques for physical measurement of radionuclides in environmental matrices, radiochemical methods for the separation and identification of different chemical forms of radionuclides, modelling and simulation, and sampling procedures.

References

[1] Commission of the European Communities, **"Annual Progress Reports of the Programme Radiation Protection,"** ECSC-EEC-EAEC, Brussels, Belgium.
[2] F. W. Whicker and V. Schultz, **"Radioecology: Nuclear**

Energy and the Environment," CRC Press: Boca Raton, Florida, 1982.
[3] Commission of the European Communities, **Methodology for Evaluating the Radiological Consequences of Radioactive Effluents Released in Normal Operations,**" Doc. No. V/3865/79-EN,FR, 1979.
[4] International Commission on Radiological Protection, **"Implication of Commission Recommendations that Doses be Kept as Low as Readily Achievable,"** Publication ICRP-22, Pergamon Press: New York, 1973.
[5] International Commission on Radiological Protection, **"Recommendations of the International Commission on Radiological Protection,"** Publication ICRP-26, Pergamon Press: New York, 1977.
[6] International Commission on Radiological Protection, **"Radionuclide Release into the Environment: Assessment of Doses to Man,"** Publication ICRP-29, Pergamon Press: New York, 1978.

XV

SPECIATION OF PLUTONIUM IN SEAWATER AND FRESHWATER

Gregory R. Choppin

Department of Chemistry
Florida State University
Tallahassee, Florida 32306 USA

The distribution of dissolved species of plutonium(III), Pu(IV), Pu(V), and Pu(VI) in freshwater at pH 6, 7, and 8, and in seawater were calculated with consideration of the plutonium complexes formed with hydroxide, chloride, carbonate, hydrogen carbonate, and humate, and of the redox equilibria pertaining to the various oxidation states of plutonium. The calculations showed Pu(VI) to be the dominant dissolved species in seawater when the redox equilibria are established rapidly. Humic materials not only complex but also reduce Pu(VI) to Pu(IV). When the free-ion redox equilibria are not reached rapidly after Pu(VI) has been reduced to Pu(IV), Pu(V) becomes the dominant dissolved species. In seawater Pu(V) was observed to be the dominant species. Plutonium(III) is predicted to be the dominant species in freshwater at pH 6, whereas Pu(IV) becomes predominant at pH 7 and 8 in agreement with observation.

Sources of Plutonium in the Environment

The injection of plutonium (Pu) into the environment began in 1945 with the detonation of the first atomic bombs. Approximately 360 kilocuries (kCi) of Pu-239 and Pu-240 corresponding to 2.3×10^{14} disintegrations per minute (dpm) from the atmospheric explosion of fission and fusion weapons, and 17 kCi of Pu-238 from a high-altitude burnup of a satellite power source (SNAP-9, April 1964) [1] have thus far been disbursed in the environment. The total mass of this plutonium is 3300 kg. The plutonium isotopes 238, 239 and 240 decay by alpha emission. The release of plutonium to the environment by the nuclear power fuel cycle has thus far been negligible except for a few locations such as the Irish Sea near the British reprocessing plant at Windscale. At Windscale approximately 100 Ci Pu have been discharged per month.

As the amount of nuclear fuel that is reprocessed increases, the contribution to the total plutonium in the environment from these operations could become more significant. Such releases could also involve the actinide elements Th, U, Np, Am, and Cm. Approximately 93 kg of Am-241 are already present in the ecosphere from the decay of the 14-year half-life Pu-241, which was released during nuclear tests. Also, over a longer time, migration from nuclear waste disposal sites could add actinide elements to the ecosphere. Whatever the source of Pu and the other actinides, their presence represents anthropogenic contamination of the environment by extremely toxic materials. An understanding of the factors involved in their retention and migration in the ecosphere is highly desirable. Comparison of the species of plutonium found in seawater and freshwater with those predicted by model calculations can serve as a test of the parameters used in a model for the behavior of Pu in the environment.

Most of the plutonium from weapons testing was initially injected into the stratosphere. The plutonium in the weapon that survived the explosion was converted to a high-fired oxide which is expected to remain insoluble after return to earth. Such insoluble particles would have sunk in a rather short time to the bottom of lakes, rivers, and oceans or would have become incorporated in soils below the surface layer. However, approximately two thirds of the plutonium was generated during the explosion via U-238 (n,γ) reactions and subsequent beta decay of nuclei such as U-239, U-240, and U-241 formed in these reactions. The plutonium from these reactions would exist as single atoms, which were never converted into high-fired particulates. The plutonium formed in this manner should be more soluble and more reactive and behave more like the plutonium released from reprocessing plants and nuclear waste repository sites.

Present Levels of Plutonium in the Ecosphere

Near test sites and reprocessing facilities the concentration of plutonium in the soil and water is much higher than in more distant locations [2]. Generally, most of the plutonium is associated with subsurface soils or sediments or with suspended particulates in water columns. For example, when vegetation, animals, litter, and soils are compared, more than 99 percent of the plutonium is present in the soil [3]. Similarly, in shallow bodies of water, more than 96 percent of the plutonium is found in the sediments. However, plutonium is mainly transported in groundwater in the form of soluble species. The calculations reported in this paper are concerned only with the chemistry of dissolved plutonium. Table 1 presents

Table 1. Concentrations of Dissolved Plutonium in Water Samples* from Lakes, Rivers and the Ocean [3, 4]

Location	Plutonium Concentration
Lake Michigan	2.0×10^{-17} M
Great Slave Lake, Canada	1.5×10^{-17} M
Okefenokee River, Florida	1.5×10^{-16} M
Hudson River, New York	1.0×10^{-17} M
Irish Sea:	
1 km from Windscale	1.6×10^{-14} M
110 km from Windscale	1.1×10^{-15} M
Mediterranean	2.6×10^{-18} M
North Pacific (surface)	3.0×10^{-17} M
South Pacific (surface)	1.0×10^{-17} M

* Samples were passed through 0.45 μm filters.

plutonium concentrations found in seawater and freshwater [3, 4]. The solubility of plutonium added to filtered seawater was measured to be 1.3×10^{-11} M 30 days after addition [5]. Approximately 60 percent was in non-ionic form. The concentration of ionic plutonium was, therefore, only 5×10^{-12} M. When humic material was added to the plutonium containing seawater samples, the solubility of plutonium had increased six-fold after 30 days. Humic materials might be responsible for the higher than normal concentrations of plutonium in the Okefenokee River (Table 1).

It is difficult to obtain reliable values for plutonium concentrations in natural aquatic systems. The concentrations are very low, providing only 0.001 dpm per liter of seawater. Moreover, the amount of plutonium associated with suspended particulates may be more than an order of magnitude greater than the plutonium in true solution. In the Mediterranean, the plutonium activity in unfiltered seawater was 25 times larger than that of a sample filtered through a 0.45 μm filter [6].

Chemical properties of plutonium. Before describing the model calculations, some of the pertinent features of the chemistry of plutonium in aqueous solution [7] are required. Plutonium can exist in aqueous solutions of pH and E_h ranges found in nature in the four oxidation states III, IV, V, and VI. The III and IV states may be present as the simple hydrated cations, whereas the V and VI states form the dioxo cations PuO_2^+ and PuO_2^{2+}, respectively. As expected, the lower oxidation states are stabilized by more acidic conditions and the higher states by more basic conditions. Various factors such as complexation can reverse the relative stability of the different oxidation states. For example, hydrolysis reactions cause the standard reduction potential of the Pu(III/IV) couple to change from +0.98 V in 1 M perchloric acid to -0.63 V in a neutral solution. Disproportionation reactions are also a common feature of the IV and V states but are unlikely to be important at environmental concentrations. Nevertheless, plutonium may exist in several oxidation states in the same solution.

Whatever the oxidation state, plutonium cations are hard acids and interact with anionic species by ionic bonding. The effective charges of plutonium in PuO_2^+ and PuO_2^{2+} are approximately +2.2 and +3.2, respectively. Therefore, the stability of plutonium complexes decreases in the order Pu(IV) > Pu(VI) > Pu(III) > Pu(V). Hydrolysis is significant at the pH values of natural waters for all the oxidation states except V (Fig. 1). Hydrolysis removes Pu(IV) above pH 2 from solution by adsorption of $Pu(OH)_4$ to suspended particles and to surfaces. The stability constants for complexes of plutonium in any particular oxidation state with anionic ligands often correlate with the ligand pK values for a series of related ligands of similar charge and dentation. Because it is difficult to maintain plutonium in only one oxidation state in solution, reliable values for stability constants of plutonium with many anions of environmental interest are lacking. However, with proper caution, stability constants for plutonium can be estimated from pertinent data for other actinides such as Am(III), Th(IV), Np(V), and U(VI). Table 2 lists the standard reduction potentials at zero ionic strength obtained by critical evaluation of literature data [8, 9]. To illustrate the effects of pH and ionic strength, the

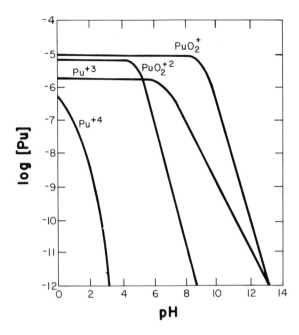

Figure 1. The concentrations of free plutonium cations as a function of pH.

formal potentials at unit ionic strength are also presented in Table 2. The stability constants used for the calculation of the species distribution are listed in Table 3. The listed values are judged to be the most reasonable available in the literature. The constants for the humate

Table 2. Standard and formal reduction potentials for Pu

pH	Ionic Strength	Reduction Potentials, E^0 or E (V)			
		$Pu^{4+} + e^- \rightleftharpoons Pu^{3+}$	$PuO_2^+ + 4H^+ + e^- \rightleftharpoons Pu^{4+} + 2H_2O$	$PuO_2^{2+} + e^- \rightleftharpoons PuO_2^+$	$PuO_2^{2+} + 4H^+ + 2e^- \rightleftharpoons Pu^{4+} + 2H_2O$
0	0	1.006±0.003*	1.10 ±0.06*	0.96+0.04*	1.032±0.037*
0	1	0.982±0.001	1.170±0.001	0.91±0.00	-1.043±0.001
8	1	-0.39 ±0.15	0.70 ±0.12	0.60+0.04	0.65 ±0.08
14	1	-1.04 ±0.24	0.52 ±0.24	0.65±0.08	0.34 ±0.12

* Standard Reduction Potentials

Table 3. Stability Constants for the Plutonium Complexes PuL_n^+

Oxidation State	Ligand L	Stability Constants				Ref.
		$\log\beta_1$	$\log\beta_2$	$\log\beta_3$	$\log\beta_4$	
Pu(III)	OH^-	6.0 (0)	11.7*			[10]
		6.2 (1)	12.0*			[11]
	Cl^-	0.1 (1)				[12]
	CO_3^{--}	5.81 (1)				[13]
	Humate	14.1 (0.1)				[14]
Pu(IV)	OH^-	13.5 (0)	9.72 (1)			[10]
		12.4 (0.7)	14.5 (0.1)			
			25.7 (0)	36.7 (0)	46.5 (0)	
			24.4 (0.7)	35.2 (0.7)	44.8 (0.7)	
	Cl^-	0.9 (1)				[15]
	CO_3^{--}	6.4* (1)	10.7* (1)			
	Humate	17.5 (0.1)	24. (0.1)			
Pu(V)	OH^-	4.3 (0)				[16]
		4.2 (0.7)				[10]
	HCO_3^-	1.9 (0.25)				[10]
Pu(VI)	OH^-	8.4 (0)	16. (0)			[15]
		7.8 (2)	15.* (1)			[17]
	Cl^-	0.1 (2)				[12]
	HCO_3^-	2.67 (0.1)**				[18]
	CO_3^{--}		13.1 (0.1)			[18]
			12.* (0.1)			
	Humate	7.* (0.1)				[19]

† Numbers in parentheses give ionic strength.
* Estimates made for the calculation of species distribution.
** For reaction $PuO_2(OH)_2 + HCO_3^- \rightleftharpoons [PuO_2(OH)_2(HCO_3)]^-$.

complexes were obtained by extrapolation of literature data to the appropriate degree of ionization for humic acid (pH>8, $\alpha = 1.0$; pH 7, $\alpha = 0.9$; pH 6, $\alpha = 0.8$).

Calculations of the Concentrations of Plutonium Species in Seawater and Freshwater

The reduction potentials (Table 2) were used to calculate the mole fractions of the free cationic plutonium species (eqn. 1-4). The standard potentials were employed for

$$\log[\text{Pu}^{3+}]/[\text{Pu}^{4+}] = (E_{4,3} - E_h)/RT \qquad (1)$$

$$\log[\text{PuO}_2^+]/[\text{Pu}^{4+}] = (E_h - E_{5,4})/RT + 4\text{pH} \qquad (2)$$

$$\log[\text{PuO}_2^{2+}]/[\text{Pu}^{4+}] = 2(E_h - E_{6,4})/RT + 4\text{pH} \qquad (3)$$

$$\log[\text{PuO}_2^{2+}]/[\text{PuO}_2^+] = (E_h - E_{6,5})/RT \qquad (4)$$

freshwater assumed to have zero ionic strength. For seawater the formal potentials were used under the assumption that these values are valid for an ionic strength of 0.7. Seawater was assumed to have a pH of 8.2 and an E_h value of 0.8 V. For freshwater, the calculations were performed for pH vaues of 6, 7, and 8 with E_h estimated for the oxic freshwater systems by equation 5. In all systems

$$E_h = 8.0 - 0.059\text{pH} \qquad (5)$$

the total pressure was 1 atm, the partial pressure of oxygen 0.21 atm, and the partial pressure of carbon dioxide $10^{-3.5}$ atm. The mole fractions of "free" plutonium cations in solution are listed in Table 4. The total mole fraction of

Table 4. *Calculated Mole Fractions of "Free" Plutonium Cations in a Solution Devoid of Complexing Agents*

Cation	Mole Fraction of Plutonium Species			
	Seawater	Freshwater		
	pH 8.2	pH 6	pH 7	pH 8
Pu^{3+}	3.56×10^{-24}	3.22×10^{-4}	3.45×10^{-6}	3.76×10^{-8}
Pu^{4+}	2.9×10^{-27}	1.04×10^{-13}	1.08×10^{-16}	1.12×10^{-19}
PuO_2^+	0.985	1.00	1.00	1.00
PuO_2^{2+}	1.35×10^{-2}	2.27×10^{-9}	2.18×10^{-10}	2.10×10^{-11}

plutonium(IV) present in a solution was obtained as shown in equation 6, using the mole fraction of free Pu^{4+}

$$\{Pu(IV)\}_t = \{Pu^{4+}\}_f \left(1 + \Sigma\beta_i^{OH}[OH^-]^i + \Sigma\beta_j^{CO_3}[CO_3^{2-}]^j + \Sigma\beta_k^{Hu}[Hu]^k + \Sigma\beta_m^{Cl}[Cl^-]^m + \Sigma\beta_n^{HCO_3}[HCO_3^-]^n\right) \quad (6)$$

from Table 4, the stability constants from Table 3, and the free uncomplexed anion concentrations from Table 5. The mole fractions of total Pu(III), Pu(V) and Pu(VI) were similarly calculated.

Table 5. Free, Uncomplexed Anion Concentrations in Seawater and Freshwater [20]

Anion	Concentration in Moles/Liter			
	Seawater	Freshwater		
	pH 8.2	pH 6	pH 7	pH 8
OH^-	$10^{-5.6}$	$10^{-8.0}$	$10^{-7.0}$	$10^{-6.0}$
HCO_3^-	$10^{-2.76}$	$10^{-5.4}$	$10^{-4.4}$	$10^{-3.4}$
CO_3^{--}	$10^{-4.86}$	$10^{-9.5}$	$10^{-7.6}$	$10^{-5.7}$
Cl^-	$10^{-0.25}$	–	–	–
Hu*	$10^{-6.7}$	$10^{-5.5}$	$10^{-5.5}$	$10^{-5.5}$

*Hu is the sum of fulvate and humate: the concentrations are expressed in equivalents/liter.

Humic acid is known to reduce Pu(VI) to Pu(IV) [21, 22]. The effective stability constant for the interaction of Pu(VI) with humic acid (eqn. 9) can be obtaind from the constants for reactions 7 and 8, and the redox potential for the Pu(VI)/Pu(IV) couple (Table 2). The ratio [Pu(IV) Hu]/[PuO_2^{2+}] calculated from the equilibrium expression for reaction 9 and the humic acid concentrations in Table 5 assumes a value of $10^{13.2}$ for seawater at pH 8.2 and 10^{23} (pH 6), 10^{19} (pH 7), 10^{15} (pH 8) for freshwater.

$$PuO_2^{2+} + 4H^+ \rightleftharpoons Pu^{4+} + 2H_2O \qquad \log K = 35.0 \quad (7)$$

$$Pu^{4+} + Hu \rightleftharpoons Pu(IV)\ Hu \qquad \log K = 17.5 \quad (8)$$

$$PuO_2^{2+} + Hu + 4H^+ \rightleftharpoons Pu(IV)\ Hu + 2H_2O \qquad \log K = 52.5 \quad (9)$$

The potential for the reduction of the Pu(VI)-humate complex to the Pu(IV)-humate complex (eqn. 10) can be

$$Pu(VI)O_2 \text{ Hu} + 4H^+ + 2e^- \rightleftharpoons Pu(IV) \text{ Hu} + 2H_2O \quad (10)$$

estimated to be approximately 1.2 V. The humic half cell should then have a potential of 0.2 V. A value of 0.35 ± 0.03 has been reported for peat humate [23].

The stability constants (Table 3) were used to calculate the major species of Pu present in seawater and freshwater. All of the Pu(VI) complexed with humate was considered to have been reduced, and was, therefore, not included in the reported Pu(VI) concentrations.

Discussion

When plutonium is introduced into an aqueous system the equilibria demanded by thermodynamics may not be achieved because one or more reactions may be kinetically hindered. Such reactions with unfavorable kinetics could be those linking the cations of plutonium in its various valence states. When Pu(VI) is reduced by humic acid the redox equilibria are disturbed. The system could then re-establish equilibrium rather rapidly or remain in the non-equilibrium condition. The species distribution of plutonium was determined for both scenarios.

In all systems Pu(III) exists almost exclusively as the 1:1 humate complex. Almost all of Pu(IV) is present in solution as $Pu(OH)_4$. The predominant species of Pu(V) is the free, aquated cation PuO_2^+. Pu(VI), which has not reacted with humic acid, forms a series of complexes with hydroxide, hydrogen carbonate or carbonate as ligands (Table 6). Variations of an order of magnitude in the values of the stability constants would not significantly change the mole fractions of the dominant Pu(III), Pu(IV) and Pu(V) species. The species distributions of Pu(VI) would, however, change considerably.

In the absence of humate, Pu(III) would be present at pH 6 as Pu^{3+} and at pH 8 at $Pu(OH)^{2+}$. The mole fractions calculated from the $E°$, E_h, β, and the concentration values of Tables 2 to 5, are listed in Table 7. The total Pu(III) concentration in the presence of humate would be 10^7 times the Pu(III) concentration in the absence of humate.

Table 6. Complexes of PuO_2^{2+} present in seawater and freshwater

Species	Percent of Total Pu(VI)			
	Seawater	Freshwater		
	pH 8.2	pH 6	pH 7	pH 8
$PuO_2(OH)^+$	0	75%	15%	0
$PuO_2(OH)_2$	60%	25%	60%	0
$PuO_2(OH)_2(HCO_3)^-$	10%	0	25%	20%
$PuO_2(CO_3)_2^{--}$	30%	0	0	80%

Table 7. Calculated Mole Fractions of Total Dissolved Pu(III), Pu(IV), Pu(V) and Pu(VI)

Cation	Seawater	Freshwater		
	pH 8.2	pH 6	pH 7	pH 8
Redox equilibrium not established after Pu(VI) reduction by humic acid.				
Pu(III)	1.4×10^{-16}	~1.0	3.3×10^{-1}	3.4×10^{-3}
Pu(IV)	7.3×10^{-7}	3.3×10^{-2}	6.7×10^{-1}	~1.0
Pu(V)	~1.0	1.0×10^{-3}	2.0×10^{-3}	3.0×10^{-4}
Pu(VI)	8.5×10^{-12}	9.0×10^{-35}	7.1×10^{-30}	5.7×10^{-26}
Redox equilibrium established after Pu(VI) reduction by humic acid.				
Pu(III)	1.0×10^{-8}	~1.0	3.3×10^{-1}	3.4×10^{-3}
Pu(IV)	5.3×10^{-9}	3.3×10^{-2}	6.7×10^{-1}	~1.0
Pu(V)	7.3×10^{-3}	1.0×10^{-3}	2.0×10^{-3}	3.0×10^{-4}
Pu(VI)	~1.0	9.0×10^{-12}	7.1×10^{-11}	5.7×10^{-11}

The principle dissolved plutonium species in seawater is Pu(V) if redox equilibrium is not established after reduction of Pu(VI) by humic acid (Table 7). However, if the system maintains redox and complexation equilibrium, Pu(VI) is predicted to be the dominant oxidation state of dissolved plutonium if the effective E_h has a value of 0.8. Experimentally, Pu(V) is found in marine waters as the predominant dissolved species, whereas Pu(IV) is the main form sorbed on particulates. These observations support either the non-equilibrium model or a lower effective E_h for the plutonium half-cell.

In freshwater in the presence of $10^{-5.5}$ eq./L humate, Pu(III) is the principal species at pH 6. However, in the absence of humate at pH 6, and even in its presence at pH 7 and 8, Pu(IV) is the dominant dissolved species. It is expected that other actinides mimic the behavior of plutonium. Americium(III) should resemble Pu(III) and should be present in solution as a humate complex in all systems considered. In the absence of humate, $AmOH^{2+}$ should be the dominant species in a medium of pH 8, whereas aquated Am^{3+} is expected at pH 6 and 7. Neptunium should be present as NpO_2^+ in all systems considered.

Acknowledgement. This research was supported by a contract with the Office of Health and Environmental Research, U.S. Department of Energy.

References

[1] R. W. Perkins and C. W. Thomas "World Wide Fallout", in **"Transuranic Elements in the Environment,"** W. C. Hanson, Ed., DOE/TIC-22800, National Tech. Inform. Serv., Springfield, VA., p. 53.

[2] J. A. Heatherington, D. F. Jeffries, N. T. Mitchell, P. J. Pentreath and D. S. Woodhead, "Environmental and Public Health Consequences of the Controlled Disposal of Transuranic Elements to the Marine Environment", in **"Transuranium Nuclides in the Environment,"** Symposium Proceedings, San Francisco, 1975, Intern. Atomic Energy Agency, STI/PUB/410, Vienna, p. 139.

[3] R. L. Watters, D. N. Edgington, T. E. Hakonson, W. C. Hanson, M. H. Smith, F. W. Whicker and R. E. Wildung, "Synthesis of the Research Literature," in **"Transuranic Elements in the Environment,"** W. C. Hanson, Ed., DOE/TIC-22800, National Tech. Inform. Serv., Springfield, VA., p. 1.

[4] Y. Miyake and Y. Sugimura, "The Plutonium Content of Pacific Ocean Waters," in **"Transuranium Nuclides in the Environment,"** Symposium Proceedings, San Francisco, 1975, Intern. Atomic Energy Agency, STI/PUB/410, Vienna, p. 81.

[5] K. C. Pillai and E. Mathew, "Plutonium in the Aquatic Environment," in **"Transuranium Nuclides in the Environment,"** Symposium Proceedings, San Francisco, 1975, Intern. Atomic Energy Agency, STI/PUB/410, Vienna, p. 25.

[6] T. M. Beasley and F. A. Cross, "A Review of Biokinetic and Biological Transport of Transuranic Radionuclides

in the Marine Environment," in **"Transuranic Nuclides in the Environment,"** W. C. Hanson, Ed., DOE/TIC-22800, National Tech. Inform. Serv., Springfield, VA., p. 524.
[7] G. R. Choppin, "Aspects of Plutonium Solution Chemistry," **ACS Symp. Series, 216** (1983) 213.
[8] J. Fuger and F. L. Oetting, **The Chemical Thermodynamics of Actinide Elements and Compounds,** Part 2. "The Actinide Aqueous Ions," Intern. Atomic Energy Agency, STI/PUB/424/2, Vienna, 1976.
[9] B. Allard, H. Kipatsi and J. O. Liljensin, "Expected Species of U, Np and Pu in Neutral Aqueous Solutions," **J. Inorg. Nucl. Chem., 42** (1980) 1015.
[10] C. F. Baes and R. E. Mesmer, **"The Hydrolysis of Cations,"** J. Wiley & Sons: New York, 1976.
[11] M. S. Caceci and G. R. Choppin, "The Determination of the First Hydrolysis Constant of Eu(III) and Am(III)," **Radiochem. Acta, 33** (1983) 101.
[12] R. M. Smith and A. E. Martell, **"Critical Stability Constants,"** Vol. 4, Plenum Press: New York, 1976.
[13] R. Lundquist, "Hydrophilic Complexes of the Actinides," **Acta Chem. Scand, A36** (1982) 741.
[14] R. A. Torres and G. R. Choppin, "Europium(III) and Americium(III) Stability Constants with Humic Acid", **Radiochim. Acta, 35** (1984) 143.
[15] S. L. Phillips, "Hydrolysis and Formation Constants at 25°," **Lawrence Berkeley Lab. Rept LBL-14B13,** 1982.
[16] P. M. Shanbhag and G. R. Choppin, "Binding of Uranyl by Humic Acid," **J. Inorg. Nucl. Chem., 43** (1981) 3369.
[17] B. Allard, "Solubilities of Actinides in Neutral or Basic Solutions," in **"Actinides in Perspective,"** N. Edelstein, Ed., Pergamon Press: New York, 1982, p. 553.
[18] J. C. Sullivan, M. Woods, P. A. Bertrand and G. R. Choppin, "Thermodynamics of Pu(VI) Interaction with Bicarbonate," **Radiochim. Acta, 31** (1982) 45.
[19] K. L. Nash and G. R. Choppin, "Interaction of Humic and Fulvic Acids with Th(IV)," **J. Inorg. Nucl. Chem., 42** (1980) 1045.
[20] W. Stumm and J. J. Morgan, **"Aquatic Chemistry,"** 2nd Ed., Wiley-Interscience: New York, 1981.
[21] E. A. Bondietti, S. A. Reynolds and M. H. Shanks, "Interaction of Plutonium with Complexing Substances in Soils and Natural Waters", in **"Transuranium Nuclides in the Environment,"** Symposium Proceedings, San Francisco, 1975, Intern. Atomic Energy Agency, STI/PUB/410, Vienna, p. 273.
[22] K. Nash, S. Fried, A. M. Friedman and J. C. Sullivan, "Redox Behavior, Complexing, and Adsorption of Hexavalent Actinides by Humic Acid and Selected Clays,"

Environ. Sci. Tech., 15 (1981) 834.
[23] S. A. Visser, "Oxidation-Reduction Potentials and Capillary Activities of Humic Acids," **Nature, 204** (1964) 581.

DISCUSSION

S. Ahrland (University of Lund): Does the plutonium concentration decrease with time as suggested by the Pu(VI)-Pu(IV) redox model?

G. R. Choppin: Although the data are fragmentary, they do seem to indicate a decrease in the concentration of dissolved plutonium.

S. Ahrland (University of Lund): Significant kinetic effects associated particularly with adsorption and precipitation processes might influence the distribution of plutonium species.

G. R. Choppin: Kinetic effects are certainly important. However, the concentrations do not exceed any solubility product constants, not even the constant for $Pu(OH)_4$. The slow processes, therefore, would be adsorption processes, redox changes particularly between Pu(IV) and Pu(V) or Pu(VI) states, and possibly the disssociation of the Pu(IV)-humate complexes.

K. C. Raymond (University of California, Berkeley): With a Pu concentration of 10^{-17} M in the ocean, the total ocean should contain much more dissolved Pu than 3300 kg.

G. R. Choppin: The data on Pu concentrations in the ocean are insufficient. Depth profiles show a decrease of Pu concentration with increasing depth. This decrease could perhaps account for a total of 30 to 300 kg Pu dissolved in the ocean.

J. J. Morgan (California Institute of Technology): Did your two limiting cases relate to rapid redox and complexation reactions leading to equilibrium, and to slow readjustment of redox equilibria after reduction of Pu(VI) by humic material?

G. R. Choppin: Yes. The real situation must be between these two extremes but sufficiently far from equilibrium to make Pu(V) the dominant dissolved species in seawater.

XVI

BIOTURBATION AND FATE OF RADIONUCLIDES IN BENTHIC MARINE ECOSYSTEMS

E. H. Schulte

Contribution No. 2023 of the Radiation Protection Programme Commission of the European Communities—DG XII/F/1

c/o ENEA, Centro Ricerche Energia Ambiente, I-19100 La Spezia, Italy

In the marine environment many contaminants, natural or man-made, associated with inorganic and organic particulates will finally become incorporated into the sediments that might be the ultimate pollutant reservoir. Evidence exists that in addition to physical processes, biological activities through the process of bioturbation remobilize, transfer, and recycle sediment-associated pollutants through benthic ecosystems. Deposit- and detritus-feeders are the most important and abundant organsims at the sediment and surface layers among the sediment infauna that redistribute contaminants and radio-nuclides from sediment to the overlying water. High population densities and turnover rates of benthic organisms may considerably enhance vertical diffusion and transport of compounds adsorbed on sediment particles. Changes in the biogeochemical regime, induced by bioturbation, may accelerate the remobilization of radionuclides from deeper sediment strata. Finally, the effect of bioturbation on the stability and composition of marine sediments influences their role as sinks, possible burial sites, and/or sources of radionuclides, and may have important implications with respect to the global cycling of radionuclides.

The process of bioturbation - mixing of surface sediments by benthic organisms - is widespread. The feeding, burrowing, irrigation, and physiological activities of benthic organisms can have marked effects on the physical and chemical properties of marine sediments and are important agents of particle and fluid transport near the sediment-water interface [1]. Perturbations of the stratigraphy and geochemical properties of the upper sediment strata can occur as a consequence of the vertical redistribution of the sediments by benthic fauna [2]. Feedback mechanisms induced by these perturbations may alter the original structure and diversity of the benthic communities [3]. The irrigation activities transport oxygenated and nutrient-rich bottom water into deeper sediment layers via bottom constructions such as holes and tubes, and increase the surface area of the transition zone between aerobic and anaerobic conditions. Remobilization of radionuclides from deeper sediment strata may be accelerated by changes in the biogeochemical regime as a consequence of bioturbation [4, 5]. The effect of bioturbation on the stability and composition of marine sediments influences their role as sinks, possible burial sites, and/or sources of radionuclides, and may have important implications with respect to the cycling of radionuclides in the marine environment.

Definition of Bioturbation

Bioturbation is the stirring of sediments by the activity of burrowing benthic organisms. The animals, usually understood to be involved, are invertebrates such as oligochaete worms, crustaceans, and bivalve molluscs as well as large decapods and certain fish. A number of other organisms probably participate as well, but because their activity is rather unimportant, they and the meiofauna [6] are usually neglected.

Sediment Infauna

In addition to physical effects, the remobilization and consequent cycling of radionuclides from the sediment to the water column is a result of the biological activities of sediment infauna and benthic organisms of the benthic boundary layer that extends from the lower limit of biological activity in the sediments through the overlying 20 to 50 m of the water column [7]. Radionuclides are mobilized from sediments by bacteria and infaunal organisms, whereas benthic organisms at the sediment-water interface and in the water column are responsible for the vertical transport from deep waters to surface waters.

Feeding strategies. Among the marine benthos, deposit-feeders, feeding on deposited detritus, and suspension-feeders, feeding on suspended seston, are the most important and frequently encountered in benthic invertebrates [3]. These two types of feeding activities are quantitatively important in changing sediment properties.

Deposit-feeders may feed at the sediment surface or at depths within the sediment. Feeding may be selective or nonselective regarding particle composition and size. Normally, deposit-feeders are ingesting a complex food resource consisting of dead organic detritus and mineral grains that may be coated with bacterial films. Meiofauna and microbial organisms living on particles and in the pore water may also be ingested as well as dissolved nutrients.

Suspension-feeders actively or passively entrap suspended seston on ciliated tentacles, mucus nets, and mucus-covered respiratory surfaces. Dense populations of suspension-feeders may effectively remove most of the suspended seston from the water column.

Vertical distribution. The vertical distribution of sediment infauna will be determined by the nature of available food and the amount of oxygen in the habitat. In particular, the distribution of deposit-feeders and suspension-feeders is related to the grade of deposit, the organic content of the sediment, and the re-suspension of the sediment surface by currents [8]. Generally, the upper 30 to 40 cm of the sediment are inhabited by macro- and meiofauna. Some big macrofaunal species may burrow down maximally to 3 m depth [9]. However, more than 90 percent of the meiofauna worms, crustaceans, and bivalve molluscs [6] are found not deeper than 15 cm. Small meiofauna such as nematodes do not penetrate more than 1 cm into the sediment. No feeding by infauna was observed below 20 cm depth, whereas the maximal abundance of small macro- and meiofauna was found in the upper 5 cm of the sediment. A good measure of the vertical distribution of sediment infauna is given by the disturbance of the stratigraphy of the sediment. The most intensive disturbance of the sediment was encountered in depths less than 12 cm but slight disturbance of layering was evident to a depth of 20 cm.

Population densities. The standing crops of sediment infauna vary considerably according to grain size, organic matter content, oxygen content, and stability of the sediment surface. There is a general trend of decreasing population densities with increasing depth and distance from the continent. Benthic faunal densities range from 10^3 to

10^4 on the upper continental slope to 10 to 100 individuals per square meter in the abyssal sea deeper than 4000 m [10]. Generally, *Polycheata*, *Crustacea*, *Pelecopoda* and *Sipuneuloidea* combined comprise 85 to 100 percent of the fauna. The dominance of polychaetes tend to decrease with distance from the continent, whereas crustaceans correspondingly increase in importance. In temperate and boreal bays macrofaunal population densities commonly range from 10^2 to 10^4 individuals per square meter [3], whereas in deep sea sediments of the North Atlantic below 3000 m densities of 106 to 123 individuals of macrofauna per square meter were found. For the North Pacific under similar conditions 84 to 160 and 36 to 268 individuals per square meter were reported [11, 12]. Many of the possible taxa were not present and more than 50 percent of the species were found only once. Polychaetes and Tanaidaceans together accounted for 75 percent of the individuals with polychaetes representing 55 percent, bivalves seven percent, and isopods six percent. Although meiofaunal taxa are difficult to sample quantitatively, they are 1.5 to 4 times more abundant than the macrofaunal taxa with nematodes numerically dominating [13]. Also Foraminifera seem to comprise an important portion of the community. Faunal diversity is normally extremely high with deposit-feeders comprising the overwhelming majority. In terms of biomass for the North Atlantic and the Mediterranean Sea values of 0.7 to 1.0 g biomass per square meter were reported [14].

Reworking and turnover rates of sediments. The feeding, burrowing, and irrigation activities of sediment infauna are important ways of particle and fluid transport near the sediment-water interface. The reworking and manipulation of sediments during feeding result in changes of grain size [13]. Although the processes of biological reworking of particles are difficult to measure directly, biological mixing rates due to bioturbation of 10^3 to 10^6 cm²/1000 years were reported for densely populated, near-shore sediments. These rates decrease at abyssal depths to 1 to 10^3 cm²/1000 years reflecting reduced biological activity in the deep sea [15]. Turnover rates - the amount of time required by a species population to completely rework all the sediment in a given area to the depth accessible to individuals - were used to quantify biological mixing of sediments. Sediment turnover rates by deposit-feeding polychaetes vary from once every 15 years to once every 10 weeks. Most rates, however, fall in the range of 1 to 5 years for a population of individual species [16]. The annual rate of sediment turnover for polychaetes from the continental slopes (250 m) was reported to be about 10 kg dry sediment/m²/year. Assuming a medium sediment reworking depth of 5 cm (5 x 10⁴ mL sediment/m²), a steady state population of polychaetes would completely turnover the sediment in 4 to 5 years [16]. In shallow areas, however, it is not surprising to find that the surface of the muddy sea floor is passed through the benthos at least once, and in some cases, several times a year [3].

Sediment reworking activities such as ingestion and egestion by infaunal species contribute significantly to the transfer of nutritive material through the sediment-water interface. The amount of organic matter in sediments can affect sediment turnover rates. The turnover rates decrease with increasing food supply. Deposit-feeders and suspension-feeders entrap biological particles from the water column and the sediment, aggregate the ingested material in the gut, and void the feces as discrete pellets of fecal strings. This process of biodeposition is known to be important to biogeochemical cycles especially in intertidal areas. Many infaunal species in coastal areas are capable of biodepositing several kilograms of feces or pseudofeces per square meter per year [15]. Deposit-feeders probably play the quantitatively most important role in "pelletizing" marine sediments. In some areas between 40 to 100 percent of the particles of the surface of the muds are in the form of pellets produced by polychaetes and bivalves [3], resulting sometimes in an upper, 1-cm thick surface layer of pellets. However, with increasing depth of burial the pellet content of muds usually decreases to reach values of five percent fecal pellets at 30-cm depth in the sediment.

Intensive mixing and irrigation activities of benthic organisms are important in accelerating vertical diffusion and transport of ions or compounds adsorbed on particles or in solution in pore water. Thus, the process of sediment reworking by benthic organisms contributes significantly to the cycling and return of dissolved organic materials buried in the sediment to the water column [16]. In particular, sediment infaunal activities influence the rate of exchange of dissolved and adsorbed ions, compounds, and gases across the sediment-water interface; the oxidation and transfer of reduced compounds from below the interface to the surface; and the cycling of carbon, nitrogen, sulfur, phosphorus, and other elements. In bottom sediments the rates of molecular diffusion are limited to the upper few millimeters, and are several orders of magnitude smaller than the vertical transport by infaunal benthos. Vertical mass transport by populations of deposit-feeding bivalves may reach values of 60 kg to 120 kg sediment/m²/year [3]. In this way organic detritus from the sediment reaches the sediment-water interface. Nutrient-rich bottom muds are resuspended into the water column serving as a potential source of food for suspension-feeders. Thus, these processes of reworking and turnover of sediment by infaunal benthos play an important role in recycling of organic matter buried in the sediment and supply nutrients to the water column.

Sediment-animal interactions. The local distribution of deposit- and suspension-feeders is related to grade of deposit, organic content of the sediment, and re-suspension of the sediment by currents. Deposit-feeding benthos are more abundant in fine grained sediments characterized by high organic content [8]. Seasonal variations in the organic content of the sediment are caused largely by variations in the standing crop of benthic microorganisms. It is known that the abundance of microorganisms is related to the surface area of the sediment. The abundance of deposit-feeders is partly dependent on the abundance of microorganisms [17]. Thus, a feedback mechanism should exist in this system.

The abundance of microorganisms is dependent on the physical characteristics of the sediment, the water content of the sediment (up to 80 to 98 percent), the depth of the aerobic zone, the vertical transport of nutrients, and the breakdown of organic matter. Bioturbation increases the surface area in the sediment available for bacterial colonization. Biodeposits such as feces and pseudofeces add new surfaces for colonization and represent a readily available nutrient source for such organisms. Thus, the more microorganisms are present as food for deposit-feeders the higher the abundance of deposit-feeders, which in turn produce new nutrients in form of fecal pellets for microorganisms [8].

Uptake of Transuranics by Sediment Infauna and Benthic Organisms

The effects of bioturbation on the redistribution of radionuclides has had little quantitative study, although there are indications that bioturbation strongly influences the recycling of radionuclides through the benthic boundary layer especially in highly productive areas [15]. Recent data suggest, that the feeding activities of infaunal benthic organisms transfer radionuclides from the overlying water to the sediment and fix them at some depths within the sediment [18]. Pu-239, Pu-240 and Am-241 fixed in sediments by this mechanism were suggested to be resolubilized in the presence of anionic organic complexes that were released deep in the sediment by the metabolic activities of sediment infauna. Once in soluble form the radionuclides are free to slowly migrate upwards through the sediment via interstitial waters or rapidly through infaunal burrows for release into the overlying waters.

Uptake from sediments. Transuranium nuclides in waters of the benthic boundary layer will become readily associated with a variety of benthic fauna living in or on the sediment [19]. Benthic organisms may play an important role in the redistribution of Pu and Am in near-shore and deep-ocean sediments. Accumulation of these isotopes by benthic biota will occur by direct sediment ingestion [20].

Epifaunal detrivores and predators can also accumulate radionuclides in their search for food in or near the sediment surface. Perhaps most important for the uptake is the sediment ingestion by deposit-feeders. It was reported that tissues of various invertebrate infauna contained concentrations of plutonium 5- to 10-times those in the sediment. This finding suggests that sediment-associated transuranics are available to biota [4]. In laboratory experiments polychaetes derived 83 to 98 percent of the plutonium from a source other than sediment, possibly from deposited material or interstitial pore water [20]. The sediment pathway accounts for three to four percent of the total body burden of plutonium in polychaetes [21]. Generally, concentration factors based on direct uptake of transuranics from sediments are very low (0.001) [20, 21]. Considering the high concentration factors of 100 to 1000 for direct uptake from water, one may conclude that water, possibly pore water, is the predominant pathway for uptake of transuranics by sediment infauna. Furthermore, the fraction of sediment-sorbed isotopes taken up by benthic infauna depends on the element involved, its physical-chemical form, and the type of the sediment. Whether uptake of transuranics from sediments by infauna occurs via ingested sediments or organic matter in the sediment, by absorption from the surface via mechanical action of infauna moving through the sediment, or by direct uptake of dissolved isotopes from interstitial water in the sediments [20] must be clarified.

Transfer factors. Although field studies clearly showed the importance of sediments in transferring radioisotopes to benthic infaunal organisms, the exact mechanism for uptake is still unclear and may vary with the different isotopes. Very little effort has gone into quantifying these processes [22, 23]. Transfer factors - radioactivity per gram of wet animal divided by the radioactivity per gram of wet sediment - were determined to measure the biological availability of radionuclides bound to the sediment for the benthic infaunal species living in contact with the sediment [22]. The transfer factors for contaminated sediments and whole bodies of infaunal organisms, that ingest sediment particles continuously, were found to be very low for plutonium, americium, and technetium. The values obtained range from 0.1 to 0.001 for polychaetes, molluscs, and crustaceans [20, 22, 24]. Transfer factors little higher than ten were reported only for some internal organs [23]. Thus, the net uptake of plutonium, americium, and technetium from contaminated sediments by infaunal benthic organisms is relatively small. Part of the radionuclides transferred to the infaunal organisms will come from the sediment's interstitial pore water. This transfer would be described by concentration factors rather than transfer factors. Another part of the radionuclides retained by the animals would, therefore, be obtained by direct transfer from sedimentary particles to the organisms [22].

Conclusion

The experimental results now available show that benthic infaunal species can mobilize transuranic elements within biodisturbed sediment layers. However, considering the relatively high concentration factors of 1000 determined for the uptake from water and the very low transfer factors of 0.001 for sediments, one may conclude that interstitial or pore water may be the predominant source of transuranics for deposit-feeding organisms of the sediment-water interface.

References

[1] D. C. Gordon, Jr., "The Effects of the Deposit-Feeding *Polychaete Pectinaria gouldii* on the Intertidal Sediment of Barnstable Harbour," **Limnol. Oceaongr.**, **11** (1966) 327.

[2] J. S. Gray, "Animal-Sediment Relationships," **Oceaongr. Mar. Biol. Ann. Rev.**, **12** (1974) 223.

[3] D. C. Rhoads, "Organism-Sediment Relations on the Muddy Sea Floor," **Oceaongr. Mar. Biol. Ann. Rev.,** 12 (1974) 263.

[4] V. T. Bowen, H. D. Livingston and J. C. Burke, "Distribution of Transuranium Nuclides in Sediment and Biota of the North Atlantic Ocean," in **"Transuranium Nuclides in the Environment,"** International Atomic Energy Agency: Vienna, IAEA-SM-199/96, 1976, p. 107.

[5] J. N. Smith and C. T. Schafer, "Bioturbation of Superficial Sediments on the Continental Slope, East of Newfoundland," in **"Marine Radioecology,"** Proceedings of the Third NEA Seminar, Tokyo, October 1-5, 1979, NEA, OECD, Paris: 1980, p. 225.

[6] T. Petr, "Bioturbation and Exchange of Chemicals in the Mud-Water Interface," in **"Interactions Between Sediments and Fresh Water,"** H. L. Golterman, Ed., W. Junk B. V. Publishers: The Hague and Centre for Agricultural Publishing and Documentation, Wageningen, 1977, 473 p.

[7] L. S. Gomez, R. R. Hessler, D. W. Jackson, M. G. Marietta, K. L. Smith, Jr., D. M. Talbert and A. A. Yayanos, "Biological Studies of the United States Subseabed Disposal Program," **Sandia Report No. SAND79-2073** (1980).

[8] E. G. Discroll, "Sediment-Animal-Water Interactions, Buzzard Bay, Massachusetts," **J. Mar. Res.,** 33 (1975) 275.

[9] G. S. Pemberton, M. J. Risk and D. E. Buckley, "Supershrimp: Deep Bioturbation in the Strait of Canso, Nova Scotia," **Science,** 192 (1976) 790.

[10] H. L. Sanders, R. R. Hessler, G. R. Hampson, "An Introduction to the Study of Deep-Sea Benthic Faunal Assemblages along the Gay-Head-Bermuda Transect," **Deep Sea Res.,** 12 (1965) 845.

[11] D. R. Anderson, Ed., "Seventh International NEA/Seabed Working Group Meeting," La Jolla, March 15-19, 1982; **Sandia Report No. SAND82-0460.**

[12] R. R. Hessler and P. A. Jumars, "Abyssal Community Analysis from Replicate Box Cores in the Central North Pacific," **Deep Sea Res.,** 21 (1974) 185.

[13] M. M. Mullin and L. S. Gomez, "Biological and Related Chemical Research Concerning Subseabed Disposal of High-Level Nuclear Waste," Jackson Hole, January 12-16, 1981; **Sandia Report No. SAND81-0012.**

[14] L. S. Gomez, R. R. Hessler, D. W. Jackson, M. B. Marietta, K. L. Smith and A. A. Yayanos, "Environmental Studies Data Base Development and Data Synthesis Activities of the US-Subseabed Disposal Program," in

"Impact of Radionuclide Releases into the Marine Environment," International Atomic Energy Agency: Vienna, No. IAEA-SM-248/142, 1980, p. 607.

[15] S. W. Fowler, "Biological Transfer and Transport Processes," in **Pollutant Transfer and Transport in the Sea, Vol. II,** G. Kullenberg, Ed., CRC Press: Boca Raton, Florida, 1982, p. 1.

[16] F. H. Nichols, "Sediment Turnover by a Deposit-Feeding Polychaete," **Limnol. Oceanogr., 19** (1974) 945.

[17] J. S. Levinton and T. S. Bianchi, "Nutrition and Food Limitation of Deposit-Feeders. I. The Role of Microbes in the Growth of Mud Snails *(Hydrobiidae),*" **J. Mar. Res., 39** (1981) 531.

[18] H. D. Livingston and V. T. Brown, "Pu and Cs in Coastal Sediments," **Earth Planet. Sci. Lett., 43** (1979) 29.

[19] M. C. Grillo, J. C. Guary and S. W. Fowler, "Comparative Studies on Transuranium Nuclide Biokinetics in Sediment-Dwelling Invertebrates," in **"Impact of Radionuclide Releases into the Marine Environment,"** International Atomic Energy Agency: Vienna, IAEA-SM-248/114, 1980, p. 273.

[20] T. M. Beasley and S. W. Fowler, "Plutonium and Americium: Uptake from Contaminated Sediments by the Polychaete *Nereis deversicolor,*" **Mar. Biol., 38** (1976) 95.

[21] C. N. Murray and W. Renfro, "Uptake of Plutonium from Seawater and Sediment by a *Polychaete* Worm," **J. Oceaongr. Soc. Japan, 32** (1976) 249.

[22] P. Miramand, P. Germain and H. Camus, "Uptake of Americium and Plutonium from Contaminated Sediments by Three Benthic Species: *Arenicola marina, Corophium volutator and Scrobicularia plana,*" **Mar. Ecol. Prog. Ser., 7** (1982) 59.

[23] S. W. Fowler, S. R. Aston, G. Benayoun and P. Parsi, "Bioavailability of Technetium from Artificially Labelled North-East Atlantic Deep-Sea Sediments," **Mar. Environ. Res., 8** (1983) 87.

[24] M. Masson, G. Aprosi, A. Laniece, P. Guegueniat and Y. Belot, "Approches Experimentales de l'Etude des Trasferts du Technetium a des Sediments et a des Especes Mariens Benthiques," in **"Impacts of Radionuclide Releases into the Marine Enviroment,"** International Atomic Energy Agency: Vienna, IAEA-SM-248/124, 1980, p. 341.

XVII

SPECIFIC SEQUESTERING AGENTS FOR IRON AND ACTINIDES

Kenneth N. Raymond

*Department of Chemistry, Materials and Molecular Research Division
Lawrence Berkeley Laboratory, University of California
Berkeley, California 94720 USA*

Transuranium actinide ions associated with the waste from the nuclear power industry have unique environmental hazards. Plutonium is a major component of this waste. Because of plutonium's relative abundance and chemical and biological properties, it is potentially hazardous. The preparation of metal-ion-specific complexing agents for ions such as plutonium(IV) is a challenge to the synthetic chemist. Such agents may solve many of the problems associated with nuclear waste. This paper describes a rational approach to the synthesis of such chelating agents. The structures of naturally-occurring complexing agents such as enterobactin that are highly specific for iron(III) serve as the basis for the design of agents specific for plutonium(IV) because of the similarities in the properties of these two metal ions. Ligands containing the same coordinating groups present in the iron complexes but capable of forming octadentate complexes, which are favored by plutonium(IV), were prepared. Preliminary tests showed these ligands to be very effective in removing plutonium from animals.

For several years I and my research group have been interested in the coordination chemistry of metal ions in biological systems and the extension of this chemistry to the preparation of new sequestering agents for ferric ion and actinide(IV) ions. The thesis, that complexing agents specific for iron in the +3 oxidation state might also serve as specific ligands for plutonium in the +4 oxidation state, is central to our approach to the coordination chemistry of the actinides. I will describe in this paper the train of thought that has led to the preparation of a number of compounds useful by themselves or as precursors to derivatives for ameliorating the environmental problems presented by plutonium. Plutonium is a major byproduct of nuclear power reactors. The treatment and storage of nuclear waste from power reactors remains a serious problem which has become a matter of substantial public concern in many countries. Because of its relative abundance and long half-life, Pu-239 is a major environmental hazard of long-term waste after the first 100 years of decay. In addition, the biological properties of plutonium make it a particular hazard once human contamination has occurred. In this paper the following questions are addressed:

- What is the aqueous chemistry of Pu(IV) and what limits does this chemistry set for the design of plutonium-selective complexing agents?
- What are the biological properties of plutonium? Where is it stored in mammals and what damage does it cause?
- How is plutonium transported once it has been incorporated into higher animals?
- What chemical reagents are used presently in the treatment of plutonium contamination in man?
- How can we design better complexing agents?

Chemical and Biological Properties of Plutonium

The plutonium cation, Pu^{4+}, which is present in solution at pH ~0, hydrolyzes with increasing pH to form dark-green colloidal plutonium hydroxides, $[Pu(OH)_x]_n$, with molecular weights in the range 100 to > 1,000,000 [1]. The hydrogen ion concentration, above which polymerization does not occur, increases with increasing plutonium concentration and is 0.12 M for 0.004 M Pu^{4+}, 0.20 M for 0.021 M Pu^{4+}, and 0.32 M for 0.062 M Pu^{4+}. The redissolution of the plutonium hydroxide can take a very long time even in strong acid once extensive polymerization has occurred either at elevated temperature or on prolonged standing (Table 1). For

Table 1. Half-Times for the Depolymerization of Polymeric Plutonium(IV) Hydroxides in Nitric Acid Solutions

Formation Temp. of Precipitate	Half Time of Depolymerization in Nitric Acid			
	2 M	4 M	6 M	10 M
25°	20 days	400 min	100 min	30 min
80°	1 year		15 days	

instance, a freshly precipitated polymeric plutonium(IV) hydroxide had a half-time of depolymerization of 20 hours in 5 M nitric acid at 25°. This half-time increased to 320 hours after the precipitate had aged for several months at room temperature. These data illustrate dramatically the insolubility of plutonium even at low pH. Other structural, thermodynamic and biological aspects of plutonium [2-4] and its elimination from higher animals [5] have recently been reviewed. The plutonium(IV) concentration in the blood plasma of beagle dogs contaminated with plutonium decreased to one-half of the initial concentration during the first ten hours. Approximately 90 percent of the plutonium remaining in the plasma is not excreted but retained. A large fraction of the plutonium(IV) is complexed by transferrin, the serum protein that transports iron in higher animals. The binding of plutonium(IV) to transferrin - just as the binding of iron(III) - requires bicarbonate ion. Plutonium(IV) cannot replace iron(III) from the iron-transferrin complex, but transferrin-bound plutonium(IV) can be replaced by iron(III), indicating that both ions bind to the same site. The plutonium(IV)-transferrin complex can be dissociated by excess citrate at pH 7.5. This complex carries plutonium(IV) from the blood to the eventual deposition sites of plutonium in the bone marrow, the bone surfaces, and the liver. In the cells of organs such as the liver, plutonium(IV) eventually is concentrated in iron storage sites such as the protein ferritin. The ferritin complex of plutonium is highly stable.

Americium behaves quite differently than plutonium. Americium is present only in the trivalent state, which is more rapidly transferred and excreted than plutonium(IV). The design of an americium-specific complexing agent poses a very different problem than the design of a reagent for plutonium.

Plutonium contamination in man is presently treated either with calcium or zinc diethylenetriaminepentaacetate (DTPA). DTPA increases urinary output of plutonium ten- to fifty-fold even months or years after exposure. However, estimates indicate that the total body burden of plutonium is decreased by not more than 25 percent by this treatment. Systemic contamination and contamination by inhalation are treated substantially in the same way. Plutonium contamination resulting from wounds is reduced by surgical removal of tissue around the wound and by washing the affected area with DTPA or ethylendiaminetetraacetic acid (EDTA). The side effects of chelation therapy administered by injection of relatively high doses of EDTA or DTPA may include sometimes quite serious kidney damage, although this is largely ameliorated by administration as the Zn or Ca salts.

Some of the chemical properties of iron(III) and plutonium(IV) are quite similar. The charge/ionic radius ratio is 4.2 for Pu^{4+} and 4.6 for Fe^{3+} [2]. These ratios, which correspond to the electrostatic potentials at the surface of a hard-sphere model of the metal ions, correlate well with many acid-base and complex-formation properties of metal ions. The solubility product per hydroxide in $Fe(OH)_3$ and $Pu(OH)_4$ (eqns. 1, 2) is approximately the same. The hydrolysis constants for the aquo ions (eqns. 3, 4) are also

$$Fe(OH)_3 \longrightarrow Fe^{3+} + 3OH^- \quad K \approx 10^{-38} \ (10^{-13} \text{ per } OH^-) \quad (1)$$

$$Pu(OH)_4 \longrightarrow Pu^{4+} + 4OH^- \quad K \approx 10^{-55} \ (10^{-14} \text{ per } OH^-) \quad (2)$$

$$Fe^{3+} + H_2O \longrightarrow Fe(OH)^{2+} + H^+ \quad K = 0.0009 \quad (3)$$

$$Pu^{4+} + H_2O \longrightarrow Pu(OH)^{3+} + H^+ \quad K = 0.031 \quad (4)$$

quite similar. The most important similarity, however, is the affinity of plutonium(IV) and iron(III) for the same site in transferrin, the iron(III) transport agent. These marked similarities are the starting point for the design of ligands specific for plutonium(IV).

Properties of specific sequestering agents include not only a high formation constant for the metal ion to be sequestered but also a low affinity for most other metal ions. Examination of the properties of biological complexing agents for a metal reveals that their specificity is usually quite high, because

- the coordination number and geometry most favorable to the metal ion is achieved;
- the cavity formed by the ligand exactly matches the size of the metal ions; and
- the electronic properties of the ligand atoms and the metal match.

The siderophore ligands for iron(III) are hexadentate and capable of providing octahedral coordination with a cavity size just right for high-spin iron(III). The ligating oxygen atoms are strongly basic and complement the high

Lewis acidity of the hard ferric ion. Siderophores [6] are good examples of metal-ion-specific complexing agents produced by organisms. These low molecular-weight compounds are synthesized by bacteria, molds, yeasts, and other microbes to obtain the iron essential for microbial growth. An example of a very stable, six-coordinate, high-spin iron(III)-siderophore complex is ferric enterobactin **1** (M = Fe) which is produced by enteric bacteria such as *E. coli*. The cyclic, trigonally symmetric triester backbone of enterobactin **2**, composed of three serine residues, acts as a template to which are appended three catechol groups. The six oxygen atoms of the catechols provide the octahedral cavity and coordination for iron(III).

Synthetic Iron(III) Chelating Agents Modeled after Enterobactin

Enterobactin **2**, desferrichrome **3**, and desferrioxamine B **4** (which is clinically used to treat human iron overload) are three siderophores, the structures of which should serve as guides for the preparation of sequestering agents for iron. The essential structural features of desferrichrome and enterobactin are very similar. Desferrichrome has a cyclic hexapeptide as a backbone to which are bonded three hydroxamate groups. In desferrioxamine B the three hydroxamate groups are arranged in a linear chain. In spite of the differences in the structures of desferrichrome and desferrioxamine B, both compounds provide essentially identical coordination sites for iron. The UV-visible spectra of the two complexes are indistinguishable. For siderophores, for which the large enthalpy change associated with the acid-base neutralization during the coordination of the metal ion is the driving force for the reaction, the ability to form the cavity required for coordination during the reaction is more important than the existence of the cavity before the reaction.

The formal stability constants for the complexes ferric enterbactin (10^{52}), ferrioxamine B ($10^{30.6}$), and ferrichrome ($10^{29.1}$) [7], in which iron(III) is coordinated to the fully deprotonated ligands, do not adequately describe the relative stabilities of these complexes under physiological conditions, because the very weakly acidic ligands make the complex-forming reaction highly pH dependent. The equilibrium concentration of free ferric ion in aqueous solution at a physiological pH of 7.4 can be calculated using pertinent formation and acid dissociation constants. The negative logarithms of the concentrations of the hexaaquoferric ion (pM) are 35.5 for enterobactin, 26.6 for ferrioxamine B_1 and 25.2 for ferrichrome assuming a total iron(III) concentration of 1.0×10^{-6} M and a total ligand concentration of 1.0×10^{-5} M. These concentrations - although too low to have a real chemical meaning - give pM values which are well defined in a thermodynamic sense and give an absolute ranking of the relative free energies of the iron complexes. The much higher relative stability of the enterobactin complex suggested it as a model for the synthesis of ligands. Compounds **5, 6** and **7** are representative examples of synthetic tricatecholate analogs of enterobactin. These ligands are all catechoyl amides, hence the acronym CAM. The prefixes ME (for mesitylene), LI (for linear), and CY (for cyclic) identify the backbone. The suffix S indicates the presence of a sulfonic acid group in the 5-position of the catechol ring. Details about the synthesis of these compounds are presented in reference 2.

The catechoyl amides are deprotonated at relatively high pH. The pK_a values for the deprotonation reactions of N,N-dimethyl-2,3-dihydroxybenzamide (eqns. 5, 6), a simple model for the ligating groups in enterobactin, are 8.4 and

Specific Sequestering Agents

MECAM 5a, R = CH₂—NH—C(=O)—(catechol: OH, OH)

MECAMS 5b, R = CH₂—NH—C(=O)—(catechol-SO₃⁻)

TRIMCAMS 5c, R = C(=O)—NH—CH₂—(catechol-SO₃⁻)

(structure **5**: 1,3,5-trisubstituted benzene with R groups)

3,4-LICAMS (**6**): NH—CH₂—CH₂—CH₂—N(C=O—Ar)—CH₂—CH₂—CH₂—CH₂—N(C=O—Ar)—CH₂—CH₂—CH₂—NH—(C=O—Ar), where Ar = sulfonated catechol (SO₃⁻, OH, OH)

4: macrocycle with (CH₂)₃, (CH₂)₃, (CH₂)₄ bridges between N—R groups

CYCAM (**7**): R = C(=O)—(catechol: OH, OH)

CYCAMS: R = C(=O)—(catechol-SO₃⁻)

$$\text{[N(CH}_3\text{)}_2\text{-C(=O)-C}_6\text{H}_3\text{(OH)}_2\text{]} \;\underset{H^+}{\overset{pK=8.4}{\rightleftharpoons}} \; \text{[deprotonated form]} \quad (5)$$

$$\text{[mono-deprotonated]} \;\underset{H^+}{\overset{pK=12.1}{\rightleftharpoons}} \; \text{[di-deprotonated]} \quad (6)$$

12.1. It has been found that all catechoyl amides, including the multidentate ligands, have pK$_a$ values that cluster around the first and second pK$_a$ of the N,N-dimethyl-2,3-dihydroxylbenzamide. The titration curve of the sulfonated tricatechoyl amide MECAMS **5b** has two plateaus

(Fig. 1). The first plateau corresponds to the removal of

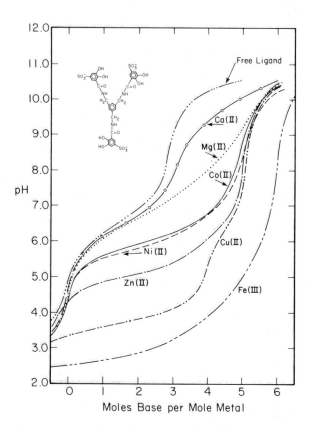

Figure 1. Titration curves of MECAMS as the free ligand and in the presence of several metal ions.

three protons, one proton from each of the three catechol groups. The second plateau corresponds to the three remaining catechol protons. The presence of divalent metal cations such as Ca(II) and Mg(II) causes a relatively small depression of the titration curve, indicating only weak complexation. The most strongly complexed of the divalent cations is copper(II). In the copper complex only two of the three catechol groups in the ligand are bonded to the metal ion. Iron(III) strongly depresses the titration curve of MECAMS, indicating very strong and relatively specific complexation.

Synthesis of Actinide(IV)-Specific Complexing Agents

Strong complexing agents that are relatively specific for plutonium(IV) and other actinide ions are urgently needed. Unfortunately, the available agents such as DTPA are relatively non-specific. A rational approach to the synthesis of complexing agents specific for plutonium(IV) has been based on the following facts.

- Plutonium(IV) is chemicaly similar to iron(III) and has similar biological transport and distribution properties in mammals.
- The most powerful naturally occurring sequestering agents which are specific for iron(III) are produced by microbes. The strongest of these is the tricatecholate ligand enterobactin **2**.
- The requirement of plutonium(IV) for a higher coordination number compared to iron(III) means that four catechol groups should be incorporated into a plutonium(IV)-specific complexing agent to make it octadentate.

To determine the preferred coordination geometry for tetracatecholate complexes of actinide(IV) ions that contain ligands with no geometric constraints, a series of $M(cat)_4^{4-}$ complexes (cat = catecholate dianion) were prepared and structurally characterized [8]. The catecholate groups surround the metal ion in a roughly tetrahedral arrangement with the eight-coordinate geometry very close to the D_{2d} symmetry of a trigonal-faced dodecahedron. Molecular models of a series of compounds with four catechole ligands appended to a backbone were constructed. The synthesis of one series of these compounds is shown in eqn. 7. The sulfonated derivatives **8** were found to be effective agents for the removal of plutonium in animal tests but had a significant toxicity at the administered doses [1, 5]. Sulfonate groups are needed in these ligands because the ligands without sulfonate groups are rather insoluble in water, are relatively unstable in air, and have an acidity too low to complex metals in a weakly acidic medium. Instead of sulfonate groups carboxylate groups can be used. The ligands with carboxylate groups can be prepared according to eqn. 8. These carboxylated ligands are as

effective as the sulfonated derivatives, but do not have the toxic side effects. The ligand 3,4,3-LICAMC **9** was tested in mice and dogs. A small amount of Pu(IV)-238 citrate was injected. One hour later the 3,4,3-LICAMC in saline solution was injected. The animals were sacrificed after 24 hours to evaluate the removal of plutonium. During the 24-hour period only 6.4 percent of the administered plutonium would have been naturally eliminated in the urine and feces. After administration of 3,4,3-LICAMC 75 percent of the plutonium was eliminated [5], a percentage substantially larger than achieved with any other chelating agent (Table 2).

Table 2. Distribution of Plutonium in Mice after Administration of Pu(IV)-238* and Elimination of Plutonium from Mice in the Urine and Feces in the Absence and Presence of Polycatechoyl Carboxylate Ligands**

Organ, Tissue or Excreta	Control 1 hr†	Control 24 hr†	Percent of Injected Plutonium-238 3,4,3-LICAMC 24 hr†	Poly-LICAMC 24 hr††	Dioctyl-LICAMC 24 hr†††
Liver	31.	49.	8.4	29.	30.
Skeleton	24.	32.	9.6	10.	15.
Soft Tissues	38.	10.	6.8	16.	16.
Urine	1.3	4.2	54.	20.	3.3
Feces	5.1	4.7	21.	15.	36.
Total Percent	100.	100.	100.	100.	100.
Percent in Urine and Feces	6.4	8.9	75.	35.	39.3
No. of Mice	13	84	15	5	5

*1.5 μCi/kg Pu(IV)-238 citrate i.v.
**30 μmol/kg ligand in saline i.p. given 1 hr after injection of Pu-238.
†Time elapsed between administration of Pu-238 and sacrifice.
††A LICAM derivative based on a polymeric amine backbone.
†††A LICAM derivative in which the two terminal amine groups of the backbone have been alkylated.

Recently an octadentate ligand based on the naturally-occurring chelating agent desferrioxamine B, which is presently used under the trade name Desferal to treat human iron overload, was synthesized [9]. Catechol groups were appended to the terminal, primary amino groups of the ferrioxamine B (eqns. 9, 10). Whereas the 2,3-dihydroxybenzoylferrioxamine B **10** is relatively insoluble, the carboxylate-substituted ligand **11** is sufficiently soluble in water. Preliminary results of experiments to remove plutonium with these chelating agents are encouraging.

Another new approach to the synthesis of actinide-specific sequestering agents, which addresses the need for the metal ions to be complexed under weakly acidic conditions, uses hydroxypyridones. Whereas catechol groups are weakly acidic and confer a second-order dependence on the hydrogen ion concentration per catechol group, the hydroxypyridonate ligands are much more acidic than catechols and are monoprotic ligands. Structural and acid-base properties of hydroxypyridonate (HOPO) ligands are summarized in Table 3. An octadentate ligand based on

Table 3. *Structures and Acid-Based Properties of Hydroxypyridones, Catechol and Acetohydroxamic Acid*

Compound	Structure	pK_a
1-Hydroxy-2(1H)-pyridinone 1,2-HOPO		-0.9
		5.78
3-Hydroxy-2(1H)-pyridinone 3,2-HOPO		0.11
		8.65
3-Hydroxy-4(1H)-pyridinone 3,4-HOPO		3.34
		9.01
Catechol		9.22
		13.0
Acetohydroxamic acid		9.36

Specific Sequestering Agents

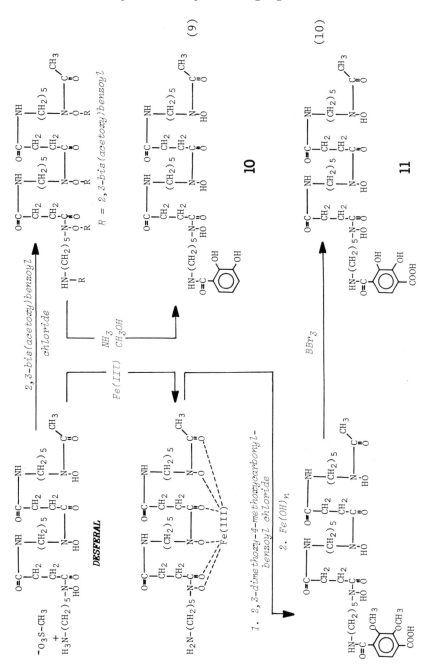

hydroxypyridone will show at most an inverse fourth-order dependence on the hydrogen ion concentration for the stability of a complex, whereas the tetracatecholate ligands will show an inverse eighth-order dependence. The hydroxypyridones can be viewed as derivatives of catechol in which a carbon of the catechol ring has been replaced by a nitrogen atom. Only 1,2-HOPO has been incorporated into multidentate structures thus far as shown in eqn. 11. The

(11)

carbonate anhydride **12** formed as an intermediate from phosgene and 6-carboxy-1,2-HOPO **13** allows the preparation of multidentate chelating agents from simple amines or polyamines (eqn. 11). The coordination and biological properties of these novel ligands are presently under investigation.

Acknowledgements. I would like to acknowledge the many co-workers, past and present, whose work I have summarized in this short paper. These include Dr. Patricia Durbin, Dr. Frederick Weitl, Dr. Wesley Harris, Mr. Steven Rodgers, Mr. Rob Scarrow, Dr. David White, Dr. Mary Kappel, Dr. Carl Carrano, Dr. Vincent Pecoraro, Dr. Stephen Sofen, and Dr. Alex Avdeef.

This work was supported by the Director, Office of Energy Research, Office of Basic Energy Sciences, Chemical Sciences Division of the U.S. Department of Energy under Contract Number DE-AC03-76SF00098.

References

[1] P. W. Durbin, "Metabolism and Biological Effects of the Transplutonium Elements," in **"Handbook of Experimental Pharmacology, Vol. 36, Uranium, Plutonium, Transplutonic Elements,** H. C. Hodge, J. N. Stannard and J. B. Hursh, Eds., Springer-Verlag: Berlin, 1973, p 739; and references cited therein.

[2] K. N. Raymond and W. L. Smith, "Actinide-Specific Sequestering Agents and Decontamination Applications," in **"Struct. Bonding (Berlin),"** Vol. **43**, J. B. Goodenough, et al., Eds., Springer Verlag: Berlin, 1981, p. 159.

[3] K. N. Raymond, M. J. Kappel, V. L. Pecoraro, W. R. Harris, C. J. Carrano, F. L. Weitl and P. W. Durbin, "Specific Sequestering Agents for Actinide Ions," in **"Actinides in Perspective,"** N. M. Edelstein, Ed., Pergamon Press: New York, 1982, p 491.

[4] K. N. Raymond, V. L. Pecoraro, W. R. Harris and C. J. Carrano, "Actinide Coordination and Discrimination by Human Transferrin," in **"Environmental Migration of Long-Lived Radionuclides,"** Proceedings, International Symposium, Atomic Energy Agency, Vienna, 1982, p. 571.

[5] R. D. Lloyd, F. W. Bruenger, C. W. Mays, D. R. Atherton, C. W. Jones, G. N. Taylor, W. Stevens, P. W. Durbin, N. Jeung, E. S. Jones, M. J. Kappel, K. N. Raymond and F. L. Weitl, "Removal of Pu and Am from Beagles and Mice by 3,4,3-LICAM(C) or 3,4,3-LICAM(S)," **Radiat. Res.**, **99** (1984) 106; and references cited therein.

[6] K. N. Raymond, K. Abu-Dari and S. R. Sofen, "Stereochemistry of Microbial Iron Transport Compounds," in **"Stereochemistry of Optically Active Transition Metal Compounds,"** B. E. Douglas and Y. Saito, Eds., **ACS Symp. Ser.**, **119** (1980) 133.

[7] A. Avdeef, S. R. Sofen, T. L. Bregante and K. N. Raymond, "Coordination Chemistry of Microbial Iron Transport Compounds. 9. Stability Constants for Catechol Models of Enterobactin," **J. Am. Chem. Soc.**, **100** (1978) 5362.

[8] S. R. Sofen, S. R. Cooper and K. N. Raymond, "Crystal and Molecular Structures of Tetrakis(catecholato)-hafnate(IV) and -cerate(IV). Further Evidence for a Ligand Field Effect in the Structure of Tetrakis-(catecholato)uranate(IV)," **Inorg. Chem.**, **18** (1979) 1611.

[9] S. J. Rodgers and K. N. Raymond, "Ferric Ion Sequestering Agents. 11. Synthesis and Kinetics of Iron

Removal from Transferrin of Catechoyl Derivatives of Desferrioxamine B," **J. Med. Chem.**, **26** (1983) 439.

DISCUSSION

E. H. Abbott (Montana State University): You pointed out that the plutonium chelates do not remove all of the plutonium from poisoning victims and that the effectiveness of these drugs depends on the type of tissue in which the plutonium resides. Does the plutonium redistribute itself in the tissues after such a treatment and can additional plutonium be extracted with a subsequent treatment?

K. N. Raymond: I do not believe that plutonium is significantly redistributed in the tissues after chelation therapy through the action of the chelating agent. It is well known that incorporated plutonium changes its distribution significantly over the course of time. Much of the plutonium that eventually is deposited in bone is, of course, inaccessible to administered chelating agents. Additional plutonium can be extracted with subsequent treatment but with decreasing effectiveness.

P. J. Sadler (Birkbeck College, University of London): Do you think new standards are required for the determination of very large metal-ligand affinity constants? You mentioned a pM scale (negative logarithm of the free ferric concentration) for Fe(III) catechols. How did you determine these concentrations?

K. N. Raymond: I think the issue is not so much new standards for the measurement of very large metal-ligand affinity constants, but rather the fact that such large constants normally are associated with ligands which are relatively strong bases and hence extensively protonated at neutral pH (or the relevant conditions for the metal complexation reaction). It is for this reason that we have used the pM scale to compare the relative effectiveness of complexing agents under a prescribed set of experimental conditions as they compete for ferric ion. When all of the relevent protonation and complexation equilibrium constants are known, a total ligand and total ferric ion concentration is sufficient to calculate the free ferric ion concentration at any given pH. The pM value is then defined as $-\log [Fe(OH_2)_6]^{3+}$. Even though this may be a number that corresponds to fractions of a free ferric ion per liter in concentration units, it is well defined in a thermodynamic sense, because it is a direct measure of the solution free energy of the metal complex. Thus, a ligand with a higher pM value under the specified conditions will always form a more stable iron complex than a second ligand with a lower pM value under those conditions. When doing electrochemical experiments, often the change in redox potential of the ferric complex from the free ferric/ferrous couple is a direct measure of the pM value. However, normally this is a quantity calculated from the measured equilibrium constants. Unlike the equilibrium constants, however, it has a built-in correction for the effects of proton competition for the ligand.

R. E. Sievers (University of Colorado, Boulder): When we were designing a new hexadentate tris(β-diketonate) ligand, we made models and tried to pick the optimal number of bridging atoms to form complexes with the highest stabilities. Did you observe a strong dependence of stability on the number of atoms in the backbone to which the catecholate groups are bound?

K. N. Raymond: Yes, there is a strong dependence.

A. E. Martell (Texas A&M University): We do not have enough examples to see a trend in the dependence of stability on the structure of the backbone. I think the number of atoms required in the backbone would depend on the number of atoms in the bridges attaching the bidentate units to the backbone.

XVIII

MERCURY ACCUMULATION IN A PELAGIC FOODCHAIN

Michael Bernhard

Centro Studi Ambiente Marino, ENEA
I-19100 La Spezia, Italy

Mercury, a trace element of concern in the Mediterranean, is ingested by man predominantly with marine foods. The mercury concentrations in marine foods from the Mediterranean are often above the legal limits and are known to reach 4500 µg Hg per kg fresh weight in large tunas. The mercury levels in Mediterranean pelagic fishes were found to be higher than the levels in specimens of the same species in the Atlantic. These differences are probably caused by lower natural mercury concentrations in the Atlantic than in the Mediterranean. A model was developed for the mercury transport through the food chain seawater - plankton - sardines - tuna. The model predicts that small differences in the mercury concentration of seawater are sufficient to explain the large differences in mercury levels observed for tunas (Thunnus thynnus) from the Atlantic and the Mediterranean, and shows, that the fraction of total mercury present as methyl mercury increases with the age and the trophic level of an organism. Accurate and precise concentrations of total mercury and methyl mercury in seawater and pelagic organisms are needed to validate the model.

More than 80 percent of the mercury intake by man stems from marine foods [1]. Concentrations of mercury several times higher than the legal limits of 0.5 to 0.7 mg/kg (fresh weight) in force in several Mediterranean countries were observed in certain marine foods from the Mediterranean. These high mercury levels in commercially important sea foods cause legal and perhaps health problems. Especially high levels well above the legal limits were detected in large tunas. The mercury levels were found to be higher in many Mediterranean pelagic fishes than in the same species in the Atlantic [2-4].

Mercury Levels in Tuna

As part of the Med Pol II pilot project, the mercury levels in medium to large tunas caught in land-based tuna traps (tonnare) in Sicily and Sardinia [4] were determined. A plot of the mercury level versus living weight (filled and open circles in Fig. 1) revealed two distinct populations: a

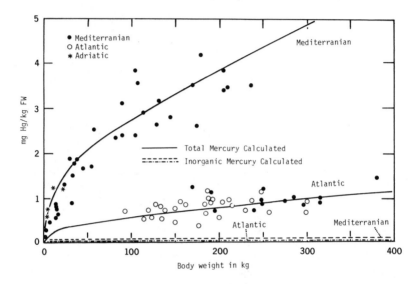

*Figure 1. Concentrations of inorganic mercury and total mercury versus body weight in **Thunnus thynnus** from the Mediterranean and the Atlantic.*

low-mercury population and a high-mercury population. The migration pattern of bluefin tuna can explain these observations. Biologists studying these migration patterns have maintained for some time that Atlantic tunas enter the Mediterranean for spawning and return to the Atlantic through the Strait of Gibraltar [5]. "Tonnare" set to trap tunas entering the Mediterranean, catch them from April to the beginning of May. "Tonnare" set in Sicily and Sardinia catch tunas from May to June, and "tonnare" in the Strait of Gibraltar catch tunas leaving the Mediterranean from July to August. Records kept for more than 150 years attest to the regularity of this migration. All tunas trapped while entering or leaving the Mediterranean belonged exclusively to the low-mercury population [4, 6]. Tunas caught during March along the north-east coast of Spain were part of the high-mercury population [7], whereas both populations were represented in tunas from Sicily and Sardinia [4, 8]. All small tunas collected north of Sicily, in the Adriatic, and in the Ligurian Sea belonged only to the high-mercury population [4].

Similar differences in mercury levels were observed in anchovy, mackerel, and sardines [2, 3]. These species are also pelagic but migrate over much more restricted areas than tunas. Therefore, anchovy, mackerel, and sardines will have mercury levels characteristic of small regions of the sea, without showing excessive coastal or benthic influences. Atlantic specimens of *Sardina pilchardus* have lower mercury concentrations than Mediterranean specimens of the same size (Fig. 2). However, published mercury concen-

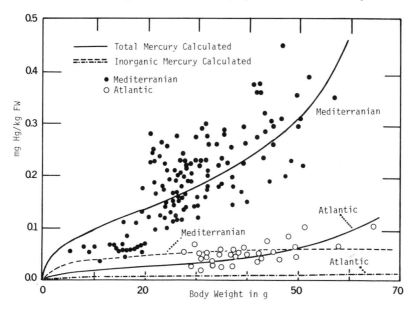

*Figure 2. Concentrations of inorganic mercury and total mercury in **Thunnus thynnus** from the Mediterranean and the Atlantic versus body weight.*

trations for seawater and plankton in the Atlantic and the Mediterranean did not show these differences [9]. The higher mercury concentrations in the Mediterranean pelagic fishes must be caused by the higher than average natural mercury levels in the Mediterranean. Evidence for such a higher mercury levels is only available for the western Mediterranean. The Mediterranean area is a well known geochemical mercury anomaly. Its significant mercury deposits belong to the orogenic and volcanic Mediterranean-Himalaya belt [10]. Although a sytematic survey of the mercury deposits in the Mediterranean region has not been carried out, 65 percent of the world's mercury resources are estimated to be located in the Mediterranean region which occupies only one percent of the earth's surface [11].

Model of Mercury Uptake in Tunas

To better understand the differences in mercury levels between Mediterranean and Atlantic biota a simple model of mercury uptake and loss for the food chain of the tunas was constructed. Plankton is the typical food of sardines. Tunas feed principally on clupeids, anchovies, and mackerel. To simplify the model, sardines were assumed to be the sole diet and the sole source of mercury from food for tunas. Approximately 90 percent of the mercury in fish is methyl mercury. Methyl mercury is preferentially retained over inorganic mercury and released at a much slower rate than inorganic mercury. The biological half-life of methyl mercury is in the range of 300 to more than 1000 days, whereas inorganic mercury has a half-life of approximately 100 days [12-14]. Methyl mercury is probably formed by organisms at the base of the foodchain. Attempts to demonstrate the biological transformation of mercury in higher marine organisms have thus far been unsuccessful [15]. Methyl mercury has not yet been identified with certainty in seawater. A few reports [16-18] claim that 0.5 to 6.0 percent of the total mercury in seawater is methyl mercury.

The accumulation of mercury in an organism (the change in its body burden), can be expressed as the difference between mercury intake and mercury loss (eqn. 1). The intake of mercury from the water, I_w, depends on the amount of water passing through the gills (the filtration volume), the assimilation efficiency or fraction, E, of methyl mercury or inorganic mercury taken up from the water, and the concentration of mercury in the water (eqn. 2). The intake of mercury from the food, $I_f(t)$, depends on the consumption rate of food containing mercury, and on the absorption of mercury from the gut (eqn. 3).

To model the accumulation of mercury in the foodchain plankton - sardines - tuna, the assumption was made that the mercury concentration in plankton is 5000-times the concen-

$$\frac{dQ(t)}{dt} = [I_w + I_f(t)]W(t) - LQ(t) \qquad (1)$$

t   : the fish's age in days
Q(t) : the mass of mercury in μg (body burden of a specimen) in form of methyl mercury or inorganic mercury at time t
W(t) : the fish's body weight in kg at time t
I_w : the intake in μg of methyl mercury or inorganic mercury per kg body weight/day from seawater
$I_f(t)$: the intake of methyl mercury or inorganic mercury per kg of body weight/day from food at time t expressed in μg mercury
L   : loss coefficient equal to log2 divided by the biological half-life (in days) of the loss rate for methyl mercury or inorganic mercury

$$I_w = EVC_w \qquad (2)$$

E : the fraction of methyl mercury or inorganic mercury taken up from the water (assimilation efficiency)
V : the volume of water passing through the gills in liters per kg body weight per day (filtration volume)
C_w: the concentration of methyl mercury or inorganic mercury in the water in μg Hg/L

$$I_f = FrC_f(t) \qquad (3)$$

F : the fraction of methyl mercury or inorganic mercury of the total mercury absorbed from the ingested food through the gut (assimilation efficiency)
r : the relative consumption rate of foods by the fish expressed in grams of food per kg body weight per day
$C_f(t)$: the concentration of mercury in food in $\mu g/kg$

tration of mercury in seawater. The growth of sardines [19] was modeled with Bertalanffy's negative exponential growth curve [20]. The growth of bluefin tunas could also have been calculated with Bertalanffy's equation [20], but a better fit to the growth data reported by Scaccini [21] and Mather [22] was obtained with equation 4.

$$W(t) = pt + \frac{p}{q} \log \frac{1 + r^{-qt}}{1 + r} \qquad (4)$$

p = 27.5 kg/year; q = 0.5/year; r = 10

Numerical Simulations

Preliminary calculations of the uptake of mercury by tuna were performed neglecting uptake of methyl mercury from seawater and of inorganic mercury from seawater and food. Thus, the sole source of mercury for tuna was considered to be the methyl mercury in sardines, which was assumed to be 90 percent of the sardines' body burden of mercury. After several sensitivity calculations 1.5 percent of the tuna's body weight was selected as its daily consumption rate of sardines. Small tunas are expected to feed on small sardines which have lower mercury concentrations than larger (older) sardines (Fig. 2). During the first three months of their life, tunas will feed on sardines, the methyl mercury concentration of which will increase from 2.5 µg/kg to 50 µg/kg in Atlantic sardines and from 2.5 µg/kg to 225 µg/kg fresh weight in Mediterranean sardines. The concentration of methyl mercury was assumed to remain constant thereafter at the levels of 50 and 225 µg/kg, respectively. The loss of methyl mercury was based on a biological half-life of methyl mercury in tunas of 7.5 years. This half-life is longer than the 300 days reported by Pentreath [12]. However, half-lives for mercury longer than 1000 days were observed in extended experiments with freshwater species [13]. Despite the extreme simplicity of the model, the total mercury levels in tunas (solid lines in Fig. 1) agree with the experimental total mercury values quite well. The model predicts that the fraction of the total mercury present as methyl mercury increases with body weight (age) of the organisms. Organisms at higher trophic levels (e.g., tunas) are predicted to have a lower body burden of inorganic mercury than organisms at a lower trophic level (e.g., sardines) (compare dashed lines in Fig. 1 and Fig. 2).

A similar calculation could not be performed for sardines, because the published mercury concentrations in plankton, the food source of sardines, were practically the same for Atlantic and Mediterranean species [9]. Therefore, the observed body burdens of mercury for sardines (Fig. 2) were used to determine the expected differences of total mercury levels between Mediterranean and Atlantic plankton. The calculations - assuming a reasonable three percent of a sardine's body weight as its daily consumption rate of plankton [23] - indicated that Atlantic plankton should reach and maintain 2.5 µg methyl mercury per kg of fresh weight and Mediterranean plankton 12.5 µg. Assuming that approximately 50 percent of the total mercury in plankton is methyl mercury [24], the average total mercury concentrations in Atlantic and Mediterranean plankton should be 5 and 25 µg per kg of fresh weight, respectively. These concentration differences are too small to be discernable in the available literature data, which pertain to mixtures of species and suffer from inaccuracies associated with the determination of mercury. Only very careful experiments and comparisons of the mercury levels in the same planktonic species might reveal statistically significant differences.

If one assumes that the concentration of mercury in plankton is 5000-times higher than in seawater, the Atlantic seawater should have 1 ng total mercury per liter and the Mediterranean 5 ng/L. At these concentrations the accuracy of the determinations of mercury is not good at all, and the difference between these concentrations is again much too small to be discernable in literature data [9].

Using the estimated mercury concentrations in seawater and plankton, the average mercury concentrations presented in Table 1 were obtained. These concentrations must be

Table 1. Concentrations of Methyl Mercury and Inorganic Mercury in Seawater, Plankton, and Sardines* in the Atlantic Ocean and the Mediterranean Sea

Mercury Species	Mercury Concentrations/Atlantic Ocean		
	Seawater µg Hg/L	Plankton µg Hg/kg FW	Sardines µg Hg/kg FW
CH_3HgX	0.0002	2.5	45.
Hg inorg.	0.0008	2.5	5.
Total	0.0010	5.0	50.

Mercury Species	Mercury Concentrations/Mediterranean		
	Seawater µg Hg/L	Plankton µg Hg/kg FW	Sardines µg Hg/kg FW
CH_3HgX	0.0005	12.5	225.
Hg inorg.	0.0045	12.5	25.
Total	0.0050	25.0	250.

*Relations used in calculations: 98% of Hg(total) in seawater is inorganic Hg. 2% is CH_3HgX. In plankton inorganic Hg is equal to CH_3HgX. Food consumption per day in percent of body weight - tunas 1.5%, sardines 3%. Liters of seawater through the gills per kg of body weight per day in tunas and sardines - 300 L. Biological half-life for CH_3HgX (inorg. Hg) in tunas and sardines - 2700 (60) days. Absorption efficiency for CH_3HgX (inorg. Hg) for tunas and sardines - 0.9 (0.1).

verified. Small changes in the assumptions, which had to be made in the absence of reliable data, will change considerably the predicted concentrations of methyl mercury and inorganic mercury in seawater and plankton. Recent mercury measurements [25] indicate that the total mercury concentration in Mediterranean seawater is in the range 1 to 5 ng/L. The model presented in this paper predicted 5 ng/L (Table 1), a value which is strongly influenced by the seawater/plankton concentration factor. This unknown factor was assumed to be 5000, but it could be different. Consequently, different seawater concentrations would result. It is important to remember that the model predicts the Mediterranean mercury concentration to be five times higher than the mercury concentration in the Atlantic.

In conclusion it can be said that small changes in the values of the parameters used may affect the prediction of seawater and plankton concentrations as well as the relative distribution of inorganic mercury versus total mercury. However, the conclusion that the total mercury concentration in the seawater from the Mediterranean must be higher than in the Atlantic remains valid. Of course, a higher proportion of methyl mercury in the Mediterranean seawater will result in higher methyl mercury concentrations and hence higher total mercury concentrations in Mediterranean biota. More precise data both for methyl mercury and total mercury capable of showing the small concentration differences are needed to validate the predictions of the model.

References

[1] F. K. Piotrowski and F. Inskip, **"Health Effects of Methylmercury,"** Progress Report 1975-1979, Monitoring and Assessment Research Centre, London.

[2] M. Stoeppler, M. Bernhard, F. Backhaus and E. Schulte, "Comparative Studies on Trace Metal Levels in Marine Biota. I. Mercury in Marine Organisms from the Western Italian Coast, the Strait of Gibraltar and the North Sea," **Sci. Total Environ., 13** (1979) 209.

[3] F. Baldi, A. Renzoni and M. Bernhard, "Mercury Concentrations in Pelagic Fishes (Anchovy, Mackerel and Sardines) from the Italian Coast and the Strait of Gibraltar. **IV Journess Etud. Pollutions, Antalya, CIESM, Monaco,** pp. 251-254 (1979).

[4] A. Renzoni, M. Bernhard, R. Sara and M. Stoeppler, "Comparison Between the Hg Concentration of *Thunnus thynnus* from the Mediterranean and the Atlantic," **IV**

Journess Etud. Pollutions, Antalya, CIESM, Monaco, pp. 255-260 (1979).

[5] R. Sara, "Sulla Biologia dei Toni (Thunnus thynnus L.), Modelli di Migrazione ed Osservazioni sui Meccanismi di Migrazione e di Comportamento," **Boll. Pesca Piscic. Idrobiol., 28** (1973) 217.

[6] R. Establier, "Concentration de Mercurio en los Tejidos de Algunos Peces, Moluscos y Crustaceos del Golfo de Cadiz y Calderos de la Costa Africana," **Invest. Pesq., 36** (1972) 355.

[7] A. Ballester, L. Cros and R. Ras, "Contenido en Mercurio de Algunos Organismos Marinos Comerciales des Mediterraneo Catalan," **Ist. Investigaciones Pesqueras (Barcelona) Minograph,** (1978) 27.

[8] B. Paccagnella, L. Prati and A. Bigoni, "Studio Epidemiologico sul Mercurio nei Pesci e la Salute Umana in un'Isola Italiana del Mediterraneo," **L'Igiene Moderna, 66** (1973) 479.

[9] M. Bernhard and G. Buffoni, "Mercury in the Mediterranean: an Overview," **Proc. Intern. Conf. Environ. Pollution, Thessaloniki,** Sept. 21-25, 1981, published by Univ. Thessaloniki, Greece, 1981, p. 458.

[10] J. O. Nriagu, "Production and Uses of Mercury," in **"The Biogeochemistry of Mercury in the Environment,"** J. O. Nriagu, Ed.; Elsevier: Amsterdam, 1979, p. 23.

[11] M. Bernhard and A. Renzoni, "Mercury Concentrations in Mediterranean Marine Organisms and their Environment: Natural or Anthropogenic Orgin?" **Thalassia Jugoslavica, 13** (1977) 265.

[12] R. J. Pentreath, "The Accumulation of Mercury from Food by the Plaice, Pleuronectes platessa L.," **J. Exp. Mar. Biol. Ecol., 25** (1976) 51.

[13] J. W. Huckabee, J. W. Elwood and S. G. Hildebrand, "Accumulation of Mercury in Freshwater Biota," in **"The Biogeochemistry of Mercury in the Environment,"** J. O. Nriagu, Ed.; Elsevier: Amsterdam, 1979, p. 277.

[14] H. L. Windom and D. R. Kendall, "Accumulation and Biotransformation of Mercury in Coastal and Marine Biota," in **"The Biogeochemistry of Mercury in the Environment,"** J. O. Nriagu, Ed.: Elsevier: Amsterdam, 1979, p. 303.

[15] H. S. K. Pan-Hou and N. Imura, "Biotransformation of Mercurials by Intestinal Microorganisms Isolated from Yellowfin Tuna," **Bull. Environ. Contam. Toxicol., 26** (1981) 359.

[16] M. Fujita and K. Iwashima, "Estimation of Organic and Total Mercury in Seawater around the Japanese Archipelago," **Environ. Sci. Technol., 15** (1981) 929.

[17] G. Topping and I. M. Davies, "Methylmercury Production in the Marine Water Column," **Nature, 290** (1981) 243.
[18] J. Yamamoto, Y. Kaneda and Y. Hikasa, "Picogram Determination of Methylmercury in Seawater by Gold Amalgamation and Atomic Absorption Spectrophotometry," **Int. J. Environ. Anal. Chem., 16** (1983) 1.
[19] B. Andreu and M. L. Fuster de Plaza, "Estudio de la Edad y Crecimiento de la Sardina (Sardina pilchardus Walb.) del NW de Espana," **Inv. Pes., 21** (1962) 49.
[20] L. von Bertalanffy, "A Quantitative Theory of Organic Growth (Inquiries on Growth Laws. II)," **Human Biol., 10** (1934) 181.
[21] A. Scaccini, "Biologia e Pesca die Tonni nei Mari Italiani," **Ministero Marina Mercantile, Roma, Memoria No. 12** (1965) 1.
[22] F. J. Mather, "The Bluefin Tuna Situation," **Proceedings, 16th Annual Intern. Game Fish Res. Conf.**, Oct. 29-30, 1973, New Orleans, Louisiana, 1974, p. 93.
[23] K. von Broecker, "Eine Methode zur Bestimmung des Kaloriengehaltes von Seston," **Kieler Meeresforschung, 29** (1973) 34.
[24] G. A. Knauer and J. H. Martin, "Mercury in a Marine Pelagic Food Chain," **Limnol. Oceanogr., 17** (1972) 868.
[25] M. Stoeppler, **KFA Julich, German Federal Republic,** private communication, 1983.

DISCUSSION

J. M. Wood (University of Minnesota): Your mercury survey in tuna fits exactly what would be predicted for a diffusion model for methyl mercury chloride. Because methyl mercury chloride diffuses through membranes at a rate of 20×10^{-9} seconds, it appears that this is the rate-determining step for transport into biota.

XIX

CHEMICAL STUDIES OF AQUATIC POLLUTION BY HEAVY METALS IN CHINA

Liu Ching-I and Tang Hongxiao

Institute of Environmental Chemistry, Academia Sinica

Beijing, China

Heavy metal pollution is an environmental problem in China. Rivers, coastal waters, and soils were polluted by industrial activities. Chinese scientists study the distribution of heavy metals in the environment, develop and use methods for the identification and determination of trace element compounds, elucidate the dynamics of trace element transport, and find ways to alleviate and prevent pollution. This paper presents a review of some of the work on mercury, cadmium, and other heavy metals in the Xiang River, Ji Yun River, Jin Sha River, and marine coastal waters. The adsorption of heavy metals on suspended river sediments and their constituents was investigated experimentally and modeled as a multi-component adsorbent system.

Heavy metal pollution is a major environmental problem in the world. Some regions of China have large deposits of nonferrous metals. Mining and industrial operations discharged large quantities of effluents into water bodies. Thus, rivers, lakes, and estuaries were polluted with heavy metals to different degrees. A few water bodies were polluted rather seriously. In Northeastern China, the Secondary Songhua River was heavily polluted by wastewater containing inorganic mercury and methylmercury that were discharged from an acetaldehyde plant and a dyestuff plant. Near the city of Tianjin, a chloralkali plant seriously polluted the Ji Yun River that flows into the Bo Hai Bay with mercury. In South China, the Xiang River, a tributary of the Yangtze River, was polluted with heavy metals particularly cadmium from mines and smelters. Huge amounts of polluted sediments cover the bottom of these rivers. Coastal waters, especially those near the estuaries, were polluted slightly by some heavy metals. Some surface waters, groundwaters, and soils in urban areas were polluted by industrial wastewater or inappropriate irrigation with wastewater. Numerous field investigations and laboratory experiments have been conducted and some control measures have been taken. The status of some studies of the aquatic pollution by heavy metals in China is briefly reported in this paper.

Distribution of Heavy Metal Species

Heavy metals polluting water bodies exist in a number of different chemical species in water, suspended matter, and sediments. Speciation of heavy metals is important for their control and for the elucidation of their fate in the environment. Different physicochemical forms cause different biological effects. Several studies on the distribution and speciation of trace metals have been carried out during the past few years in China.

A number of pretreatment and analysis procedures were adopted for the separation and enrichment of different metal species. Membrane filters, Chelex resins, and ultraviolet irradiation were employed to conduct a systematic analysis. The analytical procedures of Florence-Batley [1, 2], Tessier [3], and Eganhouse [4] were used for water and sediment samples. Sequential chemical extraction producing five to seven fractions was preferred for solids. Anodic stripping voltammetry (ASV) and atomic absorption spectrometry (AAS) were employed for the determination of metal contents. Ultrafiltration, ultrasonic filtration, and ultracentrifugation were used to separate metals of different particle sizes and specific gravities.

The mercury in the Ji Yun River was partitioned by sequential chemical extraction into five fractions [5] according to Eganhouse's procedure [4] that was slightly modified in our laboratory [6] (Table 1). Most of the mercury was associated with humic acids and other organic substances, and with the acid-insoluble fraction [7]. The mercury in the fractions was determined with a cold atomic fluorescence mercury analyzer. The humic material of the Ji Yun sediments was rich in acidic functional groups such as hydroxyl and carboxyl groups [8, 9]. The presence of other organic compounds such as amino compounds was proven by gas chromatography/mass spectrometry. These organic compounds and humic acids form relatively strong complexes with mercury. With increasing distance downstream from the

Table 1. Procedure for the Sequential Chemical Extraction of Mercury from River Sediments

Step	Procedure	Type of Mercury Extracted
1	0.5g wet sediment extracted with 20 mL deionized, distilled water by shaking for 30 min; centrifuged; filtered; bromine added to filtrate.	Hg in interstitial water and water-soluble Hg.
2	Residue extracted with 0.5 M $MgCl_2$ at pH = 7 by shaking for 30 min; centrifuged.	Ion-exchangeable Hg.
3	Residue extracted with 20 mL 0.3 M HCl. Extract treated as in step 1.	Hg associated with carbonates, oxides of Fe, Mn.
4	Residue extracted with 20 mL 1% KOH by shaking for 30 min; kept overnight; extract treated as in step 1.	Hg associated with humic acids.
5	Residue digested with 3% H_2O_2 at pH = 2 at 80° for 4 hrs; centrifuged; diluted with distilled water, filtered; filtrate treated as in step 1.	Hg associated with readily oxidizable organic matter; some Hg sulfides.
6	Residue digested with 30% H_2O_2 at pH = 2; extract treated as in step 5.	Hg associated with organic matter that is difficult to degrade; some Hg sulfides.
7	Residue digested with aqua regia at 85° for 4 hrs; centrifuged; filtered.	Hg sulfides

source of the pollution, the percentage of mercury associated with humic acids and easily degradable organic compounds increased, whereas the percentage of aqua regia-soluble mercury and mercury associated with organic matter rather resistant to degradation decreased (Table 2).

Table 2. Distribution of Various Types of Mercury in Ji Yun River Sediments Collected during September 1981

Loc.**	Type of Mercury* by Extraction with							Total Hg†, mg/kg	
	0.5 M H_2O	0.3 M $MgCl_2$	1% HCl	3% KOH	3% H_2O_2	30% H_2O_2	Aqua Regia	All Types	Aqua Regia
− 1.0	0.16	0.01	0.16	4.20	4.42	9.67	1.84	20.46	20.90
	0.78	**0.05**	**0.78**	**20.5**	**21.6**	**47.3**	**9.0**		
0.0	0.25	0.09	1.53	46.65	18.78	304.2	76.62	448.12	453.14
	0.05	**0.02**	**0.34**	**10.4**	**4.2**	**67.9**	**17.1**		
+ 0.1	0.24	0.00	0.15	2.9	0.48	15.74	0.96	20.47	20.81
	1.2	**0.00**	**0.7**	**14.2**	**2.3**	**76.9**	**4.7**		
+ 3.2	0.20	0.00	0.20	4.81	1.04	12.68	1.38	20.31	20.47
	1.0	**0.00**	**1.0**	**23.7**	**5.1**	**62.4**	**6.8**		
+ 6.5	0.19	0.00	0.19	5.15	3.14	19.08	1.70	29.45	31.94
	0.65	**0.00**	**0.65**	**17.5**	**10.7**	**64.8**	**5.8**		
+13.0	0.15	0.00	0.19	0.80	9.10	1.23	0.80	12.27	−
	1.22	**0.00**	**1.55**	**6.52**	**74.2**	**10.0**	**6.5**		

*For each location the numbers in the first row give the content of mercury in a particular extract in mg/kg, the bold numbers in the second row the percentage of mercury based on the total mercury in the sediment.

**Numbers give distances in km from the source of pollution (0) with negative numbers indicating upstream locations and positive numbers downstream locations.

†Total mercury/all types: sum of the mercury concentrations listed in the preceding seven columns. Total mercury/aqua regia: mercury concentrations in the aqua regia digests of the untreated sediments.

Using Hg-203 as a tracer, the stability of synthetic, amorphous mercury(II) sulfide under simulated river conditions was studied. Humic acids present in the sediments dissolved this sulfide. Thus, mercury sulfide in the sediment might be mobilized and cycled between sediment and the water column. Attempts to differentiate mercury associated with organic substances from mercury sulfides have not yet been successful. However, the experimental results obtained thus far indicate, that the mercury in the sediments is mainly associated with organic compounds or is present as a sulfide. This mercury can be gradually transformed into soluble species, which are then transported downstream and discharged into the Bay of Bo Hai.

The cadmium, copper, lead, and zinc species in the Xiang River water were separated by the Florence-Batley procedure and determined by anodic stripping voltammetry. The metal species in sediments and suspended matter were determined by sequential chemical extraction [10]. Considerable percentages of the total metals in the sediments were associated with carbonates, hydrous metal oxides, and organic substances (Fig. 1). The metals in the residual fractions are important only in sediments near mines and smelters.

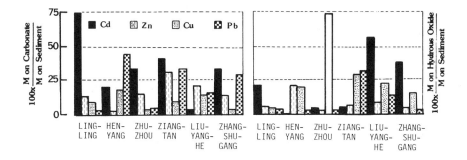

Figure 1. Association of cadmium, zinc, copper, and lead with carbonates and hydrous oxides of Xiang River sediments from the sediment/water interface.

The distribution of mercury species in Ji Yun River water was modeled considering different seasons, ligands, salinities, the adsorption of mercury on suspended matter, and the complexation of mercury by cyanide and humic acid [11]. The calculations indicated that 70 to 90 percent of the mercury in river water should be associated with suspended matter. Almost all (99.9 percent) of the dissolved mercury should be present as mercury(II) cyanide. Similarly, the distribution of copper, zinc, cadmium, cobalt, and nickel in an aquatic environment near a copper mine was modeled [12]. Included in the model were the inorganic ligands sulfate, chloride, carbonate, and hydrogen carbonate, natural humic acids, and the adsorbent kaolinite. The pertinent equilibrium constants and the adsorption isotherms of the metal ions on kaolinite were determined. The set of 81 nonlinear equations describing this system was solved on a Chinese-made TQ-16 computer. The calculations indicated, that carbonate and humic acid were the most important ligands in this system.

Adsorption, Complexation, and Transport of Heavy Metal Pollutants in Aquatic Environments

Many environmental chemical phenomena occur at the water/solid (suspended matter, sediment) interface. Most trace metal pollutants in the aquatic environment are bound to suspended matter and ultimately deposited in river sediments. Colloidal particles of clay minerals, hydrous oxides, and humic substances greatly affect water quality.

Adsorption of cadmium on freshwater particulates. The adsorption of cadmium on clay minerals, hydrous oxides of iron and manganese, humic substances, and on mixtures of these materials was systematically studied [13, 14]. The suspended sediments from the Xiang River were fractionated and the various components (clay, hydrous metal oxides, organic matter, carbonates) isolated. The experimental adsorption isotherms of cadmium on each of these components conform to the Langmuir expression (eqn. 1). At high

$$G_i = G_i^0 C / (A_i + C) \tag{1}$$

G_i: amount of cadmium adsorbed per gram of the i-th adsorbent
G_i^0: amount of cadmium per gram of the i-th adsorbent when the adsorbent is saturated
C: concentration of cadmium in solution at equilibrium
A_i: adsorption coefficient for i-th adsorbent

cadmium concentrations the adsorbents become saturated. Freundlich's adsorption isotherm describes the experimental results only at low cadmium concentrations. These experimental data were used to calculate the contributions of the sediment fractions to the total adsorption employing a multi-component adsorbent model (eqn. 2). The adsorptive

$$G_t = \sum [\alpha_i G_{i0}^0 C / (A_i + C)] \tag{2}$$

G_t: total amount of metal adsorbed per gram of sediment
G_{i0}: amount of metal adsorbed per gram of the i-th component of the sediment, when only this component is present.
G_{i0}^0: maximal amounts of metal that can be adsorbed by the i-th component in the absence of all other adsorbents.
α_i: effective coefficient for the i-th component in a mixture of adsorbents; $\alpha_i = G_i^0 / G_{i0}$

capacity of each component seems to become reduced by admixing other components (Table 3). The reduced capacities might be the result of the coating of a surface of one component by another component of the sediment. For instance, electron micrographs of Xiang River sediments show, that clay particles are partially covered by the other major constituents of the sediment. However, the exposed surfaces of the clay particles make still the dominant contribution to the total adsorption of trace elements. Metal hydroxides and organic substances are also important as adsorbents (Table 3).

Table 3. Results of Model Calculations for the Adsorption of Cadmium on Xiang River Sediments*

Component	weight % in sediment	G_i^0 for Cd mg/g	Contribution to G_t (%)	α_i
clay	80. -95.	4.7 -5.6	~60	0.85-0.90
hydrous metal oxide	0.5 - 1.2	1.2 -2.1	~20	0.50-0.75
organic matter	1.0 - 3.0	1.0 -1.7	~15	0.50-0.75
carbonates	0.03-15.0	0.01-0.5	~ 5	0.0

*The symbols are defined in eqns. 1 and 2.

Adsorption of mercury. The adsorption of mercury is considered to be the main process for the enrichment of mercury in sediments [15, 16]. The adsorptive capacity of different adsorbents for mercury was found to decrease in the sequence humic acid > MnO_2 > clay minerals > Fe_2O_3 > SiO_2. The chloride concentration of the solution influences the adsorptive capacities for mercury because of the formation of chloromercury complexes. The adsorptive capacity of sediments for mercury species decreases in the order $Hg(OH)Cl$ > $Hg(OH)_2$ > $HgCl_3^-$ > $HgCl_4^{2-}$ > $HgCl_2$. At chloride concentrations less than 1×10^{-4} M, humic acid, manganese dioxide, and illite have a considerable adsorption capacity for mercury. At a chloride concentration of 0.56 M (characteristic for seawater), at which chloromercury complexes predominate, only humic acid retains its adsorption capacity for mercury (Fig. 2).

Mercury was strongly adsorbed on the fine or suspended particulates (<5 μm) of sediments from the Ji Yun River. These particulates released mercury rather slowly. Organic substances in these sediments with thiol and amino groups could bind mercury strongly. The sulfur (1.7%) and nitrogen (9.3%) contents of the humic acids extracted from the Ji Yun

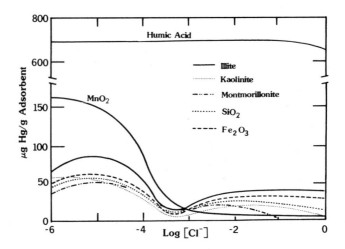

Figure 2. The adsorption of mercury(II) by humic acid, manganese dioxide, illite, iron(III) oxide, kaolinite, montmorillonite, and silicon dioxide from solutions of different chloride concentrations. Mercury was added as mercury(II) chloride.

River sediment were found to be relatively high. The humic acids and organic substances of low molecular weight were formed by biodegradation and chemodegradation of high molecular weight organic matter contributed by sewage and industrial effluents. The complexing capacity of fulvic acids and humic acids determined by gel filtration was largest for fulvic acid obtained from water, intermediate for fulvic acid from the sediments, and smallest for humic acid from the sediments [8]. Using apparent stability constants, the stability of the mercury humate in the river was found to be greater than the stabilities of the mercury chloride, hydroxide, sulfate, or phosphate complexes. Humic acids in the river water thus enhance the transport of mercury in form of soluble complexes. Chlorides and humic acids in the interstitial water of the sediments also mobilized mercury slowly from the sediments by dissolution. Adsorption and desorption, complexation, dissolution, flocculation, precipitation, and degradation of organic substances are the processes participating intermittently and interactively in the transport of mercury in the river.

The horizontal and vertical distribution of mercury in river sediments was studied using data obtained during 1976-

1977 [17], and 1980-1981 [6]. The mercury concentration in
the sediments had decreased considerably during this four-
year period. The mercury concentration on suspended sand
particles near Bo Hai Bay can be used to estimate the amount
of mercury discharged into the bay. Because the flow of
water in the Ji Yun River was much greater during 1977/78
than during 1980/81, the hydraulic process, a physical
factor, should be considered to be most important for the
transport of mercury during 1977-1979. When the flow is
small, physicochemical processes, which are specially
important when considerable amounts of organic matter are
present, might control the transport of mercury in the
river. This situation probably was characteristic for the
Ji Yun River during 1980/81. The physical and physicochemical processes interact with each other under various
environmental conditions, and thus enhance the self-
purification of the river.

Metal pollutants in the Jin Sha River. The Jin Sha River, a
tributary of the upper Yangtze River, has water with a pH of
8.0 to 8.5 maintained by a bicarbonate buffer system. The
distribution and characteristics of metal particulates, and
the adsorption of copper, zinc, cadmium, cobalt, and nickel
on the particulates in the Jin Sha River were studied
carefully [18]. The grain sizes of the suspended
particulates and the particulates of the sediment, the
concentrations of the metals, and the co-existing ions were
found to affect adsorption. The pH of this river water is
the major factor controlling the concentrations of dissolved
metal ions and their adsorption and precipitation. Each
metal reaches its maximal adsorptivity at a characteristic
pH. At higher pH values hydrolysis and precipitation will
occur. In this manner the concentration of dissolved metal
is controlled facilitating the self-purification of the
water in the Jin Sha river.

Metal pollutants in the marine environment. The transport
and transformation of trace metals in the marine environment
are largely determined by the physicochemical characteristics of the river that control the formation of metal
species. In China, marine chemists devote serious efforts
to the study of metal pollution in seawater with special
emphasis on estuarine water. Work on the speciation of
trace metals and studies of the physicochemical behavior of
some metals in seawater has been initiated. The interactions between the aqueous phase and the solids are
governed by the chemical constituents of the solids present
in the water, pH, E_h, temperature, salinity, the concentra-

tion of suspended matter, and the concentrations of metal pollutants. Wu et al. [19] studied the transport mechanism of mercury, chromium, and copper in the Yangtze River estuary. Thermodynamic and kinetic studies of the adsorption of these metals onto clays in water of different salinity showed, that these metals were largely adsorbed on inorganic and organic colloids and transported with these particulates. The order of enrichment is Cr > Cu > Hg. The presence of large organic molecules increases the rate of adsorption considerably.

Zhou et al. determined the chemical forms of arsenic, chromium, and mercury in surface seawaters from coastal sites of the Jiozhou Bay near Qingdao [20], and studied the exchange of mercury [21] and chromium [20] between seawater and sediment. Chromium(III) was the predominant form of chromium in all samples examined. This predominance is probably caused by the low pH and E_h of the sea. The chromium(III) adsorbed by particulate matter settles gradually to the bottom. This process is probably one of the main mechanisms of self-purification of seawater in estuaries and in coastal regions.

Secondary Pollution by Heavy Metals and Their Control

Field investigations and model experiments indicate that mercury contained in sediments is a real hazard. Mercury is present in sediments largely in form of sulfides and humates. Although the solubilities of these mercury compounds are low, the mercury released from the sediments cannot be ignored. In heavily polluted areas, the concentration of mercury is maintained at a few micrograms per liter of water through release of mercury from the sediments. This concentration is several times higher than the permissible level [22]. The emission of methylmercury from the sediment into the water is in the range of 0.4 to 5.0 µg Hg per m^2 per day. In addition, gaseous elemental mercury is emitted by the water body into the atmosphere.

To evaluate potential hazards assignable to heavy metal pollution, many biological experiments and epidemiological investigations have been carried out. Aquatic plants and animals, birds, soils and crops, and people were studied in polluted areas. The interesting and puzzling results thus far obtained require confirmation by additional investigations to be conducted over longer periods.

The accumulation factors for mercury in the ecosystem of the Ji Yun River are 400 to 700 for plankton and water weeds, approximately 1000 for animals other than fish, 2000 to 3000 for fishes, and 6000 to 7000 for water birds. Using these data, a mercury cycle relating to the foodchains in this ecological system was proposed. A modification of the water quality standards was suggested. The epidemiological investigations and health examinations in the area of the Secondary Songhua River discovered some phenomena that require further study.

The control and disposal of sediments polluted by heavy metals is a complex problem that must be solved to prevent long-term secondary pollution. Presently, core samples of sediments are collected frequently to investigate the transport dynamics of the polluted sediments. Attempts have been made to estimate the self-purification capacity of the river and appraise the possibility that natural processes will avert secondary pollution of the river. Dredging of the sediments or covering them with impervious material are operations that could be carried out in conjunction with river regulations and landfill projects and could remove highly polluted sediments from contact with water. However, appropriate measures must be taken to prevent the escape of mercury from the dredged sediments. Natural and chemical flocculation, dewatering and stabilization, and chemifixation have been studied to treat sediments. Model experiments with various materials for sealing the sediments in-situ were also performed. Some of these materials including natural, unpolluted sediments are effective and probably economical.

Ultimately, the sources of pollution must be controlled or eliminated. Progress toward this goal is not yet satisfactory in China. Chemical precipitation is the most popular method for the removal of heavy metal pollutants from effluents. Lime treatment and sulfide precipitation have gained wide acceptance, but the disposal of the sludges still causes great troubles. Ion exchange processes are effectively employed in some treatment plants. Chemical displacement processes with iron or red copper scraps are also used. Adsorption on activated carbon, reverse osmosis, liquid membrane extraction, macropore sulfydryl ion-exchange resins, high-gradient magnetic separation, and other procedures have been tested. Some of these processes are being tested in pilot plants but most are still explored in the laboratory. Concomitant with these attempts to remove pollutants, processes are redesigned and materials are recycled to reduce the discharge of polluting substances.

References

[1] G. E. Batley and T. M. Florence, "Determination of the Chemical Forms of Dissolved Cadmium, Lead and Copper in Seawater," **Marine Chem.**, 4 (1976) 347.
[2] T. M. Florence, "Trace Metal Species in Fresh Water," **Water Res.**, 11 (1977) 681.
[3] A. Tessier, P. G. C. Campbell and M. Bisson, "Sequential Extraction Procedure for the Speciation of Particulate Trace Metals," **Anal. Chem.**, 51 (1979) 844.
[4] R. P. Eganhouse, D. R. Young and J. N. Johnson, "Geochemistry of Mercury in Palos Verdes Sediments," **Environ. Sci. Tech.**, 12 (1978) 1151.
[5] Peng An and Wang Zijian, "Mercury in River Sediments," in **"Environmental Inorganic Chemistry,"** Proceedings, Workshop on Environmental Inorganic Chemistry, K. J. Irgolic and A. E. Martell, Eds., San Miniato, Italy, June 2-7, 1983, VCH Publishers: Deerfield Beach, Florida, 1985, p. 393.
[6] Y. H. Lin, T. M. Kang and C. I. Liu, "Speciation and Distribution of Mercury in Ji Yun River Sediment," **Huanjing Huaxue**, 2 (1983) 10.
[7] S. W. Pang, Q. K. Qui and J. F. Sun, "Sequential Chemical Extraction for Speciation of Mercury in River," **Huanjing Kexue Xuebao**, 1 (1981) 234.
[8] A. Peng and W. H. Wang, "Extraction and Characterization of Humic Acid from Ji Yun River," **Huanjing Kexue Xuebao**, 1 (1981) 126.
[9] A. Peng and W. H. Wang, "Influences of Humic Acids on the Transport and Transformation of Hg in Ji Yun River," **Huanjing Huaxue**, 2 (1982) 33.
[10] M. Z. Mao, C. H. Liu and J. X. Wei, "Studies on the Chemical States of Heavy Metals on the Surface of Sediments in Xiang Jiang River," **Huanjing Kexue**, 2 (1981) 355.
[11] K. Xu and A. Peng, "Computation of Mercury Species in Downstream Ji Yun River," **Huanjing Huaxue**, 3 (1984) 19.
[12] Z. H. Yao, H. D. Wang and P. T. Liu, "A Comprehensive Chemical Model for Heavy Metal Distribution in Aquatic Environment," **Huanjing Kexue Xuebao**, 2 (1982) 10.
[13] H. X. Tang and H. B. Xue, "Adsorption Characteristics of Cadmium Pollutants on Chinese Clay Minerals," **Huanjing Kexue Xuebao**, 1 (1981) 140.
[14] H. X. Tang and H. B. Xue, "Study on Multi-Component Adsorption Model of Aquatic Sediments with a Sequential

Chemical Separation Procedure," **Huanjing Kexue Xuebao, 2** (1982) 292.
[15] R. S. Reimers and P. A. Krenkel, "Kinetics of Mercury Adsorption and Desorption in Sediments," **J. Water Poll. Contr. Fed., 46** (1974) 352.
[16] C. S. Li and C. I. Liu, "Adsorption of Mercury in Freshwater," **Huanjing Huaxue, 1** (1982) 304.
[17] S. Zhang, Y. J. Tang and W. L. Yang, "Characteristics of Mercury-Polluted Chemical Geography of Ji Yun River," **Huanjing Kexue Xuebao, 1** (1981) 349.
[18] X. R. Wang, H. Z. Zhang, A. E. Zhou and Y. I. Liu, "Study on Heavy Metal Adsorption on the Particulates in Jin Sha River," **Huanjing Huaxue, 2** (1983) 23.
[19] Y. D. Wu and Y. W. Zheng, "On the Mechanisms of the Heavy Metals Transport in Yangtze River Estuary I," **Haiyang Yu Huzhao, 9** (1978) 168.
[20] J. Y. Zhou and W. Y. Qian, "Valency State of Cr in Seawater and Seawater-Sediment Chromium Interchange," **Haiyang Yu Huzhao, 11** (1980) 30.
[21] J. Y. Zhou and W. Y. Qian, "The Chemical Forms of Mercury in Seawater and the Seawater Sediment Mercury Interchange," **Haiyang Wenji, 4** (1981) 56.
[22] S. H. Wang, T. C. Hao and G. Q. Liu, "Release Rate of Methylmercury from River Sediment," **Huanjing Kexue, 2** (1978) 11.

XX

DETERMINATION OF THE CHEMICAL FORMS OF TOXIC ELEMENTS IN ENVIRONMENTAL AQUATIC SAMPLES FROM CHINA

Lu Zongpeng, Mao Meizhou and Wei Jinxi

Institute of Environmental Chemistry
Academia Sinica, Beijing, China

Heavy metals in seawater, river water, sediments, and soils are now thoroughly studied in China with the goal to establish the extent of pollution, to find processes suitable for cleaning heavily polluted areas, and to develop measures for preventing further pollution. Increasing emphasis is placed on speciation of trace elements. Anodic stripping voltammetry and other electrochemical techniques have gained popularity in China for the determination of metals. Sequential chemical extraction procedures are used to separate various types of metal species. The current work in this area is reviewed and information about copper, lead, cadmium, zinc, mercury, arsenic, selenium, and tellurium concentrations in coastal seawater, river waters, sediments, and soils is summarized. Background levels of metals were determined in water and snow samples collected in unpolluted areas of Tibet. Representative results are summarized in several tables.

The bioavailability and toxicity of toxic elements are critically dependent on the compounds, in which these elements occur, and not only on their "total concentrations." It is, therefore, necessary to determine the nature and the concentrations of such compounds in natural waters. Flameless atomic absorption spectroscopy (AAS), neutron activation analysis (NAA), inductively coupled plasma emission spectroscopy (ICP) and anodic stripping voltammetry (ASV) are widely used methods in trace analysis. Non-electrometric methods such as AAS, NAA, and ICP are sufficiently sensitive for the analysis of toxic trace metals in water samples, but provide only "total concentrations" unless used as element-specific detectors for gas or liquid chromatography [1]. Anodic stripping voltammetry is sensitive to the types of metal species dissolved in water, because the kinetics of an electrode process depend on the nature of the reacting species. This dependence is reflected in the measurable relationship between faradic current, electrode potential, and time. The shape of the current-potential curve may thus reveal the nature of the species present in solution [2, 3]. Anodic stripping voltammetry can distinguish between "labile" and "bound" metal species in natural waters and has been used extensively for this purpose. In recent years, the environmental chemists in China became interested in identifying chemical compounds of metals in environmental samples [4-13] by non-electric and electric methods. These studies of trace elements in aquatic environmental samples in China are summarized in this paper.

Anodic Stripping Voltammetry (ASV) for the Determination of Labile Metal Species in Seawater

During the past few years the extent of pollution of the near-shore region of China covering about 380,000 km² [14] was monitored. Samples were collected at 1,094 stations along the entire coast of China (Fig. 1). The major constituents of seawater (NaCl, $MgSO_4$, $CaSO_4$) are advantageously used as the necessary supporting electrolyte in ASV. Copper, lead, cadmium, and zinc can be determined simultaneously in a sample by ASV at the hanging mercury drop electrode or the mercury film electrode. The sample preparation is very simple. The sample is filtered through a 0.45 μm filter, and the filtrate analyzed for copper, lead, cadmium, and zinc without addition of any reagent. Tin and bismuth can be determined in the filtrate after acidification with a small volume of concentrated hydrochloric acid to make the solution 0.25 M with respect to HCl. The concentrations of trace metals were almost the same in near-shore and open-ocean regions [15-17] (Table 1). However, the concentrations of trace metal ions in the Qingdao region increased with increasing distance from the shore for zinc from 3×10^{-8} to 4×10^{-8} M, for cadmium from 2×10^{-10} to 2×10^{-9} M, for copper from 1×10^{-8} to 4×10^{-8} M, and for lead from 3×10^{-10} to 4×10^{-10} M [18]. The results of lead and cadmium determinations in seawater from different coastal regions [14] are summarized in Table 2. The concentrations of ASV-labile metals were found to decrease with increasing concentration of oil in the water. As the concentrations of complexing agents in water increase,

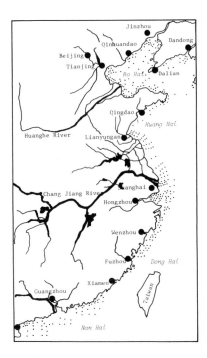

Figure 1. *Sampling stations for the near-shore monitoring of pollutants in China.*

Table 1. *The Concentrations of Labile Heavy Metals in Filtered Seawater from the Coastal Regions of China Determined by ASV on the Hanging Mercury Drop Electrode*

Element	Metal Concentration, Molarity		
	Near Shore		Pacific Ocean
	Range	Average ± Av. Dev.	Range*
Zn	$(6.0-11.0) \times 10^{-8}$	$(8.50 \pm 0.25) \times 10^{-8}$	$(0.8\ -25.4\ ) \times 10^{-8}$
Pb	$(1.5-\ 2.1) \times 10^{-10}$	$(1.80 \pm 0.30) \times 10^{-10}$	$(1.0\ -13.0\ ) \times 10^{-10}$
Cu	$(1.2-\ 2.0) \times 10^{-8}$	$(1.60 \pm 0.40) \times 10^{-8}$	$(0.6\ -15.0\ ) \times 10^{-8}$
Cd	$(3.6-\ 5.2) \times 10^{-9}$	$(4.40 \pm 0.80) \times 10^{-9}$	$(0.01-37.0\ ) \times 10^{-9}$
Sn	$(0.7-\ 1.2) \times 10^{-8}$	$(0.95 \pm 0.25) \times 10^{-8}$	$(0.10-\ 0.69) \times 10^{-8}$
Bi	$(0.7-\ 1.2) \times 10^{-11}$	$(0.95 \pm 0.25) \times 10^{-11}$	$(0.07-\ 0.96) \times 10^{-9}$

*From ref. 17, pp. 102-117.

electro-inactive metal complexes are formed. In the "oil-rich" Donghai samples, no labile cadmium and only very little labile lead were found.

Table 2. The concentrations of Heavy Metals and Oil in Seawater from Coastal Regions of China

Location	Number of Samples	Cd* Conc. µg/L Range	Cd* Conc. µg/L Average	Pb* Conc. µg/L Range	Pb* Conc. µg/L Average	Oil Conc. µg/L
Bohai	260	<DL-0.10	0.09	0.50- 17.3	2.95	0.05
Huanghai	262	<DL-0.15	0.09	0.11- 16.7	1.34	0.05
Donghai	97	<DL	<DL*	<DL- 30	**	0.067
Nanhai	138	0.03-2.57	0.27	0.09- 9.05	7.68	0.054
Overall average conc. and range µg/L		<DL-0.45	0.10	0.05-51.4	1.6	-

*Detection limits: 0.01 µg/L for Cd, 0.04 µg/L for Pb.
**Lead was detected in only 4 percent of the samples.

Speciation of Metals in Seawater

ASV is a highly sensitive technique that can distinguish between "labile" and "bound" metal species in filtered seawater. Labile species that can be determined at natural pH are the free metal ions and complexes that dissociate in the diffusion layer and liberate metal ions. Bound metal exists in relatively inert complexes and is obtained as the difference between total metal and labile metal. Chemical separations are necessary, if the bound metal associated with organic and inorganic colloidal particles or with organic ligands forming electro-inactive complexes is to be determined. The chelating resin CR-1 made in China at Nankai University was used [19, 20] in conjunction with the Batley-Florence scheme [2] to speciate heavy metals in samples from the freshwater-seawater mixing zone of the Zhujiang River and the Nanhai Sea. Non-labile organic complexes accounted for 71.7 percent of the total dissolved lead and for 60 percent of the copper (Table 3). Non-labile complexes adsorbed on organic colloidal particles (52.2%) and non-labile inorganic complexes (22.5%) are the major forms of cadmium in the seawater of Guang Zhou Bay. By a similar analytical procedure Ling et al. [20] found, that in seawater labile cadmium adsorbed on organics is the predominant species, and free cadmium and non-labile inorganic complexes are almost absent. Luo [21], He [22], and Mao [8] reported, that 80 percent of the total dissolved cadmium in the Nanhai Sea and in the Bohai Bay is present as

Table 3. Species of Heavy Metals in Filtered Seawater from the Nanhai Sea

Species*	Zn µg/L	Zn %	Cd µg/L	Cd %	Pb µg/L	Pb %	Cu µg/L	Cu %
M+MA$_1$+ML$_1$	10.0	11.7	0.089	10.0	**	**	0.84	13.3
ML$_2$	4.0	4.7	0.136	15.3	0.4	5.3	0.64	10.1
MA$_2$	4.0	4.7	**	**	0.8	10.6	**	**
ML$_3$	32.8	38.5	**	**	5.4	71.7	38.0	59.9
MA$_3$	9.2	10.8	0.20	22.5	0.4	5.3	0.27	4.3
ML$_4$	20.0	23.5	0.464	52.2	**	**	0.79	12.5
MA$_4$	5.1	6.0	**	**	0.53	7.0	**	**
Total	85.1	100.0	0.889	100.0	7.53	100.	6.34	100.

* M = free metal ions, ML$_1$ = labile organic complexes, MA$_1$ = labile inorganic complexes, ML$_2$ = labile metal adsorbed on organics, MA$_2$ = labile metal absorbed on inorganics, ML$_3$ = non-labile organic complexes, MA$_3$ = non-labile inorganic complexes, ML$_4$ = non-labile metal adsorbed on organics, MA$_4$ = non-labile metal adsorbed on inorganics.

** Below detection limits of 0.01 µg/L for Cd, 0.04 µg/L for Pb, 0.01 µg/L for Cu.

non-labile cadmium. The determination of cadmium species in seawater is very difficult especially at the low concentrations characteristic of unpolluted seawater.

The chemical forms of zinc in surface waters of the northeast Jiaozhou Bay were studied by Wang et al. [23] by ASV and extraction with dithizone/spectrophotometry. Particulate zinc (2.7 µg/L), ASV-labile zinc (4.1 µg/L), weakly bound zinc (5.5 µg/L), and non-labile zinc (8.4 µg/L) were found. The concentration of zinc in Nanhai seawater (85.1 µg/L, Table 3) is higher [20] than in the Jiaozhou Bay. It is possible, that the concentration of zinc in surface waters near the polluted shore is higher than in the open ocean. Great care must be exercized to avoid contamination of the samples.

The chemical forms of lead in seawater near the Jiaozhou Bay were reported [24] to be particulate (30%), labile (9%), weakly bound (26%), and strongly bound lead (35%). The weakly bound species, which dissociate at pH 4, are the labile complexes adsorbed on inorganic and organic particulates. The strongly bound lead present in the form of non-labile organic complexes can be determined only after digestion with nitric acid and hydrogen peroxide. The non-labile organic lead complexes were found to be the dominant form of lead in Nanhai and Bohai seawaters.

Species of Trace Metals in Freshwater

During the past ten years, the inland rivers and lakes such as the Yangtze River, the Shonghua River, the Xiang River, and the Taihu Lake were monitored. However, in most cases only total concentrations of trace metals were determined. Since 1979, Mao et al. [8] used the analytical scheme given in Figure 2 to speciate trace metals in the

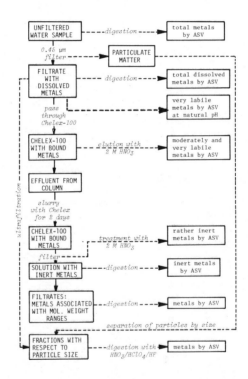

Figure 2. Scheme for the speciation of trace metals in Xiang River water by ASV.

Xiang River. This procedure divides the metals in the river water into species adsorbed on particulates and into species dissolved in the water. The dissolved species are further divided into labile and inert species. Table 4 presents the results of such distribution studies for cadmium, lead, and

Table 4. *The Distribution of Copper, Lead, and Cadmium in Xiang River Waters among Labile Dissolved Species, Inert Dissolved Species,† and Species Absorbed on Particulates*

Metal	Species	Concentration and Percent* of Species							
		Zhuzhou		Xiangtan		Changsha		Zhangshou	
		µg/L	%	µg/L	%	µg/L	%	µg/L	%
Cu	total dissolved	1.9	66	2.3	62	3.3	79	3.0	88
	labile dissolved	0.4	14	0.2	5	1.1	26	0.8	23
	inert dissolved	1.5	52	2.1	57	2.2	53	2.2	65
	adsorbed on particulates	1.0	44	1.4	38	0.9	21	0.4	12
	total metal	2.9	100	3.7	100	4.2	100	3.4	100
Pb	total dissolved	1.6	73	2.6	67	1.4	67	1.8	67
	labile dissolved	0.6	27	1.1	28	0.4	19	0.6	22
	inert dissolved	1.0	46	1.5	39	1.0	48	1.2	45
	adsorbed on particulates	0.6	27	1.3	33	0.7	33	0.9	33
	total metal	2.2	100	3.9	100	2.1	100	2.7	100
Cd	total dissolved	0.21	66	0.41	43	0.21	68	0.17	68
	labile dissolved	0.18	56	0.35	36	0.17	55	0.14	56
	inert dissolved	0.03	10	0.06	7	0.04	13	0.03	12
	adsorbed on particulates	0.11	34	0.55	57	0.10	32	0.08	32
	total metal	0.32	100	0.96	100	0.31	100	0.25	100

† Labile dissolved species are those determinable by ASV at the natural pH of the sample. The species not determinable by ASV under these conditions are called inert dissolved species.

* Percentages based on total metal concentration and rounded to the nearest percent.

copper in Xiang River water. A large fraction of the total cadmium is present in the form of labile, dissolved species including simple inorganic complexes with carbonate and hydroxyl groups as ligands, and labile organic complexes. In contrast, approximately half of the total copper was found in the inert dissolved fraction. Inert dissolved species are also the most abundant among the lead species. The percentage of inert dissolved species in the Xiang River water decreases in the sequence copper-lead-cadmium. The stabilities of complexes of these metals with organic ligands also decrease from copper to cadmium. A more detailed subdivision of the dissolved metal species by Chelex-100 chromatography and ASV is presented in Table 5 [25].

Table 5. *The Distribution of Dissolved Copper, Lead, and Cadmium among the Very Labile, Moderately Labile, Rather Stabile and Inert Species at Three Locations in the Xia Wan Region of the Xian River*

Species	Percent of Total Dissolved Metal at								
	0.0 m*			100 m**			500 m**		
	Cu	Pb	Cd	Cu	Pb	Cd	Cu	Pb	Cd
very labile	13	40	32	25	42	50	21	15	57
moderately labile	15	2	4	4	2	1	9	7	3
rather stabile	27	20	41	36	27	12	7	44	0
inert	47	36	23	34	29	37	62	34	40
TSM†	33.9	47.	1.1	17.7	8.3	0.82	8.5	7.3	0.30
MAP†	33.3	90.	0.3	3.4	12.5	0.30	1.1	0.3	0.05
TSM+MAP†	67.2	137.	1.4	21.1	20.8	1.12	9.6	7.6	0.35

*Discharge point.
**Distance in meters downstream from discharge point.
† TSM: Total Dissolved Metal; MAP: Metal Absorbed on Particles; concentrations are given in $\mu g/L$.

Analysis of Metal Complexes in Water by Anodic Stripping Voltammetry

Anodic stripping voltammetry (ASV) and differential pulse anodic stripping voltammetry (DPASV) [26] are very sensitive methods for the determination of trace metals in environmental samples. ASV is not only very sensitive, precise, and accurate, but also useful for investigating different types of metal complexes at trace levels in natural waters. The direct identification of metal complexes in natural water by determination of ligand numbers and stability constants was attempted only recently [27]. The electrochemical reduction of a metal is influenced by complexation. The amount of metal deposited as a function of deposition potential depends on the chemical nature of the complex. Reversible deposition of a metal at various potentials coupled with a sensitive ASV technique allows to measure stripping peak currents as a function of deposition potential and obtain a pseudopolarogram. Zhang et al. [28] determined cadmium-EDTA complexes and their conditional formation constants by ASV. The metal complexing ability of water in the Xuanwu Lake, Nanjing, at pH 8.0 was determined to be 5.42×10^{-7} M lead(II). The composition of the lead-humic acid complex, its apparent formation constant, and the molecular weight of humic acid were determined by square wave polarography [9]. Three complexes with two, four, and six moles of lead per mole of humic acid were formed. The average molecular weight of humic acid was calculated (eqn. 1) to be 1154. The complexes formed between lead and human albumin, human gammaglobulin, and bovine albumin were investigated similarly [29].

$$MW_{humic\ acid} = \frac{a[L]W}{b[M]E} \qquad (1)$$

E : volume of sample at the equivalence point (liters)
[L]: original concentration of ligand (g per liter)
[M]: concentration of metal ion (moles per liter)
V : volume of sample (liters)
a : number of metal atoms in the complex
b : number of ligand molecules in the complex

Chemical Forms of Arsenic in Aquatic Environmental Samples

The chemistry of arsenic in natural water is complex. Arsenic exists naturally in +5, +3, 0, and -3 oxidation states. The toxicity of arsenic varies widely with the chemical nature of the compounds in which this element occurs. The valence state and forms of arsenic can be determined by spectrophotometry [7], neutron activation analysis [30], and ASV [31, 32]. The surface seawater of Dalian Bay had arsenic concentrations in the range 5 μg/L to 11 μg/L. Using averaged concentrations, the following ratios were obtained: dissolved arsenic/particulate arsenic = 5.0; inorganic arsenic/organic arsenic = 1.2; and arsenite/arsenate = 1.1. The concentrations of arsenic in this area are higher than in normal seawater.

The determination of chemical forms of arsenic in alluvial soils was reported by Li [12]. Soil samples from ten Chinese provinces contained the following arsenates (percentage range based on total arsenic): adsorbed (0-21%), aluminum (1-9%), iron (1-16%), calcium (3-39%), and occluded (32-86%).

Chemical Forms of Mercury in Aqueous Environmental Samples

Seawater of the middle Pacific Ocean contains 3.5-12.1 ng/L of mercury at the surface and 8.6 to 30.7 ng/L at a depth of 1500 m [33]. The surface water in Bohai Wan had total mercury concentrations in the range of 11.8 to 46 ng/L [34]. The water in the Xiamen harbor had 5.52 ng/L. The estuary of the Zhujian River had 3 ng/L of inorganic mercury, 1 ng/L of organic mercury, and 30 ng/L of suspended mercury [33]. The mercury in the estuarine water was strongly associated with particulate matter [20, 35]. Rao [36] reported concentration ranges 1.5 to 6.2 ng/L of methylmercury and 36 to 55 ng/L of total mercury for Xiang River water.

The procedure of sequential chemical extraction by different reagents was used to separate the mercury species present in sediments. The extracts were analyzed with a cold atomic fluorescent mercury analyzer [37]. The uptake of mercury by plants grown on mercury-containing sediments and soils is influenced by the chemical state of mercury [38]. The uptake of mercury by rice decreases in the sequence $C_6H_5HgOOCCH_3$ > $HgCl_2$ > HgO > HgS [38]. The methylated forms of mercury are much more toxic than inorganic mercury compounds. Methylmercury can be concentrated during passage from water to and through biota in the food chain. Muscle tissue of fish was found to exceed the methylmercury concentration of water by a factor of 38,000 [10]. The organic mercury compounds present in several fish species were determined by gas chromatography with a mercury analyzer as detector [39]. Methylmercury was separated from inorganic mercury on sulfhydryl cotton fibers [10, 40, 41]. Chen [10] and Yang [11] found that organic mercury accounts for 80 to 90 percent of the total mercury in fish from Bohai Bay (Table 6).

Determination of the Valencies of Elements in Natural Water

It is known that the valence states of elements influence their toxicities. For example, the toxicity of arsenic compounds decreases in the sequence AsH_3 > AsO_3^{3-} > As_2O_3 > AsO_4^{3-} > H_3AsO_4. The valence states of elements such as arsenic [31, 32, 42, 43], selenium [44-46], antimony [47], chromium [48-50], and tellurium [51] were determined. The results for arsenic and selenium are summarized in Table 7. Pentavalent arsenic is the major form of arsenic in seawater. The arsenate/arsenite ratios varied in freshwaters.

Table 6. Chemical Forms of Mercury in Fishes from Bohai Bay

Fish	Concentration* and Percentage of Mercury						
	Total	Inorganic		Organic		methylated	
	µg/kg	µg/kg	%	µg/kg	%	µg/kg	%
Mugil soiny	17.2	3.58	20.8	13.6	79.1	13.8	80.2
Stromateoides argentens	12.4	2.4	19.4	10.0	80.6	10.3	83.1
Nibea albiflora	203.0	23.5	11.6	179.5	88.4	153.7	75.7
Lateolabrax japonicus	91.1	14.6	16.0	76.5	84.0	77.5	85.1
Coilia mystus	134.0	13.1	9.8	120.9	90.2	109.9	82.1

*In fish muscle, based on fresh weight.

Table 7. Valence States of Arsenic and Selenium in Natural Waters

Sample	Arsenic, µg/L		Ref.	Sample	Selenium, µg/L		Ref.
	III	V			IV	VI	
Quishui*	2.5	2.03	[42]	River W.	0.4	0.1	[44]
Wei*	1.6	0.9		Well W.	<0.1	0.6	
Jinling*	1.47	1.17		Sea W.	<0.1	<0.1	
Hot Spring	4.07	1.47		Hai*	0.22	0.43	[45]
Spring W.††	0.37	0.0		Kunming†	0.50	0.46	
Baoji Reserv.	0.93	0.0		Donnigu*	0.30	0.90	
Bohai Sea†††	0.8	5.65	[31]	Canal W.	0.80	0.37	
River**	8.0	7.5	[32]	Ground W.	1.0	0.84	
Spring**	3.05	5.44	[43]	Bohai Bay	<0.02	0.20	

*Name of river   **Polluted   †Lake   ††W.: Water
†††Digestion of the sample with $K_2S_2O_4$ gave an As(V) value of 5.65 µg/L, whereas digestion with sulfuric acid yielded 4.29 µg/L.

Trivalent arsenic was the predominant species in anoxic river water. The two common oxidation states of selenium are Se(VI) and Se(IV). The Se(VI)/Se(IV) ratios vary considerably in freshwaters. An agreement has not yet been reached about the predominant valence state of selenium in seawater [1, 45]. The study of the valence states of tellurium and antimony in natural waters of China has just been initiated [47, 51].

Determination of Background Concentrations of Elements in Unpolluted Waters by Anodic Stripping Voltammetry and Catalytic Wave Polarography

Anodic stripping voltammetry (ASV) and catalytic wave polarography (CWP) are techniques of sufficient sensitivity for the determination of trace element levels in unpolluted waters [52-60]. Trace metals in snow samples collected on Zhumulangmafeng, a mountainous area with elevations in the range from 3800 to 5600 m, were determined in our laboratory [61] by ASV at a mercury film electrode. The concentrations of several heavy metals in river and snow samples are reported in Tables 8 and 9. The concentration ranges in

Table 8. The Concentration of Heavy Metals in River Waters from Xizang, Tibet

Element	Method[†]	Metal Concentration, µg/L				
		Zhangmu 2300 m*	Lasa 3670 m*	Rikazo 3900 m*	Dingri 4300 m*	Base Camp 5000 m*
Cu	ASV	0.4	1.8	2.5	2.3	1.2
	AAS	0.75	1.9	1.9	2.5	1.2
Pb	ASV	0.4	2.3	3.1	2.4	3.0
	AAS	0.5	2.3	3.0	2.3	2.2
Cd	ASV	0.039	0.065	0.043	0.039	0.045
	AAS	<0.07	<0.07	<0.07	0.018	0.09
Zn	ASV	**	5.4	16.0	16.0	3.6
Se	ASV	**	0.04	0.02	0.02	**

*Elevation of sampling site.
[†]ASV: Anodic stripping voltammetry; AAS: Graphite furnace atomic absorption spectrometry.
**Not determined.

river waters were 0.4-2.5 µg/L for copper, 0.4-3.1 µg/L for lead, 0.032-0.065 µg/L for cadmium, and 3.6-16.0 µg/L for zinc. The concentrations in snow were somewhat higher. Selenium determined by ASV on a gold electrode was present in river water at concentrations of 0.02 to 0.04 µg/L, and in snow at 0.03 to 0.05 µg/L [56]. Water samples from the mountainous region near Huairou had similar selenium concentrations [56]. The concentration range of silver in samples collected near Huairou were 1.2 to 2 ng/L for rain and 3.8 to 7 ng/L for snow [62]. Molybdenum was determined by CWP in samples from Zhumulangmafeng [63]. Rain contained 0.033 µg/L, river water 0.1 to 1.4 µg/L, and snow 0.007 to 0.68 µg/L. Recently, tellurium was found in natural waters

Table 9. The Concentrations of Heavy Metals in Snow Samples Collected on Zhumulangmafeng

Elevation, meter	Cu µg/L		Pb, µg/L		Cd, µg/L		Zn, µg/L
	ASV	AAS	ASV	AAS	ASV	AAS	ASV
3800	3.3	2.2	6.2	5.0	0.36	0.30	4.0
4300	1.9	3.9	0.4	0.5	0.045	0.07	†
5000*	4.8	8.5**	11.0	10.0	0.35	0.28	1.6
5000	2.2	3.0	11.0	15.0	0.26	0.22	
7600	2.0	†	2.2	†	0.17	†	43.0
8600	5.1	3.6	12.0	7.5	0.15	†	15.0

* Fresh snow.
** Neutron activation analysis gave 4.6 µg/L.
† Not determined

using CWP [64]. The concentrations of tellurium are 0.8 ng/L in seawater from Nanhai, 0.4 to 0.7 ng/L in seawater from Donghai, 2.8 ng/L in water from the Min River, 2.3 ng/L in tap water, and 0.3 ng/L in ground water.

Conclusions

The speciation of trace metals in aquatic environmental samples is progressing rapidly in China. Knowledge about the types of trace metal compounds present in aquatic environmental samples is essential for an understanding of the toxicity and bioavailability of these metals. Information about the distribution of trace metal species is much more important than knowledge of total trace metal concentrations. The environmental analytical chemist must strive to identify and quantitate the chemical forms of trace elements and must not be satisfied with the determination of total trace element concentrations. The experimental techniques for the determination of species of trace metals must be improved to achieve higher sensitivity, prevent contamination, and preserve the species in their original distribution at the moment of sampling. Perhaps the simplest and most useful technique for speciation at this time is anodic stripping voltammetry, which distinguishes between labile and bound metals. In recent years, polarographic instruments, such as pulse polarographs, voltammetric analyzers, and Neopolarographs have been manufactured in China and have become available. Because of its sensitivity and versatility, anodic stripping voltammetry is becoming increasingly popular in China as a technique for the speciation of trace metals in aquatic samples.

Acknowledgement. The authors express their gratitude to Professor Ching-I Liu, under whose guidance this article was prepared.

References

[1] K. J. Irgolic, "Environmental Inorganic Analytical Chemistry," in **"Environmental Inorganic Chemistry,"** Proceedings, Workshop on Environmental Inorganic Chemistry, San Miniato, Italy, June 2-7, 1983, VCH Publishers: Deerfield Beach, Florida, 1985, p. 547.
[2] T. M. Florence and G. E. Batley, "Chemical Speciation in Natural Waters," **CRC Crit. Rev. Anal. Chem., 9** (1980) 219.
[3] W. Stumm and P. A. Brauner, "Chemical Speciation," in **"Chemical Oceanography," Vol. 1,** 2nd Ed., J. P. Riley and G. Skirrow, Eds., Academic Press: New York, 1975, p. 173.
[4] Xu Ouying and Zeng Canxing, "Speciation of Metals in Natural Water," **Huanjing Huaxue, 1** (1982) 329.
[5] Zhang Zuxun, "Investigations of the Chemical Speciation of Metals in Aquatic Environments by Anodic Stripping Voltammetry," **Huanjing Kexue, 2** (1981) 66.
[6] Pang Shuwei, "Analytical Techniques for the Study of Heavy Metals in Environment," **Huanjing Kexue, 1** (1980) 63.
[7] Lu Yaqin, Wang Xiuling, Zhang Tianhui and Qi Guanfa, "The Research on Removing Rule of Arsenic in the Sea Area near the No. 53 Outfall of Dalian Bay," **Haiyang Huanjing Kexue, 1** (1982) 49.
[8] Mao Miezhou, Wei Shinhsi and Liu Tsehui, "A Preliminary Study of Some Metals in Natural Waters," **Huanjing Kexue, 2** (1981) 265.
[9] Zhang Zuxun and Wang Chunming, "Estimation of the Composition of Lead-Humic Acid Complex, Conditional Formation Constant, and Determination of Humic Acid Molecular Weight with Square Wave Polarography, **Huanjing Kexue, 4** (1983) 31.
[10] Chen Xulong, Zhang Yuqi and Chen Zhiqiong, "The Chemical Forms, Accumulation and Distribution of Mercury in the Organs and Tissues of Fish in the Bohai Sea," **Haiyang Huanjing Kexue, 1** (1982) 66.
[11] Yang Huifen, Zhu Lin and Wang Huaizhou, "Determination

of Organic and Inorganic Mercury in Fish-Model 590 Acid Extraction Mercurimetry," **Zhongguo Huanjing Kexue, 2** (1982) 58.
[12] Li Xunguan, "Chemical Forms and Content of Arsenic in Some Soils of China," **Turang Xuebao, 19** (1982) 360.
[13] Fu Shaoqing and Song Jinyu, "Study on the Method of Determination of Soil Available Phosphorus in Relation to Phosphorus Forms," **Turang Xuebao, 19** (1982) 305.
[14] Shi Ehou, Han Jiangao, Huang Shuigang and Wu Chengbin, "Seawater Pollution in Coastal Areas of China," **Hainyang Huanying Kexue, 1** (1982) 13.
[15] Gu Hongkan, Liu Mingxing, Bao Wanyou, Zhang Xingjun, Wang QI., Guo Ruxin and Zeng Zhaowen, "Distribution of Trace Amounts of Various Metal Ions in China Coastal Waters," **Haiyang Kexue Jikan, 13** (1978) 1.
[16] Gu Hongkan, Liu Mingxing, Zhang Xingjun, Bao Wanyou, Guo Hongkan, Wang QI. and Zeng Zhaowen, "Concentrations of Trace Metal Ions in Sea Water," **Haiynag Kexue Jikan, 14** (1978) 23.
[17] J. A. Page and G. W. Van Loon, "A Report on the Application of Electroanalytical Methods to the Study of Trace Metals in the Marine Environment," **Report, Natl. Res. Council, Canada, Marine Anal. Standard Program, 5,** 1978, 131 pp. NRCC No. 16924.
[18] Cheng Jiemin, "Anomalous Distribution of Trace Metal Ions in Coastal Seawater," **Huanjing Huaxue, 1** (1982) 400.
[19] Zhen Jianlin, Lin Zhiqing, Wan Zhaoding, Li Zhuohong, Li Zijiang and Chen Jinsi, "Applications of China-Made Chelating Resins to the Analysis of Seawater. 1. Determination of the Species of Heavy Metals in Seawater," **Haiyang Xuebao, 4** (1982) 431.
[20] Ling Zhiging, Wang Zhaoding and Zheng Jianlu, "Determination of Chemical Forms of Copper, Lead, Cadmium and Zinc in the Water Phase near Macas," **Huanjing Kexue, 3** (1982) 37.
[21] Luo Weiquan, He Qingxi, Fang Ping and Chen Guoqing, "Preliminary Study on the Speciation of Mercury and Cadmium in the Zhujiang River (Pearl River) Estuarine Area, China," **Huanjing Huaxe, 1** (1982) 245.
[22] He Qingxi and Fang Ping, "Method for Analysis of Chemical Forms of Cadmium in the Water of Estuaries," **Huanjing Huaxue, 1** (1982) 71.
[23] Wang Shuchang, Shi Zhili, Suh Bingyi, Wang Yongchen, Yu Shengrui, Dai Guosheng, Li Liangzhong, Chang Fenghan, Ke Dongsheng and Chang Yuenli, "The Chemical Forms of Zinc and its Distribution in the Northeast Jiaozhou Bay," **Shandong Haiyang Xueyuan Xuebao, 10** (1980) 64.

[24] Sun Bingyi, Shi Zhili, Wang Shuchang, Yu Shengrui, Wang Yongchen, Dai Guosheng, Fang Ping and Ciu Liuji, "Chemical Forms and Distribution of Lead in the Northeast Jiaozhou Bay, China," **Shandong Haiyang Xueyuan Xuebao,** 10 (1980) 79.

[25] Mao Meizhou, Liu Zihui and Cui Chunguo, "Determination of Labile Metal Fractions in Polluted Water in the Xia Wan Section of the Xiang River by Chelex Resin Chromatography and Anodic Stripping Voltammetry," **Huanjing Kexue,** in press.

[26] Qi Deyao, Ying Tailin and Ying Qizhou, "Determination of Lead, Copper, Cadmium, and Zinc in Airborne Particulate Matter with Differential Pulse Anodic Stripping Voltammetry," **Huanjing Kexue,** 4 (1983) 27.

[27] S. D. Brown and B. R. Kowalski, "Pseudopolarographic Determination of Metal Complex Stability Constants in Dilute Solution by Rapid Scan Anodic Stripping Voltammetry," **Anal. Chem.,** 51 (1979) 2133.

[28] Zhang Zuxun and Liu Xianhao, "Determination of Organometallic Complexes and Conditional Formation Constants by Anodic Stripping Voltammetry," **Huanjing Kexue,** 3 (1981) 293.

[29] Zhang Zuxun and Zhan Xinming, "Estimation of Lead-Serum Protein Complex, Conditional Formation Constants, and Determination of the Molecular Weight of Serum Protein by Anodic Stripping Voltammetry," **Huaxue Xuebao,** 40 (1982) 891.

[30] Xie Zhilong and Sun Yungjun, "Separation and Neutron Activation Analysis of Different Valences of Arsenic in Water," **Zhongguo Huangjing Kexue,** 2 (1982) 50.

[31] Huang Weiwen, He Chongli, Li Furong, Wang Qinzhang and Li Jing, "Determination of Arsenic in Seawater by Anodic Stripping Voltammetry," **Shandong Haiyang Xueyuan Xuebao,** 10 (1982) 42.

[32] Wang Jinhao, Liu Jianming, Xing Huiying and Wang Zhijinag, "Determination of Total Arsenic in Natural Water by Anodic Stripping Voltammetry on Gold Film Electrodes," **Huanjing Huaxue Congkan,** 3 (1982) 32.

[33] Xu Kuncan, Wu Liqing and Wang Ruixian, "Differential Cold Atomic Absorption Method for Direct Determination of Trace Mercury in Seawater," **Haiyang Xuebao,** 4 (1982) 564.

[34] You Daoxin, Jin Zhaohui, Wang Yanfei and Liu Yinghua, "The Distribution of Mercury in Surface Seawater of Bohai Wan," **Huanjing Huaxue,** 1 (1982) 369.

[35] Qian Wanying, Zhou Jiayi, Li Jiliang, Yao Yunling and Qiu Lisheng, "Reduction-Aeration Preconcentration at Ordinary Temperature and Cold Vapor Atomic Absorption

Method for the Determination of Mercury at ppt Levels in Seawater and Natural Waters," **Haiyang Yu Huzhao, 9** (1978) 36.
[36] Rao Lili and Zhang Shen, "Study of the Distribution of Mercury and its Pattern in the Sediments of the Xiangjiang River," **Zhungguo Huanjing Kexue, 2** (1982) 36.
[37] Pang Shuwei, Qui Qangkui and Sun Jingfang, "Sequential Chemical Extraction for Speciation of Mercury in River Sediments," **Huanjing Kexue Xuebao, 1** (1981) 234.
[38] Wang Qingmin, Feng Guozhou and Cao Hongfa, "Correlation between Mercury Forms and its Absorption by Rice," **Huanjing Kexue, 3** (1982) 23.
[39] Dai Shugui, Huang Guolan and Fu Xueqi, "Determination of Organic Mercury Compounds in Different Species by Using Gas Chromatography in Combination with the Mercury Analyzer," **Huanjing Kexue Xuebao, 1** (1981) 369.
[40] Wang Shuhai, "Determination of Methylmercury in Water, Soil and Biological Materials by Concentration of Sulfhydryl Cotton Fiber-Gas Chromatography," **Huanjing Kexue, 3** (1979) 36.
[41] Zhang Xin and Guo Xiuying, "Determination of the Valency of Mercury in Industrial Wastewater," **Huanjing Huaxue, 2** (1983) 54.
[42] Jiang Yanqing, "Spectrophotometric Determination of Trace Arsenic(III) and Arsenic(V)," **Huanjing Kexue, 2** (1981) 210.
[43] Yu Muqing and Liu Guiqin, "Hydride Generation Atomic Absorption Spectrophotometric Determination of Trace Arsenic(III) and Arsenic(V) in Water by Concentration and Separation with Sulfhydryl Cotton Fibers," **Fenxi Huaxue, 10** (1982) 747.
[44] Wan Guiqin and Yu Muqing, "Determination of Trace Selenium(IV) and Total Selenium in Waters by Hydride Generation-Atomic Absorption Spectrophotometry," **Fenxi Huaxue, 11** (1983) 200.
[45] Wan Shungrong, Xu Fuzheng, Zhou Hengfu, Jin Xiulan and Li Cong, "Gas Chromatographic Determination of Selenium(IV), Selenium(VI), and Total Selenium," **Huanjing Kexue, 2** (1981) 419.
[46] Yu Muqing and Liu Guqin, "Determination of Trace Selenium of Different Valence in Water by Concentration and Separation on Sulfhydryl Cotton Fiber-Atomic Absorption Spectrophotometry after Hydride Generation," **Huanjing Huaxue, 1** (1982) 467.
[47] Liu Guiqin and Yu Muqing, "Sulfhydryl Cotton Fiber Concentration Separation-Hydride Atomic Absorption Method for Separate Determination of Antimony(III) and

Antimony(V) in Water," **Fenxi Huaxue, 10** (1982) 680.
[48] Wan Shunrong, Xu Fuzheng, Zhou Hengfu and Jin Xinlan, "Determination of Chromium(III) and Chromium(VI) by Gas Chromatography," **Huanjing Huaxue, 1** (1980) 11.
[49] Wan Lijun and Zhang Shen, "Studies on the Method of Separation and Concentration of Trace Chromium(III) and Chromium(VI) in Water with Ion Exchange," **Huanjing Kexue, 2** (1981) 84.
[50] Wang Mengxia and Wang Erkang, "Determination of Trace Chromium(VI) and Total Chromium in Plating Water by Pulse Polarography with a Rotating Glassy Carbon Electrode," **Huanjing Huaxue Congkan, 2** (1981) 38.
[51] Shan Xiaoquan and Ni Zheming, "Matrix Modification for the Differential Determination of Tellurium(IV) and Tellurium(VI) in Water Samples by Graphite-Furnace Atomic Absorption Spectrometry," **Huanjing Kexue Xuebao, 1** (1981) 74.
[52] Song Hongzi and Wang Naixing, "Cathodic Stripping Voltammetric Determination in Kelp and Laver," **Fenxi Huaxue, 10** (1982) 489.
[53] Wang Erkang, Li Aiguo and Zhang Shousong, "Pulse Polarographic Stripping Determination of Trace Mercury in Pure Water," **Huanjing Kexue, 1** (1979) 24.
[54] Wang Mengxia, Wang Erkang and Zhang Xinhua, "Determination of Trace Arsenic in Water by Pulse Polarographic Stripping Method," **Fenxi Huaxue, 9** (1981) 156.
[55] Deng Jiaqi, Zhang Qingyuan and Wang Zhaizhong, "Study of Anodic Stripping Voltammetry of Trace Germanium on Gold Film Electrode," **Huaxue Xuebao, 40** (1982) 151.
[56] Lu Zongpeng, Xie Guangguo and Zeng Xianying, "Determination of Selenium by Inverse Stripping Voltammetry on a Gold Electrode," **Fenxi Huaxue, 8** (1980) 223.
[57] Xin Xueyi, Yin Xiangchun, Duan Zhe and Wang Guiyun, "Determination of Low Level Cadmium in Seawater by Anodic Stripping Voltammetry," **Haiyang Xuebao, 4** (1982) 35.
[58] Peng Tuzhi and Lu Rongshan, "A Study of Cathodic Stripping Voltammetry of Cyanide," **Huaxue Xuebao, 40** (1982) 598.
[59] Song Hongzi and Wang Naixing, "Determination of Sulfur in Sera by Cathodic Stripping Voltametry," **Fenxi Huaxue, 10** (1982) 170.
[60] Cui Chunguo, "Simultaneous Determination of Copper and Bismuth at the Gold Electrode by Anodic Stripping Voltammetry," **Huanjing Kexue, 3** (1982) 13.
[61] Lu Zongpeng and Liu Yulan, "Determination of Copper, Lead, Cadmium and Zinc in the Environmental Samples Collected from the Zhumulangmafeng Region," in **"Report**

on Scientific Research of Zhumulanmafeng (1975): Meteorology and Environment," Daqi Wuli So and Zhongyang Qixiang Ju Yanjiu Suo, Eds., Scientific Publishing House, 1982, p. 222.
[62] Lu Zongpeng, Xie Guangguo and Zeng Xianying, "Determination of Trace Silver in Ground Water and Precipitation by Anodic Stripping Voltammetry," in **"Collected Works of the Electroanaytical Symposium 1980,"** Zhongguo Huaxue and Zhongguo Jinshu Xuehui," Eds., Jiangsu Taizhou Publishing House, 1 (1981) p. 384.
[63] Chen Xibao, "Determination of Molybdenum in Waters Collected from the Region of Zhumulangmafeng," in **"Report of Scientific Research on Zhumulangmafeng (1975): Meteorology and Environment,"** Daqi Wuli Suo and Zhongyang Qixiang Ju Yanjiu Suo, Eds., Scientific Publising House, 1982, p. 233.
[64] An Jinru and Zhang Qing, "Method for Determining Ultratrace Tellurium in Seawater and Environmental Water Samples," **Haiyang Xuebao,** 4 (1982) 555.

DISCUSSION

R. E. Sievers (University of Colorado, Boulder): How were the snow samples from the Zhumulangmafeng region collected?

Lu Zongpeng: The snow samples were collected by a mountaineer on foot.

J. J. Morgan (California Institute of Technology): Which procedures were used to obtain the nine-compartment speciation results you presented?

Lu Zongpeng: We used the procedures published by Florence and Batley.

XXI

MERCURY IN RIVER SEDIMENTS

Peng An and Wang Zijian

Institute of Environmental Chemistry, Academia Sinica
Beijing, China

Sediments from the Ji Yun and the Xia Wan River, China, were collected and extracted sequentially with solutions of sodium acetate, hydroxylamine HCl/acetic acid, sodium hydroxide/NaCl, 30 percent hydrogen peroxide, and aqua regia to determine water soluble/ exchangeable mercury, mercury associated with carbonates and oxides, mercury associated with humic materials, mercury present as organic compound and as sulfide, and finally residual mercury. Very little of the mercury was water soluble/exchangeable and associated with carbonates and oxides. Most of the mercury in the Ji Yun sediments is associated with humic materials, organics, and perhaps with sulfides, whereas the Xia Wan sediments probably contain mercury in the form of sulfides. The thermal decomposition of sediment samples spiked with various mercury compounds provided evidence for characteristic regions of decomposition temperatures for elemental mercury, mercury associated with organic materials, and inorganic mercury compounds. The results of the thermal decomposition of the river sediments provide some support for the conclusion drawn from the results of extraction experiments.

Mercury is known to accumulate in river sediments. Information about mercury compounds in such sediments is needed for a mechanistic study of the toxicity of mercury, its transport and transformation, and for the identification of its sources. Two techniques that have been used to obtain information about inorganic and organometallic compounds in environmental samples are sequential extraction [1] and thermal decomposition [2, 3]. Similar procedures suitable for mercury compounds need to be developed. Although the thermal decomposition of many mercury compounds has been studied, the information about the thermal behavior of organic mercury compounds and of coordination compounds of mercury is insufficient.

This paper reports the results of the sequential chemical extraction of mercury from contaminated sediments of the Ji Yun and Xia Wan rivers, and attempts to characterize the types of mercury compounds in these sediments by decomposition techniques.

Experimental

Samples of sediments from the Ji Yun and the Xia Wan rivers known to be contaminated with mercury were collected at the sites shown in Fig. 1.

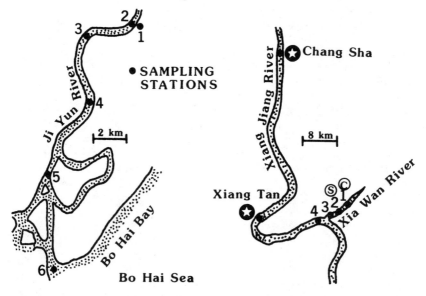

Figure 1. Sampling stations on the Ji Yun River and the Xia Wan River. The locations of the smelter and the chemical plant on the Xia Wan River are indicated by "S" and "C".

Chemical sequential extraction. The sediment samples were thoroughly homogenized and frozen as quickly as possible. All analyses were performed with the wet sediments. The results were converted to a dry weight basis using the separately determined dry weight/wet weight ratios. The accurately weighed wet sample (0.5-5.0 g) was placed into a quartz flask equipped with a stirrer and a reflux condenser.

The solution (10-20 mL/g sample), to be used as extractant, was added. The mixture, kept in a thermostat, was constantly agitated. The mixture was then centrifuged at 4000 rpm for 20 minutes, and the supernatant decanted. The residue was returned to the flask for extraction with the next reagent.

Aqueous 1.0 M sodium acetate was allowed to react with the sediment samples at room temperature for 0.5 hours to extract water-soluble and exchangeable mercury. The mercury associated with carbonates, iron oxides, and manganese oxides was dissolved with a 3:1 (v/v) solution of 0.4 M hydroxylamine hydrochloride/25% acetic acid at 90° for 2 hours. The mercury associated with humic materials was largely, but not completely, brought into solution by treating the sample for 4 hours at 90° with an aqueous solution of potassium hydroxide (1%) and sodium chloride (3%). Mercury sulfides and mercury associated with organic compounds were mineralized at pH 1-2 (HCl) with 30 percent hydrogen peroxide for 2 hours at 70°. To avoid losses of mercury the condenser was packed with cotton containing thiol groups. The residue was treated with aqua regia at 70° for 2 hours. The mercury concentrations in the extracts were determined with a PE Coleman-50 Mercury Analzyer [4].

Thermal analysis. The apparatus for the thermal analysis of mercury-containing samples is shown in Figure 2. The sample

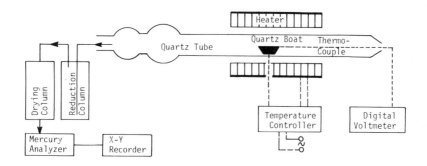

Figure 2. Diagram of the thermal decomposition apparatus.

(0.01-2.0 g) (sediments, sediment spiked with mercury compounds) was weighed and placed into a quartz boat. The loaded boat was inserted into the quartz tube and the temperature increased at a rate of 4°-5°/min in an atmosphere of flowing air (1.3-1.5 mL/min). The evolved gases passed through a solution of 0.5 percent tin(II) chloride in aqueous 5 percent hydrochloric acid and a drying tower filled with silica gel into the mercury analyzer [5].

Thermal decomposition of mercury compounds. Organic and inorganic mercury compounds were thoroughly mixed by grinding with dry river sediment that had been collected at an unpolluted upriver site. Mercury was not detectable in this sediment with a mercury analyzer. The samples were spiked with several milligrams of a mercury compound and then heated at 4°-5°/min under an atmosphere of flowing air. Several mercury complexes were synthesized as reference materials. The mercury humate was prepared with purified humic acid extracted from the river sediments.

Results and Discussion

Sequential extraction with 1.0 M aqueous sodium acetate (water soluble and exchangeable Hg), aqueous 1.0 M hydroxylamine hydrochloride/acetic acid (3:1 v/v) (Hg associated with carbonates, iron oxides, and manganese oxides), 30% hydrogen peroxide (Hg associated with humic acid), and aqua regia (residual Hg) was used to separate the various types of mercury in river sediments. Water soluble and exchangeable mercury is best extracted with aqueous 1.0 M sodium acetate. Extractions with 1.0 M magnesium chloride gave inferior results. X-ray powder patterns of the samples before and after extraction with hydroxylamine hydrochloride/acetic acid clearly showed the absence of carbonates in the residues from the extraction. Most of the manganese oxides and approximately half of the iron oxides were dissolved by this procedure.

Synthetic mercury humates were used to test the efficiencies of various reagents for the extraction of this type of mercury. At room temperature 1.0 percent aqueous potassium hydroxide or sodium hydroxide dissolved almost no mercury. After four hours at 90°, especially in the presence of sodium chloride, a considerable fraction of the mercury had dissolved. Solutions of tetrasodium diphosphate are also not very efficient. Although complete dissolution of mercury humates was not achieved with aqueous base/sodium chloride at 90°, this solution was used to extract humate mercury from the sediments. Extraction of mercury-containing river sediments with these reagents showed the KOH/NaCl solution to be the most effective extractant.

Mercury associated with organic compounds and mercury sulfides can be brought into solution with 30 percent hydrogen peroxide at 60°. Precautions must be taken to avoid losses of volatile elemental mercury. Any mercury in humates remaining from the extraction with base will be dissolved by hydrogen peroxide. This reagent quantitatively solubilized mercury from synthetic mercury humates.

The total mercury concentrations in the sediments analyzed ranged from 0.6 to 93.8 mg Hg per kg of dry sediment (Table 1). The salt sludge from location Ji Yun 1 contained 1,999 mg Hg/kg. With exception of the salt sludge, the sediments had very little water-soluble/exchangeable mercury and mercury associated with carbonates and oxides. At least 75 percent of the mercury in the Ji Yun sediments is associated with humic materials, other organic compounds, and sulfides. The good correlation of mercury with the content of total organic carbon in the sediments [6] corroborate the results obtained by sequential extraction. Such an association is to be expected because the Ji Yun river receives organic materials from urban sewage and organic wastes and mercury from a chloralkali plant.

Most of the mercury in the Xia Wan sediments could only be extracted by hydrogen peroxide and aqua regia. Relatively little of the mercury is associated with humic acids. Because a smelter is the source of the mercury, it is reasonable to assume that a large fraction of the mercury will be present in inorganic forms, perhaps as very insoluble mercury sulfides. In these sediments total mercury is not correlated with total organic carbon.

The decomposition temperature of a mercury compound is determined by its chemical and physical properties. Thermal decomposition of a mixture of mercury compounds, volatilization of the products of decompostion, and monitoring mercury in the gases escaping from the furnace might be a procedure for the identification of types of mercury compounds in sediments. Experiments using river sediments spiked with various mercury compounds showed that three distinct decomposition regions delineated by characteristic "maximal thermal decomposition temperatures" (MTDT) exist (Fig. 3). The MTDT is the temperature at the maximum of the temperature/Hg concentration curve (Figs. 4, 5). Elementary mercury belongs to region I as a representative of compounds with MTDTs below 100°. Region II contains mercury associated with organic compounds. All

Table 1. Mercury in the Sediments of the Ji Yun and Xia Wan Rivers

Location	Type of Mercury, %					Total Hg mg/kg dry sediment
	Water-Soluble & Exchangeable	Associated with Carbonates and Oxides	Associated with Humic Acid	Organic & Sulfides	Aqua Regia Soluble	
Ji Yun 1*	3.9	13.1	2.7	41.4	34.8	1,999.
Ji Yun 2†	0.06	0.09	25.7	64.3	9.6	93.8
Ji Yun 3**	0.35	0.12	47.8	34.0	13.0	23.0
Ji Yun 4**	0.09	0.0	77.4	21.0	1.8	5.7
Ji Yun 5**	0.06	0.0	60.5	41.0	0.0	4.8
Ji Yun 6††	2.4	0.0	18.0	56.0	26.3	1.2
Xia Wan 1	0.2	0.2	6.0	66.0	29.0	41.2
Xia Wan 2	0.1	0.1	47.0	21.0	24.0	21.6
Xia Wan 3	0.01	0.1	7.0	43.0	50.0	41.9
Xia Wan 4	0.03	0.02	36.0	20.0	35.0	8.9
Xiang Tan	2.4	2.4	7.0	29.0	57.0	1.2
Chang Sha	3.4	3.4	33.0	26.0	32.0	0.6

* Salt sludge, the solid waste from the chloralkali plant.
† Sample collected at the point of discharge of wastes from the chloralkali plant into the river.
** Sites are downstream from discharge point as shown in Figure 1.
†† Estuary
*** Total mercury determined in the aqua regia digests.

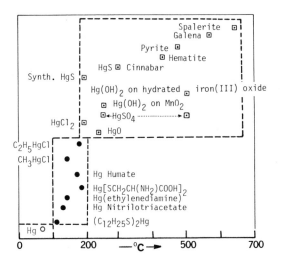

Figure 3. *Maximal thermal decomposition temperatures of mercury-containing ores and mercury compounds.*

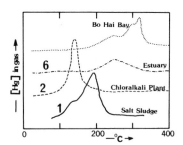

Figure 4. *Thermal decomposition curves of sediments from the Ji Yun River.*

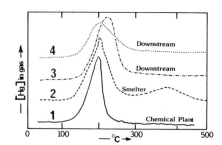

Figure 5. *Thermal decomposition curves of sediments from the Xia Wan River.*

the investigated derivatives had MTDTs between 100° and 200°. Region III, spanning the temperature range from 200° to 650°, is reserved for inorganic mercury compounds. Mercury sulfide cannot be distinguished easily from other inorganic mercury compounds.

The thermograms of the Ji Yun sediments (Fig. 4) indicate that different types of mercury compounds exist in the various samples. The salt sludge and the sediment from location 2 (Fig. 1) gave peaks indicating the presence of "organic mercury" in agreement with the results of the extraction experiments. The thermograms for the estuarine sediment and the sediment from Bo Hai Bay are characteristic of inorganic mercury compounds. It is therefore likely that most of the 56 percent of mercury extracted by hydrogen peroxide (Table 1), a reagent that does not distinguish between organic mercury and mercury sulfides, is present in the sediment in the form of sulfides. The thermograms for the Xia Wan sediments (Fig. 5) suggest that the mercury is present in organic and/or sulfidic form.

The results of the thermal investigation do not contradict the results obtained with the extraction procedures. The resolution achievable in the thermograms is not sufficient to unequivocally identify the types of mercury compounds in the sediments.

References

[1] R. P. Eganhouse, D. R. Young and J. N. Johnson, "Geochemistry of Mercury in Palos Verdes Sediments," **Environ. Sci. Tech., 12** (1978) 1151.
[2] R. L. Watling, "The Identification and Significance of Mercury Compounds in Estuarine Sediments," in **"International Conference on Heavy Metal in Environment,"** CEP Consultants, Ltd., Amsterdam (1981) 591.
[3] Hu Goulian, "Transformation and Migration of Mercury in Geological Pollution Sources," **Huanjing Kexue Xuebao, 2** (1982) 229.
[4] Wang Zijian and Peng An, "A Study on the Methodology of Speciation of Mercury in River Sediments," **Fenxi Huaxue, 11** (1983) 321.
[5] Peng An and Wang Zijian, "Speciation of Mercury in River Sediments by Thermal Decomposition," **Huanjing Huaxue, 3** (1984) 53.
[6] Lin Yuhuan, Kang Demeng and Liu Ching-I, "Species Distribution of Mercury in Ji Yun River Sediments," **Huanjing Huaxue, 2** (1983) 10.

XXII

DEPOSITION OF ATMOSPHERIC AND METAL POLLUTANTS AND THEIR IMPACT ON THE FRESHWATER ENVIRONMENT: CASE STUDIES IN SWITZERLAND

Werner Stumm

Institute for Water Resources and Water Pollution Control (EAWAG)
Swiss Federal Institute of Technology (ETH) Zürich
8600 Dubendorf, Switzerland

The hydrogeochemical cycling of metals, carbon, nitrogen, phosphorus, and sulfur is accelerated by the release of products from fuel combustion and extractive metallurgy. The two effects of major concern are the production of acid rain by the oxidation of sulfur and nitrogen compounds, and the release of metals as particles and as vapors into the atmosphere. The atmosphere - in comparison to the lithosphere and the hydrosphere - is a relatively small reservoir with small time constants for alterations and thus especially sensitive to anthropogenic emissions. Although the free acidity in rain of continental Europe is similar to that in Scandinavia, the consequences of acid rain on the freshwater environment in continental Europe have been minimal, because soils and sediments contain relatively large amounts of carbonates. Lakes in non-calcareous mountain regions are very susceptible to the effects of acidic precipitation. The ability of these lakes to neutralize acids increases with the size of the catchment area and the residence time of the waters. The chemical forms, in which metals are present in the environment, need to be known in order to understand the mechanisms that control their concentrations in waters. Chemical processes occurring at the particle/water interface, especially complex formation between oxygen donor atoms at organic and inorganic particle surfaces and metal ions, are important in determining the residence times and the ultimate fate of metals in oceans and lakes. Lakes, despite being polluted with metals 10 to 100 times more than the oceans, are nearly as depleted in trace metals as the oceans. Larger productivity and higher particle sedimentation are the efficient elimination mechanisms in lakes. The effect of technology and the consequences of energy dissipation are now being felt over increasingly larger distances, at times becoming global in character. The anthropogenic dispersion of elements into the atmosphere appears to rival and sometimes exceed natural mobilizations. For instance, acid rain is a consequence of the anthropogenic disturbance of the geochemical cycles of sulfur and nitrogen.

The Atmosphere as a Conveyor Belt of Acidic and Metallic Pollutants

The atmosphere is more susceptible to anthropogenic emissions than the terrestrial or the aqueous environment, because the atmosphere is much smaller than the other reservoirs. Furthermore, the time constants for atmospheric alterations are small in comparison to those in the seas and the lithosphere. When the rate of oxidation of carbon, nitrogen, and sulfur increases relative to the rate of reduction of carbon dioxide, nitrogen, oxides of nitrogen, sulfur dioxide, and sulfuric acid, the delicate natural balance is disturbed. Such a disturbance caused by the activities of civilization in the fossil fuel age has increased the concentration of carbon dioxide globally, and the concentrations of sulfur dioxide, sulfuric acid, nitrogen oxides, nitrous acid, and nitric acid regionally. The ability to counteract disturbances by these substances is much greater on land and in the oceans than in the atmosphere. For example, the mixing time of the ocean is approximately 1000 years.

Genesis of Acid Precipitation

Most of the major and minor gaseous components of the atmosphere (O_2, N_2, CO_2, CO, NO, NO_2, SO_2, CH_4) participate in elemental cycles that are governed by oxidation-reduction reactions of biological origin. Photosynthesis and respiration are major reactions in these cycles. These oxidation and reduction reactions consume and produce the gases of the atmospheric reservoir. A steady-state concentration is established over a period of millions of years for each component. In oxidation-reduction reactions, electron transfers are coupled with the transfer of protons to maintain charge balance. A modification of the redox balance elicits a modification of the acid-base balance. The oxidation of carbon, sulfur, and nitrogen during the combustion of fossil fuels increases the concentrations of sulfur oxides, nitrogen oxides, and carbon dioxide in the atmosphere. The oxides are converted to sulfuric acid, nitric acid, and carbonic acid. These acids and the oxides become associated with aerosol particles, rain drops, snow flakes, and fog. The compounds mainly responsible for acid rain are the strong acids sulfuric acid and nitric acid (Fig. 1). Because carbonic acid is a weak acid, its influence on the composition of precipitation is insignificant in spite of the increase in atmospheric carbon dioxide during the past 100 years. The importance of the interactions of sulfur and nitrogen oxides with atmospheric gases, aerosols, and water is dependent on several factors. The assumption seems reasonable, that 50 percent of the sulfate and nitrate, 80 percent of the ammonia, and 100 percent of the hydrogen chloride are transferred from the atmosphere to the rain droplets. Figure 1 summarizes the formation of acid rain in terms of oxidation and acid-base reactions and illustrates interactions of acid rain with the terrestrial and aquatic environment [1].

A considerable part of the strong acids originate from the oxidation of sulfur during combustion of fossil fuels and the conversion of atmospheric nitrogen to NO and NO_2 during the combustion of gasoline in motor vehicles and other combustion processes of sufficiently high temperature. Hydrogen chloride is formed during the combustion and decomposition of organic chlorine compounds, for example, during incineration of polyvinyl chloride plastics. Natural sources of these pollutants are volcanic activity, the oxidation of hydrogen sulfide formed in anaerobic sediments, and the oxidation of dimethyl sulfide and carbon oxysulfide generated in the ocean. Bases in the atmosphere are carbonates in wind-blown dust, and ammonia that is generally of natural origin. The ammonia is formed from ammonium ions and by the decomposition of urea in soil environments. Redox and acid-base reactions occur in the gas phase, in aerosols, in rain drops, cloud droplets, and fog.

Sulfur dioxide has a maximal residence time in the atmosphere of several days. During this time the sulfur dioxide may be transported several hundred to one thousand kilometers. Nitric acid is formed more rapidly than

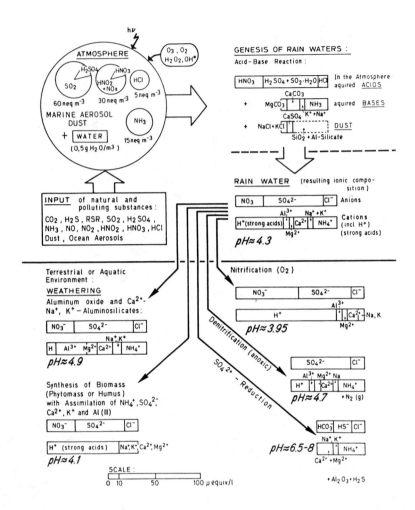

Figure 1. The genesis of acid rain and the composition of Swiss rain water. Reproduced with permission from ref. 1.

sulfuric acid resulting in more restricted dispersion of nitric acid. Sulfuric acid may react with ammonia to form ammonium hydrogen sulfate or diammonium sulfate aerosols. In addition, the ammonium nitrate aerosols are in equilibrium with gaseous ammonia and gaseous nitric acid. The importance of gas and aerosol scavenging by atmospheric condensates and rain drops depends on many factors.

Acid Rain in Switzerland and its Influence on Some Alpine Lakes

Although the concentration of free acidity in Swiss rain (pH 4.3) has approached that of Scandinavia, the consequences of acid rain in Switzerland have been minimal compared to those in Scandinavia and parts of North America, because Swiss soils and sediments contain relatively large amounts of carbonates. These carbonates neutralize excess acids rapidly. This neutralization does not occur in lakes in non-calcareous mountainous regions. South of the Alps, the mountain lakes in the vicinity of the watersheds of the Maggiatal in the Verzasca Valley are in an area consisting exclusively of crystalline rocks. Weathering of these crystalline rocks (granite, gneiss and mica schist) produces basic materials, but their reactions with free acidity is much slower than those of carbonates. Therefore, these lakes have a low pH, although the precipitation entering these lakes was shown not to be as acidic as the precipitation in the non-mountainous areas of Switzerland. These lakes become acidic, when the residence time of acid rain and melt water from snow in soils is relatively short. Because the soils in this area are very thin, little fine material is available, and many weathered rocks are exposed, the water has little time to react with the minerals. Also, the buffering effect of trees or thick vegetation is lacking at these altitudes. Analyses by the Office of Water Protection of the Canton of Tessin [2] identified 20 small mountain lakes with pH <6. Among them ten lakes had pH <5.5. The lowest pH measured was 4.6. The acidity was highest in lakes with very small surface and catchment areas, resulting in the shortest residence times of runoff water in the watershed. The acidity of lakes that receive runoff and melt water from snow very rapidly is not greatly different from the acidity of the snow.

The composition of some typical waters in the Maggia Valley and the successive reactions of the source water during its flow to the lakes are shown in Figure 2. The pH of the water increases with increasing catchment area and concomitantly increasing residence time of the water. During this flow the concentrations of cations (sodium, potassium, calcium) and of silicic acid increase also [3]. The concentration of aluminum ion that can increase only through dissolution of aluminum oxides and hydroxides is higher in the lakes with low pH than in the lakes with high pH. That fish cannot reproduce in acidic lakes, is probably a result of high concentrations of aluminum ion and other

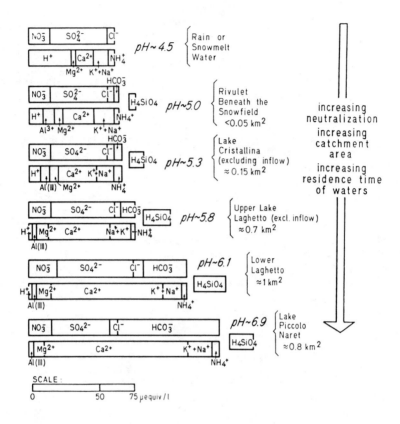

Figure 2. *The composition and acidity of Swiss mountain lakes at the end of July 1982.*

heavy metal ions rather than of low pH. These lakes are not sterile. However, the supply of nutrients is extremely limited. With less than 5 µg phosphorus per liter, the productivity is small. The rate of chemical weathering in the watersheds of these lakes can be estimated from the difference between the acidities of the precipitation and the lakes. The calculated weathering rate is 500 equivalents per hectar per year, approximately half of the weathering rate of alumino-silicate minerals in the alpine valley of the Rhine [4].

The condition in the Tessin mountain lakes is compared to the conditions in other acidic lakes in Scandinavia and North America [5] in Figure 3. The acid deposited in the drainage areas of the lakes reacts with the basic substances

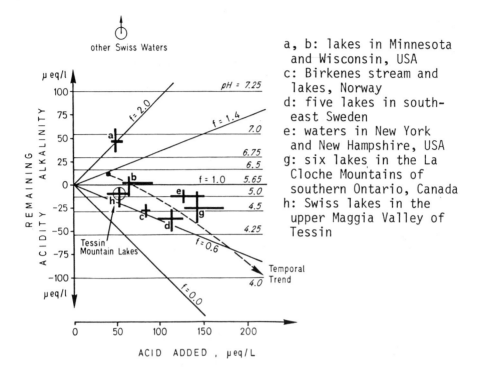

Figure 3. *Alkalinity of natural waters in Canada, Norway, the United States of America, and Switzerland as a function of the acid load. The acid added to the lakes is given by (Lake Inflow x Acidity of Precipitation)/Lake Outflow. Modified with permission from ref. 1.*

in rocks and minerals. The parameter, f, representing the equivalents of acid neutralized by chemical weathering, generates curves (Fig. 3) that express the trend in acidity or alkalinity of a water body. When f equals zero, no acid is neutralized. When f is one, all the added acid is neutralized and the pH does not change. When f is larger than one, the water will become with time increasingly alkaline, for instance, through the reaction of carbon dioxide with calcium carbonate. The historical change of pH for a lake would produce a "titration" curve (Fig. 4) with progressive increase of an acid load. Figure 3, in which the ranges for added acid and remaining acidity or alkalinity are given by horizontal and vertical bars, respectively, indicates that the Tessin lakes ceased to be able to compensate for an increasing acid load only five to

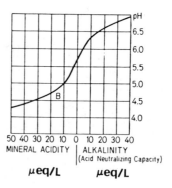

Figure 4. *The change of pH of a lake with an increasing acid load. Modified with permission from ref. 5.*

ten years ago. The acidities determined for similar lakes a decade ago appear to confirm the trend toward higher acidity (dashed line in Fig. 3) in the Tessin mountain lakes. It is feared that the acidity of precipitation will rise further. If the concentration of the hydrogen ions in precipitation were to double, most of the lakes in the crystalline rock regions will problably have pH values less than 5 in the near future [3]. Lakes in the Swiss lowlands are not endangered by acid rain because of their alkalinity of 1000 eq/L.

Early warning systems of nature. The atmosphere is not only an important vehicle for the deposition of acid pollutants but also for a number of other substances that affect the ecology of aquatic and terrestrial systems. Atmospheric deposition accounts for approximately 100 g per hectare per year of polyaromatic hydrocarbons in rural environments. In populated regions, rain waters contain concentrations of heavy metals considered excessive for surface water [6].

Metal Pollutants in Lakes

The concentrations of metal pollutants in lakes increase 10- to 100-times faster than in the oceans. Much of this pollution is "fallout" from the atmosphere [7, 8] (Table 1). Anthropogenic fluxes were calculated by various authors [9, 10]. That pollution of coastal and inland waters by trace elements has increased markedly over the last decades is recorded in the sediments.

Table 1. *Atmospheric Deposition and Sedimentation Rates of Heavy Metals in the North Atlantic, in Lake Constance, and in Swiss Rivers*

Location	Particulate Flux, ng cm^{-2} yr^{-1}			
	Cu	Pb	Cd	Zn
North Atlantic [8]				
Atmospheric Deposition	25	310	–	130
Sinking Suspended Particles	234	330		1040
Concentrations in Seawater	30–300* (100)	1–15* 3	1–120* 10	10–600* (100)
Lake Constance [9]				
Atmospheric Deposition	714	11,000	20	8,400
Sinking Suspended Particles	6500	9500	(100)	36,000
Concentration in Lakes	300–800*	50–100*	6–20*	1000–4000*
Swiss Rivers				
Concentration	1000–3000*	300–1000*	20–100*	10,000*

*Range of concentrations in ng/L. The values in parentheses are averages.

The need for chemical speciation. To understand the factors that control the concentrations of metals in natural waters, the chemical forms, in which metals are present, their chemical reactivities, their biological availability and toxicity, and their ultimate fate must be known. Usually, the evidence for the occurrence of a particular chemical form of a metal is indirect. Analytical techniques are needed that distinguish between the various dissolved and "adsorbed" species, and identify, for instance, sites on organic surfaces, on iron(III) oxide, and aluminum silicates, to which metals are or can be bound [11]. As an example, Figure 5 categorizes copper species that could be present in natural waters and summarizes techniques for their characterization. Ion selective electrodes, if they were sufficiently sensitive, would permit the measurement of the activities of free metal ions. No other simple method permits the unequivocal determination of free metal ions. Bioassays with algae were used to determine free copper(II) ion. It is operationally very difficult to distinguish between concentrations of dissolved metal species and metal species adsorbed on particulates. Colloids are often

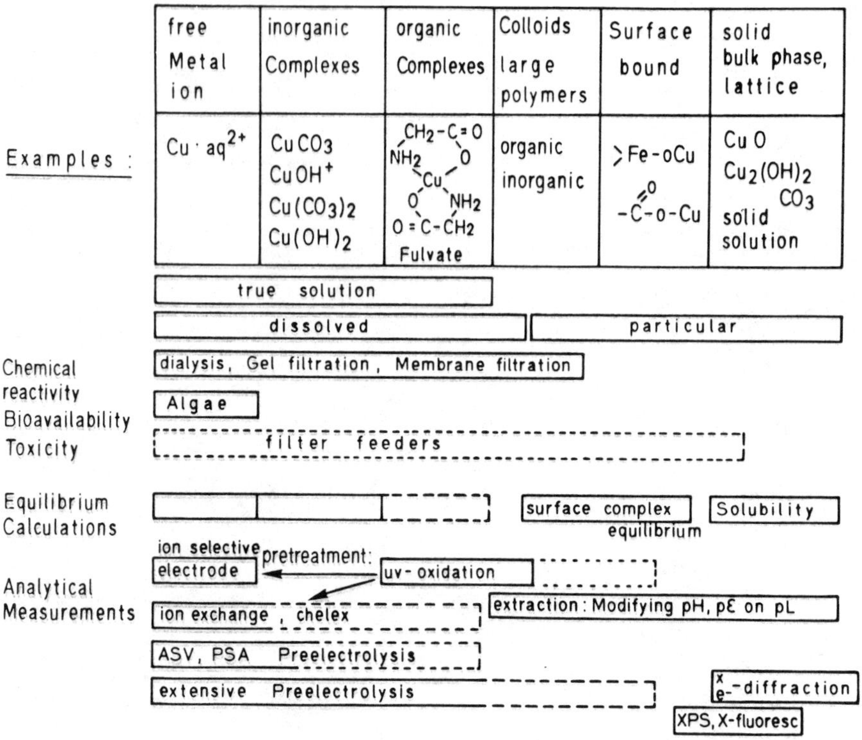

	free Metal ion	inorganic Complexes	organic Complexes	Colloids large polymers	Surface bound	solid bulk phase, lattice
Examples :	$Cu \cdot aq^{2+}$	$CuCO_3$ $CuOH^+$ $Cu(CO_3)_2$ $Cu(OH)_2$	$CH_2-C=O$ $NH_2 \quad O$ $\diagdown Cu \diagup$ $O \quad NH_2$ $\diagup \quad \diagdown$ $O=C-CH_2$ Fulvate	organic inorganic	$>Fe-oCu$ $-\overset{O}{\overset{\|}{C}}-o-Cu$	CuO $Cu_2(OH)_2$ $\quad CO_3$ solid solution

| true solution |
| dissolved | particular |

Chemical reactivity
Bioavailability
Toxicity

| dialysis, Gel filtration, Membrane filtration |
| Algae |
| filter feeders |

Equilibrium Calculations

| | | surface complex equilibrium | Solubility |

Analytical Measurements

| ion selective electrode | pretreatment: | uv-oxidation |
| ion exchange , chelex | extraction: Modifying pH, pE on pL |
| ASV, PSA   Preelectrolysis |
| extensive Preelectrolysis | | e^x_e-diffraction |
| | XPS, X-fluoresc |

Figure 5. *Types of copper(II) species in freshwater and techniques for their characterization.*

sufficiently small to pass through the pores of membrane filters that are commonly used to separate dissolved metals from particulate metals. For instance, anodic stripping voltammetry (ASV) cannot be used for the determination of copper(II) in freshwater unless the samples are acidified to pH < 6. Any measurements of complexation capacity in acetate-buffered samples by ASV may not be representative of the complexing capacity of a water at its natural pH. Although ASV is an extremely sensitive technique, it measures only the concentration of electrochemically labile metal ions and the electrochemically available metal ions present within the diffusion layer. Copper complexes with carbonate or glycine as ligands are examples of electrochemically labile species. At low pH ASV measures usually the concentrations of total soluble metal ion.

Multimetal, multiligand equilibrium models. Models for equilibria among complexes often provide information about the complexes of a metal present in a body of water with certain reacting ligands and competing metals. Table 2 presents an example of the distribution of metal species in a typical freshwater with several added organic substances at concentrations of less than 10^{-5} M [11]. Two amino acids, citrate, tartrate, and phthalate, ligands that have functional groups similar to those in fulvic acids, were chosen as organic substances. The distribution of all the species was computed by considering all of the more than 100 equilibria. Although these organic substances are not representative of those found in natural waters, general conclusions can be drawn from the results of these model calculations.

- Conventional complex-forming organic ligands at concentrations equal to or greater than those encountered in open surface waters will affect the distribution of trace metal species only to a limited extent.
- Only the distribution of copper is markedly affected by complex formation with organic ligands considered in the calculations (Table 2).
- Approximately one third of the organic complex-forming donor groups will become bound to cations, mostly to Ca and Mg.
- Concentrations of individual amino acids as high as 100 µg/L are not sufficient to cause significant interactions with trace metals.
- At the concentration levels considered, monodentate complexing agents such as acetate are less efficient than chelate formers in sequestering trace metals.

Copper(II) is most likely to be affected by organic complex formers. Because algal growth is strongly dependent on the concentration of free copper(II) ions, organic matter may indeed exert a pronounced influence on the physiological response of algae by regulating the copper concentration. Other soluble trace metals are primarily present in natural waters (seawater and freshwater) as inorganic species.

The role of particles. Particles play an important role in regulating the concentration of most reactive elements in natural waters and especially in freshwaters. More than 95 percent of the heavy metals that are transported from land to the sea are associated with particulate matter (Table 3). The concentration of metals is typically much larger in the solid phase than in the solution phase. Thus, the buffering

Table 2. *The Distribution of Metals in Aerobic Freshwater* in the Absence and the Presence of Organic Ligands** Corresponding to 2.3 mg/L of Soluble Organic Carbon. Reproduced with Permission from Ref. 11*

Metal	Freshwater[†] -Log Concentration			Major Inorg. Species	Freshwater with Organic Ligands[†] -Log Concentration**							
	M_T	M_F	I_S		M_F	Acet.	Cit.	Tart.	Glyc.	Glu.	Phthal.	
Ca	2.7	2.72	4.6	$CaHCO_3$	2.72	7.0	5.2	5.6	9.1	8.6	5.7	
Mg	3.7	3.72	5.1	$MgSO_4$	3.72	8.0	7.0	7.1	8.0	9.1	††	
Fe(III)	Satd.	17.70	8.7	$Fe(OH)_2$	17.73	19.0	7.2	††	15.1	††	††	
Mn(II)	7.0	7.04	8.5	$MnSO_4$	7.04	11.3	9.7	††	11.5	11.1	††	
Cu(II)	7.0	7.46	7.2	$CuCO_3$	9.93	13.1	7.0	11.3	9.4	9.4	11.4	
Zn(II)	6.7	6.72	8.2	$ZnSO_4$	6.72	10.3	10.5	8.9	9.6	8.6	9.1	
Cd(II)	7.7	7.73	9.2	$CdSO_4$	7.76	11.5	9.2	9.6	11.5	10.3	9.7	
Pb(II)	7.0	8.02	7.1	$PbCO_3$	8.04	11.0	8.9	8.8	10.1	††	9.2	
Ag(I)	9.0	9.19	9.5	$AgCl$	9.19	13.8	17.5	††	13.3	††	††	
Free org. ligand						7.55	6.91	8.02	5.16	5.16	6.99	
Percent of ligand bound to metal						1.5	98.2	35.2	0.3	0.4	29.0	

*-Log concentration of free anion: sulfate 3.4, hydrogen carbonate 3.1, carbonate 6.1, chloride 3.3; pH 7.0, 25°.
**Organic ligands: 7×10^{-6} moles of each ligand are added to one liter. The concentrations of the metal complexes refer to the sum of the concentrations of all complexes formed by a particular metal ion, e.g., CuCit, CuHCit, $CuCit_2$.
[†] M_T: total metal concentration. These concentrations are higher than those typically found in unpolluted waters. The relative effects of complex formation are independent of the total concentrations. The waters are assumed to be in equilibrium with $Fe(OH)_3(s)$. M_F: free metal ion; I_S: major inorganic species.
[††] Stability constants were not available.

Table 3. Distribution of Reactive Elements Among Particulate Matter in Natural Waters [12]

Particulate Matter	Distribution Coefficient*, m^3/kg											
	Al	Si	Ca	Fe	Mn	P	Na	Ni	Pb	Cu	Cd	Zn
Average World Rivers	1900	40	1.6	1200	130	30	1.4	180	1000	66	50	8
Deep Sea Clays	1.9×10^5	140	0.06	3×10^4	3×10^4	23	0.002	1000	7×10^4	2000	23	1200
Settling Particles in Lake Constance [8]	—	—	2.3	1×10^4	3600	21	—	—	1000	36	180	100
Percent of Metal Present in Particulate Form in the Average World River**	99.9	100	39	100	98	80	35	99	99.8	86.5	95	77

*Distribution coefficient: D = metal concentration in particles [g/kg]/ concentration of dissolved metal [g/m^3].
**100 [metal concentration in particles [g/kg] × g particles per m^3]/ (concentration of dissolved and particulate metal [g/m^3]); average concentration of particulate matter: 400 g/m^3.

of metals is much higher in the presence of particles than in their absence. Particles are the major sinks for metals and other reactive elements.

The oxygen-metal ion bond. Surfaces of hydrous oxides, surfaces coated with organic materials, and surfaces of organic materials possess functional groups such as OH and COOH that act as coordinating sites. Formation constants for surface complexes are available. These constants can be used to estimate the extent of surface binding, the degree of adsorption, and the magnitude of the coefficients for the distribution of metal ions between particles and the solution as a function of pH and solution variables [13-15]. These equilibrium constants are related to the oxygen-metal ion bond strength as indicated, for example, by an electronegativity function, or to corresponding equilibrium constants in solution. For example, the tendency to form a metal-O-Si(particle) bond is related to the tendency to form a metal-OH complex in solution. Recent evidence from EPR and ENDOR measurements confirm, that such metal ion adsorption occurs typically via "inner sphere" surface complexes.

Sequestering surfaces and complex formation in solution. With the help of equilibrium constants, simple models can be constructed to evaluate the competition for metal ions between the coordination sites of soluble ligands and surface ligands. An example of the estimated distributions for cadmium, copper, and lead among free ions, carbonato complexes, salicylato complexes and (adsorbed) surface complexes is given in Figure 6. This example illustrates that surfaces can scavenge significant proportions of trace metals even in the presence of an organic chelating agent such as salicylate. This calculation showed again, that organic ligands considerably influence the species distribution of copper. In a real lake and the ocean, the effects of particles on the regulation of metal ions are even more pronounced, because the continuously settling particles such as phytoplankton and particles introduced by rivers act as a conveyor belt in transporting reactive elements. Indeed, the partitioning of metals and other reactive elements between particles and water is the key parameter in establishing the residence times and the residual concentrations of these elements in the ocean and in lakes.

Thus, the geochemical fate of metals is controlled by the chemical processes occurring between the solid surfaces and

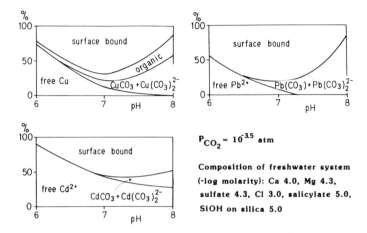

Figure 6. The calculated distribution of metal ions in freshwater among carbonato complexes, surface complexes and free metal ions as a function of pH.

the water [16]. The more reactive an element is in a lake or the ocean, the more of it will be bound to particles, the more rapidly will it be removed, and the shorter will be its residence time.

The scavenging of metals by particles in lakes. Lakes, despite being polluted with metal ions ten to a hundred times as much as oceans from riverine and atmospheric inputs, are nearly as much depleted in these trace metals as are the oceans. Therefore, the elimination mechanisms in lakes must be more efficient than those in oceans. Larger productivities and higher sedimentation rates for particles are primarily responsible for the more efficient scavenging in lakes through adsorption of metals on phytoplankton and to a lesser extent by other particles. The input of phosphorus into a lake, influencing the production of biogenic particles, is a major factor controlling the sedimentation rate of biogenic particles and in removing "biophile" heavy metals. Other potential scavenging and metal regeneration cycles operate near the sediment/water interface. Subsequent to early epidiagenesis in the partially anoxic sediments, iron(II) and manganese(II), and other elements depending on redox conditions are released by diffusion from the sediments to the overlying water, where iron and manganese are oxidized to insoluble iron(III) and manganese(III, IV) oxides. These oxides are important conveyors of heavy metals.

References

[1] W. Stumm, J. J. Morgan and J. L. Schnoor, "Saurer Regen, eine Folge der Störung hydrogeochemischer Kreisläufe," **Naturwissenschaften, 70** (1983) 216.

[2] G. Righetti, "Controllo sui Laghi Alpini del Cantone Ticino," **Aquicoltura Ticinese, 3,** Sept. 1981, p. 10.

[3] W. Stumm and G. Righetti, "Saurer Regen, saure Schweizer Bergseen," **Neue Zürcher Zeitung, Beilage, "Forschung und Technik," Nr.** 232, Okt. 6, 1982.

[4] J. Zobrist and W. Stumm, "Chemical Dynamics of the Rhine Catchment Area in Switzerland: Extrapolation to the Pristine Rhine River Input into the Ocean," in **"River Inputs to Ocean Systems,"** Proceedings of a Review Workshop at FAO Rome; United Nations: New York, 1981, p. 52.

[5] J. L. Schnoor, W. D. Parmer, J. M. Eilers and G. E. Glass, "Modeling Impact of Acid Precipitation for Northeastern Minnesota," in **"Modeling of Total Acid Precipitation Impacts,"** J. L. Schnoor, Ed., Butterworth Publishers: Stoneham, MA; Acid Precipitation Series, Vol. 9, 1984, p. 155.

[6] L. Sigg, M. Sturm, W. Stumm, L. Mart and H. W. Nürnberg, "Schwermetalle im Bodensee; Mechanismen der Konzentrationsregulierung," **Naturwissenschaften, 69** (1982) 546.

[7] P. Buat-Menard and R. Chesselet, "Variable Influence of the Atmospheric Flux on the Trace Metal Chemistry of Oceanic Suspended Matter," **Earth Planet Sci. Lett., 42** (1979) 399.

[8] L. Sigg, M. Sturm, J. Davis and W. Stumm, "Metal Transfer Mechanisms in Lakes," **Thalassia Jugoslavica, 18** (1982) 293.

[9] R. L. Lantzy and F. T. Mackenzie, "Global Cycles and Assessment of Man's Impact," **Geochim. Cosmochim. Acta, 93** (1979) 511.

[10] H. W. Georgii and J. Pankrath, Eds., **"Deposition of Atmospheric Pollutants,"** D. Reidel Publishing Co.: Dordrecht, Netherlands, 1982.

[11] W. Stumm and J. J. Morgan, **"Aquatic Chemistry,"** Wiley Interscience: New York, 1981, p. 393.

[12] J.-M. Martin and M. Whitfield, "The Significance of the River Input of Chemical Elements to the Ocean," in

"**Trace Metals in Seawater,**" C. S. Wong, E. Boyle, K. W. Bruland and J. D. Burton, Eds., Plenum Press: New York, 1983, p. 265.
[13] L. Sigg, W. Stumm and B. Zinder, "Chemical Processes at the Particle/Water Interface; Implications Concerning the Form of Occurrence of Solute and Adsorbed Species," in **"Metals in Natural Waters,"** C. J. H. Kramer and J. Duinker, Eds., Martinus Nijhoff/Dr. W. Junk Publishers: The Hague, Netherlands, 1984, p. 251.
[14] W. Stumm, H. Hohl and F. Dalang, "Interaction of Metal Ions with Hydrous Oxide Surfaces," **Croat. Chem. Acta, 48** (1976) 491.
[15] W. Stumm, R. Kummert and L. Sigg, "A Ligand Exchange Model for the Adsorption of Inorganic and Organic Ligands," **Croat. Chem. Acta, 52** (1980) 291.
[16] Y. H. Li, "Ultimate Removal Mechanisms of Elements from the Ocean," **Geochim. Cosmochim. Acta, 45** (1981) 1659.

DISCUSSION

R. E. Sievers (University of Colorado, Boulder): Did you attempt to distinguish between metals adsorbed on the surfaces of particulates in rain drops and metals dissolved in the rain drops?

W. Stumm: Not yet. I would welcome your help in showing us how to design such experiments.

R. E. Sievers: We observed that potential organic ligands tend to be associated with particulates in melted snow rather than with the aqueous phase.

XXIII

ATMOSPHERIC ACIDITY: ANALYTICAL PROBLEMS RELATED TO ITS DETERMINATION

A. Liberti, I. Allegrini, A. Febo and M. Possanzini

Istituto Inquinamento Atmosferico C.N.R., Area della Recerca di Roma
Via Salaria Km. 29,300 – C.P. 10
I-00016 Monterotondo Stazione (Roma), Italy

Analytical problems related to the determination of atmospheric acidity are outlined. The need for analytical procedures that supply information on the status of the environment while preventing artifacts and erroneous sampling is stressed. Two lines of investigation are presented; one on pollutants sampling and the other on the determination of gaseous compounds in the presence of aerosols and particulate matter. High precision samplers for ambient, emission, and indoor monitoring can be built based on the constant flow principle. Constant flow is maintained by an electronic device equipped with a flow rate transducer. The standard flow rate is measured with a differential pressure transducer. The adiabatic effusion of a gas through an orifice maintains a constant sampling gas flow, which might be applied to a variety of sampling systems. Gases can be determined in the presence of aerosols by means of diffusion tubes called denuders. They consist of glass tubings coated with suitable media that selectively absorb gases before particles are collected. New denuders were designed that can operate with a flow rate up to 30 L/min with a recovery efficiency greater than 95 percent. The air is forced through the annular space of two coaxial glass cylinders. This device permits the determination of acidic and basic compounds and sulfur dioxide present in the atmosphere.

Atmospheric acidity has become one of the major current environmental concerns since the discovery of rainfall in several areas with an acidity higher than expected from normal meteorological events. Atmospheric acidity, responsible for acid rain, is caused by primary pollutants such as sulfur dioxide, nitrogen oxide, NO, and hydrogen chloride, that are emitted into the atmosphere, and by secondary pollutants such as nitrogen dioxide, nitric acid, sulfuric acid, and sulfate and nitrate aerosols formed through reactions with normal components of the atmosphere.

Combustion of fossil fuels in stationary and mobile power units generates most of these pollutants. After release, the gases disperse into the atmospheric boundary layer, from whence they are ultimately removed either by dry deposition on the earth's surface or by incorporation into precipitation.

Because the atmosphere is an oxidizing environment, the primary pollutants are oxidized as they are dispersed. Oxides of nitrogen and sulfur are converted to low volatility oxidation products, sulfuric acid and nitric acid, which form aerosols. These strong acids as well as hydrochloric acid are readily incorporated into the various forms of precipitation. The pH of cloud droplets, rain, snow, dew, and fog is largely determined by these acids.

The complexity of the reactions in the atmosphere between species emitted through anthropogenic activity and components of the atmosphere requires the development of analytical procedures and methodologies suitable to supply representative information. Commonly, the monitoring of the atmospheric environment and the determination of atmospheric acidity requires the measurement of species present in small concentrations. This situation demands long sampling times and enrichment of the species to be determined. Two lines of investigations have been pursued: sampling of pollutants and the determination of gaseous compounds in the presence of aerosols and particulate matter. High precision samplers were needed to monitor pollutants and trace components in the atmosphere with a high reliability. This need was met by operating the sampling system at a constant air flow automatically maintained with a suitable transducer. For the determination of gaseous compounds high performance difussion tubes, called denuders, were developed. In the denuders the commonly used air filtration system was replaced because enrichment by this technique is plagued by many artifacts. Theoretical considerations relative to these new methods are outlined on the following pages.

Flow Rate and Volume Transduction in the Sampling of Air-Pollutants: Theoretical Considerations

Flow rates and sample volumes are usually measured in air sampling equipment with various mechanical devices such as rotameters and volume meters. These instruments are in most cases less precise and accurate than the analytical methods employed to determine atmospheric pollutants. In addition, temperature and pressure must be continuously recorded because the final data are reported for standard thermodynamic conditions. To avoid these limitations, a flow rate transducer was developed to keep the flow rate constant and obtain the sample volume directly at standard conditions [1].

The transducer consists of an orifice, located downstream of the pump, across which a pressure drop develops during sampling. The flow rate through the orifice, Q_O, is a complex function of several parameters such as absolute pressure, temperature, and fluid density (eqn. 1). To

$$Q_O = f(K, P_1, P_2, \rho_s, T) \qquad (1)$$

K: constant dependent on size and geometry of orifice
P_1: pressure upstream of the orifice
P_2: pressure downstream of the orifice
ρ_s: density of the gas
T: absolute temperature

determine the standard flow rate, several quantities have to be measured, complicating the use of the standard flow rate transducer. If, however, the reference pressure is set at atmospheric pressure, P', relationship 1 is simplified (eqn. 2). The standard flow rate is thus determined by the

$$Q_O = f[K, (P_1 - P'), T] \qquad (2)$$

P_1: pressure upstream of orifice
P': atmospheric pressure

pressure drop, the absolute temperature, and the size and geometry of the orifice. Therefore the flow rate can be continuously measured by means of a simple transduction of the pressure drop and the absolute temperature.

The measurement of the standard flow rate does not require a critical orifice or a special geometry for the orifice. The physical principle that controls the measurement is the adiabatic effusion of a gas through an orifice. This principle has never been used before. The standard flow rate is thus measured by using a differential pressure transducer with the reference arm kept at atmospheric pressure, and an absolute temperature transducer, that senses the absolute temperature of the gas flowing through the orifice.

Figure 1 shows a diagram of the standard flow rate transducer. The air from the sampling pump is drawn into a ballast that minimizes the pressure fluctuations in the

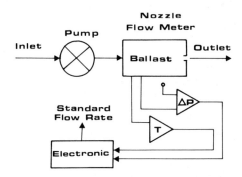

Figure 1. Diagram of the standard flow rate transducer.

sampling line. The orifice is directly drilled through the ballast. Its size depends on the flow rate range to be measured and on the characteristics of the pressure transducer used. The signals from the transducers are processed by an electronic circuit to avoid thermal instabilities and offset fluctuations. An electronic circuit performs the algorithm of equation 2 by using either discrete devices or a microprocessor. The signal for the standard flow rate can be read on a seven-segment display or processed by an additional electronic curcuit, that integrates the signal over the sampling time to obtain the standard volume of the air sample. The flow rate transducer is calibrated according to reference methods with an accuracy and precision better than one percent.

The outlined measurement of the standard flow rate is rather simple and extremely reliable. This system might be used to sample air pollutants with instruments that must operate unattended for long times. The principle of the adiabatic effusion of a gas through an orifice, which assumes a constant gas flow during sampling, was applied to the instruments described in this report.

Applications of Constant Flow Systems

Constant flow is important for a variety of sampling devices such as

- precision, low-volume samplers (adsorption traps for organics and impingers);
- samplers for total suspended particulate matter;
- samplers for particulate matter with size classification;
- samplers for diffusion tubes; and
- samplers for stacks and ducts.

High performance denuders: theoretical considerations. Air components are in a dynamic equilibrium that might be disturbed by air filtration processes used for sampling. This disturbance might cause changes in the composition of gas and particle phases. Gaseous compounds can be retained on the filter through adsorption or reactions with particulate matter. Because of the non-equilibrium conditions during sampling, ammonium salts in the aerosols will partially dissociate and volatilize from the filter. Several problems associated with gas-particle discrimination were overcome by using diffusion tubes, that selectively remove gaseous pollutants before particulate matter. However, the cylindrical denuders used so far have some limitations. They have a low adsorption capacity because of their reduced surface area and cannot operate at high air flows. Thus, glass tubings of considerable length are needed to achieve an acceptable sorption efficiency [2]. To operate with high flow rates, that are required to obtain significant information about the environment and its variation with time, a new denuder geometry was studied and a high performance denuder was developed. It consists of two coaxial cylinders. Air is forced to pass through the annular space [3].

During the laminar passage of air through a cylindrical denuder, the molecules of a reactive gas will diffuse to the wall of the tube while the particles will proceed unaffected. By calling C_o the gas concentration entering the tube and C the average gas concentration leaving the tube, the ratio of the concentrations of a pollutant in the air entering the tube (C_o) and leaving the tube (C) is expressed approximately by equation 3. The term Δ is defined by equation 4. For an annular denuder Δ_a is given by equation 5.

$$\frac{C}{C_o} = e^{-14.6272\,\Delta} \qquad (3)$$

$$\Delta = \frac{DxL}{\nu Rd} = \frac{\pi DL}{4F} \qquad (4)$$

D: diffusion coefficient of the gaseous pollutant
L: length of the coated tube
γ: kinematic viscosity of air
d: internal diameter of the tube
R: Reynolds number
F: air flow rate

$$\Delta_a = \frac{\pi DL(d_1 - d_2)}{4F(d_1 - d_2)} \qquad (5)$$

d_1: inside diameter of annulus
d_2: outside diameter of annulus

Thus, in a laminar flow through an annulus Δ_a may be optimized by varying not only the tube length and the air flow rate but also the width of the annular section and the diameter of the inner tube. As a consequence, values of Δ_a much larger than those corresponding to Δ at given F/L ratios are obtained. The ratio C/C_o for an annular denuder is given by equation 6. The coefficients A and α must be determined by experiment.

$$\frac{C}{C_o} = Ae^{-\alpha\Delta_a} \qquad (6)$$

Operation of the Annular Denuder

To determine the values of the coefficients A and α, and to check the performance of denuders that have annular geometry three sulfur dioxide glass (Pyrex) denuders of difference size were tested. The width of the annulus and the length of the denuders were constant at 1.6 mm and 200 mm, respectively. As outside diameter of the inner tube 10, 20 or 30 mm were chosen. A cross sectional drawing of such a denuder is shown in Figure 2. The walls of the annulus were chemically frosted to increase their surface area and then covered with a sodium tetrachloromercurate-maleic buffer solution.

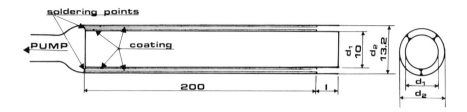

Figure 2. Diagram of an annular denuder with dimensions given in millimeters. Reprinted with permission from ref. 3.

The ratio C/C_o was determined at various flow rates by passing through the denuder standard mixtures of 100 to 500 µL/L of sulfur dioxide and measuring the exit concentration with a fluorescent sulfur dioxide monitor (Monitor Labs Model 8850). Permeation tubes were used as a source of sulfur dioxide. Air was drawn in by means of a high precision electronic sampler developed in our laboratory. The experimental set-up is outlined in Figure 3. The relative humidity during the experiment was between 60 and 80 percent.

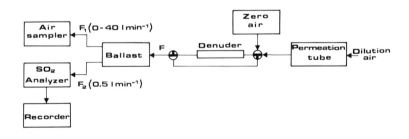

Figure 3. Experimental apparatus for the testing of denuders. Reprinted with permission from ref. 3.

The exit concentration, C, was recorded after the full response time of the analyzer had passed (eight minutes from the start of sulfur dioxide exposure). It has to be pointed out, that equation 3 describes the behavior of diffusion tubes of infinite absorbing capacity, whereas their actual capacity is limited. Consequently, C increases with sampling time. Equation 3 is only valid as long as the cumulative amount of absorbed sulfur dioxide is negligible compared with the loading capacity of the denuder.

Test results are summarized in Table 1. Values for Δ_a were calculated from equation 5 and the sorption efficiency, E, from the expression $1-(C/C_o)$. Figure 4 shows $-\ln C/C_o$ as a function of Δ_a. Each point corresponds to the average

Table 1. *Determination of the Sulfur Dioxide Sorption Efficiency in Several TCM-Coated Annular Denuders of 3.2 mm Equivalent Diameter (d_2-d_1)*

Denuder	Coated Length L, mm	Reynolds Number R	Efficiency E%	Flow Rate, F L/min	Δ_a
1	200	233	>99	3.9	0.218
	200	299	98.7	5.0	0.187
	200	347	98.0	5.8	0.161
	200	407	96.5	6.8	0.137
	200	448	94.8	7.5	0.123
	200	526	92.5	8.8	0.106
	200	627	88.5	10.5	0.089
	200	713	85.0	11.9	0.078
	200	807	82.4	13.5	0.069
	200	854	81.3	14.3	0.065
	120	119	>99	2.0	0.280
	120	221	97.3	3.7	0.151
	120	329	91.7	5.5	0.102
	70	72	>99	1.2	0.272
	70	167	94.0	2.8	0.117
	70	240	85.1	4.0	0.081
2	200	218	>99	7.8	0.223
	200	367	97.5	11.4	0.152
	200	589	90.2	18.3	0.095
	200	818	81.4	25.4	0.068
3	200	340	98.0	15.4	0.165
	200	512	93.8	23.2	0.112
	200	662	89.3	30.0	0.087
	200	883	81.6	40.0	0.064

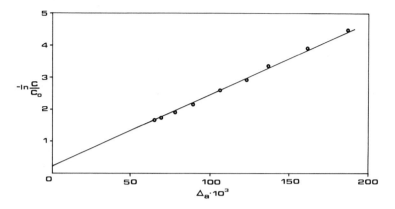

Figure 4. Plot of the experimental values for $ln(C/C_o)$ versus calculated values for Δ. From the regression line A (0.82) and α (-22.53) were calculated. Reprinted with permission from ref. 3.

value of C/C_o from five replicate measurements at each flow rate.

Deposition of atmospheric particles on the denuder walls may interfere with the operation of the denuder. In the laminar flow regime two major effects have to be considered: sedimentation and Brownian diffusion. Whereas sedimentation is eliminated by setting the denuder in a vertical position, Brownian diffusion depends on the coefficients for particle diffusion. These coefficients range from 5.2×10^{-4} to 2.7×10^{-7} cm³/s for aerosols with 0.01 to 1 µm diameter. These values are several orders of magnitude lower than those of gaseous species. Because the equation describing fractional penetration of a monodisperse aerosol through an annular denuder is of the same form as the equation for a reactive gas, we may conclude that the particle mass deposition is negligible for ambient atmospheric samples.

To obtain experimental verification of these theoretical conclusions, the mass deposition of particles in a vertically mounted annular denuder was investigated. A sodium sulfate polydisperse aerosol, generated by an atomizer, was sampled at 15 L/min from a dilution chamber. The particles were collected on a back-up filter (Gelman, type TF450) placed at the denuder outlet. The amounts of deposited sulfate were found by the Thorin colorimetric method. Five replicate measurements showed that the average mass (1.1 µg) of deposited sulfate on the denuder walls was negligible at less than 0.4 percent of the sulfate collected on the filter (292 µg).

The transmission of aerosols larger than 0.3 µm in diameter was measured in a laboratory atmosphere by connecting at alternate times the outlet of the denuder to an aerosol counter (Royco Model 225). The particles were detected in five size ranges (0.3-0.5 µm, 0.5-0.7 µm, 0.7-1.4 µm, 1.4-3.0 µm and >2.0 µm). Sampling was extended to eight hours at a flow rate of 2.8 L/min. At the end of each preset measurement period (10 minutes) with and without denuder, the data accumulated in the selected channels were automatically logged on a digital printer. The results relative to the total sampling period are shown in Table 2. As expected, no significant changes in the particle size distribution were observed with the denuder inserted.

Table 2. Transmission of Ambient Aerosols Through an Annular Denuder as Measured by Optical Counting

Sampling Time min	Size Range of Aerosol µm	Particle Count		Difference %
		Without Denuder	With Denuder	
150	0.3 - 0.5	12,653,254	12,630,257	- 0.2
150	0.5 - 0.7	3,153,602	3,140,253	- 0.4
150	0.7 - 1.4	1,273,351	1,264,950	- 0.7
150	1.4 - 3.0	470,706	466,940	- 0.8
150	> 3.0	15,632	15,413	- 1.4

Discussion and Conclusion

The high performance denuder that has been developed can be a very valuable tool to solve difficult analytical problems related to the environment and can help to obtain a better understanding of the interactions that might occur during wet and dry deposition. The denuder provides the possibility of operating with the highest efficiency even at flow rates as high as 20 L/min. This denuder is currently the simplest device for sampling different gases present in the atmosphere, if suitable absorbing media are available.

By setting three denuders in series [4] (Fig. 5), most species responsible for atmospheric acidity can be sampled from air: acids such as hydrochloric and nitric acids,

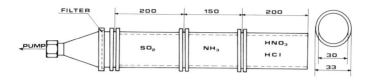

Figure 5. Denuder assembly for sampling strong acids, ammonia and sulfur dioxide with dimensions given in millimeters. Reprinted with permission from ref. 4.

basic compounds such as ammonia, and gaseous oxides such as sulfur dioxide. Sulfur dioxide is separated by diffusion to the wall of a tube coated with sodium tetrachloromercurate, ammonia by diffusion to the oxalic acid-coated tube, and hydrochloric acid and nitric acid by diffusion to a sodium fluoride-coated tube. Sulfuric acid and other aerosols are collected on the filter and determined separately. This three-stage denuder minimizes artifacts and permits the determination of each one of these species when all of them are present in air (Table 3). The same device permits the evaluation of artifacts, that might occur during sampling of ammonia and ammonium aerosols from the environment. The

Table 3. Concentration of Atmospheric Gases Sampled with a Multiple Denuder in a Rural Area

Sampling Date	SO_2 µg/m^3	NH_3 µg/m^3	HNO_3 µg/m^3	HCl µg/m^3
8/10/82	0.5	7.5	*	*
9/10/82	1.1	8.7	*	*
11/10/82	0.5	18.0	*	*
30/10/82	1.2	12.3	*	*
1/11/82	0.5	8.2	*	*
4/2/83	4.7	3.7	*	*
9/2/83	6.1	*	*	0.3
10/2/83	*	1.4	*	0.7
14/2/83	3.2	*	0.8	0.5
15/2/83	*	6.8	0.2	*
22/2/83	3.7	0.2	*	6.4
7/3/83	8.0	2.7	*	*
3/5/83	*	2.7	0.4	*

*not detectable.

constant flow principle can now be applied to air samplers, because pressure and temperature transducers to measure flow rates and volumes are now available. The measurements are easily performed as individual differential pressure and temperature transducers are set across an orifice that is located downstream of the pump.

The use of constant flow monitors realized by means of electronic devices may have a noticeable impact upon the development of instrumentation for the sampling of gases and aerosols.

References

[1] I. Alegrini, A. Liberti, A. Febo, M. Salmi and M. Beretta, "Automatic Beta Gauge Instrument for Atmospheric Dust Monitoring," in **"Proceedings, 2nd European Symposium on Physico-Chemical Behavior of Atmospheric Pollutants,"** B. Versino and H. Ott, Eds., Varese, Italy, September 29 - October 1, 1981, Reidel Publishing Co.: Dordrecht, Holland, 1982, p. 20.

[2] M. Ferm, "Method for Determination of Atmospheric Ammonia," **Atmospheric Environment, 13** (1979) 1385.

[3] M. Possanzini, A. Febo and A. Liberti, "New Design of a High-Performance Denuder for the Sampling of Atmospheric Pollutants," **Atmospheric Environment, 17** (1983) 2605.

[4] F. de Santis, A. Febo, M. Possanzini and A. Liberti, "Determination of Gaseous and Aerosolic Acidity in the Atmosphere," **Ann. Chim. (Rome), 74** (1984) 135.

XXIV

RECOVERY AND REUSE OF INORGANIC MATERIALS IN THE TANNERY INDUSTRY

M. Acampora, U. Croatto, G. Piovan and P. A. Vigato

Universita-Consiglio Nazionale delle Ricerche-Centro Tecnochimico
I-35100 Padova, Italy

A new method for the removal of inorganic and organic pollutants from tannery wastewaters is reported. Environmentally acceptable technologies for the recovery and reuse of chromium and sulfides were investigated in pilot plants and then scaled up to commercial plants. The chromium-containing wastewater is treated separately. Addition of sodium hydroxide to the wastewater precipitates chromium(III) hydroxide. The hydroxide is collected, dissolved in sulfuric acid, and the resulting solution recycled to the tanning operation. The sulfide-containing wastes are fed slowly into the aerobic treatment tank, where sulfide is oxidized by bacteria to sulfate. The sludges produced by this method can be used as fertilizers and for the production of biogas.

The tannery industry uses various inorganic compounds such as sodium sulfides, chromium(III) salts, ammonia, and water for the production of leather, and iron(II) sulfate, aluminum(III) sulfate, manganese(II) salts and nickel(II) salts for the purification of wastewaters. To protect the environment and maximize recycling of materials, especially sulfide and chromium compounds, new purification systems for wastewater had to be developed. In the old method chromium-containing suspended solids are separated from the basic wastewater by preliminary sedimentation. The solids thus collected are rich in organic materials but contaminated by chromium and are, therefore, often placed into a waste dump. Alternately this material is treated with bases that solubilize and hydrolyze the proteins. The products of the hydrolysis can be used for feed production. Chromium is recovered as a residue. The wastewater ⁻ after primary sedimentation ⁻ contains sulfides, which are then precipitated by addition of iron(II) sulfate. Suspended solids and dissolved organic materials coprecipitate with the iron sulfide. The precipitate must be placed into a waste dump, because the high content of iron sulfide makes this solid unusable. This method produces large quantities of non-reusable sludges and pollutes the atmosphere with hydrogen sulfide.

Finally, the residual wastewater must be purified by biological oxidation. The precipitation of sulfides by iron(II) sulfate is now being replaced by a new process based on the oxidation of sulfide to sulfate by atmospheric oxygen in the presence of manganese(II) or nickel(II) sulfate as catalyst. The old methods of treatment of tannery wastes have several disadvantages:

- Sludges contain high percentages of sulfides and of metals such as chromium, iron, and aluminium.
- The sludges are difficult to filter and difficult to reuse.
- The sludges must be stored in safe sites such as controlled dumps to avoid contamination of groundwaters, soils, and rivers.

The limited availability of mineral and energy resources and environmental concerns demand new industrial technologies that are frugal with resources and benign to the environment. The tannery industry must change to meet these new demands.

The toxicity of chromium, the dwindling supplies of chromium ores, and the ever-increasing cost of chromium compounds make the recovery and reuse of chromium from tannery wastes unavoidable and economic. Sulfides are also toxic and must be eliminated from wastes. Sulfides are currently removed from liquid wastes that were treated in lime pits by diffusion through permeable membranes. The recovery of sulfides is less economical than the recovery of chromium.

New Treatment Processes for Tannery Waste Waters

We investigated more economic and efficient processes for the recovery and reuse of chromium and for the disposal of sulfides. The wastewater from the chrome-tanning operation is not allowed to mix with other wastes. The chromium-containing wastes are treated to recover chromium(III) hydroxide that is dissolved in sulfuric acid and then subsequently reused as tanning material. The recovered chromium can be further purified for the tanning of particularly valuable hides. The recovery of chromium as chromium(III) hydroxide is almost quantitative. We studied the biological removal of sulfides via their incorporation into amino acids and their bacterial conversion to sulfur and sulfates. Because of the high toxicity of sulfides toward bacteria, the sulfide-containing wastewaters must be fed slowly into the tank, in which the organic materials are removed by biological action. The concentration of sulfides in the tank must never exceed 400 mg sulfur per liter. Therefore, the limepit water must be stored in a separate tank and fed into the tank for biological oxidation at a carefully controlled rate. All of the sulfide wastes can be treated in this manner. These purification processes increase the sulfate concentration in the discharge and thus contribute to the load of inorganic materials of the water discharged into rivers. The discharge waters may contain 2900 mg/L chloride and 2000 mg/L sulfate.

Removal and recycling of chromium. The acidic, chromium-containing wastewater [0.7-0.8% chromium(III) oxide] is passed through a screen into a storage tank (Fig. 1). The

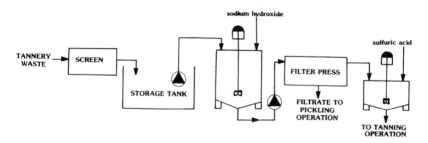

Figure 1. Flow sheet for the recovery and reuse of chromium.

solution is made alkaline by addition of sodium hydroxide. The alkaline mixture is filtered in a filter press. The filtrate containing 3-5 mg/L chromium is used to prepare the pickle bath. The chromium(III) hydroxide retained by the filters is purified and dissolved in concentrated sulfuric acid. The resulting solution of basic chromium(III) sulfate is reused in the tanning bath after any necessary adjustment.

Removal of sulfides. Before tanning hair, epidermis and subcutaneous tissue must be removed from the hides. This operation is carried out in limepits through the action of several chemicals, the most important of which are sodium sulfide and lime. The effluents from the limepits have a chemical oxygen demand of 70,000 to 80,000 mg/L, total nitrogen at 4,000 to 5,000 mg/L, and sulfides at 4,000 mg/L. The limepit water is added in controlled quantities to the other wastewaters undergoing biological degradation in treatment tanks (Fig. 2). Appropriately selected and

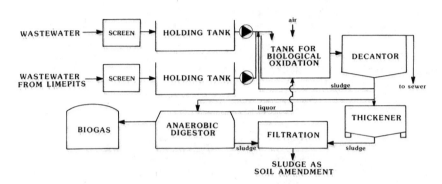

Figure 2. Flow sheet for the biological treatment of tannery wastewaters.

acclimatized bacteria oxidize sulfides to sulfates at 25° provided that the sulfide concentration in the tank does not exceed 400 mg/L. Ninety-five percent of the sulfides are removed after five hours and 98 percent after ten hours. The pH of the wastewater is reduced from 11 to 7.5 concomitant with the oxidation of the sulfides. When the limepit wastewaters are appropriately diluted with wastewaters from other processes the aerobic biological activity in the treatment tanks is not diminished by the toxic substances in the limepit water.

Commercial treatment plants. The pilot plants (Fig. 2) were scaled up and commercial treatment plants were built at tanneries. These plants performed well in removing pollutants (Table 1). The COD, BOD and sulfide values are approximately the same as those in the wastewater treated conventionally. The new method removes total nitrogen and ammonium-nitrogen more efficiently than the old technique. However, the efficiency of nitrogen removal varies in these new plants. The reasons for these variations are now being explored.

Table 1. *Analytical Data for Untreated and Treated Wastewaters from Tanneries*

	Concentration, mg/L	
Parameter	Untreated Wastewater*	Treated Wastewater
COD	6500	249
BOD_5	3500	54
BOD_{20}**	4117	78
Chloride	2950	2750
Sulfate		1950
Phosphorus	2	3
Total chromium	7.5	0
Total nitrogen	280	25
Ammonium-nitrogen	75	19
Phenols	7	3
Hydrogen sulfide	350	0.8
Sediments †	6	0.2
pH	11.5	7.2

*Substances extractable with diethyl ether: 750 mg/L.
**Value obtained by extrapolation.
†Concentration in mL/L.

Analytical data for the sludges obtained by this process are reported in Table 2. The new treatment process produces

Table 2. *Analytical Data for the Sludges from the Biological Treatment Tanks*

Parameter	Percent of Sludge Dried at 105°	Parameter	Percent of Sludge Dried at 105°
Organic material	55.	Phosphorus (P_2O_5)	0.05
Mineral matter (at 550°)	45.	Chromium(III)	0.4
Organic carbon	26.6	Aluminium	0.22
Proteins	37.8	Iron	0.28
Organic nitrogen	6.	Calcium	16.16
Ammonia-nitrogen	0.07	Sulfate	1.95

only 25 percent of the sludge formed in the old process. The new-process sludge - because of its composition - is easy to dispose.

Conclusions

The treatment methods for wastewater from tanning operations including the new method described in this paper are not optimal. The concentrations of several constituents in the discharged treated waters are still higher than the legal limits, a situation very common for wastewater treatment plants in Italy. The salinity of effluents from wastewater plants is generally very high. The systems available for the removal of chloride and sulfate are too expensive for wide application. Recycling of treated water within an industry is, therefore, strongly recommended. Water-recycling does not seem to be practiced in the tannery industry. The tanning plants obtain the needed, high quality water, that requires no treatment, at low cost from groundwater reservoirs. However, the practice of recovery and reuse of chromium not only prevents environmental deterioration, reduces demand for chromium compounds, allows the sludge to be used for biogas production and as a fertilizer, but makes the tanning process also more economical.

Bibliography

W. W. Eckenfelder, Jr. and D. J. O'Conner, **"Biological Waste Treatement,"** Pergamon Press: Oxford, 1961.
W. W. Eckenfelder, Jr. and D. L. Ford, **"Water Pollution Control,"** Pembleton Press: New York, 1970.
J. C. Lamb, W. C. Westgarth, J. L. Rogers and A. P. Vermimmen, "A Technique for Evaluating the Biological Treatability of Industrial Wastes," **J. Wat. Pollut. Control Fed., 36** (1964) 1263.
E. Metcalf, **"Wastewater Engineering: Collection Treatment Disposal,"** McGraw-Hill: New York, 1972.
J. Monod, "The Growth of Bacterial Cultures," **Ann. Rev. Microbiol, 3** (1949) 371.
R. W. Ramalho, **"Introduction to Wastewater Treatment Processes,"** Academic Press: New York, 1977.

XXV

ASSESSMENT OF EXPOSURE OF HUMAN POPULATIONS TO HEAVY METALS THROUGH DIETARY INTAKES

Gianni Tomassi

Istituto Nazionale della Nutrizione
Via Ardeatina 546, I-00179 Roma, Italy

The contamination of food by heavy metals and particularly by those metals that accumulate in organisms is a potential hazard for human populations. The risk associated with the ingestion of contaminants can be evaluated by comparing the actual or estimated dietary intake with the provisional tolerable weekly intake. Appropriate estimates of dietary intake of contaminants can be obtained from surveys on actual food consumption and the concentrations of contaminants in foodstuffs. Specific fish consumption surveys conducted in Italian coastal villages indicate that people consuming a diet rich in seafood, with which mercury is generally associated, are likely to exceed the recommended tolerable intake of mercury. The calculated cadmium and lead intakes of the Italian population show that the tolerable level of intake for cadmium is exceeded and that the calculated lead intake almost equals the tolerable level. Continuous surveillance of the contamination of food, especially of cereals and vegetables that contribute heavily to the estimated intakes is recommended.

Many heavy metals, although they occur in the earth's crust and are present in organism, have no known biological function. When their concentrations exceed background levels and their bioavailabilities are increased as a consequence of human activity, they are regarded as environmental contaminants and present a potential hazard to man. Acute and chronic intoxications by heavy metals are well documented. Their effects are described at the clinical and, when possible, at the biochemical level. For example, the first sign of cadmium intoxication from long-term exposure is a disfunction of the kidneys expressed by a decreased tubular absorption of proteins [1]. Chronic lead intoxication affects the nervous and hematopoietic systems, reducing the nervous conduction velocity and decreasing the biosynthesis of hemoglobin [2]. Irreversible neurological disturbances are well documented for mercury poisoning, particularly after exposure to highly toxic organic mercury compounds [3].

In chronic, non-occupational exposure, foodstuffs generally contribute most of the heavy metal to the total body burden. The risk from dietary exposure may vary according to a person's physiological and nutritional state. For instance, only ten percent of ingested lead are absorbed by the gastrointestinal tract of adults, whereas 50 percent are taken up by children and malnourished adults [4]. Approximately five percent of cadmium is absorbed by normal populations, whereas people with low body burdens of cadmium or those on a calcium-deficient diet resorb twenty percent [5]. In the case of mercury, fetal and infant exposures are believed to be the most critical, because mercury compounds readily cross the placental barrier and are frequently present at elevated levels in human milk. These conditions lead to high exposures of unborn and young children during periods of maximal growth of the nervous system [3].

Methodologies for Dietary Surveys

Dietary intakes of contaminants are determined by their concentration in the food and the quantities of food consumed. It is difficult to accurately estimate the dietary intake of contaminants with the food for individuals and for the general population, because the concentrations of contaminants and the consumption of food vary widely. Concentrations of organochlorine pesticides, lead, cadmium, and aflatoxins in selected foodstuffs from 20 countries were collected by the Global Environmental Monitoring System [6]. These concentrations serve as a basis to estimate dietary intake.

Data on individual and collective food consumption should reflect typical food consumption patterns of a population. Ideally, the data should come from current national food consumption surveys based on standard statistical methods. There are two general approaches to obtain information on the dietary habits of a population: those that use indirect data on the movement and disappearance of foodstuffs in a region or a home, and those that use data on the actual

Assessment of Exposure to Heavy Metals

amounts of food consumed by an individual [7]. The methods most often used are summarized in Table 1. The "food diary

Table 1. Methods for Collecting Data on Food Intake from Population Groups and Individuals [7]

Assessment	Approach	Method
Individual	Direct	Food diary
		Duplicate portion
	Indirect	Dietary recalls
		Diet histories (food frequency)
Populaton	Direct	Food diary
	Indirect	Food disappearance method national or for a household
		Dietary recall
		Diet histories (food frequency)

method" requires an individual or a household to keep a diary on the weight or estimated amount of foods consumed over a certain period. This approach is useful to determine the food consumption of individuals and of large populations. For large surveys, a one-day record is believed to be sufficient to obtain reasonable data on the average consumption of foods, whereas a seven-day record obtained by weighing all foods should give a reasonable estimate of actual intake.

The "duplicate portions method" requires the purchase and the preparation of twice the usual portions consumed by an individual at home or away from home. The duplicate of the 24-hour intake is weighed, refrigerated or frozen, and subsequently homogenized and analyzed. This method is suitable for institutional foods and for small surveys, but is not appropriate for large scale food consumption studies because of cost and time. Because foods are often collected and mixed in large containers, individual foodstuffs cannot always be analyzed. Unless a diary is kept in conjunction with the sample collection, no written record of eaten or collected foods will be available.

In the "recall method", individuals or families are requested to recall the types and amounts of food they consumed during the previous 24-hour period. The method requires a trained interviewer and is more suitable for studies on populations, because the individual day-to-day variations in food consumption are very large.

The "food frequency method" uses a list of commonly consumed foods. The participants indicate the number of times per day, per week or per month a particular food is consumed. Generally, information about the quantity of food consumed is not requested on such a form. Data on average serving sizes from previous diary or recall surveys are used with the frequency data to produce the desired information on dietary intake. This approach is particularly useful for obtaining information about the consumption of specific types of food that are known to be contaminated and that are likely to contribute predominantly to the overall intake.

The "household disappearance method" estimates individual intake by assessing the amount of food disappearing from a home kitchen during a given period per family member. A similar approach is useful at the national level when resources are insufficient to conduct a national food consumption survey. The food balance sheets provide data on food consumption by subtracting food exported, food used for animals, food used for non-eating purposes, food lost during storage and food wasted from food produced, food imported, and food taken from stocks. The "food disappearance method" provides a rather crude measure of food consumption and should be used only, when all other approaches are not feasible. It has been recognized at the 1982 meeting of FAO/WHO experts on Guidelines for the Study of Dietary Intakes of Contaminants in Rome that such theoretically deduced intakes can be misleading and must be used with caution. Therefore, estimates of heavy metals and pesticide intake should be based on the actual dietary intake. To assess consumer risk the estimates should be compared with the acceptable daily intakes. Such an estimate also helps to establish the relationship between the level of exposure to a contaminant and any health effect in humans identified by epidemiological studies. The estimate of actual dietary intakes of heavy metals as a measure of exposure is indispensable for risk assessments.

The selection of an appropriate method is influenced by many factors such as the objective of the study, the financial resources available for the survey, and the behavior of the contaminant in the environment and in the human body.

Evaluation of Risk

The concepts of acceptable daily intake for food additives and admissible daily intake for pesticides - defined by the Joint FAO/WHO Expert Committee on Food Additives as the amounts that may be ingested daily during a life time without risk - do not seem to be suitable for certain food contaminants. For these cases the concept of provisional tolerable weekly intake was developed [8]. A distinction must be made between contaminants with cumulative properties (cadmium, mercury, lead) and contaminants without cumulative properties. Thus, for copper and zinc, which are essential for humans but are toxic at higher levels of intake, the Committee established Provisional Maximum Tolerable Daily Intakes, expressed by two numbers, one indicating the level of essentiality and the other the level of safety (Table 2) [9]. A provisional

Table 2. Evaluations of Tolerable Intakes of Elements [9]

Element	Tolerable Intake		
	mg/person	mg/kg	µg/kg*
Cadmium**	0.5	0.0083	8.3
Copper†	–	0.05 –0.5	50. – 500.
Lead**	3.0	0.05***	50.0
Mercury, Total**	0.3	0.005	5.0
Mercury, Methyl**	0.2	0.0033	3.3
Phosphorus††	–	70.0	70,000.
Tin, Inorganic†	–	2.0	2,000.
Zinc†	–	0.3 –1.0	300. –1000.

*Body weight
**Provisional Tolerable Weekly Intake
***Not applicable to infants and children
†Provisional Maximum Tolerable Intake
††Maximum Tolerable Daily Intake
†††Applicable to diets that are nutritionally adequate with respect to calcium. If the calcium intake were high, the intake of phosphate could be proportionately higher, and vice versa.

maximum daily intake was established for tin, a metal contaminant without cumulative properties, and a maximum tolerable daily intake was approved for phosphorus, an essential nutrient and an unavoidable constituent of food.

To properly protect human health, the fact must be considered, that certain contaminants occur only in one or a few food items (mercury in seafoods), whereas other contaminants are found in a great number of foods (lead and cadmium). Critical foods for contaminants such as mercury occurring in a few items only, are easily identified. Appropriate and specific dietary surveys will give accurate estimates of the intakes of the contaminant by population groups. Critical populations can be identified and monitored for signs of intoxication. The identification of critical foods associated with lead or cadmium is difficult, because of the wide variations in the concentration of these elements in various food categories.

Fishermen and their families, and employees at fish restaurants and fish processing plants are the groups most exposed to mercury. The determination of the size and distribution of these "hot spots" of exposure to mercury is an acceptable approach to risk evaluation provided that the general population is not at risk. A pilot study of fish consumption by fishermen and non-fishermen and their families was recently conducted with questionnaires (Fig. 1)

RECORD OF FISH CONSUMPTION BY A FAMILY AT HOME

DATE	TYPE OF FISH OR SHELLFISH	INDICATE WHETHER fresh or processed (canned, frozen, etc.)	QUANTITY OF EACH TYPE PREPARED PER MEAL in grams	PORTION NOT CONSUMED 1/4, 1/2, 1/4	NAME	Members of the family present at the meal 01 02 03 04 05 06 07 08 09 10	FISH IDENTIFICATION CODE Leave blank

Figure 1. Questionnaire used to survey the fish consumption of families in Italian coastal villages.

in coastal areas of Italy. Data on the mercury concentration in seafoods were collected separately. The combined results of these investigations showed, that a high percentage of the surveyed peopled exceeded the tolerable limits for mercury intake (Table 3). The people in Bagnara

Table 3. Percentage of Persons at the Five Percent Risk Level of Exceeding the Provisional Tolerable Weekly Intake in Population Groups in Italy*

Group	Number of panelists	Percentage of persons at >5% risk
RAVENNA		
All panelists	184	17.9
Women of childbearing age (16-45 yrs.)	41	12.2
Children (1-10 yrs.)	14	42.9
Youths (11-18 yrs.)	29	31.0
Aged persons (>60 yrs.)	27	14.8
FIUMICINO		
All panelists	211	46.0
Women of childbearing age (16-45 yrs.)	44	47.6
Children (1-10 yrs.)	32	62.6
Youths (11-18 yrs.)	41	49.0
Aged persons (>60 yrs.)	14	29.0
BAGNARA CALABRA		
All panelists	243	84.8
Women of childbearing age (16-45 yrs.)	53	79.2
Children (1-10 yrs.)	40	97.5
Youths (11-18 yrs.)	38	86.8
Aged persons (>60 yrs.)	22	86.4

*Adapted from reference [10].

Calabra not only consumed more fish than the people in Ravenna and Fiumicino, the other localities surveyed, but also ate more species such as swordfish that have elevated mercury levels [10]. It is therefore not surprising that the people of Bagnara Calabra had the highest risk of exceeding the limit of mercury intake (Table 3). The risks of various age groups and of women of childbearing age were estimated. Children and women of childbearing age might face special risks. Because of their low body weight, children consume mercury frequently in excess of the recommended maximal level. Mercury intake by women of childbearing age might affect prenatal life especialy during the period of brain formation. In some countries legal action levels exist for certain foods that help to reduce contamination levels, for instance, by technological improvements in food production. In Italy the legal limit for lead in wine is 0.3 mg/L.

Dietary intakes of metals calculated from their mean concentrations in food and the average food consumption might not reflect the exposure of the population very accurately. Nevertheless, these calculated intakes are useful indicators of exposure for continuous surveillance programs and are helpful, when preventive measures to reduce contamination must be selected. When the concentration of a contaminant in a food or food category is known, various food consumption patterns can be simulated, the risk associated with the contaminant estimated, and the contribution of a particular food to the dietary intake calculated.

Estimated dietary intakes of cadmium and lead for the Italian population are reported in Table 4. Seafoods contribute only six percent to the total dietary intake of lead and cadmium. Meat and meat products, cereals and fruit, and vegetables are the major contributors. The

Table 4. Calculated Dietary Intakes of Lead and Cadmium in Italy

Food	Av. Food Consumption in Italy kg/per week per person	Mean conc. in Foods, mg/kg		Dietary Intakes, mg/week*	
		Cd	Pb	Cd	Pb
Meat and meat products	1.5	0.15	0.30	0.23(20)	0.45(16)
Cereal products	3.75	0.10	0.20	0.62(53)	0.75(26)
Fruits and vegetables	6.25	0.03	0.15	0.19(16)	0.94(33)
Fish	0.14	0.2	1.0	0.03(3)	0.13(5)
Crustacean	0.01	1.0	1.0	0.01(1)	0.01(<1)
Molluscs	0.05	1.0	1.0	0.03(3)	0.04(1)
Milk and milk products	1.75	0.03	0.1	0.05(4)	0.17(6)
Wine	1.88	—	0.2	—	0.38(13)
			Total	1.16	2.87

*Numbers in parentheses: Contribution to Weekly Intake in percent.

provisional tolerable weekly intake for adults is 0.4 to 0.5 mg for cadmium and 3 mg for lead. The calculated dietary intakes (Table 4) exceed the provisional tolerable values for cadmium and are slightly less than the tolerable value for lead. These results indicate the necessity for epidemiological studies and for the surveillance of the food with respect to contamination by lead and cadmium.

Available data on food contamination by heavy metals and dietary intake of heavy metals indicate the need for continued surveillance. Accurate values for dietary heavy metal intakes in populations should be obtained from specific dietary surveys that pay particular attention to the high risk groups such as women of childbearing age, infants, and children. A better understanding of the biochemical reactions that damage the organism after ingestion of heavy metals is required to establish definitive safety limits, choose appropriate preventative measures, and identify valid early indicators of chronic exposure in population groups.

References

[1] M. Vahter, Ed., **"Global Environmental Monitoring System (GEMS): Assessment of Human Exposure to Lead and Cadmium Through Biological Monitoring,"** National Swedish Institute of Environmental Medicine and Department of Environmental Hygiene, Karolinska Institute: S-10410, Stockholm, Sweden (P.O. Box 60208) 1982.
[2] Environmental Health Criteria, Vol. 3, **"Lead,"** World Health Organization: Geneva, 1977.
[3] J. K. Piotrowski and H. J. Inskip, "Health Effects of Methylmercury," **Monitoring and Assessment Research Center (MARC) Report, 24,** Chelsea College, University of London, 1981.
[4] K. R. Mahaffey, "Nutritional Factors in Lead Poisoning," **Nutr. Rev., 39** (1981) 353.
[5] Gesamp Working Group, **"Review of Potentially Harmful Substances: Report on Hazard Evaluation of Cadmium,"** Geneva, 1983.
[6] Global Environmental Monitoring System: Joint FAO/WHO Food and Animal Feed Contamination Monitoring Programme, **"Summary of Data Received from Collaborating Centres 1977 to 1980,"** World Health Organization, Geneva, 1981.
[7] Global Environmental Monitoring System: Joint FAO/WHO

Food Contamination Monitoring Programme, **"Guidelines for the Study of Dietary Intakes of Contaminants,"** Meeting in Rome, December 16-21, 1982.

[8] G. Vettorazzi, "Quantification in Food Regulatory Toxicity," in **"Health Evaluation of Heavy Metals in Infant Formulas and Junior Food,"** E. H. F. Schmidt and A. G. Hildebrand, Eds., Springer Verlag: Berlin, Heidelberg, 1983, p. 13.

[9] G. Vettorazzi, "Lead as a Food Contaminant," **Riv. Soc. Ital. Alim., 11** (1982) 303.

[10] C. Nauen, G. Tomassi, G. P. Santaroni and H. Josupeit, "Results of the First Pilot Study on the Chance of Italian Seafood Consumers Exceeding their Individual Allowable Daily Intake of Mercury," **VIth Workshop on Marine Pollution of the Mediterranean,** December 2-4, 1982, Cannes, France, International Commission for the Scientific Exploration of the Mediterranean Sea, UNESCO, 1984, p. 571.

XXVI

METALS AS GENOTOXIC AGENTS: THE MODEL OF CHROMIUM

Vera Bianchi and Angelo Gino Levis

Istituto Biologia Animale, Universita di Padova
Via Loredan 10, I-35000 Padova, Italy

Genotoxic effects include induction of tumors, anomalies of embryonic development, and mutations. Genotoxicity is an important part of general environmental toxicity. The principles of genetic toxicology and short-term tests for mutagens and carcinogens are briefly reviewed. The genotoxicity of chromium compounds is discussed and the hypotheses for the genotoxic actions are presented.

Human populations in industrialized countries are exposed to an impressive number of hazardous chemicals. The number of different chemicals is estimated to be approximately four million. More than 60,000 of these chemicals are produced in large enough quantities to constitute an environmental risk [1]. Every year approximately 25,000 new substances are synthesized, and two percent of them are used commercially.

Environmental pollutants produce a wide range of toxic effects, some of which become immediately evident as physiological or metabolic disorders. However, genotoxic effects, which are caused by alterations of the genetic material, may become evident only many years, or even generations, after genetic alterations were induced. During the past 15 years it has been demonstrated that some chemicals used as food additives, drugs, cosmetics, and pesticides, and industrial products, by-products and wastes may be or contain genotoxic agents [2]. Increasing evidence is accumulating that genotoxicity plays an important role in environmental toxicity. Genotoxic effects include induction of tumors (carcinogenesis), anomalies of embryonic development (teratogenesis), and mutations (mutagenesis). The importance and social burden of environmental carcinogenesis and teratogenesis is easily recognized, because their consequences directly affect the exposed individuals or their offsprings. The risk arising from mutations is less widely recognized and requires a few comments.

Principles of Genetic Toxicology

Mutations are transmissible alterations of the genetic material occurring at the level of genes (gene mutations) or chromosomes (chromosomal aberrations). Both types of mutations can occur in germ cells (germ cell mutations) or in somatic cells (somatic mutations). Whereas germ cell mutations can be transmitted to offsprings, somatic mutations are expressed only in the affected individuals. These two classes of mutations have mostly negative consequences, although their effects can differ markedly [3, 4]. Germ cell mutations reduce the reproductive ability through sterility or abortions, cause perinatal mortality and genetic diseases in newborns, accumulate in subsequent generations and impair the genetic pool of the population (Fig. 1). The somatic mutations that cause cell death or

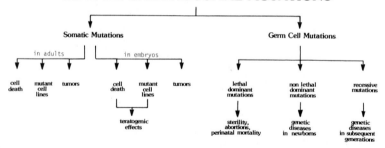

Figure 1. Consequences of mutations in humans.

produce mutant cell lines have negligible effects in adults, but contribute to teratogenesis if occurring in embryos. The major danger associated with somatic mutations in adults and embryos is neoplastic cell transformations. It is widely accepted that carcinogenesis can be initiated by structural or functional alterations of the genetic material in somatic cells [5, 6].

The somatic cell mutation hypothesis for the induction of tumors is based not only on indirect evidence, but also on the observation that most carcinogens and mutagens cause common biological effects when tested in a variety of short-term assays [2-5, 7]. Such effects include

- DNA damage: alterations of the macromolecular structure of DNA;
- DNA repair: stimulation of enzymatic systems which mend damaged DNA;
- chromosomal effects: structural alterations of chromosomes;
- gene mutations: alterations of the informational content of genes producing morphological or physiological changes of gene-controlled traits; and
- in-vitro cell transformation: stable changes of growing patterns of cultured mammalian cells which, once transformed, become able to produce tumors in experimental animals.

Short-Term Tests for Mutagens and Carcinogens

Since the late 1960s a great number of short-term tests for the detection of genotoxic agents has been developed and validated to identify mutagens among environmental contaminants [5-8]. The development of such methods was stimulated by the observed correlation between mutagenic and carcinogenic power of genetically active compounds [2, 5-9]. In 1979 Hollstein et al. [8] reviewed more than 100 different tests (Table 1) capable of detecting genotoxic effects ranging from DNA damage to in-vitro cell transformation. These relatively inexpensive tests use a variety of organisms and conditions. Some of the tests that are based on the selection of induced mutants are strictly objective. Because the estimation of genotoxic risk to humans is the goal of short-term tests, the most predictive assays are those performed in-vivo on mammals. Testing in whole animals makes it possible to detect the influence of metabolism that activates or inactivates xenobiotics or genotoxicity. To simulate in-vivo conditions, tests on microorganisms and mammalian cells in-vitro were modified by adding actively metabolizing cell fractions such as liver microsomes. As an alternative, host-mediated assays can be performed. In these tests, compounds are injected into host mammals and then recovered from the host. Microorganisms are then exposed to the recovered compounds and the genetic effects caused by these compounds determined.

Table 1. Short Term Tests for Mutagens and Carcinogens*

Type of Test	Number of Tests
DNA Damage	
binding to DNA	2
DNA fragmentation and other damages	9
inhibition of DNA replication	2
induction of errors in DNA replication	3
DNA Repair	
killing of repair deficient bacteria	2
stimulation of DNA repair synthesis in mammalian cells	7
Chromosomal Effects	
in lower Eukaryotes (yeast and fungi)	12
in plants and insects	9
in mammalian cells in-vivo and in-vitro	13
Gene Mutations	
in bacteria	7
in lower Eukaryotes (yeasts and fungi)	21
in mammalian cells in-vivo and in-vitro	13
In-Vitro Cell Transformation	
morphological or physiological transformations of mammalian cells	17
enhancement of virus-induced transformation of mammalian cells	4

*Adapted from Hollstein et al. [8].

The results of short-term assays are useful also to predict carcinogenic potential in first-level screening studies and to plan subsequent long-term in-vivo tests and epidemiological investigations [8, 9]. Results of short-term tests for mutagens and carcinogens led to regulatory actions and proposals by governmental committees and agencies in several countries designed to protect the public from hazardous chemicals [7]. The correlation between mutagenicity and carcinogenicity is higher, when tests responsive to different aspects of genotoxicity give concordant results. Therefore, guidelines for environmental protection are usually based on a series of tests that are constructed to identify the broadest possible spectrum of genetic damage. Each series contains a minimal number of tests that must be performed before a final decision about the hazards of a compound is made. The number of tests in a series is determined by the agencies and depends on the kind of compounds under examination [7] (Tables 2, 3).

Table 2. *Guidelines Proposed by the U.S. Environmental Protection Agency for Industrial Wastes**

Requires a minimum of one test from each group capable of detecting:
DNA Damage and Repair
  DNA repair in bacteria;
  Unscheduled DNA repair synthesis in human diploid cells;
  Mitotic recombination and/or gene conversion in yeast;
  Sister chromatid exchanges in mammalian cells.
Gene Mutations in Prokaryotes
  Point mutations in bacteria.
Gene Mutations in Eukaryotes
  Point mutations in mammalian somatic cells in culture;
  Point mutations in fungal micro-organisms.

---

*Federal Register, December 18, 1978.
Adapted from reference 7

Table 3. *Guidelines Proposed by the U.S. Environmental Protection Agency for Registering Pesticides**

Requires tests capable of detecting:
DNA Damage and Repair: a minimum of 2 tests selected from
  DNA repair test in bacteria;
  Unscheduled DNA repair synthesis in mammalian cells;
  Mitotic recombination and/or gene conversion in yeast;
  Sister chromatid exchange in mammalian cells.
Chromosomal Aberrations: a minimum of 3 tests selected from
  In-vivo cytogenetic test in mammals;
  Insect tests for heritable chromosomal effects;
  Dominant lethal tests in rodents;
  Heritable translocation test in rodents.
Gene Mutations: a minimum of 3 tests selected from
  Bacteria;
  Eukaryotic micro-organisms;
  Insects (e.g. sex-linked recessive lethal test);
  Mammalian somatic cells in culture;
  Mouse specific locus test.

---

*Federal Register, August 22, 1978.
Adapted from reference 8

Metals as Genotoxic Agents

Metals and their compounds are widely used and thus give rise to occupational and environmental exposure. For instance, the world production of chromite, the principal ore of chromium, was 8.6 million tons in 1976. The United States used 912,000 tons of chromite in 1974. One hundred and four occupations, in which exposure to chromium(VI) can occur, were identified in 1975 in the United States, and an estimated 175,000 persons could have been exposed [10]. The genotoxic properties of metals have been extensively studied [11-13] and were recently reviewed by the International Agency for Research on Cancer [2]. This agency evaluated the existing data about mutagenicity, and carcinogenicity to humans and animals (Table 4). Because the mutagenicity of

Table 4. Evidence of Genotoxic Activity of Metal Compounds [2]

Metal	Carcinogenicity to Humans	Carcinogenicity to Animals	Mutagenicity
Arsenic	sufficient	inadequate	limited
Beryllium	limited	sufficient	inadequate
Cadmium	limited	sufficient	inadequate
Chromium	sufficient	sufficient	sufficient
Lead	inadequate	sufficient	inadequate
Nickel	limited	sufficient	inadequate

metals is difficult to assess, its correlation with the carcinogenicity of metals is weak. Only for chromium is "sufficient" evidence available for mutagenicity, and carcinogenicity in humans and animals.

The Model of Chromium: Mechanisms of Genotoxic Action

Chromium is peculiar, because its two common oxidation states, chromium(VI) and chromium(III), differ markedly in a number of their biological properties [14]. In water, chromium(III) exists at pH values less than four as the hexaaquo cation, $[Cr(H_2O)_6]^{3+}$, or at sufficiently high chloride concentration as $[Cr(H_2O)_4Cl_2]^+$. At physiological pH values and in basic medium polymeric, rather insoluble hydroxo complexes, $[Cr(H_2O)_m(OH)_{6-m}]^{m-3}$, are formed that are unable to cross cell membranes. In solutions of chromium(VI), chromate ions, CrO_4^{2-}, predominate at physiological and basic pH values, whereas the dichromate ion, $Cr_2O_7^{2-}$, is favored under acidic conditions. Chromate and dichromate ions are strong oxidizing agents. Chromate ions easily cross cell membranes. Chromium(III) cannot be oxidized to chromium(VI) by biological systems. Chromium(III) is devoid of cytotoxic and genotoxic activity. Chromium(VI) is easily reduced to chromium(III) inside and outside a cell. Extracellular reduction abolishes the cytotoxicity and genotoxicity of chromium(VI). Intracellular reduction of chromium(VI) occurs either in the cytoplasm, where chromium(III) is then sequestered, or in the nucleus, where chromium(VI) may interact with genetic materials, for instance, with nucleophilic sites in DNA such as the nitrogen and oxygen atoms of the cyclic bases and the negatively charged oxygen atoms of phosphate groups.

Chromium(VI) compounds, unlike most other metal compounds, are actively transported through cell membranes and thus can reach genetic targets. Consequently, the genotoxic effects of chromium have been studied with a variety of short-term tests covering the entire range of genetic damages [2, 10-19]. The results (Table 5) indicate that chromium(VI) is an active genotoxic agent in intact cell systems, whereas chromium(III) is inactive. Chromium(III) gives a positive mutagenic response only when tested in-vitro on purified DNA. Chromium(VI) gives contradictory results under these conditions.

Several hypotheses were formulated for the mechanisms of the genotoxic action of chromium. Schoental [20] proposed that chromium(VI) acts indirectly producing, on account of its oxidizing power, reactive aldehydes, which are the final genotoxic agents (eqn. 1). For instance, glycerol, obtained

$$\underset{glycerol}{\overset{OH\ \ OH\ \ OH}{\underset{|\ \ \ \ \ |\ \ \ \ \ |}{CH_2-CH-CH_2}}} \xrightarrow{Cr(VI)} \underset{acroleine}{\overset{O}{\underset{||}{HC}}-CH=CH_2} \xrightarrow{Cr(VI)} \underset{glycidal}{\overset{O\ \ \ \ \ O}{\underset{||\ \ \ /\backslash}{HC-CH-CH_2}}} \quad (1)$$

$$\underset{\substack{cross-linked\ cellular\\macromolecule}}{\equiv\!-N\!=\!CH\!-\!\overset{OH}{\underset{|}{CH}}\!-\!CH_2\!-\!S\!-\!\equiv} \longleftarrow \equiv\!-NH_2 \quad HS\!-\!\equiv$$

from triglycerides, is oxidized via acroleine to glycidal, which may crosslink macromolecules of chromatin, interfere with cell division, and produce genotoxic effects. This mechanism may be operative only in a few in-vivo carcinogenicity assays. The alternate hypotheses share the

Table 5. *Critical Evaluation of Genotoxic Effects of Chromium in Short-Term Tests**

Genetic Effect	Number of Published Reports	Activity** Cr(VI)	Cr(III)
DNA Damage			
Alterations of purified DNA	11	±	+
Errors in the replication of purified DNA	5	±	+
Alterations of DNA in treated cells	10	+	−
Inhibition of DNA replication	8	+	−
DNA Repair			
Killing of repair deficient bacteria	5	+	−
Stimulation of DNA repair synthesis	4	+	−
Chromosomal Effects			
in lower Eukaryotes (yeasts)	2	+	n.t.
in mammalian cells in-vitro	22	+	±
in mammalian cells in-vivo	4	+	−
in exposed humans	4	+	n.t.
Gene Mutations			
in bacteria	23	+	−
in lower Eukaryotes (yeasts)	1	+	n.t.
in mammalian cells in-vivo and in-vitro	7	+	−
In-Vitro Cell Transformation			
Morphological or physiological transformation of mammalian cells	9	+	−
Enhancement of virus-induced transformation of mammalian cells	1	+	n.t.

*Adapted from reference 14.
**+: assessed positive activity; −: assessed negative activity; ±: contradictory results; n.t.: not tested.

assumption that the activity of chromium(VI) is caused by the ability of chromate ions to enter the cell. The mechanism [14] stresses the mutagenic potential for the imbalance in the nucleotide pools. This imbalance, one of the cytotoxic effects of chromium(VI), may be caused by the cytoplasmic reduction of chromium(VI) and the subsequent impairment of enzymes involved in nucleotide metabolism. Other mechanisms consider the cytoplasmic reduction of chromium(VI) to be a detoxification reaction, and stress that chromium(VI) becomes genetically active only after it has reached the nucleus (Fig. 2). These hypotheses differ

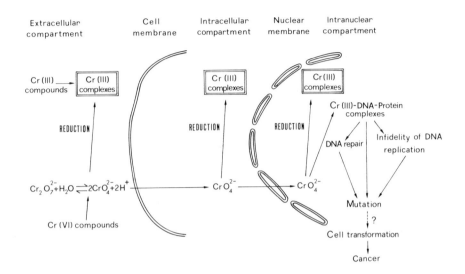

Figure 2. Proposed mechanisms for the genotoxicity of chromium (adapted from reference 14).

in the final step of the process. Mutations can arise through direct oxidation of DNA-protein complexes [21], through the interaction of intranuclearly reduced chromium(III) with DNA [22], or, more indirectly, through the interaction of chromium(III) with precursor nucleotides or DNA polymerases that diminish the fidelity of DNA replication [23]. All but Schoental's mechanisms are supported by experimental data. These mechanisms are not mutually exclusive [14]. Chromium(III) is confirmed as the final genotoxic agent by the finding that the targets identified through exposure of purified DNA to chromium(III) are the same as those affected by treatment of whole cells with chromium(VI) [14]. The affected DNA regions are rich in guanine and cytosine, contain nucleophilic sites, and are the common targets of many electrophilic mutagens. Chromium(III) interacts with such sites on DNA and produces interstrand DNA crosslinks. DNA-protein cross-linking was also detected [24]. Cross-linking is observed, when isolated cell nuclei are treated with chromium(III) or chromium(VI) in the presence of reducing systems such as microsomes. The requirement for a reducing system demonstrates that the reduction of chromium(VI) to chromium(III) is crucial for the induction of genotoxic effects [24, 25]. It has been suggested, that chromium(VI) can be reduced stepwise in a cell with chromium(V), a reactive intermediate, acting as the genotoxic agent [26].

The mechanisms of chromium genotoxicity derived from short-term tests explain also some features of chromium carcinogenicity such as the site of tumors in experimental animals and exposed humans. Cell extracts and microsomes markedly differ in their ability to prevent the mutagenicity of chromium(VI) [27]. Human erythrocyte lysate, rat liver microsomes, and mouse liver microsomes have the highest preventive ability; human gastric juice, human liver microsomes, and rat suprarenal microsomes possess intermediate activity; and human lung microsomes are least active. Human serum and plasma, mouse and rat lung microsomes, and rat muscle microsomes are inactive. Erythrocytes, liver microsomes, and gastric juice, agents with good preventive abilities, are also the major sites for in-vivo reduction of chromium(VI). Lung and muscles do not reduce chromium(VI), are unable to prevent the mutagenic effects of chromium(VI), and are preferred targets for the induction of tumors.

Chromium Genotoxicity in Relation to Environmental Exposure

In Table 5 the activities are attributed to chromium(VI) and chromium(III), although a variety of chromium(III) and chromium(VI) compounds were tested. The compound-specific results of the genetic tests are summarized in Table 6. All chromium(VI) compounds - with the exception of lead chromate - are active genotoxic agents. Most chromium(III) compounds are inactive [14]. Lead chromate, the only insoluble chromium(VI) salt investigated for DNA damage and repair in bacteria, was found to be inactive in a few tests.

Chromium(VI) compounds produce chromate ions when dissolved under physiological conditions. Chromate ions are known to cross cell membranes. Therefore, the genotoxic activity of chromium(VI) is attributable to the chromate ion acting via one or more of the proposed mechanisms.

Genotoxicity is not only associated with reagent grade laboratory compounds of chromium, but also with chromium-

Table 6. Critical Evaluation of the Genotoxic Effects of Chromium Compounds in Short-Term Tests [14]

Compound	Water Solubility	Number of Tests with Positive/Negative Response for									
		DNA Damage and Repair		Chromosomal Effects		Genetic Mutations		Cell Transformation		Total	
		+	−	+	−	+	−	+	−	+	−
Chromium(VI)											
K_2CrO_4	+	3	0	3	0	3	0	1	0	10	0
Na_2CrO_4	+	*	*	3	0	1	0	1	0	5	0
$K_2Cr_2O_7$	+	1	0	3	0	3	0	1	0	8	0
$Na_2Cr_2O_7$	+	1	0	3	0	1	0	*	*	5	0
CrO_3	+	2	0	3	0	2	0	*	*	7	0
$CaCrO_4$	±	*	*	2	1	2	0	1	0	5	1
$PbCrO_4$	−	1	2	2	0	2	1	1	0	6	3
$PbCrO_4 \cdot PbO$	−	*	*	2	0	1	0	*	*	3	0
$ZnCrO_4$	−	*	*	*	*	1	0	1	0	2	0
$ZnCrO_4\ Zn(OH)_2$	−	*	*	2	0	1	0	*	*	3	0
$(NH_4)_2CrO_4$	+	*	*	*	*	1	0	*	*	1	0
$SrCrO_4$	−	*	*	*	*	1	0	*	*	1	0
CrO_2Cl_2	+	*	*	*	*	1	0	*	*	1	0
Chromium(III)											
$CrCl_3$	+	1	2	0	1	1	0	1		1	5
$Cr(NO_3)_3$	+	1	0	0	2	1	*	*		1	3
$Cr_2(SO_4)_3$	−	0	1	0	2	1	*	*		0	4
$KCr(SO_4)_2$	+	0	1	1	1	1	*	*		1	3
$Cr(OH)SO_4$	±	*	*	0	2	1	*	*		0	3
$Cr(CH_3COO)_3$	+	1	0	1	1	3	*	*		2	4

*No test performed

containing industrial materials such as tannins used in the leather industry, the volatile chromyl chloride used in a variety of processes, fume particles emitted during welding processes [14, 28], and pigments used in paints. These results consistently identify chromium(VI)-containing compounds as the only active genotoxic agents. Positive results occasionally obtained with chromium(III) compounds could be caused by chromium(VI) present as a contaminant. Chromium(VI) could be formed in chromium(III) wastes by oxidation. This reaction was observed to occur at pH 11 at 45° in the presence of an oxygen source [29].

Soluble chromium(VI) compounds are the most active mutagens in short-term tests, whereas poorly soluble chromium(VI) compounds are the most powerful carcinogens in long-term assays on mammals. Soluble chromium(VI) compounds are so toxic that cells at the site of implantation in animal assays are killed. The rapid transport of soluble chromium(VI) compounds in the blood stream and their rapid inactivation prevents the occurrence of long term effects. The short-term genotoxic effects are easily detected even when acute symptoms of chromium(VI) toxicity are present [14]. Only chronic exposure to chromium(VI) provided by poorly soluble chromium(VI) compounds sets up local conditions suitable for the expression of genotoxicity. Several short-term tests use experimental mammals for in-vivo exposure to chromium(VI) to check for gene mutations, chromosomal effects, and cell transformations, or exposed workers to investigate chromosomal damage in blood cells. Chromium(VI) was demonstrated to cross the placenta in mice. The effects of chromium(VI) were detected in the cells of litters born to mice that had been exposed to chromium(VI). A synergistic interaction of chromium(VI) with oncogenic viruses and with other metals such as zinc was discovered during tests for genetic damage [14].

It appears, that in the case of chromium(VI) the extrapolation of genotoxic risk to man is based on sound evidence. However, a quantitative evaluation of risk is necessary to regulate environmental exposure. Such an evaluation cannot be obtained by simply extrapolating data from short-term tests, but must be based on an assessment of the actual conditions of exposure. Although no threshold dose can be defined for mutagens and carcinogens [30], the final regulation cannot avoid to consider the risk/benefit ratio.

Acknowledgements. This work was supported by grants from the National Research Council (C.N.R.) of Italy (Progetto Finalizzato "Medicina Preventiva e Riabilitativa", Sottoprogetto "Rischio Tossicologico"), the Italian Ministry of Public Education, and the Venetia Region ("Centro di Alta Specializzazione in Cancerogenesi Ambientale").

References

[1] T. H. Maugh, "Chemicals: How Many Are There?", **Science, 199** (1978) 162.
[2] International Agency for Research on Cancer, **"Mono-**

graphs on the Evaluation of the Carcinogenic Risk of Chemicals to Humans," Suppl. 4; "Chemicals, Industrial Processes and Industries Associated with Cancer in Humans," Lyon, France, 1982.

[3] D. Brusick, **"Principles of Genetic Toxicology,"** Plenum Press: New York, 1980.

[4] K. C. Bora, G. R. Douglas and E. R. Nestmann, Eds., **"Chemical Mutagenesis, Human Population Monitoring and Genetic Risk Assessment,"** Elsevier Biomedical Press: Amsterdam, 1982.

[5] International Agency for Research on Cancer, **"Monographs on the Evaluation of the Carcinogenic Risk of Chemicals to Humans,"** Suppl. 2; "Long-Term and Short-Term Screening Assays for Carcinogens: a Critical Appraisal," Lyon, France, 1980.

[6] H. F. Stich and R. H. C. San, Eds., **"Short-Term Tests for Chemical Carcinogens,"** Springer-Verlag: New York, 1981.

[7] G. M. Williams, "Batteries of Short-Term Tests for Carcinogen Testing," in **"The Predictive Value of Short-Term Screening Tests in Carcinogenicity Evaluation,"** G. M. Williams, R. Kroes, H. W. Waaijers and K. W. Van de Poll, Eds., Elsevier: Amsterdam, 1980, p. 327.

[8] M. Hollstein, J. McCann, F. A. Angelosanto and W. W. Nichols, "Short-Term Tests for Carcinogens and Mutagens," **Mut. Res.**, **65** (1979) 133.

[9] F. J. De Serres and J. Ashby, Eds., "Evaluation of Short-Term Tests for Carcinogens," **Progr. Mut. Res. Vol. 1,** Elsevier: Amsterdam, 1981.

[10] International Agency for Research on Cancer, **"Monographs on the Evaluation of the Carcinogenic Risk of Chemicals to Humans,"** Vol. 23: "Some Metals and Metallic Compounds," Lyon, France, 1980.

[11] F. W. Sunderman, "Carcinogenic Effects of Metals," **Fed. Proc.**, **37** (1978) 40.

[12] A. Leonard, "Carcinogenic and Mutagenic Effects of Metals (As, Cd, Cr, Hg, Ni): Present State of Knowledge and Needs for Further Studies," in **"Trace Metals. Exposure and Health Effects,"** E. Di Ferrante, Ed., Pergamon Press: New York, 1979, p. 199.

[13] C. P. Flessel, A. Furst and S. B. Radding, "A Comparison of Carcinogenic Metals," in **"Metal Ions in Biological Systems,"** H. Sigel, Ed., Marcel Dekker AG: Basel, 1980, p. 23.

[14] A. G. Levis and V. Bianchi, "Mutagenic and Cytogenetic Effects of Chromium Compounds," in **"Biological and Environmental Aspects of Chromium,"** S. Langård, Ed., Elsevier Biomedical Press: Amsterdam, 1982, p. 171.

[15] T. Norseth, "Health Effects of Nickel and Chromium," in **"Trace Metals. Exposure and Health Effects,"** E. Di Ferrante, Ed., Pergamon Press: New York, 1979, p. 135.
[16] S. Langård, "Chromium," in **"Metals in the Environment,"** H. A. Waldron, Ed., Academic Press: New York, 1980, p. 111.
[17] A. Leonard and R. R. Lauwerys, "Carcinogenicity and Mutagenicity of Chromium," **Mut. Res., 76** (1980) 227.
[18] R. B. Hayes, "Carcinogenic Effects of Chromium," in **"Biological and Environmental Aspects of Chromium,"** S. Langård, Ed., Elsevier Biomedical Press: Amsterdam, 1982, p. 221.
[19] V. Bianchi, L. Celotti, G. Lanfranchi, F. Majone, G. Marin, A. Montaldi, G. Sponza, G. Tamino, P. Venier, A. Zantedeschi and A. G. Levis, "Genetic Effects of Chromium Compounds," **Mut. Res., 117** (1983) 279.
[20] R. Schoental, "Chromium Carcinogenesis, Formation of Epoxyaldehydes and Tanning," **Br. J. Cancer, 32** (1975) 403.
[21] F. L. Petrilli and S. De Flora, "Metabolic Deactivation of Hexavalent Chromium Mutagenicity," **Mut. Res., 54** (1978) 139.
[22] A. G. Levis, M. Buttignol, V. Bianchi and G. Sponza, "Effects of Potassium Dichromate on Nucleic Acid and Protein Syntheses and on Precursor Uptake in BHK Fibroblasts," **Cancer Res., 38** (1978) 110.
[23] L. K. Tkeshelashvili, C. W. Shearman, R. A. Zakour, R. M. Koplitz and L. A. Loeb, "Effects of Arsenic, Selenium, and Chromium on the Fidelity of DNA Synthesis," **Cancer Res., 40** (1980) 2455.
[24] A. J. Fornace, D. S. Seres, J. F. Lechner and C. C. Harris, "DNA-Protein Cross-Linking by Chromium Salts," **Chem. Biol. Interactions, 36** (1981) 345.
[25] J. D. Garcia and K. W. Jennette, "Electron-Transport Cytochrome P-450 System is Involved in the Microsomal Metabolism of the Carcinogen Chromate," **J. Inorg. Biochem., 14** (1981) 281.
[26] K. Wetterhahn-Jennette, "Microsomal Reduction of the Carcinogen Chromate Produces Chromium(V)," **J. Am. Chem. Soc., 104** (1982) 874.
[27] S. De Flora, "Biotransformation and Interaction of Chemicals as Modulators of Mutagenicity and Carcinogenicity," in **"Environmental Mutagens and Carcinogens,"** T. Sugimura, S. Kondo and H. Takebe, Eds., Alan R. Liss: New York, 1982, p. 527.
[28] P. Venier, A. Montaldi, F. Majone, V. Bianchi and A. G. Levis, "Cytotoxic, Mutagenic and Clastogenic Effects of Industrial Chromium Compounds," **Carcinogenesis, 3**

(1982) 1331.
[29] S. A. J. Shivas, "Factors which can Influence the Oxidation State of Chromium in Tanning Wastes," **J. Amer. Leather Chem. Ass.**, **75** (1980) 45.
[30] E. Freese, "Thresholds in Toxic, Teratogenic, Mutagenic and Carcinogenic Effects," **Environ. Health Perspect.**, **6** (1973) 171.

DISCUSSION

E. H. Abbott (Montana State University): I would like to point out that most of the chromium(III) salts you tested for genetic toxicity contain the same cation, $Cr(H_2O)_6^{3+}$. I think that this cation may not pass through the cell membrane but rather may bind to the exterior of the cell. Do you have information that shows that this cation actually gets into the cell?

V. Bianchi: Recent data show that chromium(III) cations enter the cell and produce DNA damage, when bacteria are incubated in a liquid medium of pH 7.4 with chromium trichloride for at least six hours. A few reports indicate, that with animal cells endocytosis of extracellular chromium(III) is followed by induction of chromosomal damage. Therefore, the generally stated lack of genetic activity of chromium(III) compounds seems to be related to the difficulty of crossing the membrane rather than to the inability of chromium(III) complexes to cause genetic damage.

P. J. Sadler (Birkbeck College, University of London): It is usually assumed that chromate rapidly enters cells and is reduced to chromium(III). Do you know how rapid the reduction is? Does any chromium(VI) remain in the cell membrane?

V. Bianchi: Chromium(VI) uptake by the cell and reduction inside the cell was studied in detail in mammalian erythrocytes, in which hemoglobin activity reduces chromium(VI). After six hours approximately five percent of the cellular chromium were still in the form of free, anionic chromium(VI). No chromium(VI) was found in the cell membrane. It must be emphasized that erythrocytes are a specialized cell system, and the results obtained with erythrocytes may not be characteristic of other types of cells.

P. J. Sadler: Although your chromium(III) complexes were not taken up by cells, presumably some chromium(III) complexes such as the glucose tolerance factor are specificially taken up by cells. Have you thought of extending your studies to chromium(III) complexes of amino acids?

V. Bianchi: In addition to the glucose tolerance factor other chromium(III) complexes permeate the cell membrane provided that the complexes have lipophilic ligands. Such complexes were shown to be genetically active.

J. M. Wood (University of Minnesota, Navarre): I have real reservations about an oxidation state of six for chromium inside cells. The reducing conditions in a cell would reduce chromium(VI) to chromium(III). Could it be possible that complexes of chromium(III) with biological ligands act as the mutagenic factor of chromium(III)? Chromium(VI) may be required for the transport of chromium across membranes. After entering the cell chromium(VI) is reduced to chromium(III), chromium(III) forms complexes inside the cell, and the complexes cause mutations.

V. Bianchi: With the exception of the direct genetic effects of chromium(VI) ascribable to its oxidizing power most of the damage to DNA by chromium is produced by intracellular chromium(III). The reduction of chromium(VI) within the cell can be regarded as a mechanism of inactivation of chromium, if stable, not diffusible chromium(III) complexes are produced. However, this process can also produce chromium(III) complexes that are genetically active, because they involve genetically important molecules or carry ligands exchangeable for nucleophilic sites in DNA.

XXVII

URINARY CHROMIUM AS AN ESTIMATOR OF EXPOSURE TO DIFFERENT TYPES OF HEXAVALENT CHROMIUM-CONTAINING AEROSOLS

A. Mutti*, C. Minoia†, C. Pedroni*, G. Arfini*, G. Micoli†, A. Cavalleri§, and I. Franchini*

*Cattedra di Medicina del Lavoro, Università di Parma, I-43100 Parma
†Centro Ricerche di Fisipatologia e Sicurezza del Lavoro, Fondazione Clinica del Lavoro, Università di Pavia, I-27100 Pavia
§Cattedra di Medicina del Lavoro, Università di Modena, I-41100 Modena
Italy

Chromium concentrations in the air were measured in seven different workroom environments in which exposure to water-soluble, hexavalent or trivalent chromium compounds was suspected. Total chromium concentrations were determined in urine spot samples collected at the beginnings and at the ends of the same arbitrarily chosen workshifts. End-of-shift urinary chromium concentrations and their increases above the pre-shift levels were closely related to concentrations of water-soluble, hexavalent chromium in the air. An end-of-shift value of 29.8 µg Cr/g creatinine and a difference of 12.2 µg Cr/g creatinine between pre-shift and end-of-shift values were estimated to be characteristic for an exposure to 50 µg Cr/m^3 of air in form of water-soluble, hexavalent chromium. Urinary chromium in workers exposed to water-soluble chromates or to water-soluble chromium(III) sulfate cannot be recommended as a test for short-term exposure to chromium and for the quantitative evaluation of job-related hazards.

Industrial pollutants may contain chromium in various oxidation states and in compounds with different solubility. Chemically different chromium compounds are thought to have different metabolic and biological activities [1]. The complexity of most industrial operations causing occupational exposure to chromium compounds leads to a wide range of possible interactions of chromium with the human body. The knowledge of the metabolism and toxicity of chromium compounds in humans is mostly drawn from investigations that are complicated by a number of confounding variables. Most of these variables such as the interactions with other toxic metals and the characteristics of the populations involved are difficult or even impossible to control. The adverse health effects caused by exposure to chromium are believed to be attributable to hexavalent chromium. Trivalent chromium compounds are thought to be less toxic or, at least, less hazardous to occupationally exposed workers.

The total chromium concentration and the concentration of chromium(VI) in the workroom air are reasonably measures of exposure and related hazards [2]. For the assessment of individual exposure, a biological test would provide additional information and would have advantages over air monitoring. The concentration of chromium in biological specimens would give "integrated" information about the overall exposure resulting from different sources of contamination, absorption routes, and physiological characteristics of each individual. However, the usefulness of a biological test is mainly related to effective sampling strategy and proper sampling procedure. Chromium is usually present in trace amounts in biological specimens. Disposable steel needles may contaminate blood and serum samples with chromium during the collection procedure [3]. Contamination can easily be avoided with urine samples, which have also several other practical advantages.

Urinary chromium has been successfully used to assess exposure to chromium(VI) in chrome-plating and stainless steel welding industries [4-11]. During plating and welding operations, chromium(VI) is taken up by the respiratory tract in the form of readily soluble chromic acid mists and chromium(VI)-containing welding fumes. Chromium(VI) is quickly reduced in the human organism to the trivalent form, which is the only oxidation state of chromium in biological specimens [12]. The kinetics of urinary excretion of chromium is rather complex [13-15]. However, much of the chromium is excreted fast enough to permit evaluation of very recent exposure. A single-compartment mathematical model based on quickly excreted chromium was developed to describe the short-term release of urinary chromium following one week of exposure to welding fumes [16]. In spite of the relationship between urinary chromium and exposure to hexavalent, water-soluble chromium in airborne particulates, urinary excretion also depends on the total body burden of chromium. Among workers with the same current exposure those with higher body burdens show higher urinary chromium values than workers with low body burdens [9].

So far, no information is available about the biological fate and urinary excretion rate of chromium compounds reaching the respiratory tract in forms different from those in welding fumes and plating mists. The aim of the present study was to evaluate the influence of the solubility and valency of chromium compounds and of body factors on the absorption and urinary excretion of chromium compounds.

Subjects. The study involved 137 workers employed in plants, in which exposure to hexavalent or trivalent chromium compounds was expected. Exposure to chromium(VI)-containing welding fumes was evaluated for 36 manual metal-arc/stainless-steel (MMA/SS) welders using high-chromium alloyed electrodes and for 12 workers exposed to metal inert-gas/stainless-steel (MIG/SS) welding fumes. Subjects from chrome-plating plants comprised 24 workers at "hard" plating baths, in which the physical characteristics of a metal are improved with a thick (5-10 µm) layer of chromium, and 16 workers at "bright" plating plants, in which the appearance of a metal is improved with a thin (0.5-1.0 µm) layer of chromium. The subjects from a dichromate producing factory were divided into two groups according to their exposure to chromium compounds: 15 workers were mainly exposed to chromium(III) in the form of basic chromic sulfate containing measurable amounts of chromium(VI), and 22 workers were mainly exposed to chromium(VI) in the form of chromium trioxide and potassium dichromate. Twelve spray painters exposed to water-insoluble chromium(VI) compounds as lead and zinc chromates and 38 non-exposed subjects were also examined.

Methods. Urine samples were collected before and after the shift of an arbitrarily chosen working day in acid-washed polyethlene bottles. Particular care was taken to avoid contamination. Total chromium was determined by electrothermal atomic absorption spectroscopy [17] in two different laboratories. An interlaboratory comparison was carried out on 15 urine samples with chromium contents in the range of 5 to 35 µg/L. The results were in good agreement with a correlation coefficient of 0.99 and a regression line of negligible intercept with a slope very close to unity. Urinary creatinine was measured by Jaffe's reaction using a Technicon AutoAnalyzer [18].

Air samples near welding operations were collected on PCV or cellulose nitrate filters with mean pore sizes of 0.8 to 0.45 μm with personal Dupont 4,000 or stationary Gelman EC 3,000 samplers. Flow rates ranged from 10 to 15 L/min for stationary pumps and 3 to 4 L/min for personal samplers. The sampling times varied from 30 to 60 min.

Total chromium in the airborne particulate was measured by X-ray fluorescence and atomic absorption spectroscopy. Chromium(VI) assays were performed according to the the method No. P&CAM 319 of 8.29.80 "Hexavalent Chromium" reported in the NIOSH Manual of Analytical Methods [19]. The water- and acid-soluble chromium compounds in welding fumes were determined as reported [9].

Results and Discussion

The mean values of urinary chromium and exposure levels in seven groups of workers occupationally exposed to different chromium compounds and in a control group are summarized in Table 1. All of the exposed groups show urinary levels definitely higher than those observed in occupationally non-exposed subjects.

Both valency and solubility seem to play a role in the metabolism of chromium compounds. Total chromium concentrations in urine spot samples collected at the end of the workshift and the daily increases above the pre-shift levels correlate with the water-soluble fraction of hexavalent chromium in the air. Heavy exposure to poorly soluble hexavalent chromium or to water soluble trivalent chromium caused only a slight increase in urinary chromium. Painters exposed to water-insoluble zinc and lead chromates showed urinary chromium levels much lower than expected on the basis of airborne chromium(VI) concentrations. Chromate workers heavily exposed to water-soluble trivalent chromium present in the workroom environment as chromic sulfate showed urinary chromium values rather low and consistent with exposure to water-soluble hexavalent chromium present as impurities in the trivalent chromium compound.

For a small group of workers, mostly welders, the chromium concentrations in the air and in the urine were available. The end-of-shift urinary chromium was closely related to the concentration of water-soluble, hexavalent chromium in the air (Fig. 1) in spite of the great difference in aerodynamic properties of inhaled aerosols.

Table 1. *Chromium Concentrations in the Workroom Environment and in Urine of Workers Exposed to Different types of Chromium-Containing Compounds*

Group	No. of Subjects	8-h TWA* airborne chromium (μg/m³)				Urinary chromium (μg/g creatinine)	
		Hexavalent Cr		Insoluble Median (range)	Total Cr Median (range)	End-of-Shift Urinary Cr Mean (SD)	Daily Increase Urinary Cr Mean (SD)
		Water-soluble Median (range)					
Welders							
MMA/SS	36	65(10-152)		n.a.**	94(12- 224)	33.3 (6.9)	8.8 (3.8)
MIG/SS	12	12(3- 35)		n.a.	62(15- 348)	12.1 (4.9)	2.6 (1.3)
Chrome-Platers							
"HARD"	24	31(4-146)		n.a.	47(6- 160)	15.3 (9.5)	6.1 (3.1)
"BRIGHT"	16	5(0- 31)		n.a.	7(0- 39)	5.8 (4.2)	1.9 (1.5)
Chromate Workers							
$K_2Cr_2O_7$	22	43(8-212)		n.d.***	89(18- 312)	19.5 (10.1)	8.0 (6.4)
$Cr_2(SO_4)_3$	15	6(2- 23)		n.d.	512(48-1710)	16.7 (12.5)	6.4 (4.8)
Lead and Zinc Chromate							
Painters	12	n.a.		990(450-1450)	1221(536-1720)	13.2 (10.4)	1.3 (1.2)
Non-Exposed Subjects	38	0		0	0	0.8 (0.4)	-0.1 (0.2)

*TWA: Time Weighted Average   **n.a. = not analyzed   ***n.d. = not detected

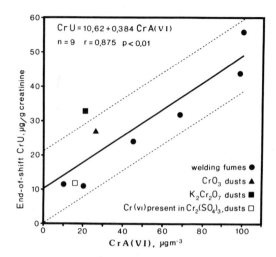

Figure 1. The relationship between the time-weighted average, eight-hour concentration of water-soluble, hexavalent chromium, CrA(VI), in the air of the breathing zone in the workroom environment and the chromium concentration in end-of-shift urine, CrU. Reprinted with permission from. ref. 20.

The correlation improved (Fig. 2) when the daily increases in urinary chromium concentrations - the differences between morning (pre-exposure) and afternoon (end-of-shift) values - were compared with the concentrations of water-soluble, hexavalent chromium(VI) in the air. The workers involved in this survey probably had heavy body burdens of chromium acquired during long-term exposures to water-soluble, hexavalent chromium compounds. An end-of-shift value of 29.8 µg Cr/g creatinine and a daily increase in urinary chromium of 12.2 µg Cr/g creatinine were estimated from the regression lines for an exposure to 50 µg Cr/m^3 of air in form of water-soluble, hexavalent chromium, the current TLV in most countries. The lower limits for these values at the 95 percent confidence level are 18.5 and 7.1 µg Cr/g creatinine. The subject mainly exposed to chromic sulfate was not included in establishing these correlations because he had high, pre-shift urinary chromium values probably caused by the slow absorption of trivalent chromium through the digestive tract.

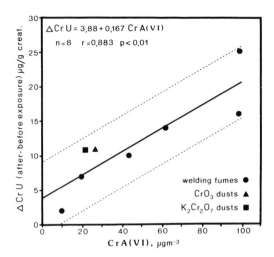

Figure 2. The relationship between the time-weighted average, eight-hour concentration of water-soluble, hexavalent chromium, CrA(VI), in the air of the breathing zone in the workroom and the difference in chromium concentrations between pre-shift and end-of-shift urine, ΔCrU. Reprinted with permission from ref. 20.

The analysis of urine samples of exposed subjects by two techniques selective for chromium(VI) failed to show detectable amounts of chromium(VI) [21]. Once absorbed, hexavalent chromium is quickly reduced to the trivalent form. Therefore, the differences in the metabolism of chromium compounds can be ascribed only to their absorption rates [22]. Urinary chromium was suggested as a short-term exposure test for welders and chrome platers exposed to water-soluble, hexavalent chromium compounds that are readily absorbed through the respiratory tract [4-11]. Urine chromium levels account not only for current exposure but also for the total chromium burden [8, 9]. Moreover, at the same current exposure levels, higher urinary total chromium should be expected in workers with heavy past exposure than in newly exposed subjects [9].

The present study involved workers with long-term exposure to different types of chromium-containing aerosols and evaluated the role of valency and solubility in the absorption and urinary excretion of chromium. The results are consistent with the hypothesis that only hexavalent and water-soluble chromium compounds are readily absorbed. Urinary chromium is useful as an exposure test only when workers are exposed to hexavalent and water-soluble chromium compounds. Neither water-soluble, trivalent compounds such as chromium(III) sulfate, nor hexavalent, insoluble compounds such as lead and zinc chromates are absorbed quickly enough to cause any exposure-related change in total chromium concentrations in urine. Chromium levels in urine of workers heavily exposed to chromium(III) sulfate or to lead and zinc chromates were similar or lower than those of subjects exposed to much lower concentrations of water-soluble chromium(VI) compounds. The slight increase in urinary chromium after exposure to insoluble hexavalent or to trivalent chromium compounds might be ascribed to trace amounts of soluble, hexavalent chromium. In spite of the differences in the aerodynamic properties of inhaled particles, soluble, hexavalent chromium compounds showed a similar biological availability and urinary excretion. In subjects with long-term exposure, the very recent absorption of chromium(VI) resulted in a rapid increase in urinary chromium levels proportional to the airborne chromium concentrations. No difference could be detected in urinary excretion of chromium between workers exposed to welding fumes and the few subjects exposed to chromate and dichromate dusts at the same exposure levels. Therefore, exposure to water-soluble chromates and dichromates might be monitored by the procedure recommended for exposure to chromium(VI)-containing welding fumes and plating mists [22]. This procedure is not applicable to trivalent or less soluble, hexavalent compounds.

Because of the great differences in bioavailability and toxicity of the various chromium compounds, these compounds must be identified before a biological monitoring program can be initiated and its results evaluated. The distribution of chromium among valence states and solubility fractions must be known. The aerodynamic properties of the particulates seem to be less important with respect to the absorption rate of water-soluble, hexavalent chromium. The aerodynamic properties might play a role in the accumulation of poorly soluble chromates in the lung.

References

[1] International Agency for Research on Cancer, Lyon, France, "Some Metals and Metallic Compounds," **IARC Monographs on the Evaluation of Carcinogenic Risk of Chemicals to Man,"** 230 (1980) 205.

[2] National Institute for Occupational Safety and Health, **"Criteria for a Recommended Standard: Chromium(VI),"** U.S. Department of Health, Education and Welfare (NIOSH) Publ. No. 76/129, National Institute for Occupational Safety and Health: Cincinnati, Ohio, 1975.

[3] J. Versieck, J. Hoste, F. Barbier, H. Steyaert, J. De Rudder and H. Michels, "Determination of Chromium and Cobalt in Human Serum by Neutron Activation Analysis," **Clin. Chem., 24** (1978) 303.

[4] A. Borghetti, A. Mutti, A. Cavatorta, M. Falzoi, F. Cigala and I. Franchini, "Indices Renaux d'Exposition Aigue et d'Impregnation Chronique par le Chrome," **Med. Lavoro, 68** (1977) 355.

[5] I Franchini, A. Mutti, F. Gardini and A. Borghetti, "Excretion et Clearance Renale du Chrome par Rapport au Degre et a la Duree de l'Exposition," **Rein et Toxiques,"** J. Traeger and L. Roche, Eds., Masson: Paris, 1975, p. 271.

[6] M. P. Guillemin and M. Berode, "A Study of the Difference in Chromium Exposure in Workers in two Types of Electroplating Processes," **Ann. Occup. Hyg., 21** (1978) 105.

[7] B. Gylscth and S. Langård, "Evaluation of Chromium Exposure Based on a Simplified Method of Urinary Chromium Determination," **Scand. J. Work, Environ. Health, 3** (1977) 28.

[8] P. L. Källiomäki, E. Rahkonen, V. Vaaranen, K. Källiomäki and K. Aittoniemi, "Lung-Retained Contaminants, Urinary Chromium and Nickel among Stainless-Steel Welders," **Int. Arch. Occup. Environ. Health, 49** (1981) 67.

[9] A. Mutti, A. Cavatorta, C. Pedroni, A. Borghi, C. Giaroli and I. Franchini, "The Role of Chromium Accumulation in the Relationship between Airborne and Urinary Chromium in Welders," **Int. Arch. Occup. Environ. Health, 43** (1979) 123.

[10] B. Sjogren, L. Hedstrom and U. Ultrarson, "Urine Chromium as an Estimator of Air Exposure to Stainless Steel Welding Fumes," **Int. Arch. Occup. Environ. Health, 51** (1983) 347.

[11] S. Tola, J. Kilpio, M. Virtamo and K. Haapa, "Urinary Chromium as an Indicator of the Exposure of Welders to Chromium," **Scand. J. Work, Environ. Health, 3** (1977) 192.

[12] C. Minoia, M. Colli and L. Pozzoli, "Determination of Hexavalent Chromium in Urine by Flameless Atomic Absorption Spectrophotometry," **At. Spectrosc., 2** (1981) 163.

[13] R. J. Collins, P. O. Fromm and W. D. Collings, "Chromium Excretion in the Dog," **Am. J. Physiol., 201** (1961) 795.
[14] A. Mutti, A. Cavatorta, L. Borghi, M. Canali, C. Giaroli and I. Franchini, "Distribution and Urinary Excretion of Chromium," **Med. Lavoro, 70** (1979) 171.
[15] C. Onkelinx, "Compartment Analysis of Chromium(III) in Rats at Various Ages," **Am. J. Physiol., 232** (1977) 478.
[16] A. Tossavainen, M. Nurminen, P. Mutanen and S. Tola, "Application of Mathematical Modelling for Assessing the Biological Half-Times of Chromium and Nickel in Field Studies," **Br. J. Ind. Med., 37** (1980) 285.
[17] K. H. Schaller, H. G. Essing, H. Valentin and G. Schake, "The Quantitative Determination of Chromium in Urine by Flameless Atomic Absorption Spectroscopy," **At. Abs. Newsl., 12** (1973) 147.
[18] R. J. Henry, **"Clinical Chemistry: Principles and Technics,"** Harper & Row: New York, 1965, p. 287.
[19] P. M. Eller and D. Molina, "Hexavalent Chromium," in **"NIOSH Manual of Analytical Methods,"** D. G. Taylor, Manual Coordinator, U.S. Department of Health, Education and Welfare (NIOSH) Publ. 80/125, NIOSH: Cincinnati, Ohio, 1980, p. 319.
[20] A. Mutti, C. Pedroni, G. Arfini, I. Franchini, C. Minoia, G. Micoli and C. Baldi, "Biological Monitoring of Occupational Exposure to Different Chromium Compounds at Various Valency States," **Intern. J. Environ. Anal. Chem., 17** (1984) 35.
[21] C. Minoia, A. Cavalleri and F. D'Andrea, "Urinary Excretion of Total and Hexavalent Chromium," in **"Trace Element Analytical Chemistry in Medicine and Biology,"** P. Bratter and P. Schramel, Eds., Walter de Gruyter: Berlin, 1983, p. 623.
[22] I. Franchini and A. Mutti, "Metabolism and Toxicity of Chromium Compounds," in **"Environmental Inorganic Chemistry,"** K. J. Irgolic and A. E. Martell, Eds., Proceedings, Workshop on Environmental Inorganic Chemistry, San Miniato, Italy, June 2-7, 1983, VCH Publishers: Deerfield Beach, Florida, 1985, p. 473.

XXVIII

METABOLISM AND TOXICITY OF CHROMIUM COMPOUNDS

I. Franchini and A. Mutti

Istituto di Clinica Medica e Nefrologia, Cattedra di Medicina del Lavoro
Universita di Parma
I-43100 Parma, Italy

Water-soluble chromium(VI) compounds are easily absorbed in the respiratory tract. Once absorbed, chromium(VI) is quickly reduced to the trivalent form and excreted in the urine. Chromium(III) and water-insoluble chromium(VI) compounds are cleared slowly from the lungs. Adverse effects of chromium are mainly caused by hexavalent compounds, which can cause skin and respiratory lesions. Many epidemiological studies indicate that chromium(VI) is carcinogenic in humans. The chromium concentration in urine provides information about the current exposure to water-soluble chromium(VI) and about the body burden of chromium. End-of-shift urinary chromium can be used to estimate the total body burden whereas the difference between end-of-shift and pre-shift values permits evaluation of the mean daily exposure. Urinary chromium appears to be a useful tool for the biological monitoring of job-related hazards caused by water-soluble chromium(VI), but is not applicable to water-insoluble chromium(VI) compounds, which are thought to be the main species responsible for human cancer in the chromate and dichromate industry.

A U.S. Public Health Service Publication [1] lists 104 occupations in which exposures to chromium occur. Chromium compounds are widely used in industrial processes such as the production of steels, chromium plating, leather tanning, pigment production, preparation of graphics, and production of refractory materials. In many occupations exposure to chromium is possible but not proven. Exposure may often come from unrecognized sources. For most of the proven cases the concentration of chromium in the workroom environment, the oxidation state of the chromium, and the identity of the chromium compound(s) are not known. Several reviews cover the toxicity, metabolism [2-5], and carcinogenicity [6-8] of chromium compounds, and the biological methods to monitor occupational exposure to chromium [9]. This paper presents general information about the absorption, distribution, and excretion of chromium, discusses aspects of the toxicity of chromium compounds, and reviews advances in monitoring occupational exposure to selected chromium(VI) compounds.

Metabolism

Absorption. During occupational exposure, chromium compounds are almost exclusively absorbed in the respiratory system. Sometimes the skin and the gastrointestinal tract may also absorb chromium. Hexavalent, trivalent or elemental chromium may reach the respiratory tract as vapors, mists, fumes or dusts. Water-soluble chromium(VI) compounds inhaled as fumes or mists are known to be absorbed in the respiratory tract including the upper airways. More information is needed about the absorption of chromium(III) compounds and of water-insoluble chromium(VI) compounds [10]. Very little chromium(III) is absorbed by the gastrointestinal tract. For instance, only one percent of the orally administered dose of chromium trichloride was taken up by animals [11]. Almost all of the chromium was eliminated in the feces shortly after administration [12]. Only two percent of chromates given orally to man were absorbed in the intestines. The gastric juice reduced chromates to poorly absorbing trivalent chromium [13]. Chromium(VI) readily penetrates intact skin [14] and may be retained in human skin for a considerable time. Percutaneous absorption is negligible [15].

Distribution. In man chromium(VI) is reduced to trivalent chromium, which binds to plasma proteins. Siderophilin [16, 17] and a low molecular weight, chromium-binding protein [18] seem to be the chromium(III) carriers in blood. Water-soluble chromates and dichromates may escape reduction by incorporation into red blood cells [19, 20]. However, the fraction of diffusible chromium in the blood i.e., chromium circulating in ionic form or bound to carriers with molecular weight less than 50,000, is very small even in workers occupationally exposed to chromium(VI) [21]. In non-occupationally exposed subjects the tissue concentrations of chromium are very low and appear not to increase with age [22]. The chromium concentrations in lung, spleen, liver, and kidney tissue of chromate workers with lung cancer were much higher than in normal subjects [23, 24]. Because chromium(VI) is rapidly reduced to chromium(III) in biological materials, the distribution of hexavalent and trivalent chromium in tissues is rather similar. The greater accumulation of chromium(VI) is probably caused by the higher absorption rate in the gastrointestinal tract [25].

Excretion. Absorbed chromium is eliminated from the body almost exclusively in the urine. The renal elimination of chromium was studied in dogs and rats. After filtration through the kidney glomeruli, chromium in the diffusible fraction of serum is largely re-absorbed by the renal tubules. Depending on the dose and the duration of exposure, 65 to 99 precent of the chromium are re-absorbed [20]. The kinetics of chromium excretion after administration of trace amounts can be described by a three-compartment model. Fifty percent of the chromium are eliminated in urine. Fecal clearance accounts for less than ten percent. A very low residual clearance corresponds to an apparently irreversible loss into body sinks of approximately 40 percent of the administered dose [26]. Chromium may accumulate in kidneys. As a result the renal clearance of diffusible chromium is enhanced proportionally to the chromium concentration in the renal cortex [20]. Accumulation may, therefore, play a role in the relationship between external and internal exposure [27]. Workers currently exposed to the same chromium levels may show different urinary excretion rates for chromium dependent on the history of exposure. Workers with heavy past exposure excrete much more chromium in the urine than workers with low past exposure.

Toxicity

Chromium(III) is an essential element for man and animals [2, 28] required as part of the glucose tolerance factor for the initiation of peripheral insulin action. The exact structure of the glucose tolerance factor is still unknown. Acute and chronic adverse effects of chromium are associated mainly with hexavalent chromium compounds, which are highly toxic to humans. The major lesions caused by acute and generally accidental poisoning with chromium(VI) involve the kidneys and the liver [9]. Cases of acute bronchopneumonia were also reported [29]. Lesions caused by prolonged occupational exposure to chromium(VI) are generally local.

The most affected organs are the skin, the respiratory tract, and the digestive system. Skin lesions include chronic ulcers and irritative dermatitis attributable to the oxidizing properties of chromium(VI), and eczematous dermatitis caused by cutaneous sensitization. Contact hypersensitivity seems to be related to the chromium(VI) content of the materials used. Effectual concentrations of chromium(VI) can easily be reached in non-occupational environments [30].

In workers exposed to chromium(VI) a mild to severe rhinopathy is the most common ailment. Severe lesions including hypertrophy, atrophy, ulcerous atrophy, and asymptomatic perforation of the nasal septum may occur even in subjects with brief occupational exposure [31]. Cross-sectional epidemiological investigations showed that lung function may be impaired in workers exposed to chromium(VI) compounds [32-39]. The oxidizing action of chromium(VI) may injure the digestive system causing sometimes very severe gastroenterocolitis [40, 41]. A working group at the International Agency for Research on Cancer judged the epidemiological evidence of respiratory carcinogenicity in men exposed during chromate production to be sufficient, whereas data on lung cancer risks in other chromium-associated occupations were considered to be insufficient [6]. Early reports suggested that only some insoluble chromium(VI) compounds formed during roasting of chromite ore could cause cancer [3, 6, 7]. It has not been proven that chromium(VI) must be water soluble to be carcinogenic. The importance of water solubility of hexavalent chromium compounds to carcinogenicity may have been overemphasized as suggested by a recent review [7]. In fact, the highly soluble chromium trioxide may also cause cancer in humans [42]. The incidence of respiratory cancer was found to be higher than expected in workers associated with the manufacture of chromate pigments, the spraying of zinc chromate paints, the welding of stainless steel, and the operation of chrome-plating baths [7, 42]. All of these occupations involve exposure to mostly water soluble chromium(VI) compounds.

Monitoring the Exposure to Chromium Compounds

To prevent adverse effects resulting from exposure to chromium or any other pollutant, the work environment and the workers must be monitored. A few definitions will facilitate the following discussion.

- **Exposure:** the contact between the outer and inner surfaces of an organism and a pollutant, which may be present in air, water, and/or food. Quantitative statements about exposure use appropriate concentration units and provide information about the duration of the contact. Exposure is often called "external dose".
- **Dose:** the amount or concentration of a pollutant at the site of an effect. Dose is often specified as "internal exposure" or "internal dose".
- **Biological Effect:** any change noticable in an organism resulting from an exposure.
- **Ambient Monitoring:** measurement and assessment of harmful agents in the working environment.
- **Biological Monitoring:** measurement and assessment of toxic substances, their metabolites, and their specific early effects in biological samples (tissues, secreta, excreta, expired air, blood).
- **Health Surveillance:** any periodic physiolgical, biochemical, or clinical examination of workers with the objective of protecting health and preventing occupationally related disease.

Biological monitoring is an important tool for the evaluation of exposure and for the prevention of adverse effects of toxic chemicals for individuals that are in contact with pollutants. Ambient and biological monitoring are not mutually exclusive but complementary tools. Meaningful biological monitoring requires the existence of indicators such as the concentrations of a pollutant and its metabolites in tissues or body fluids, and the availability of practical and valid analytical methods for the determination of these concentrations. Biological monitoring is dependent on knowledge about dose-effect, dose-response, and exposure-dose relationships, and factors that might modify such relationships; and presupposes that the data can be unambiguously interpreted in terms of risk. Interpretation in terms of risk is crucial for the establishment of safe levels. The risk that can or cannot be accepted is a matter of debate and ultimately of choice. The risk assessment must be based on the threshold level for each effect. Whether or not a threshold exists for chemical carcinogens is not known with certainty. If such threshold levels do exist, they probably will be very low. Under these conditions, every exposure might increase the risk in a statistical sense. Exposure to chemical carcinogens including chromium(VI) compounds should be kept as low as possible.

Health surveillances do not need to be as frequent as biological and ambient monitoring. Health surveillance programs, for which biological monitoring is not an alternative but a required part, should consider those indicators that require medical judgements and are needed for a meaningful evaluation of risk to an individual.

Biological monitoring of exposure to chromium compounds. The metabolism of chromium varies considerably with the physicochemical properties of the chromium compounds. The solubility of chromium compounds and the valence state of chromium greatly influence the rates of absorption and excretion of chromium. Data are available only for a limited number of industrial processes that cause exposure to water-soluble chromium(VI) compounds. Several papers [21, 27, 43-52] addressing the exposure to chromium(VI) during plating and welding operations report the usefulness of the chromium concentration in urine as an indicator of the uptake of water-soluble chromium(VI).

Indicators of internal dose. The time dependence of the chromium concentration in the urine of platers exposed to chromic acid mists indicated that the end-of-the-shift chromium concentration in urine is correlated with exposure intensity [21]. In addition to current exposure, urinary chromium reflects also past exposure and accumulation. In spite of a relatively short biological half-life with the fast phase ranging from seven hours to 15-38 hours [48], chromium is not completely eliminated during two consecutive days free of additional exposure to chromium. Chromium, therefore, must accumulate in the body during prolonged exposure [45]. Kidneys are important sinks for chromium. With increasing concentration of chromium in the renal cortex, the re-absorption of diffusible chromium by the renal tubules diminishes [20].

The renal accumulation of chromium was indirectly confirmed in platers, in whom the renal clearance of diffusible chromium was closely related to the duration of exposure [46]. The chromium content of the cortex determines the degree of renal clearance of diffusible chromium and contributes together with chromium from current exposure to the concentration of chromium in the urine. Workers with heavy past exposure to chromium and accumulation of chromium, excrete much more chromium in the urine than workers without such exposure even when both types of workers are exposed to the same chromium-containing environment in their present occupation. The additional urinary chromium in the workers with past exposure comes from the high renal clearance of diffusible chromium [27]. The chromium concentration in end-of-shift urine cannot discriminate between current heavy exposure of workers with a low chromium body burden and current low exposure of workers with a high chromium body burden. The difference between urinary chromium levels before and after exposure is dependent not only on the actual exposure but also on renal clearance of chromium and, therefore, on past exposure and accumulation. Thus, neither end-of-shift values nor daily-increase-above-pre-shift values permit to draw any conclusion when considered separately.

Absolute urinary chromium values might be useful as indicators of chromium accumulation as an alternative to measurements of renal clearance of diffusible chromium. Urinary chromium remains high for a long time after removal from a chromium-contaminated environment only in workers with heavy body burdens [27]. The good relationship between urinary chromium and the average remanent magnetic field - an indirect measure of chromium accumulation in the lungs of welders [46] - support the use of urinary chromium values for the assessment of body burdens. The estimation of current and past exposures might be possible with a two-point, simplified sampling strategy. The absolute end-of-shift urinary chromium value would quantify the body burden of chromium replacing the more complicated and impractical determination of renal clearance of diffusible chromium [27]. The daily increase in urinary chromium above the pre-shift values would reflect the exposure and uptake during the working day [9].

Urinary chromium as a tool for biological monitoring. Chromium levels in urine are related partly to the body burden characterized by the end-of-shift chromium values, and partly to the current exposure characterized by the daily increase of chromium above pre-shift levels. Which of the two levels is more important for the assessment of risk, depends on the type of exposure. In the case of exposure to water-soluble chromium(VI) compounds, acute effects are the most common, and the risk is related mainly to current levels of exposure, which is measured by the daily increase of chromium above pre-shift values. To assess the body burden of chromium, the levels in the end-of-shift urine should be used. Guidelines for the estimation of exposure to water-soluble chromium(VI) compounds and for the determination of the body burden of chromium are summarized in Table 1.

Table 1. Guidelines for the biological monitoring of workers occupationally exposed to water-soluble chromium compounds

Parameter	1	2A	2B	3A	3B	Other Tests
Indicators of Internal Dose						
ΔCr-U* (μg Cr/g creatinine)	0	<5.0	<5.0	>5.0	>5.0	Renal clearance of diffusable chromium
End-of-Shift Urinary Cr	<5.0	>15.0	5.0–15.0	>15.0	5.0–15.0	
Exposure Estimation						
Current exposure as water-soluble Cr(VI) in air (μg/m^3)	very low	<25	<50	>25	>50	
Past exposure (body burden)	very low	heavy	rather heavy	rather heavy	rather low	
Preventive Measures						
Individuals	annual detn. of urinary Cr	Semestral** check of urinary Cr and ΔCr-U		Removal from exposure and health examinations		Kidney dysfunction indicators: proteinuria, β_2-microglobulin or Retinol-binding protein in urine; β-glucuronidase in urine or other enzymes in urine
Group	None	Monitoring of the working environment		Technical control of exposure (exhaust, ventilation, etc.)		

* Cr-U = Urinary chromium after exposure minus urinary chromium before exposure: 1 μg Cr/g creatinine = 2.17 μmol Cr/mol creatinine.

** A trimestral check is recommended for workers exposed to chromic acid mists.

The risk classes (Table 1) provide health-based biological limits without considering possible carcinogenic effects, for which the threshold is not known. Risk class 1 includes values within the normal reference range. Classes 2 and 3 indicate increasing current exposure levels as measured by the increase in urinary chromium (ΔCr-U). Subclasses A and B refer to past exposure and accumulation as reflected by the end-of-shift urinary chromium.

A daily increase in urinary chromium less than 5 µg/g creatinine indicates a low current exposure, especially when such an increase coexists with end-of-shift values higher than 15 µg/g creatinine, a value suggestive of heavy past exposure and body burden (Table 1, risk class 2A). End-of-shift urinary chromium is generally lower than 15 µg/g creatinine in newly exposed workers. For these subjects, a daily increase in urinary chromium lower than 5 µg/g creatinine suggests low current exposure (Table 1, risk class 2B). Values within the class 2B, however, should be interpreted with caution, because the current exposure could be underestimated. High exposure to more than 50 µg/g Cr/m^3 of air in form of water-soluble hexavalent compounds would cause a daily increase higher than 5 µg/g creatinine. The exposure is proportionally lower for subjects with end-of-shift values higher than 15 µg/g creatinine.

The guidelines in Table 1 are applicable only to exposure to water-soluble chromium(VI) compounds and must be modified for other chromium compounds. This strategy for biological monitoring is based on a simplified procedure. One must guard against misinterpretations, which are possible whenever the different behavior of the biological indices under various conditions of exposure and body burden are not considered.

Acknowledgement. The editorial help of Ms. G. Intini is gratefully acknowledged.

References

[1] T. H. Milby, M. M. Key, R. L. Gibson and H. E. Stockinger, "Chemical Hazards," in **"Occupational Diseases. A Guide to Their Recognition,"** M. W. Gafafer, Ed., U.S. Public Health Service Publication, U.S. Government Printing Office, Washington, D.C., 1966, p. 120.

[2] National Academy of Sciences, **"Chromium,"** Washington,

D.C., 1974.
[3] National Institute for Occupational Safety and Health, "Criteria for a Recommended Standard: Chromium(VI)," Department of Health, Education and Welfare (NIOSH) Publication No. 76-129, Cincinnati, Ohio, 1975.
[4] S. Langård, Ed., "Biological and Environmental Aspects of Chromium," Elsevier: Amsterdam, 1982, p. 149.
[5] D. Burrows, Ed., "Chromium: Metabolism and Toxicity," CRC Press: Boca Raton, Florida, 1983.
[6] International Agency for Research on Cancer, Lyons, France. "Some Metals and Metallic Compounds," **IARC Monographs on the Evaluation of Carcinogenic Risk of Chemicals to Man, 23** (1980) 205.
[7] S. Langård, "The Carcinogenicity of Chromium Compounds in Man and Animals," in "Chromium: Metabolism and Toxicity," D. Burrows, Ed., CRC Press: Boca Raton, Florida, 1983, p. 13.
[8] A. G. Levis and V. Bianchi, "Mutagenetic and Cytogenetic Effects of Chromium Compounds," in "Biological and Environmental Aspects of Chromium," S. Langard, Ed., Elsevier: Amsterdam, 1982, p. 171.
[9] I. Franchini, A. Mutti, A. Cavatorta, C. Pedroni and A. Borghetti, "Chromium," in "Human Biological Monitoring of Industrial Chemicals Series," L. Alessio, A. Berling, R. Roi and W. G. Town, Eds., Commission of the European Community: Luxembourg, in press.
[10] P. L. Källiomäki, M. L. Juntilla, K. K. Källiomäki and R. Kivela, "Comparison of the Behavior of Stainless and Mild-Steel Manual Metal Arc Welding Fumes in Rat Lung," **Scand. J. Work, Environ. Health, 9** (1983) 176.
[11] W. J. Visek, I. B. Whitney, U. S. G. Kuhn and C. L. Comar, "Metabolism of 51-Cr by Animals is Influenced by Chemical State," **Proc. Soc. Exp. Biol. Med., 84** (1953) 610.
[12] M. T. Fisher, P. R. Atkins and G. F. Joplin, "A Method for Measuring Faecal Chromium and its Use as a Marker in Human Metabolic Balance," **Clin. Chim. Acta, 41** (1972) 109.
[13] R. M. Donaldson and R. F. Barreras, "Intestinal Absorption of Trace Quantities of Chromium," **J. Lab. Clin. Med., 68** (1966) 419.
[14] B. Baranowska-Dutkiewicz, "Absorption of Hexavalent Chromium by Skin in Man," **Arch. Toxicol., 47** (1981) 47.
[15] N. B. Pederson, "The Effects of Chromium on the Skin," in "Biological and Environmental Aspects of Chromium," S. Langård, Ed., Elsevier: Amsterdam, 1982, p. 249.
[16] L. L. Hopkins and K. Schwarz, "Chromium(III) Binding to Serum Proteins, Specifically Siderophilin," **Biochem.**

Biophys. Acta, **90** (1964) 484.
[17] R. Jett, J. O. Pierce and K. M. Stemberg, "Toxicity of Alloys of Ferrochromium," **Arch. Environ. Health, 17** (1968) 29.
[18] A. Yamamoto, O. Wada and T. Ono, "A Low-Molecular Weight, Chromium-Binding Substance in Mammals," **Toxicol. Appl. Pharmacol.**, **59** (1981) 515.
[19] S. Langård, N. Gundersen, D. L. Tsalev and B. Glyseth, "Whole Blood Chromium Level and Chromium Excretion in the Rat after Zinc Chromate Inhalation," **Acta Pharmacol. Toxicol.**, **42** (1978) 142.
[20] A. Mutti, A. Cavatorta, L. Borghi, M. Canali, C. Giaroli and I. Franchini, "Distribution and Urinary Excretion of Chromium. Studies on Rats after Administration of Single and Repeated Doses of Potassium Dichromate," **Med. Lavoro, 70** (1979) 171.
[21] I. Franchini, A. Mutti, F. Gardini and A. Borghetti, "Excretion et Clearance Renale du Chrome par Rapport au Degre et a la Duree de l'Exposition," in **"Rein et Toxique,"** L. Roche and J. Traeger, Eds., Masson: Paris, 1975, p. 271.
[22] H. A. Schroeder, J. J. Balassa and J. H. Tripton, "Abnormal Traces of Metals in Man: Chromium," **J. Chron. Dis., 15** (1962) 941.
[23] A. M. Baetjer, C. Damron and V. Budacz, "The Distribution and Retention of Chromium in Men and Animals," **Arch. Ind. Health, 20** (1959) 136.
[24] H. Teraoka, "Distribution of 24 Elements in the Internal Organs of Normal Males and the Metallic Workers in Japan," **Arch. Environ. Health, 36** (1981) 155.
[25] R. D. Mackenzie, R. R. Byerrum, C. F. Decker, C. A. Hopper and R. F. Langham, "Chronic Toxicity Studies," **Arch. Ind. Health, 18** (1958) 232.
[26] C. Onkelinx, "Compartment Analysis of Metabolism of Chromium(III) in Rats at Various Ages," **Am. J. Physiol., 232 E** (1977) 478.
[27] A. Mutti, A. Cavatorta, C. Pedroni, A. Borghi, C. Giaroli and I. Franchini, "The Role of Chromium Accumulation in the Relationship between Airborne and Urinary Chromium," **Int. Arch. Occup. Environ. Health, 43** (1979) 123.
[28] W. Mertz, "The Essential Trace Elements," **Science, 213** (1981) 1332.
[29] J. B. Meyer, "Acute Pulmonary Complications Following Inhalation of Chromic Acid Mist: Preliminary Observation of two Patients who Inhaled Massive Amounts of Chromic Acid," **Arch. Ind. Hyg., 2** (1950) 742.

[30] J. Oleffe, "Presence du Chrome dans l'Environnement de Travail. Ses Repercussions dans la Reparation es Dermatoses," **Arch. Belge. Derm. Siphil.**, **27** (1971) 159.

[31] M. Cavazzani and A. Viola, "Aspetti Clinici della Patologia da Cromo," **Med. Lavoro**, **61** (1970) 168.

[32] P. Lerza, "Contributo allo Studio delle Alterazioni dell'Apparato Respiratorio nel Cromismo Professionale," **Folia Medica**, **60** (1957) 100.

[33] E. Capodaglio, G. Pezzagno, A. Salvadeo and G. Catenacci, "Le Alterazioni della Funzione Respiratoria in Cromatori," **Lav. Umano**, **16** (1964) 57.

[34] A. Reggiani, M. Lotti, E. De Rosa and B. Saia, "Alterazioni Funzionali Respiratorie in Soggetti Esposti al Cromo. Nota I. Alterazioni Spirografiche," **Lav. Umano**, **25** (1973) 23.

[35] A. Reggiani, M. Lotti, E. De Rosa and B. Saia, "Alterazioni Funzionali Respiratorie in Soggetti Esposti al Cromo. Nota II. Alterazioni del Transfer del CO in Steady State," **Lav. Umano**, **25** (1973) 56.

[36] P. Bovet, M. Lob and M. Grandjean, "Spirometric Alterations in Workers in the Chromium Plating Industry," **Int. Arch. Occup. Environ. Health**, **40** (1977) 25.

[37] A. Cavatorta, M. Falzoi, A. Mutti, C. Frigeri and C. Pedroni, "Alterazioni Funzionali Respiratorie in Saldatori Professionalmente Esposti a Cromo," **L'Ateneo Parmense, Acta Biomedica**, **51** (1980) 299.

[38] A Cavatorta, A. Mutti, G. Frigeri, M. Falzoi, F. Cigala and I. Franchini, "Esposizione a Cromo ed Alterazioni Funzinali Respiratorie in Lavoratori dell'Industria Galvanica," **L'Ateneo Parmense, Acta Biomedica**, **51** (1980) 289.

[39] S. Langård, "Chromium," in **"Metals in the Environment,"** H. A. Waldron, Ed., Academic Press: London, 1980, p. 111.

[40] S. Tara, G. Delplace and A. Cavigneau, "Lesions Gastriques Chez les Chromeurs," **Arch. Mal. Prof.**, **12** (1951) 578.

[41] H. Wohlenberg and J. Lenhard, "Die Chrom-Enteropathie," **Deutsch. Med. Wochenschr.**, **22** (1970) 1224.

[42] I. Franchini, F. Magnani and A. Mutti, "Mortality Experience among Chromeplating Workers," **Scand. J. Work, Environ. Health**, **9** (1983) 247.

[43] B. Gylseth, N. Gundersen and S. Langård, "Evaluation of Chromium Exposure Based on a Simplified Method for Urinary Chromium Determination," **Scand. J. Work, Environ. Health**, **3** (1977) 28.

[44] S. Tola, J. Kilpio, M. Virtamo and K. Haapa, "Urinary Chromium as an Indicator of the Exposure of Welders to

Chromium," **Scand. J. Work, Environ. Health,** 3 (1977) 192.
[45] A. Borghetti, A Mutti, A. Cavatorta, F. Cigala, M. Falzoi and I. Franchini, "Indices Renaux d'Exposition Aiguë et d'Impregnation Chronique par le Chrome," **Med. Lavoro, 68** (1977) 355.
[46] P. L. Källiomäki, E. Harkonen, V. Vaaranen, K. Källiomäki and A. Sittionieni, "Lung-Retained Contaminants: Urinary Chromium and Nickel among Stainless Steel Welders," **Int. Arch. Occup. Environ. Health, 49** (1981) 67.
[47] I. Franchini, A. Mutti, A. Cavatorta, A. Cosi, A. Corradi, G. Olivetti and A. Borghetti, "Remarks on an Experimental and Epidemiological Investigation," in **"Toxic Nephropathies,"** L. Migone, Ed., Karger: Basel, 1978, p. 98.
[48] A. Tossavainen, M. Nurminen, P. Mutanen and S. Tola, "Application of Mathematical Modelling for Assessing the Biological Half-Times of Chromium and Nickel in Field Studies," **Br. J. Ind. Med., 37** (1980) 285.
[49] B. Sjogren, L. Hedstrom and U. Ulfvarson, "Urine Chromium as an Estimator of Air Exposure to Stainless Steel Welding Fumes," **Int. Arch. Occup. Environ. Health, 51** (1983) 347.
[50] E. Lindberg and O. Vesterberg, "Monitoring Exposure to Chromic Acid in Chrome-Plating by Measuring Chromium in Urine," **Scand. J. Work, Environ. Health, 9** (1983) 333.
[51] H. Welinder, M. Littorin, B. Gullberg and S. Skerfving, "Elimination of Chromium in Urine after Stainless Steel Welding," **Scand. J. Work, Environ. Health, 9** (1983) 397.
[52] M. Kiilunen, H. Kivisto, P. Ala-Laurila, A. Tossavainen and A. Aitio, "Exceptional Pharmacokinetics of Trivalent Chromium during Occupational Exposure to Chromium Lignosulfonate Dust," **Scand J. Work, Environ. Health, 9** (1983) 265.

XXIX

MICROBIAL RESISTANCE TO HEAVY METALS

John M. Wood and Hong-Kang Wang*

Gray Freshwater Biological Institute
University of Minnesota
Navarre, MN 55392 USA

*Beijing Agricultural University
People's Republic of China

During the 200 years following the beginning of industrialization, huge changes in the distribution of elements at the surface of the Earth have occurred. Microorganisms are adapting to these changes by evolving strategies to maintain low intracellular concentrations of toxic pollutants. An understanding of the biochemical basis for resistance to metal ion toxicity is emerging but is complicated by the different resistance mechanisms. Several strategies for resistance to metal ion toxicity have been identified:

- *The development of energy-driven efflux pumps that keep toxic element levels low in the interior of the cell. Such mechanisms have been described for Cd(II) and As(V).*
- *Oxidation (e.g., arsenite to arsenate) or reduction (e.g., Hg^{2+} to Hg^0), which can enzymatically and intracellularly convert a more toxic form of an element to a less toxic form.*
- *Biosynthesis of intracellular polymers that serve as traps for the removal of metal ions from solution. Such traps have been described for cadmium, calcium, nickel, and copper.*
- *The binding of metal ions to cell surfaces.*
- *The precipitation of insoluble metal complexes (e.g., metal sulfides and metal oxides) at cell surfaces.*
- *Biomethylation and transport through cell membranes by diffusion-controlled processes.*

Each of the mechanisms for resistance to toxicity requires inputs of cellular energy and as such represents a non-equilibrium component for the distribution of elements at the Earth's surface.

It is clear that living organisms have influenced steady-state levels of elements in the atmosphere, in the oceans, and at the surface of the Earth throughout geological time [1]. Both organic compounds and some inorganic complexes have been formed through biological activities that began about four billion years ago. Before examining some of the processes involved in element uptake by living cells, one should consider some of the selection principles involved in the chemistry of life. Several fundamental questions must be asked:

- Which elements are essential for the growth and cell division of microorganisms, plants, and animals?
- Why were these elements selected during the evolution of microorganisms more than four billion years ago?
- What is the role of the geosphere in determining the uptake of essential elements?
- What is the role of the biosphere in the selection of these elements?

Obviously, the uptake of elements and their use by living cells depend on the chemical and physical properties of each element. Of the more than 100 elements in the periodic table, 30 have been found to be required for microbial life, although not all of these elements are necessary for the growth and cell division of every microbial species. Therefore, in addition to the bulk elements - carbon, nitrogen, hydrogen, and oxygen - 26 others are required in intermediate to trace amounts. An overabundance of any of these elements can cause buildup to an intracellularly toxic level, which often results in death.

The reason for the selection of these 30 elements in the evolution of microorganisms appears to have been determined by two factors: the abundance of these elements in the Earth's crust, and their solubility in water under strictly anaerobic conditions. Table 1 lists the elements in order of their crustal abundances. Twenty-two of the 26 non-bulk elements required by microbes are also essential for higher organisms. Non-essential elements are generally of low abundance in the Earth's crust and, therefore, should not be effective in competing with essential elements in cells through their specific transport systems.

The effect of solubility properties of elements under strictly anaerobic conditions is best illustrated by the biological use of iron and the rejection of aluminum. Under anaerobic conditions, iron will be present in complexes of

Table 1. Crustal Abundances of Elements

O,[†] Si, Al,[*] Fe, Na, Ca, Mg, K, H, Mn, P, S, C, V, Cl, Cr, Zn, Ni, Cu, Co, N, Pb,[*] Sn,[*] Br, Be,[*] As, F, Mo, W, Tl,[*] I, Sb,[*] Cd,[*] Se;[†]

all the rest are less than 0.1 µg/g.

[†] Range of concentration 0 - 46.6%; Se - 0.1 µg/g.
[*] Element with no known biological function.

lower oxidation state, [Fe(II) salts], that are water-soluble and available for transport into primitive anaerobic bacteria. By contrast, aluminum will not be soluble under these conditions, unless the pH is extremely low. The same is true for the possible solubilization of lead, tin, and antimony.

Metal Ion Selection

Williams determined a number of chemical parameters that influence the uptake of metal ions by living cells [2]. These parameters are charge, ionic radius, preference of metals for coordination to certain organic ligands, coordination geometry and coordination numbers. Spin-pairing between metal ions for more stability and the degree of covalence of metal-ligand interactions, available concentrations of metal ions in the aqueous environment, kinetic controls pertinent to metal ion transport and binding, and chemical reactivities of metal ions in solution are also major factors.

The principles for the uptake of essential metal ions by primitive microorganisms can now be set forth. First of all, the availability of metal ions for transport into cells is restricted by their natural abundance and solubility in water. Solubility is profoundly influenced by pH, temperature, standard reduction potential (E^0), the presence of competing anions and cations, and the presence of surface-active substances such as particulates and macromolecules including proteins, humic acids, and clays.

Both pH and E° can vary widely from outside the living cell to inside that cell. For example, many essential transition metal ions such as iron, copper, cobalt, chromium, and nickel occur in higher oxidation states outside the cell but in lower oxidation states inside the cell. The pH can vary widely outside the cell, but is usually between 7.0 and 7.2 inside the cell. Such changes in oxidation state and pH affect the reactivity of chemical species. These changes determine, for instance, whether inorganic complexes function as nucleophiles or electrophiles. Most metal ions function as Lewis acids (electron acceptors), but depending on pH, oxidation state, and complexation, metal complexes can function also as bases. This is especially true for thiol-containing complexes.

Changes in oxidation state can profoundly affect steric factors in addition to coordination geometry and coordination number. Moreover, changes in pH and E° can have an effect on the charge of the inorganic complex, and such alterations in charge can greatly influence the transport of metals that are known to be involved in the most primitive metabolic pathways.

Outside the cell, chemical properties can be used to predict interactions between chemical species, often in quite complex situations. Pearson summarized the order for complexation of inorganic ions on the basis of his theory of "hard" and "soft" acids and bases [3]. Table 2 outlines ligand preferences for a number of essential and nonessential trace elements. It should be noted here that many of the more reactive metals are soft acids preferring coordination to bases found in living systems, such as thiolate groups, which are present in sulfur-containing amino acids. For example, the stability of Cu(II) compounds decreases with increasing hardness of the ligands establishing the following sequence: $S^{2-} > CN^- > CO_3^{2-} > OH^- > PO_4^{3-} > NH_3 > SO_4^{2-} > I^- > Br^- > Cl^- > F^-$.

However, the coordination number becomes a very important factor for the stability and the kinetics of binding. Some metal ions, once coordinate-covalently saturated in the interior of biological macromolecules, are difficult to replace by other competing metal ions. This is demonstrated clearly by the specific selection of metals in metalloproteins containing iron, copper, zinc, cobalt, nickel, manganese, or molybdenum.

Table 2. Classification of Hard and Soft Acceptors and Donors [3]

Hard	Acceptor Intermediate	Soft
H^+, Na^+, K^+, Be^{2+}, Mg^{2+}, Ca^{2+}, Mn^{2+}, Al^{3+}, Cr^{3+}, Co^{3+}, Fe^{3+}, As(III)	Fe^{2+}, Co^{2+}, Ni^{2+}, Cu^{2+}, Zn^{2+}, Pb^{2+}	Cu^+, Ag^+, Au^+, Tl^+, Hg^{2+}, CH_3Hg^+
Hard	Donor Intermediate	Soft
H_2O, OH^-, F^-, Cl^-, PO_4^{3-}, SO_4^{2-}, CO_3^{2-}, O^{2-}	Br^-, NO_2^-, SO_3^{2-}	HS^-, S^{2-}, RS^-, CN^-, SCN^-, CO, R_2S, RSH

Williams extended Pearson's ideas by recognizing factors important to the transport and partitioning of elements in cells [2, 4]. For example, the failure of a cell to transport and use sufficient amounts of an essential element could arise from low availability, excessive competition from other ions with similar chemical properties [e.g., competition between Co(II) and Ni(II), Ca(II) and Cd(II), phosphate and arsenate, Li(I) and Na(I)], inadequate synthesis of carrier molecules by the cell, excessive excretion of metal ions by the cell, or failure of the energy-driven uptake systems. Excessive uptake by elements can occur through the reversal of these factors. It is important to recognize, that living cells are not at equilibrium with the external environment. Therefore, a kinetic approach to metal ion transport, binding, toxicity, and resistance to toxicity is much more meaningful than a thermodynamic approach.

Transport in Aerobic Bacteria

The research of J. B. Neilands and his group at the University of California at Berkeley has led to a clearer understanding of the molecular biology of iron transport in certain bacteria [5, 6]. Ligands are excreted into the external environment to form extremely stable complexes with Fe(III). These ligands then bind to a specific receptor site at the cell surface. This mechanism provides a very elaborate yet efficient system for the uptake of iron, which involves the investment of considerable energy by the cell.

Figure 1 illustrates the parameters to be considered in

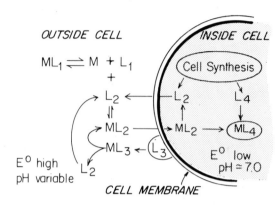

Figure 1. *A model for ion transport through cell membranes by means of ligands synthesized by the cell.*

both the transport and the control of intracellular concentrations of essential trace elements. Consider a general trace metal M and ligands L_1 through L_4 that can coordinate to M. The free external concentration of M is determined by environmental factors and so are the stability constants for ML_1 (L_1: a variety of ligands including, for example, humic acids). The internal concentrations are determined by selective carrier ligands, L_2; by ligands L_3 that can remove M from solution at the cell surface; by concentration gradients that are established between the exterior and the interior of the cell through binding to L_2 or by energy-coupled channels and pumps; and by internal concentrations of M. These internal concentrations can also be controlled through removal of M from solution by special biomolecules (L_4). An example of such a control mechanism is the removal of Cd through non-specific binding to the protein metallothionein.

It is apparent, that uptake is controlled by metabolic activity. Of crucial importance is the specific design of L_2 for transport of each specific metal ion. Siderophores, small chelating agents excreted by microbes for binding and transporting essential metal ions [2, 6], are examples of ligands designed for the transport of iron into bacteria and fungi. L_2 must be able to compete effectively with external ligands L_1; therefore, stability constants are of critical importance to element transport. Also, ML_2 must have an affinity for cell membranes by reacting at a specific membrane receptor site. This is well established for iron transport in unicellular organisms known as prokaryotes [5] that contain no nucleus. Therefore, there will be a K_D for each element in this membrane interaction. K_D is an equilibrium constant for the interaction of a metal complex with the surface of a microbial cell. Competition between similar elements, such as Co(II) and Ni(II), must be taken into account. The kinetics for interactions of ML_2 outside the cell, at the membrane, and inside the cell are very important in determining intracellular concentrations of M.

It is clear from these concepts that the accumulation of a trace metal by a cell is not at equilibrium, because metabolic activity is responsible for the synthesis of L_2, L_3, and L_4 as well as for energy-coupled channels, pH gradients, and redox potential differences. Clearly, uptake requires the investment of cellular energy. Therefore the thermodynamics of trace-metal ion transport depends on the following parameters:

- log K_{aq}: stability constants for ligand-metal ion interactions in aqueous solution;
- ΔpH: difference between internal and external pH;
- ΔE: difference between internal and external redox potentials;
- free [M]: free metal ion concentraion outside the cell;
- free $[L_1]$, $[L_2]$, $[L_3]$, $[L_4]$: concentration of free ligands inside the cell, outside the cell, and in the cell membranes.

Metal and ligand concentrations are expressed in molarities or molalities. The problem becomes even more complicated if one considers these principles for uptake in terms of kinetics. For example, one must understand the rates of input from the environment, the rates for transport through cell membranes, the rates for cellular exclusion or inclusion on cellular traps, and the rates of exit from the cell.

Outside the cell, the availability of a trace metal ion is determined by its abundance and solubility in aqueous solution. Inside the cell, redox potentials and pH gradients are determined by different environmental situations. High pH and high E^0 favor higher oxidation states for trace metals, and it follows that low solublity and availability result. Transfer through membranes requires a combination with a carrier or the presence of special channels in membranes, such as Ca channels. Carrier molecules can be specific small molecules or proteins. Nevertheless, there is still a great lack of understanding of the selectivity principles involved in trace metal ion transport and of the mechanisms that energize inward transport.

Essentiality vs. Toxicity

The biochemical basis for resistance to toxicity is complicated by the great variety of reactions at the molecular and cellular levels, even in closely related organisms and tissues. Several pathways for resistance to the effects of toxicants have been identified. Williams pointed out, that metal ion interactions in biology can be divided into three classes: ions in fast, intermediary, and slow exchange with biological ligands [4].

Examples of elements in fast exchange are the alkali metals sodium and potassium, the alkaline earth metals calcium and magnesium, and H^+. Elements that sometimes can be in intermediary exchange are Fe^{2+} and Mn^{2+}, whereas examples of elements in slow exchange are those associated with active sites of metalloenzymes (Fe^{3+}, Zn^{2+}, Ni^{2+}, Cu^{2+}). Metal-to-metal interactions and covalency predominate in the slow-exchange metals, and this provides the basis for stability. However, competition for these metal ions in fast exchange is often severe. Living cells have membranes that act as initial barriers to metal ion uptake. In prokaryotes, the external cell membrane represents the only barrier, but in eukaryotes (nucleus-containing cells of advanced organisms) there are many membranous organelles that can partition metal ions by a variety of mechanisms.

A comparative study of the metal ion resistance of blue-green algae (prokaryotes) and green algae (eukaryotes) demonstrates how important membranes are to metal ion uptake and toxicity. The green algae are much more resistant to high concentrations of toxic metal ions such as Cu(II) and Ni(II) than are the blue-green algae [7, 8]. The external cell membranes of prokaryotes carefully select ions in fast exchange as exemplified by the rejection of sodium and calcium and selection of potassium and magnesium [9]. In eukaryotes, spatial partitioning occurs even for metals in slow exchange, because metal-binding macromolecules can be partitioned in different organelles and in tissues of different cell lines. For instance, zinc, cadmium, copper, and mercury are bound by metallothionein in the kidney cortex. The regulation of metal ion transport is presented in a general way in Figure 2, which attempts to show how toxicity becomes evident once the cell buffering capacity for essential metal ions is exceeded.

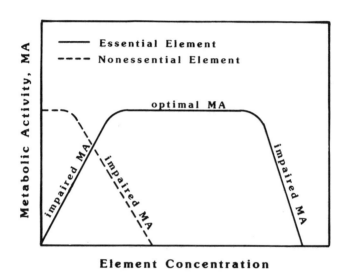

Figure 2. The dependence of metabolic activity on the concentration of essential and nonessential elements.

Toxic effects become evident at much lower concentrations for nonessential than for essential metals. In the natural environment, nonessential metals seldom reach concentrations in excess of 1 µg/g (Table 1). Examples of exceptional circumstances in which higher metal concentrations occur in nature include active volcanic regions such as deep sea vents, hot springs, and volcanic lakes.

However, during the 200 years since the beginning of industrialization, huge changes in the distribution and solubilization of metal ions at the Earth's surface have occurred. These changes are attributable to the use of new chemistry in industrial society. Activities ranging from mining to modern agriculture pose risks to health and welfare and challenge the regulatory systems for the transport of metal ions through cell membranes that took billions of years to evolve. Microorganisms that have short generation times and consequently increased evolution rates have adapted to deal with high concentrations of metal ions.

Microbial Resistance to Metal Ions

Over the past two centuries, vast quantities of energy have been used to extract and process elements at the surface of the Earth. Microorganisms adapted to these changes by evolving strategies to maintain low intracellular concentrations of toxic pollutants. This adaptation to resist toxic substances has arisen in two distinct patterns. Some microorganisms inherited the ability to resist high concentrations of toxic elements through their evolution under extreme environmental conditions [10]. Other microorganisms acquired a transferred resistance to the polluted environment relatively recently, and certainly since the start of the Industrial Revolution. Such bacteria achieved resistance through the acquisition of extrachromosomal DNA molecules (plasmids) [11].

Whether primitive or recent, each of the mechanisms for resistance to toxicity relies on inputs of cellular energy and thus represents a nonequilibrium component for the distribution of elements at the Earth's surface. Therefore, kinetic aspects of these processes are especially important to the field of environmental health and toxicity. Rates of uptake of all elements by cells determine whether they live or die. However, it should be recognized that resistance mechanisms in higher as well as lower organisms can be quite variable.

Biomethylation

Recent analyses of oil shale show, that primitive organisms actively synthesized organometals and organometalloids. Therefore, biomethylation must have given certain microorganisms selective advantages for the elimination of heavy metals such as mercury and tin and of metalloids such as arsenic and selenium. The synthesis of less polar organometallic compounds from polar inorganic ions has certain advantages for cellular elimination by diffusion-controlled processes [12, 13].

The microbial synthesis of organometallic compounds from inorganic precursors in the terrestrial environment and in the sea is well understood. Mechanisms for B_{12}-dependent synthesis of metal alkyls (requiring the presence of vitamin B_{12}) were discovered for the metals Hg, Pb, Tl, Pd, Pt, Au, Sn, and Cr, and for the metalloids As and Se [14-18]. Also, the synthesis of organoarsenic compounds was shown to occur by a mechanism involving S-adenosylmethionine as the methylating coenzyme [19].

To date, two different mechansims were identified for methyl transfer from methyl B_{12} to heavy metals: electrophilic attack by the metals on the Co-C bond of methyl B_{12}, and methyl-radical transfer to an ion pair between the attacking metal ion and the corrin-macrocycle (the part of the vitamin B_{12} molecule that sequesters the central cobalt atom). Metal ions that displace the methyl group by electrophilic attack are Hg(II), Pb(IV), Tl(III), and Pd(II). Examples of free radical transfer are the reactions with Pt(II)/Pt(IV), Sn(II), Cr(II), and Au(III). The radical mechanism provides experimental support for Kochi's ideas on charge transfer complex formation and electron transfer [20].

The ecological significance of B_{12}-dependent biomethylation is best illustrated by B_{12}-dependent and B_{12}-independent strains of *Clostridium cochlearum*. The B_{12}-dependent strain is capable of methylating Hg(II) salts to CH_3Hg^+, whereas the B_{12}-independent strain is capable of catalyzing this reaction. Both strains transport Hg(II) into cells at the same rate, but the B_{12}-independent strain is inhibited by at least a 40-fold lower concentration of Hg(II) than the B_{12}-dependent strain. This result clearly demonstrates, that the dependent strain of *C. cochlearum* uses biomethylation as a mechanism for detoxification. Biomethylation gives the organism a clear advantage in mercury-contaminated systems. This biomethylation capability was shown to be plasmid mediated [21].

Once methylmercury is released from the microbial system, it enters food chains as a consequence of its rapid diffusion rate. In the estuarine environment, the reduction of sulfate by *Desulfovibrio* species to produce hydrogen sulfide is important in reducing CH_3Hg^+ concentrations by S^{2-}-catalyzed disproportionation to volatile $(CH_3)_2Hg$ and insoluble HgS. It should be mentioned that there is overwhelming evidence to support the notion that membrane transport of methylmercury is diffusion controlled.

Fluorescence techniques and high-resolution nuclear magnetic resonance spectroscopy show that diffusion is the key to CH_3Hg^+ uptake [13]. Also, a field study of the uptake of CH_3Hg^+ by Mediterranean tuna perfectly fits the diffusion model for biota in tuna food chains [22]. An updated view of the mercury cycle is presented in Figure 3 [23].

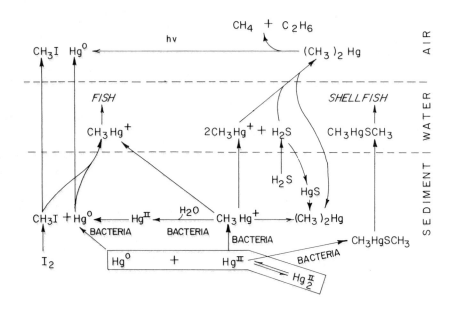

Figure 3. *The mercury cycle.*

Some details of the individual reactions of this mercury cycle will illustrate the molecular biology and biochemistry of the equilibria [23] shown in equation 1. Microorganisms

$$CH_3Hg^+ \rightleftharpoons Hg^{2+} \rightleftharpoons Hg^0 \qquad (1)$$

were isolated that catalyze these reactions both forward to Hg^0 and back to CH_3Hg^+. The enzymes carrying out the forward reactions are coded by DNA on bacterial plasmids and transposons and not by normal chromosomal genes [24-26]. Transposons are small segments of DNA transferable between closely related microorganisms. Therefore, it is not too surprising, that mercuric and organomercurial strains of bacteria were isolated from a variety of ecosystems such as soil, water, and marine sediments [27-32]. The enzymology of methylmercury hydrolysis and mercuric ion reduction is now understood in some detail [24-26]. In fact, the sequence of the active site of mercuric ion reductase was found to be identical to that of glutathione reductase [33, 34].

Clearly, the reduction of Hg^{2+} to Hg^0, which is volatile, represents a very effective detoxification mechanism. Much less is known about the reverse reaction, the oxidation of Hg^0 to Hg^{2+}. However, an enzyme critical to the oxygen cycle (catalase) will carry out this reaction. Microbial methylation of the mercuric ion is also widespread [32, 35, 36]. Biomethylation has been shown to occur in sediments and in human feces [36].

The important role played by sulfide in the biological cycle for mercury is evident from Figure 3. Hydrogen sulfide is extremely effective at volatilization and precipitation of mercury through disproportionation chemistry in the aqueous environment. The same is true for the volatilization and precipitation of lead compounds. This chemistry is important to the mobilization of metals from the aquatic environment into the atmosphere. Such reactions occur only in polluted lakes, rivers, coastal zones, estuaries, and salt marshes where *Desulfovibrio* species have access to sulfate in anaerobic ecosystems (Fig. 3). The disproportionation of organometals by hydrogen sulfide is outlined by equations 2 and 3. Once in the

$$2CH_3Hg^+ + H_2S \longrightarrow (CH_3)_2Hg + HgS \quad (2)$$
$$2(CH_3)_3Pb^+ + H_2S \longrightarrow (CH_3)_4Pb + (CH_3)_2PbS \quad (3)$$

atmosphere, volatile organometallics such as dimethylmercury are unstable, because metal-carbon bonds are susceptible to homolytic cleavage by light.

Intracellular Traps for Ion Removal

The biosynthesis of intracellular traps for the removal of metal ions from solution represents a temporary measure adopted by cells to prevent metals from reaching toxic levels. This temporary measure precedes mechanisms for the expulsion of these metals from the cell by vacuoles. However, such temporary traps can be very effective. One example is the biosynthesis of metallothionein and the removal of cadmium [2] or copper [37] by this sulfhydryl-containing protein.

This strategy adopted for the biosynthesis of intracellular traps fits quite closely the predicted partitioning of elements in organic or inorganic matrices. For instance, Na, K, Mg, Ca, Al, P, Si, and B prefer to react with an oxygen-donor matrix, but Cu, Zn, Fe, Ni, Co, Mo, Cd, and Hg perfer a nitrogen- and sulfur-donor matrix. A recent example that we discovered in our laboratory came about through the selection of nickel-tolerant mutants of the cyanobacterium *Synechococcus*. Mutants that would tolerate up to 20×10^{-5} M nickel sulfate were selected [38] and were found to synthesize large quantities of an intracellular polymer [39]. Nickel analysis and electron microscopy of sections of these nickel mutants showed that this polymer effectively removes nickel from solution by providing an intracellular mechanism to prevent nickel toxicity. Figure 4 shows an electron micrograph of a nickel mutant of *Synechococcus* in which large granules are present.

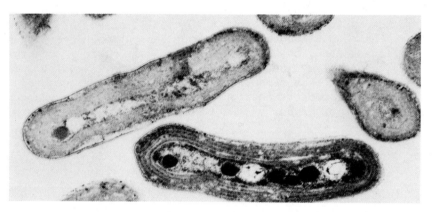

Figure 4. Synechococcus, *a nickel-tolerant microorganism grown in 20×10^{-6} M nickel sulfate, with dense intracellular granules (magnification 21,500).*

We failed to isolate similar mutants of *Synechococcus* showing tolerance to 10^{-5} M copper sulfate. However, we found that our nickel-tolerant mutants were also resistant to copper. This was presumably caused by the stronger coordination of copper to the granules. Mutants with intracellular trapping mechanisms tend to bioconcentrate the toxic metal intracellularly to approximately 200 times the external concentration. However, whereas this strategy works quite well for some organisms, it does not compare favorably with the precipitation or binding of metals outside the cell. Intracellular concentrations of metal ions can be controlled by deposition on a solid surface, as in the crystallization of calcium salts in blood platelets or the removal of copper and nickel by intracellular granules.

The concentration of free metal ions can also be controlled by the biosynthesis of ligands in the form of small molecules with high stability constants. The removal of iron by siderophores fits this category. Energy may be expended by the cell to pump the metal ion out of the cell (for example, by the sodium-potassium ATPase pump). The cell may synthesize ligands that bind metals strongly at the cell surface or use the activities of surface-bound enzymes to precipitate metals extracellularly.

Efflux Mechanisms

The resistance of microbial cells to the toxic elements arsenic, antimony, and cadmium was shown to be caused by cellular exclusion mechanisms. In *Staphylococcus aureus* and in *Escherichia coli*, resistance to arsenate, arsenite, and antimony(III) salts is induced by an operon-like system [40]. An operon is a DNA region that codes for several enzymes in a reaction pathway. Each of the three ions will induce resistance to ions of the other two elements. Arsenate is transported through cell membranes in competition with the transport of phosphate ions, but arsenite is transported through a phosphate-independent mechanism [41]. The genetics of arsenate efflux from resistant cells is quite well understood at the genetic level and is the subject of two recent reviews [11, 24]. The toxicity of arsenic and antimony salts resides in the reactivity of these compounds with sulfhydryl groups [42]. Sulfhydryl-containing enzymes are inactivated by arsenicals and antimonials.

In the case of cadmium, resistance was shown to be mediated by a plasmid. Resistant cells of *S. aureus* have a very efficient chemosmotic efflux system specific for Cd^{2+} ions [24]. Two separate plasmid genes are responsible for cadmium resistance. These genes also prevent zinc toxicity. The cad-A gene codes for proteins that are involved in the efflux of Cd^{2+} on the inside and for two H^+ from the outside of the cell. The cad-B gene is believed to be responsible for the synthesis of a cadmium-binding protein, which may be similar to metallothionein [24, 44].

Arsenate, cadmium ions, and antimony ions are accidentally taken up by normal transport systems designed for the transport of ions such as phosphate and zinc. Therefore, the uptake of these elements is nonselective. However, resistance to toxicity by these microorganisms resides in their ability to remove these elements selectively by energy-driven efflux systems.

Surface Binding of Metal Ions

Microorganisms, including algae, synthesize extracellular ligands that complex metals and prevent their cellular uptake. The research groups of Francois Morel and Pamela Stokes at MIT and the University of Toronto, respectively, carried out extensive work on the complexation of copper with a variety of extracellular substances [44-49]. Considerable literature also exists on the toxicity of nickel and copper to algae [50-57]. Cyanobacteria, brown and green algae, and yeast cels all bioconcentrate nickel [57-59].

Recently, we used the isotope nickel-63 in a study of nickel binding by seven different strains of nickel-tolerant algae [60]. The cyanobacteria were found to be more sensitive to nickel toxicity than the green algae, which points out differences in transport mechanisms between prokaryotes and eukaryotes. Both, cyanobacteria and green algae, could concentrate nickel primarily at the cell surface to 3000-times the concentration in the culture medium.

Both nickel binding and nickel toxicity were shown to be very pH dependent. The optimum pH for binding was between 8 and 8.5. Binding was shown to be rather specific with charge, ionic radius, and coordination geometry being the predominant factors. The only significant competing cation for Ni(II) binding was Co(II). The orientation of ligands at the cell surface must be important, because only surface-active substances such as humic acids could compete effectively for nickel binding [60]. Table 3 shows the ability of these seven strains of nickel-tolerant algae to bioconcentrate nickel simply through ion exchange.

Table 3. *The Dependence of the Ni-63 Concentration Factor on pH for Different Strains of Algae**

Algae	Concentration Factor** at pH					
	4	5	6	7	8	9
Scenedesums ATCC 11460	†		7.9×10^2	1.8×10^3	2.2×10^3	2.0×10^3
Scenedesums B-4	†	90	2.8×10^2	6.6×10^2	1.0×10^3	4.0×10^2
Synechococcus ATCC 17146	†	†	4.2×10^2	3.0×10^3	3.3×10^3	3.1×10^3
Synechococcus Nic	†	†	2.2×10^2	4.4×10^2	5.5×10^2	†
Oscillatoria UTEX 1270	†	29	9.5×10^2	1.1×10^3	1.1×10^1	†
Chlamydomonas UTEX 89	27	38	5.4×10^2			†
Euglena UTEX 753	†	†	†	1.7×10^1	6.9×10^2	†

* 0.34 µM Ni(II)-63 (0.02 µg/mL) incubated for 6 hours at 20° under light.
** Concentration factor = (µg of Ni-63 removed per gram of algae)/(µg of Ni-63 in culture medium).
† No significant uptake at the 95 percent confidence level.

Precipitation at the Cell Surface

The precipitation of insoluble metal complexes occurs through the activities of membrane-associated sulfate reductases [60] or through the biosynthesis of oxidizing agents such as oxygen or hydrogen peroxide [23]. The reduction of sulfate to sulfide and the diffusion of O_2 and H_2O_2 through the cell membrane provide highly reactive agents by which metals can be complexed and precipitated. This process depends on the metabolic activity of the cell and is closely tied to heavy-metal resistance.

Several bacteria were found that precipitate silver as Ag_2S at the cell surface. Also, certain fungi are very efficient at recovering uranium [61]. However, the most interesting organisms are certain strains of green algae that grow under acidic conditions and at high temperatures [10]. *Cyanidium caldarium* is such an organism; it has the remarkable ability to grow in media containing either 1 N H_2SO_4 or 1 M HCl.

A strain of *C. caldarium* was adapted to grow at 45° in acid mine water from a copper-nickel mine. This thermophilic alga grows extremely well in sulfuric acid over the pH range from 0 to 4 and removes toxic metal ions from solution by their precipitation as metal sulfides at the cell surface. Batch cultures of *C. caldarium* were grown that remove very high concentrations of metals from solution - 68 percent of the iron, 50 percent of the copper, 41 percent of the nickel, 53 percent of the aluminum, and 76 percent of the chromium. Because *C. caldarium* is both thermophilic and acidophilic, it is easily grown free of other contaminating autotrophs without sterilization of the acid mine water medium. This alga has great promise as a biological agent for the recovery of metals and for cleaning metal-polluted wastewaters.

In the early 1950's. Allen [62] was the first to report the isolation of *C. caldarium* from acidic hot springs in California. This was subsequently confirmed by Seckbach et al. [63]. Since that time, substantial research has been done on the microbiology, life cycle, structure, and ecology of this unusual alga [10]. This organism has an optimal growth temperature of 45° and grows very well in temperatures ranging from 35° to 55°. *C. caldarium* adapts to fit a whole range of temperatures and pH without any apparent selection of specialized strains for specific ecosystems. Cultures grow faster when provided with 5 percent carbon dioxide, and excellent cell yields are obtained under strictly autotrophic conditions [10]. However, more than double the cell yield is obtained when cultures are provided with one percent of a soluble carbon source such as glucose. Therefore, *C. caldarium* grows heterotrophically in the dark on a variety of organic substrates, including monosaccharides, disaccharides, mannitol, glycerol, ethanol, succinate, glutamate, lactate, and acetate. The organism is easily maintained on acid media with viabilities of stock cultures lasting longer than one year [62].

A culture of *C. caldarium* isolated from the Waimangu Caldron Outlet, North Island, New Zealand, was slowly adapted to growth in acid mine water by adding ten percent incremental increases of mine water to a culture grown under the conditions described by Allen [62]. Using this procedure, one can select strains of *C. caldarium* that grow very well on a culture medium consisting of unfiltered acid mine water provided with five percent carbon dioxide.

Table 4 presents data on the elemental composition of acid mine water at pH 2.1 and also shows the removal

Table 4. *Precipitation of Metal Ions by Cyanidium caldarium from Acid Mine Water**

Metal	Concentration, mg/L		Percent of Element Removed from Solution	Conc. µg/mL Culture Supernatant 5% CO_2	Percent of Element Removed from Solution
	Acid Mine Water, pH 2.1	Culture Supernatant 5% CO_2 1%$(NH_4)_2SO_4$			
Ca	342	219	36	277	19
Mg	456	228	50	342	25
Fe	632	205	68	386	39
Cu	119	60.6	50	95	20
Al	329	155	53	245	25
Cr	1.31	0.31	76	0.38	71
Na	28.4	13.2	54	20.9	26
Ni	4.32	2.55	41	2.69	28
P	26.0	13.5	50	19.1	29

*Cyanidium caldarium (one liter inoculum) grown for one week in eight liters acid mine water. The solutions were analyzed by plasma emission spectrometry.

efficiencies for stationary-phase cultures of *C. caldarium* in the presence and absence of glucose plus ammonium sulfate. Figure 5 shows a section of *C. caldarium* examined under the electron microscope. Microcrystals of metal sulfides are found to adhere to the external cell membrane. Cells contain approximately 20 percent metal on the basis of dry weight.

Clearly, toxic metals such as copper, nickel, and chromium are prevented from entering the cell through an extracellular precipitation mechanism. This result suggested to us that *C. caldarium* possesses a membrane-associated sulfate reductase system. Heterotrophic cultures such as this alga's that are allowed to attain anaerobic conditions in the dark produce hydrogen sulfide gas quite efficiently. Therefore, sulfide precipitation of metals can be regarded as a cellular detoxification mechanism.

Of special interest are the removal efficiencies of chromium and nickel, which are present in very low concentrations in the acid mine water. These metals are potential carcinogens and their selective removal from wastewaters is highly desirable. Organisms such as *C. caldarium* may well be effective in treating polluted waters

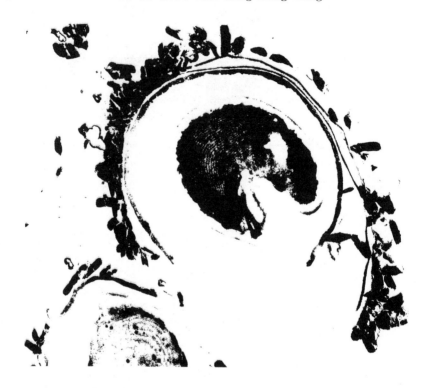

Figure 5. *Cyanidium caldarium* - *grown in acid mine water with five percent carbon dioxide at pH 2.1 and 45° - with precipitates of metal compounds on the cell surface (magnification 19,900).*

so that effluents can meet federal standards for such toxic elements.

Genetic Engineering

The examples previously given suggest that microorganisms can be selected from extreme environments or can be manipulated genetically to recover toxic elements from industrial wastewaters. This is certainly true for those organisms with resistance mechanisms coded by the cell genome. However, a word of caution should be said with regard to the uses of plasmids to engineer special organisms genetically for this purpose. It is more than likely that the toxic conditions found in industrial wastewaters will cause plasmid losses from organisms, even though they can be shown to function extremely well in the laboratory. Also, at ambient pH and temperature, organisms introduced into industrial wastewaters must compete for nutrients with those organisms already established in this special environment. The use of biotechnology looks much more promising for those organisms - such as *C. caldarium* - that grow under extreme conditions of pH and temperature and that use light as the major energy source.

Acknowledgement. We wish to acknowledge the research of postgraduate colleagues Y.-T. Fanchiang and F. K. Gleason. Francis Engle and Steve Michurski also are thanked for their technical assistance. We wish to dedicate this paper to R. J. P. Williams, F.R.S., for his encouragement, insight, and contributions to our knowledge of this subject. Some of the research reported in this paper was supported by grants from the National Institutes of Health AM 18101 and by a grant from Atlantic Richfield Company.

References

[1] J. E. Lovelock, **"Gaia. A New Look at Life on Earth,"** Oxford University Press: Oxford, U.K., 1979.
[2] R. J. P. Williams, "Physicochemical Aspects of Inorganic Element Transfer Through Membranes," **Phil. Trans. Roy. Soc. London, Ser. B, 294** (1981) 57.
[3] R. G. Pearson, "Hard and Soft Acids and Bases, HSAB, Part II," **J. Chem. Educ., 45** (1968) 643.
[4] R. J. P. Williams, "Structural Aspects of Metal Toxicity ," in **"Changing Metal Cycles and Human Health,"** J. O. Nriagu, Ed., Dahlem Workshop, March 20-25, 1983, Berlin, Springer Verlag: Berlin, 1984, p. 251.
[5] J. B. Neilands, "Continuous Synthesis of Iron-Regulated Membrane-Proteins during the Division Cycle of *Escherichia coli* K12," **Chem. Scripta, 21** (1983) 123.
[6] J. B. Neilands, "Biomedical and Environmental Significance of Siderophores," in **"Trace Metals in Health and Disease,"** N. Kharasch, Ed., Raven Press: New York, N.Y., 1979, p. 27.
[7] J. W. Wood and H.-K. Wong, "Microbial Resistance to Heavy Metals," **Environ. Sci. Technol., 17** (1983) 582A.
[8] H.-K. Wong and J. M. Wood, "Bioaccumulation of Nickel by Algae," **Environ. Sci. Tech., 18** (1984) 106.
[9] R. J. P. Williams, "Inorganic Elements in Biological Space and Time," **Pure Appl. Chem., 55** (1983) 1089.
[10] T. D. Brock, **"Thermophilic Microorganisms and Life at**

High Temperatures," Springer Verlag: Heidelberg, Germany, 1978.
[11] S. Silver, "Bacterial Interactions with Mineral Cations and Anions: Good Ions and Bad," in **"Biomineralization and Biological Metal Accumulation,"** P. Westbroek and E. W. De Jong, Eds., D. Reidel Publishing Co.: Dordrecht, Holland, 1983, p. 439.
[12] J. M. Wood, "Biological Cycles for Elements in the Environment," **Naturwissenschaften, 62** (1975) 357.
[13] J. M. Wood, A. Cheh, L. J. Dizikes, W. P. Riley, S. Rakow and J. R. Lakowicz, "Mechanisms for the Biomethylation of Metals and Metalloids," **Fed. Proc., 37** (1978) 16.
[14] W. R. Ridley, L. J. Dizikes, A. Cheh and J. M. Wood, "Biomethylation of Toxic Elements in Environment," **Science, 197** (1977) 329.
[15] W. R. Ridley, L. J. Dizikes, A. Cheh and J. M. Wood, "Recent Studies on Biomethylation and Demethylation of Toxic Elements," **Environ. Health Perspect., 19** (1977) 43.
[16] J. M. Wood, "Biochemistry of Toxic Elements," **Quart. Rev. Biophys., 11** (1978) 467.
[17] J. M. Wood, Y.-T. Fanchiang and W. R. Ridley, "Kinetic and Mechanistic Studies on B_{12}-Dependent Methyl Transfer to Certain Toxic Metal Ions," **ACS Symp. Series, 82** (1978) 54.
[18] P. J. Craig and J. M. Wood, "Biological Methylation of Lead: An Assessment of the Present Position," in **"Environmental Lead,"** D. R. Lynam, L. E. Piatanida and J. F. Cole, Eds., Academic Press: New York, N.Y., 1981, p. 333.
[19] B. C. McBride, H. Meriless, W. R. Cullen and W. Pickett, "Anaerobic and Aerobic Alkylation of Arsenic," **ACS Symp. Ser., 82** (1978) 94.
[20] J. K. Kochi, "Mechanism for Alkyl Transfers in Organometals," **ACS Symp. Ser., 82** (1978) 205.
[21] H. S. Pan Hou and N. Imura, "Involvement of Mercury Methylation in Microbial Mercury Detoxification," **Arch. Microbiol., 131** (1982) 176.
[22] R. Buffoni, M. Bernhard and A. Penzoni, "Mercury in the Mediterranean Tuna. Why are their Levels Higher than in the Atlantic Tuna? A Model," **Thalassia Jugoslavica, 18** (1982) 231.
[23] J. M. Wood, "Selected Biochemical Reactions of Environmental Significance," **Chem. Scripta, 21** (1983) 155.
[24] S. Silver, "Bacterial Transformations of and Resistance to Heavy Metals," in **"Changing Metal Cycles and Human Health,"** J. O. Nriagu, Ed., Dahlem Workshop, March 20-

25, 1983, Berlin, Springer Verlag: Berlin, 1984, p. 199.
[25] W. J. Jackson and A. O. Summers, "Polypeptides Encoded by the mer-Opernon," **J. Bacteriol., 149** (1982) 479.
[26] N. N. Bhriain, S. Silver and T. J. Foster, "Tn5 Insertion Mutations in the Mercuric Ion Resistance Genes Derived from Plasmid R100," **J. Bacteriol., 155** (1983) 690.
[27] D. A. Friello and A. M. Chakraborty, "Transposable Mercury Resistance in *Pseudomonas putida* Plasmids and Transposons: Environmental Effects and Maintenance Mechanisms" **Proceedings, 4th Annual Symposium on the Scientific Basis of Medicine,** C. Stuttard and K. R. Rozee, Eds., Academic Press: New York, N.Y., 1980, p. 249.
[28] B. H. Olson, T. Barkay and R. R. Colwell, "Role of Plasmids in Mercury Transformation by Bacteria Isolated from the Aquatic Environment," **Appl. Environ. Microbiol., 38** (1979) 478.
[29] G. J. Olson, W. P. Iverson and F. E. Brinckman, "Volatilization of Mercury by *Thiobacillus ferrooxidans*," **Curr. Microbiol., 5** (1981) 115.
[30] A. J. Radford, J. Oliver, W. J. Kelly and D. C. Reaney, "Translocatable Resistance to Mercuric and Phenylmercuric Ions in Soil Bacteria," **J. Bacteriol., 147** (1981) 1110.
[31] J. F. Timoney, J. Port, J. Giles and J. Spanier, "Heavy-Metal and Antibiotic-Resistance in Bacterial Flora of Sediments of New-York Bight," **J. Appl. Environ. Microbiol., 36** (1978) 465.
[32] J. W. Vonk and A. K. Sjipesteijn, "Studies on the Methylation of Mercuric Chloride by Pure Cultures of Bacteria and Fungi," **Antonie van Leeuwenhoek, 39** (1973) 505.
[33] C. H. Williams, Jr., L. D. Arscott and G. E. Schultz, "Amino Acid Sequence Homology between Pig Heart Lipoamide Dehydrogenase and Human Erythrocyte Glutathione Reductase," **Proc. Natl. Acad. Sci., USA, 79** (1982) 2199.
[34] N. L. Brown, R. D. Pridmore and D. C. Fritzinger, Medical School, University of Bristol, Bristol, England, personal communication, 1983.
[35] M. K. Hamdy and O. R. Noyes, "Formation of Methyl Mercury by Bacteria," **Appl. Microbiol., 30** (1975) 424.
[36] B. C. McBride and T. L. Edwards, "Role of the Methanogenic Bacteria in the Alkylation of Arsenic and Mercury," in **"Biological Implications of Metals in the Environment,"** Proceedings, 15th Annual Hanford Life Sciences Symp., Sept. 29-Oct. 1, 1975, Richland,

Washington, **ERDA Symposium Series, 42,** 1977, p. 1; NTIS CONF-750929.
[37] K. Lerch and M. Beltranini, "Neurospora Copper Metallothionein-Molecular Structure and Biological Significance," **Chem. Scripta, 21** (1983) 109.
[38] F. K. Gleason, Gray Freshwater Biological Institute, University of Minnesota, Navarre, personal communication, 1983.
[39] R. D. Simon and P. Weathers, "Determination of Structure of Novel Polypeptide Containing Aspartic Acid and Arginine which is Found in Cyanobacteria," **Biochim. Biophys. Acta, 420** (1976) 165.
[40] S. Silver, K. Budd, K. M. Leahy, W. V. Shaw, D. Hammond, R. P. Novick, G. R. Willsky, M. H. Malamy and H. Rosenberg, "Inducible Plasmid-Determined Resistance to Arsenate, Arsenite, and Antimony(III) in Escherichia coli and *Staphylococcus aureus*," **J. Bacteriol., 146** (1981) 983.
[41] R. P. Novick, E. Murphy, T. J. Gryczan, E. Baron and I. Edellman, "Penicillinase Plasmids of *Staphylococcus aureus* - Restriction - Deletion Maps," **Plasmid, 2** (1979) 109.
[42] A. Albert, **"Selective Toxicity,"** 5th Ed., Chapman and Hall: London, U.K., 1973, p. 392.
[43] D. P. Highan, P. J. Sadler and M. D. Scawen, "Bacterial Cadmium-Binding Proteins," **Inorg. Chim. Acta, Bioinorg. Art. Lett., 79** (1983) 140.
[44] D. M. McKnight and F. M. M. Morel, "Release of Weak and Strong Copper-Complexing Agents by Algae," **Limnol. Oceanogr., 24** (1979) 823.
[45] K. C. Swallow, J. C. Westall, D. M. McKnight and F. M. M. Morel, "Potentiometric Determination of Copper Complexation by Phytoplankton Exudates," **Limnol. Oceanogr., 23** (1978) 538.
[46] N. M. L. Morel, J. G. Rueter and F. M. M. Morel, "Copper Toxicity to *Skeletonema costatum (Bacillariophyceae)*," **J. Phycol., 14** (1978) 43.
[47] P. M. Stokes and S. I. Dreier, "Copper Requirement of a Copper-Tolerant Isolate of *Scenedesmus* and the Effect of Copper Depletion on Tolerance," **Can. J. Bot., 59** (1981) 1817.
[48] P. M. Stokes, "Metal Tolerance Mechanisms in Algae," 13th Intern. Botanical Congress, August 21-28, 1981, Sidney, Australia, **Proc. Intern. Bot. Congr., 13** (1981) 84.
[49] G. Mierle and P. M. Stokes, "Heavy Metal Tolerance and Metal Accumulation by Planktonic Algae," in **"Trace Substances in Environmental Health,"** Proceedings, 10th

Annual Conference on Trace Substances in Environmental Health, June 8-10, 1976, University of Missouri, Columbia, Missouri, University Missouri: Columbia, Missouri, 1976, p. 113.

[50] D. F. Spencer, "Nickel and Aquatic Algae," in **"Nickel in the Environment,"** J. D. Nriagu, Ed., Wiley Interscience: New York, N.Y., 1980, p. 339.

[51] J. M. Hassett, J. H. Jennett and J. E. Smith, "Microplate Technique for Determining Accumulation of Metals by Algae," **Appl. Environ. Microbiol., 41** (1981) 1097.

[52] J. S. Fezy, D. F. Spencer and R. W. Greene, "Effect of Nickel on the Growth of the Freshwater Diatom *Navicula pelliculosa*," **Environ. Poll., 20** (1979) 131.

[53] G. W. Stratton and C. T. Corke, "Effect of Mercuric, Cadmium and Nickel Ion Combinations on a Blue-Green Algae," **Chemosphere, 8** (1979) 731.

[54] D. F. Spencer and R. W. Green, "Effects of Nickel on 7 Species of Fresh-Water Algae," **Environ. Poll. Ser. A, Ecol. Biol., 25** (1981) 241.

[55] P. T. S. Wong, Y. K. Chan and P. L. Luxon, "Toxicity of a Mixture of Metals on Freshwater Algae," **J. Fish Res. Board Can., 35** (1978) 479.

[56] P. Foster, "Concentrations and Concentration Factors of Heavy-Metals in Brown Algae," **Environ. Poll., 10** (1976) 45.

[57] R. Fuge and K. H. James, "Trace Metal Concentrations in Brown Seaweeds, Cardigan Bay, Wales," **Mar. Chem., 1** (1973) 281.

[58] A. Ballester and J. Castelvi, "Bioaccumulation of V and Ni by Marine Organisms and Sediments," **Investigacion Pesquera, 44** (1980) 1.

[59] M. Itoh, M. Yuasa and T. Kobayashi, "Absorption of Metal Ions on Yeast Cells at Varied Cell Concentrations," **Plant Cell Physicol., 16** (1977) 1167.

[60] J. M. Wood, F. Engle and R. Nice, Gray Freshwater Biological Institute, University of Minnesota, Navarre, Minnesota, unpublished results.

[61] M. Galun, P. Keller, D. Malki, H. Feldstein, E. Galun, S. M. Siegel and B. Z. Siegel, "Removal of Uranium(VI) from Solution by Fungal Biomass and Fungal Wall-Related Biopolymers," **Science, 219** (1983) 285.

[62] M. B. Allen, "Studies with *Cyanidium caldarium*, an Anomalously Pigmented Chlorophyte," **Archiv. Mikrobiol., 32** (1959) 270.

[63] J. Seckbach, F. A. Baker and P. M. Shugarman, "Algae Thrive under Pure Carbon Dioxide," **Nature (London), 227** (1970) 744.

DISCUSSION

P. J. Sadler (Birkbeck College, University of London): What are the nickel deposits in your nickel-resistant cells? They appeared to have hexagonal symmetry in your electron micrographs? Are the deposits crystalline?

J. M. Wood: The deposits are not crystalline. They are most likely cyanophycin granules (polyaspartylarginine) or carboxysomes (ribulose-5-phosphate carboxylase).

P. J. Sadler: You mentioned that *Cyanidium* can reduce metal ions in hydrochloric acid solution. What is the reducing agent? Which metals have you tried? Which potential range is available?

J. M. Wood: Reduction occurs at the membrane surface presumably by electron transfer from the photosynthetic apparatus. We examined the reduction of insoluble ferric hydroxide to produce increasing concentrations of ferrous ions in 0.5 M HCl solution.

XXX

BACTERIAL TRANSFORMATIONS OF AND RESISTANCES TO HEAVY METALS

Simon Silver

Biology Department, Washington University
St. Louis, Missouri 63130 USA

Bacteria carry out chemical transformations of heavy metals. These transformations including oxidation, reduction, methylation, and demethylation are sometimes only byproducts of normal metabolism that confer no known advantage upon the organism. Sometimes, however, the transformations constitute a mechanism of resistance. Many species of bacteria have genes that control resistances to specific toxic heavy metals. These resistances often are determined by extrachromosomal DNA molecules (plasmids). The same mechanisms of resistances occur in bacteria from soil, water, industrial waste, and clinical sources. The mechanism of mercury and organomercurial resistance is the enzymatic conversion of the mercurials to volatile species (methane, ethane, metallic Hg) that are rapidly lost to the atmosphere. Cadmium and arsenate resistances are due to reduced net accumulation of these toxic materials. Efficient efflux pumps cause the rapid excretion of cadmium(II) and arsenate. The mechanisms of arsenite and of antimony resistance, usually found associated with arsenate resistance, are not known. Silver resistance is due to lowered affinity of the cells for Ag^+, which can be complexed with extra-cellular halides, thiols, or organic compounds. Sensitivity is due to binding of Ag^+ more effectively to cells than to Cl^-.

Bacterial cells divide the elements in the Periodic Table into three classes: elements that are necessary for intracellular metabolism [1]; elements that are abundant in natural environments, that are generally not used within the cell but can be associated with extracellular structural or regulatory functions; and elements that have no useful biological function [2, 3]. Potassium and phosphorus are examples for the first class; calcium and chlorine, elements not needed at all by most bacteria, are examples for the second class; and arsenic, mercury and cadmium, toxic elements without biological utility, are examples for the third class. This report discusses toxic elements and their compounds. Some elements such as mercury and arsenic can be transformed by microbes from relatively nontoxic inorganic ions into relatively toxic methylated forms. The same or other microbes can degrade organometallic compounds [3, 4]. Oxidation and reduction by microbial enzymes also affect the bioavailability and toxicity of heavy metals.

Free-living bacterial cells have evolved resistance mechanisms to cope with heavy metal pollution. These resistance mechanisms are highly specific. The genes determining these resistance mechanisms occur on small nonchromosomal DNA molecules called plasmids. These resistance plasmids also have genes controlling resistance to most known antibiotics. I have reviewed this subject periodically [2, 3, 5, 6] and most recently in 1983 [6]. This paper is an updated version of the 1983 review.

Mercury and Mercurial Transformations and Resistances

The transformation of mercury (eqn. 1) is the best known

$$CH_3Hg^+ \rightleftharpoons Hg^{2+} \rightleftharpoons Hg^0 \tag{1}$$

case of microbial metabolism affecting the chemical form of a heavy metal. Microbial activity is associated with the methylation and demethylation of mercury [4], the oxidation of elemental mercury [7], and the reduction of inorganic mercury compounds [3]. The transformations of highly toxic methyl-mercury found in fish to less toxic ionic Hg^{2+}, the predominant form in seawater, and finally to the least toxic elemental mercury are carried out by enzymes controlled by bacterial resistance plasmids and transposons (moveable DNA sequences) and not by chromosomal genes. Without the plasmid genes, the cells remain mercury sensitive.

The earliest studies of enzymatic detoxification of mercuric ion used a multiply drug-resistant *E. coli* [28] and a soil pseudomonad [8, 9]. The frequency of Hg^{2+} resistance among clinical isolates can be more than 50 percent [10-12]. Recently, mercuric- and organomercurial-resistant strains with very similar properties were found in a wide variety of bacterial species from soil, water, and marine environments [13-18]. Mercuric-resistant microorganisms are found in much higher frequencies in polluted waters than in nearby cleaner waters [19].

Bacteria with plasmids showed a small number of patterns of resistance to organomercurials [20, 21].

- In *E. coli* about 96 percent of the mercuric resistance plasmids controlled also resistance to the organomercurials merbromin and fluorescein mercuric acetate, but to no other tested organomercurial. The other four percent determined additional resistances to phenylmercuric acetate 1 and thimerosal 2 [20].

$$\text{Ph-Hg}^+ \; {}^-\text{OOCCH}_3$$
1

$$\text{(2-COO}^-\text{Na}^+\text{)C}_6\text{H}_4\text{-S-Hg-CH}_2\text{CH}_3$$
2

- The plasmids in *Pseudomonas aeruginosa* also fell into two equally populated classes with regard to resistance to organomercurials [22]. All *Pseudomonas* plasmids controlled resistance to p-hydroxymercuribenzoate, and some *Pseudomonas* showed additional resistances to methylmercuric and ethylmercuric compounds [21, 22].
- Initially a single pattern was reported with *S. aureus* plasmids [21, 23]. This pattern is different from those with the gram-negative bacteria, because *S. aureus* plasmids control resistances to phenylmercuric acetate, p-hydroxymercuribenzoate, and fluorescein mercuric acetate but not to thimerosal or to merbromin.

Recently, the first thimerosal-resistant *S. aureus* were found [24]. These new clinical isolates volatilize mercury more rapidly from thimerosal (Fig. 1) than do the previous *S. aureus* strains. Of course, the sensitive strain without a plasmid cannot volatilize mercury from thimerosal at all (Fig. 1). The new strains also showed activity for thimerosal as an inducer of synthesis of this mercuric detoxification system, which is only made after exposure to low levels of Hg^{2+} or organomercurials [23, 24]. The limited number of patterns of resistance can be understood in terms of the biochemistry of the enzymes involved.

Mercurial-detoxifying, mercurial-resistant strains have been found in every type of bacteria tested including soil

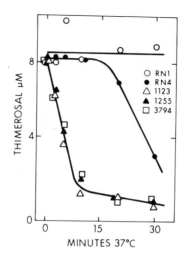

Figure 1. Detoxification of thimerosal through loss of mercury by a plasmid-less sensitive strain (RN1), by a thimerosal-sensitive plasmid-containing strain (RN4), and by new thimerosal-resistant isolates (1123, 1255, 3794). The thimerosal left in solution is plotted versus time. Reprinted with permission from ref. 24.

and marine bacteria (*Bacillus* [14, 17], *Pseudomonas* [22], and *Mycobacterium* [25]), mineral-leaching *Thiobacillus* [16, 26], and even antibiotic-producing *Streptomyces* [27a].

Enzymatic Mechanism of Mercurial Detoxification. Resistance to mercuric ion in bacteria results from enzymatic detoxificiation of mercury compounds (eqns. 2-5). The

$$Hg^{2+} \xrightarrow{mercuric\ reductase} Hg^0 \uparrow (volatile) \qquad (2)$$

$$C_6H_5Hg^+ \longrightarrow C_6H_6 \uparrow \qquad (3)$$
$$CH_3Hg^+ \xrightarrow{organomercurial\ hydrolase(s)} CH_4 \uparrow \qquad (4)$$
$$CH_3CH_2Hg^+ \longrightarrow C_2H_6 \uparrow \qquad (5)$$

mercury is volatilized [9, 28, 29] as elemental mercury. The enzyme responsible for these reactions is called mercuric reductase (eqn. 2). Several organomercurials are also enzymatically detoxified to volatile compounds (eqns. 3, 4, 5; Fig. 2). Benzene is produced from phenylmercury, methane from methylmercury, and ethane from ethylmercury. The enzymes responsible for cleaving the Hg-C bond are

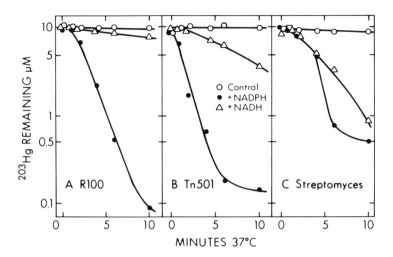

Figure 2. Volatilization of mercury from mercuric ion by enzyme preparations from cells bearing plasmid R100 (A), transposon Tn*501* (B), or a new **Streptomyces** strain (C). The experiments were performed under conditions given by Schottel [32] except 25 µM NAD(P)H were added with 10 µM Hg-203 mercuric ion. Reprinted with permission from ref. 27.

organomercurial lyases. Tezuka and Tonomura [30, 31] separated from a *Pseudomonas* strain two small soluble lyase enzymes that have molecular weights of about 19,000. One enzyme cleaved phenylmercuric acetate, p-hydroxymercuribenzoate, and methylmercury, whereas the other enzyme cleaved only phenylmercuric acetate and p-hydroxymercuribenzoate [30, 31]. An organomercurial-resistant *E. coli* gave no evidence for cleavage of p-hydroxymercuribenzoate [21]. Schottel [32] was unable to separate the two *E. coli* lyases. However, kinetic analyses indicated that two enzymes were present, both of them active toward phenylmercuric acetate, but only one active toward methyl- and ethylmercury. The general properties of the enzymes from the soil pseudomonad and *E. coli* were similar, except that the *E. coli* organomercurial lyases had a somewhat greater molecular weight [32].

Mercuric reductase has been studied in great detail [32-35]. We once thought the enzyme to be strictly NADPH-dependent, but it now appears that some mercuric reductases such as those in *Streptomyces* strains (Fig. 2C) can utilize either NADPH or NADH. The mercuric reductase enzyme [33, 38] is related to other NAD(P)H-dependent FAD-containing flavoproteins such as glutathione reductase and lipoamide dehydrogenase [36, 37]. These enzymes follow a reaction sequence starting with an NAD(P)H-reducing, enzyme-bound FAD to $FADH_2$. Next, $FADH_2$ reduces the disulfide bond between two neighboring cysteines in the active site of the enzyme. Finally, mercuric ion is bound to the two cysteine SH groups and then reduced to elemental mercury [33]. The structure and partial amino acid sequences of *E. coli*, yeast, and human glutathione reductase, and of *E. coli* and pig lipoamide dehydrogenase are known [36, 37]. The DNA sequences of two mercuric reductase genes were determined by direct analysis [38, 38a, 38b]. Mercuric reductase contains fifteen amino acids, two of which are cysteine, in a

$$\begin{array}{c}\text{Thr-Ile-Gly-Gly-Thr-Cys-Val-Asn}\\|\qquad\qquad\qquad|\\\text{Lys-Ser-Pro-Val-Cys-Gly-Val}\end{array}$$

3

sequence (formula 3) essentially identical to that at the active sites of glutathione reductase and lipoamide dehydrogenase [36, 37]. These studies were the first to apply DNA sequencing to the elucidation of heavy metal resistances. The results suggest, that the enzymes able to detoxify heavy metals might have evolved from enzymes of normal cellular metabolism.

The subunit structure of the mercuric reductase enzyme from various sources is currently under re-evaluation. Furukawa and Tonomura [39] determined the molecular weight of the enzyme from a soil pseudomonad to be 65,000. Schottel [32] found the enzyme from plasmid R831 that originated in *Serratia* to be a trimer of 170,000 molecular weight consisting of three identical monomer subunits, each approximately the size of the *Pseudomonas* enzyme. Each subunit contained a single, bound FAD for three FADs per 170,000 molecular weight. We obtained similar evidence for a trimeric structure of mercuric reductase from a variant of plasmid R100 by gel filtration and electrophoresis through gels of differing porosity [40, 41]. Yet, Rinderle et al. [42] found a mobility ratio by gel filtration/denatured gel electrophoresis consistent with a dimeric enzyme for another variant of plasmid R100. Fox and Walsh [33] reported that the molecular weight of another mercuric reductase from transposon Tn501 was approximately 125,000 and that the enzyme appeared to be dimeric. We confirmed the dimeric structure of the mercuric reductase of Tn501.

Antibodies were prepared against purified mercuric reductases coded by two plasmids in *E. coli* [40]. Reductases from all but one gram-negative source reacted with these antibodies as shown by the inhibition of enzyme activity and by the formation of precipitin bands on double-diffusion gels. The enzymes can be divided into two major sub-classes based on only partial immunological identity. The prototype of the first enzyme class is coded by transposon Tn501 [43]. This enzyme class includes also mercuric reductases governed by a variety of plasmids found in clinical, soil, and marine bacteria. One strong conclusion from studies of plasmid-determined mercuric resistance is that the same system appears widely in bacterial isolates from diverse environments. Newer mercuric ion- and in one case phenylmercuric-resistance transposons from soil microbes showed different patterns of digestion by DNA restriction endonucleases [17].

The second immunological subgroup of gram-negative mercuric reductases has as its prototype the enzyme coded by plasmid R100. The genetic structure of the mercuric resistance operon is being studied in detail with R100 [44-46]. This subgroup contains enzymes from a wide variety of sources including the enzyme determined by a second *Pseudomonas* mercury transposon Tn502 [40, 47]. All of the mercuric reductases from gram-negative bacteria were immunologically related with the single exception of the enzyme from *Thiobacillus ferrooxidans* [26]. The antibodies perpared against the two classes of gram-negative enzymes did not cross-react with mercuric reductases from *S. aureus* strains and marine and soil bacilli. The enzymes from gram-positive sources showed functional requirements otherwise similar to those from the gram-negative bacteria [23, 40] but they are immunologically distinct.

Thus, plasmid-determined mercuric and organomercurial resistances occur widely in all bacterial types tested and are the best understood of all plasmid-coded heavy metal resistances. Resistance is achieved by enzymatic detoxification of the mercurials to volatile compounds of lesser toxicity. The mercuric reductases and organomercurial lyases, the enzymes responsible for detoxification, were purified and studied from the viewpoint of enzymology. The newer tools of DNA cloning and sequencing are now being applied to these systems.

Oxidation and Methylation of Mercury. Much less is known about the oxidation of elemental mercury to mercuric ion and the subsequent methylation of mercuric ion to CH_3Hg^+ and $(CH_3)_2Hg$ than about the reverse processes. Microbial activities have been implicated in both reactions. How is the geochemical mercury cycle maintained, when most of the mercury in the atmosphere is elemental mercury and most of the mercury in the sea is mercuric ion [3]? A wide range of bacteria can oxidize elemental mercury to mercuric ion [7]. These bacteria generally do not convert mercuric ion to methylmercury [7]. The ubiquitous enzyme catalase - found in animal tissues and bacteria - can oxidize elemental mercury to mercuric ion and is probably responsible for the rapid conversion of inhaled mercury vapor into ionic mercury in the blood [48]. There appears to be general agreement that mercuric ion can be methylated abiotically and extracellularly by methylcobalamin (vitamin B_{12}) [49-51]. Cobalamin is synthesized only by microbial cells and is either excreted or released on cell lysis. Other bacteria such as *E. coli* and cells of higher organisms such as plants and animals, all of which require this vitamin for growth, can accumulate cobalamin, methylate it enzymatically [51], and thus synthesize a reagent required for methyl transfer reactions. Many common bacteria methylate mercury [52] presumably in this way. Mercury can be methylated by microbial activities in such environments as river sludge [53] and human feces [50]. Both, bacteria and fungi, can methylate mercury [52]. Although only the methylcobalamin-dependent process has been studied in detail [51], there are reasons such as lack of dependency upon added cobalamin with some but not with other bacteria [52] to believe that a second, biologically important methyl transfer process does exist. This alternate methylation process must be investigated [49]. It is not clear whether all of the methylmercury found in fish and in man was synthesized by microbes growing in intestines and on gills [3] or produced by mammalian enzymes that may also function as methylators of mercury.

Arsenic and Antimony Resistances

Arsenic and antimony resistances are governed by plasmids that also code for antibiotic and other heavy metal resistances [54-56]. Arsenate, arsenite, and antimony(III) resistances are coded by an inducible operon-like system in *S. aureus* and *E. coli* [57]. Each of the three ions induces all three resistances. In *E. coli* bismuth(III) is a gratuitous inducer of arsenate resistance, even though the plasmid system does not confer bismuth(III)-resistance. *S. aureus* has a genetically separate plasmid-mediated bismuth(III) resistance determinant [55, 58] of unknown mechanism.

The mechanism of arsenate resistance is a reduced accumulation of arsenate by induced resistant cells. Arsenate is normally accumulated via the cellular phosphate transport systems, of which many bacteria appear to have two, the Pst and the Pit system (Fig. 3). The distinction

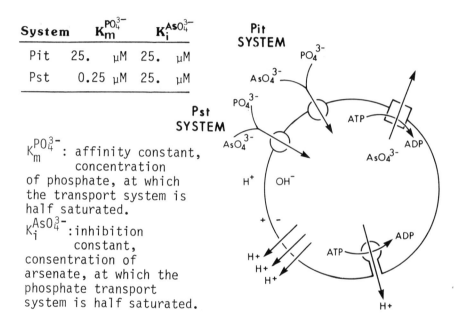

System	$K_m^{PO_4^{3-}}$	$K_i^{AsO_4^{3-}}$
Pit	25. µM	25. µM
Pst	0.25 µM	25. µM

$K_m^{PO_4^{3-}}$: affinity constant, concentration of phosphate, at which the transport system is half saturated.

$K_i^{AsO_4^{3-}}$: inhibition constant, consentration of arsenate, at which the phosphate transport system is half saturated.

Figure 3. The systems for the transport of phosphate and arsenate into the cell and the system for the efflux of arsenate.

between arsenate and arsenite resistance was shown initially by the observation that phosphate did not protect against growth inhibition by arsenite [57]. Genetic studies demonstrated that the arsenate resistance gene is different from, but closely linked to, the arsenite resistance gene. Arsenite and antimony resistances may be determined by separate genes [58]. The presence of the resistance plasmid does not alter the kinetic parameters of the cellular phosphate transport systems. Even the K_i for arsenate, a competitive inhibitor of phosphate transport (Fig. 3), is unchanged. This finding, along with direct evidence for the plasmid-governed energy-dependent efflux of arsenate, indicated that the block on net uptake of arsenate resulted from rapid efflux. The energy dependence of the efflux process was shown by its sensitivity to "uncouplers" and ionophore antibiotics such as Nigericin and Monesin [59]. Mercuric ion also inhibited the efflux system. This inhibition by mercuric ion was readily reversed by mercaptoethanol [59]. Mobley and Rosen [60] demonstrated, that glucose but not succinate could energize arsenate efflux in an *E. coli* strain incapable of synthesizing ATP from respiratory substrates. Although all of these data are indirect, arsenate efflux appears to directly involve an ATPase-linked transport system (Fig. 3). An interesting question about the arsenate efflux system concerns its specificity. Arsenate generally functions as a phosphate analog and is accumulated by bacteria via phosphate transport systems [1]. The arsenate-resistance efflux system should not excrete phosphate, because the cells would then become phosphate starved, a situation no more advantageous than being arsenate inhibited. That toxic heavy metals often enter cells by means of transport systems for normally required nutrients [1, 3] is a basic conclusion from our work on arsenate and cadmium(II) resistance. Energy-dependent efflux systems functioning as resistance mechanisms should be highly specific for the toxic anion or cation to prevent loss of the required nutrient.

The mechanism(s) of plasmid-determined arsenite and antimony resistances are not known. Arsenite is not oxidized to the less toxic arsenate by plasmid-bearing *E. coli* or *S. aureus* [57], although *Alcaligenes* strains were isolated that do have an inducible oxidizing system for arsenite [61, 62]. Arsenicals and antimonials are toxic by virtue of inhibiting thiol-containing enzymes [63]. Plasmid-containing resistant cells did not excrete soluble thiol compounds into the medium [57]. The absence of "detoxificiation" - measured by experiments inoculating sensitive cells into media "preconditioned" by growth of resistant cells - eliminates all other mechanisms of extracellular detoxification involving changes in extracellular chemical states. It is possible that arsenite resistance results from a change in cell membrane transport as does arsenate resistance. Arima and Beppu [64, 65] isolated an arsenite resistant *Pseudomonas pseudomallei* that had an inducible decrease in the energy-dependent accumulation of arsenite. No further work has been done on this system.

Oxidation and Methylation of Arsenic. Oxidation and methylation of arsenic compounds by microbes was known well before the modern era of molecular genetics. Arsenite-resistant bacteria capable of oxidizing the more toxic arsenite ion to somewhat less toxic arsenate were first isolated in 1918 from cattle-dipping tanks in South Africa and later in Australia [3]. Similar strains have appeared over the years. These arsenite-oxidizing strains contain an inducible, soluble arsenite dehydrogenase enzyme [3]. There is no evidence of involvement of plasmids in the oxidation of arsenite by these bacteria. Presumably chromosomal genes are involved.

The arsenic compound produced by fungi growing on an arsenic-containing medium was identified as trimethylarsine in 1932 [3, 50]. The process of methyl donation to arsenic seems to require methylcobalamin as does the methylation of mercuric ion [50]. However, experiments with *Scopulariopsis brevicaulis*, *Gliocladium roseum* [66], and *Candida humicola* [66, 67, 68] indicate, that S-adenosylmethionine or some related sulfonium compound might also serve as a methyl donor. An enzymatic reaction sequence of reduction of arsenate to arsenite, methylation of arsenite to methylarsonic acid, reduction of methylarsonic acid to a trivalent methylarsenic species followed by its methylation to dimethylarsinic acid, and reduction of dimethylarsinic acid to dimethylarsine was postulated [50] and later demonstrated [69, 70]. Clearly this sequence is more complex than the abiotic methylation of mercuric ion by methylcobalamin. However, under some conditions dimethylmercury was the only volatile product of the methylation of mercuric ion [71]. How different or how similar the methylation of heavy metals and metalloids is at the biochemical level [49-51] is not yet known.

Cadmium and Zinc Resistance

Plasmid-determined cadmium resistance was found only in *S. aureus* [55]. In some collections of organisms from hospitals, cadmium resistance is the most common plasmid resistance in *S. aureus* and is more common than mercury or penicillin resistance [10]. Gram-negative cells without plasmids are just as resistant to cadmium as staphylococci with plasmids [11], probably because of relatively reduced cadmium uptake. However, occasionally gram-negative bacteria are sensitive or even "hypersensitive" to cadmium [15, 72]. The basis of cadmium sensitivity and resistance in other bacterial species is not known. There is no evidence, however, for chemical transformations associated with resistance. In staphylococci, cadmium is accumulated by a membrane transport system utilizing the cross-membrane electrical potential [73-75] (Fig. 4). This uptake system

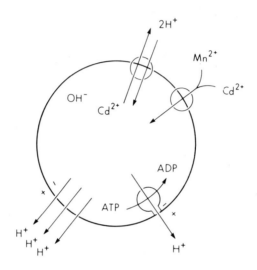

Figure 4. *Model for the cadmium(II) uptake and efflux systems. Redrawn from ref. 77.*

is highly specific for cadmium and manganese(II) with respective affinity constants, the concentrations at which the transport system is half saturated, of 10 µM and 16 µM in whole cells [75] and 0.2 µM and 0.95 µM in membrane vesicles [73].

Two separate plasmid genes are responsible for the cadmium resistance of *S. aureus* strains [55, 58] (Fig. 5). The cadA and cadB genes confer, respectively, a large and a small increase in cadmium resistance (Fig. 5). When both genes are present, the effect of cadA masks the cadB effect; cadA$^+$ cadB$^+$ strains are no more resistant than cadA$^+$ and B$^-$ strains. Both genes confer increased zinc resistance (Fig. 5). These observations and the genetic linkage of cadmium and zinc resistance [58] indicate that the cadA and cadB genes are also responsible for zinc resistance. Cadmium

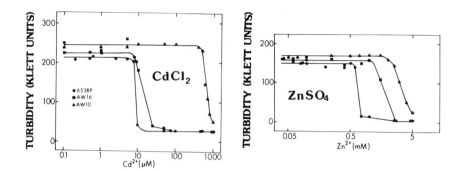

Figure 5. Cadmium(II) and zinc(II) resistances of strains with and without plasmids (sensitive strain 6538P, cadA+-resistant AW10, cadB+-resistant AW16). Overnight cultures were diluted 1:200 into broth containing cadmium chloride or zinc sulfate. The turbidities of the cultures were measured after 7 hours of growth at 37⁰. Redrawn from ref. 73.

resistance is conferred through decreased accumulation of cadmium by resistant cells [74, 75]. The cadA gene product causes this lessened cadmium accumulation [73, 75] (Fig. 6).

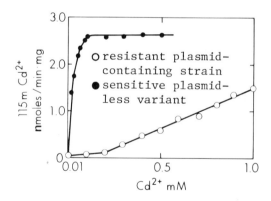

Figure 6. Cadmium(II) uptake by sensitive and by resistant S. aureus. The initial rate of accumulation of Cd-115m is shown as a function of the cadmium concentration. Reproduced with permission from ref. 74.

A plasmid-encoded efflux system rapidly excretes cadmium from resistant cells (Fig. 7). It seems plausible, but has not yet been directly demonstrated, that the cadA gene

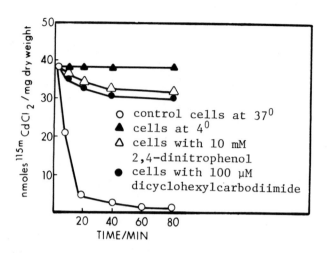

Figure 7. Retention of Cd-115m by cadmium-resistant S. aureus. The cell suspensions were pre-incubated for 10 minutes with 1 mM Cd-115m at 37^0. The cells were washed free of cadmium(II). The efflux of Cd-115m was then assayed. Reproduced with permission from ref. 77.

product also causes zinc efflux. However, the presence of the cadA$^+$ gene neither reduces manganese(II) uptake nor causes rapid efflux of accumulated manganese(II) [73]. CadA$^+$ resistant strains possess this efflux system, but sensitive cells and cadB$^+$ resistant strains do not [78]. The cadA-encoded efflux system is energy dependent. Cadmium efflux ceased at low temperature and after addition of dinitrophenol (Fig. 7). Inhibitor studies indicate, that the efflux system is a cadmium(II)/2H$^+$ exchange system (Fig. 4) analogous to the tetracyline efflux system [79] responsible for plasmid-mediated resistance to that antibiotic.

We do not know the mechanism of the cadB gene function. Our clues at the moment include the significant resistance to zinc conferred by cadB (Fig. 5) and an inducible cadmium-binding activity governed by this gene [78]. This binding activity is not energy-dependent, and we hypothesize, that a membrane component analogous to metallothionein, the small cadmium-binding protein made by mammals and microbes, might be responsible. Corrigan and Huang [80], seeking cadmium-resistant mutants of mammalian cells, found what appear to be transport mutants that accumulate less cadmium and more zinc than wild-type cells. Whether the metallothionein studies or mammalian transport studies will provide a basis for understanding cadB resistance is an open question.

Silver Resistance

Microbial silver toxicity is found in situations of industrial pollution, especially associated with the use of photographic film. In hospitals, silver salts are preferred antimicrobial agents for burns [81, 82]. It is thus not surprising that silver-resistant bacteria were found in urban and industrial polluted sites [83], and that silver-resistant bacteria [84-86] and silver-resistant plasmids [87] were described. The plasmid-determined resistance is very great, and the ratio of minimal-inhibitory silver concentrations (resistant:sensitive) can be greater than 100:1 [6, 78]. The level of resistance is strongly dependent on available halide ions. Without chloride, relatively little difference exists between cells with or without resistance plasmids [78]. Bromide and iodide protected both sensitive and resistant cells at halide concentrations far below those used in the chloride experiments. Sensitive and resistant cells bind silver tightly and are killed by effects on cell respiration and other cell surface functions [88, 89]. Once bound extracellularly, silver enters the cell and is found in high-speed centrifugal supernatant fluids [27]. We hypothesize that the sensitive cells bind silver so tightly allowing them to extract silver from AgCl and other bound forms, whereas cells with resistance plasmids do not compete successfully with silver-halide precipitates for silver.

Other Heavy Metal Resistances

Several other plasmid heavy metal resistances exist [3, 83, 90] but nothing is known about the mechanisms of resistance to bismuth, boron, cobalt, lead, nickel, or tellurium. Chromate resistance in a pseudomonad isolated from river sediment is caused by reduction of toxic Cr(VI) to less toxic Cr(III). This chromium resistance appears to be plasmid-determined [91]. Cupric ion resistance was found with plasmids in *E. coli* [76, 92] and in *S. aureus* [93]. The mechanism of this resistance has not been studied.

Methylation of other heavy metals. In addition to arsenic and mercury, chemical or biogenic methylation reactions were reported for palladium, thallium, selenium, lead, platinum, gold, tin, chromium, and sulfur [49, 51, 94]. Tin-resistant bacteria were isolated from polluted areas of Chesapeake Bay [95]. These bacteria were also resistant to dimethyltin. Bacteria from Chesapeake Bay were able to synthesize dimethyl- and trimethyltin [96] as previously proposed during a general discussion of methylcobalamin-dependent methylations [51]. It is not known whether the bacteria resistant to methyltin synthesize or degrade methyltin. Resistance may - as in the case of mercury - be unrelated to the ability to synthesize these organometallic compounds. Once methylated by biological processes, abiotic transmethylation reactions occur [49], and heavy metal compounds may be destroyed abiotically by oxidation [97].

During the last 13 years, our laboratory has been studying the mechanisms and genetics of plasmid-mediated heavy metal resistances. In the frequent absence of any obvious source of direct selection, one may ask why these resistances occur with such high frequencies. Selective agents in hospitals and "normal human" environments are only beginning to be examined. It has been suggested that heavy metal resistances may have been selected in earlier times, and that they are merely carried along today "for a free ride" with selection for antibiotic resistances. I doubt that there is such a thing as "a free ride" for these determinants. For example, in Tokyo in the late 1970s, both heavy metal resistances and antibiotic resistances were found with high frequencies in *E. coli* isolated from hospital patients, whereas heavy metal resistance plasmids without antibiotic resistance determinants were found in *E. coli* from an industrially polluted suburban river [98]. Selection occurs for resistances to both types of agents in the hospital, but only for toxic heavy metals in the river environment. Radford et al. [17] found mercury-resistant microbes in agricultural soil with no known human mercurial input. In such settings the prevalence of resistant microbes may be very low, but these microbes may become prominent in response to industrial or agricultural pollution [8, 15, 18, 83]. This situation may be analogous to one in which antibiotic-resistance plasmids were found in low frequencies in antibiotic-virgin populations [99, 100].

The transformations of toxic heavy metals that occur on a global scale are generally brought about by microorganisms. Considerations of human health problems must take into account the chemical forms of heavy metals available. Microbes might be utilized to convert more toxic forms of heavy metals into less toxic forms. Other microbes might be fooled into accumulating heavy metals in order to clean up polluted sites. In both cases, understanding of the basic biochemical and genetic bases for microbial resistances to heavy metals and microbial transformations of these materials is essential for rational planning.

Acknowledgements. The work in our laboratory on these topics during the last decade was performed by many talented people whose names often appear as coauthors in the references. Support was provided most recently by National Science Foundation Grant PCM 82-055992 and by U.S. Public Health Service Grant AI15672.

References

[1] S. Silver, "Transport of Cations and Anions," in **"Bacterial Transport,"** B. P. Rosen, Ed., Marcel Dekker Inc.: New York, 1978, p. 221.

[2] S. Silver, "Mechanisms of Plasmid-Determined Heavy Metal Resistances," in **"Molecular Biology, Pathogenicity and Ecology of Bacterial Plasmids,"** S. B. Levy, R. C. Clowes and E. L. Koenig, Eds., Plenum Press: New York, 1981, p. 179.

[3] A. O. Summers and S. Silver, "Microbial Transformations of Metals," **Ann. Rev. Microbiol.**, 32 (1978) 637.

[4] T. Edwards and B. C. McBride, "Biosynthesis and Degradation of Methylmercury in Human Faeces," **Nature**, 253 (1975) 462.

[5] S. Silver, "Mechanisms of Bacterial Resistances to Toxic Heavy Metals: Arsenic, Antimony, Silver, Cadmium and Mercury," in **"Environmental Speciation and Monitoring Needs for Trace Metal-Containing Substances from Energy-Related Processes,"** F. E. Brinckman and R. H. Fish, Eds., Special Publ. 618, National Bureau of Standards: Washington, D.C., 1981, p. 301.

[6] S. Silver, "Bacterial Transformations of and Resistance to Heavy Metals," in **"Changing Metal Cycles and Human Health,"** Report of the Dahlem Workshop, March 20-25, 1983, Berlin, J. O. Nriagu, Ed., Springer-Verlag: Berlin, 1984, p. 199.

[7] H. W. Holm and M. F. Cox, "Transformation of Elemental Mercury by Bacteria," **Appl. Microbiol.**, 29 (1975) 491.

[8] K. Furukawa, T. Suzuki ad K. Tonomura, "Decomposition of Organic Mercurial Compounds by Mercury-Resistant Bacteria," **Agric. Biol. Chem.**, 33 (1969) 128.

[9] K. Tonomura and F. Kanzaki, "The Reductive Decomposi-

tion of Organic Mercurials by Cell-Free Extract of a Mercury-Resistant Pseudomonad," **Biochim. Biophys. Acta, 184** (1969) 227.

[10] H. Nakahara, T. Ishikawa, Y. Sarai and I. Kondo, "Distribution of Resistances to Metals and Antibiotics of Staphylococcal Strains in Japan," **Zentralb. Bakteriol. Parasitenkd. Infektionskr. Hyg., 1 Abt. Orig. A, 237** (1977) 470.

[11] H. Nakahara, T. Ishikawa, Y. Sarai, I. Kondo, H. Kozukue and S. Silver, "Linkage of Mercury, Cadmium, and Arsenate and Drug Resistance in Clinical Isolates of *Pseudomonas aeruginosa*," **Appl. Environ. Microbiol., 33** (1977) 975.

[12] W. Witte, N. Van Dip and R. Hummel, "Resistenz gegen Quecksilber und Cadmium bei *Staphylococcus aureus* unterschiedlicher ökologischer Herkunft," **Z. Allg. Mikrobiol., 20** (1980) 517.

[13] D. A. Friello and A. M. Chakrabarty, "Transposable Mercury Resistance in *Pseudomonas putida*," in **"Plasmids and Transposons: Environmental Effects and Maintenance Mechanisms,"** C. Suttard and K. R. Rozee, Eds., Academic Press: New York, 1980, p. 249.

[14] K. Izaki, "Enzymatic Reduction of Mercurous and Mercuric Ions in *Bacillus cereus*," **Can. J. Microbiol., 27** (1981) 192.

[15] B. H. Olson, T. Barkay and R. R. Colwell, "Role of Plasmids in Mercury Transformation by Bacteria Isolated from the Aquatic Environment," **Appl. Env. Microbiol., 38** (1979) 478.

[16] G. J. Olson, W. P. Iverson and F. E. Brinckman, "Volatilization of Mercury by *Thiobacillus ferrooxidans*," **Current Microbiol., 5** (1981) 115.

[17] A. J. Radford, J. Oliver, W. J. Kelly and D. C. Reanney, "Translocatable Resistance to Mercuric and Phenylmercuric Ions in Soil Bacteria," **J. Bacteriol., 147** (1981) 1110.

[18] J. F. Timoney, J. Port, J. Giles and J. Spanier, "Heavy-Metal and Antibiotic Resistance in the Bacterial Flora of Sediments of New York Bight," **Appl. Environ. Microbiol., 36** (1978) 465.

[19] J. D. Nelson, Jr. and R. R. Colwell, "The Ecology of Mercury-Resistant Bacteria in Chesapeake Bay," **Microb. Ecol., 1** (1975) 191.

[20] J. Schottel, A. Mandal, D. Clark, S. Silver and R. W. Hedges, "Volatilisation of Mercury and Organomercurials Determined by Inducible R-Factor Systems in Enteric Bacteria," **Nature, 251** (1974) 335.

[21] A. A. Weiss, J. L. Schottel, D. L. Clark, R. G. Beller

and S. Silver, "Mercury and Organomercurial Resistance with Enteric, Staphylococcal, and Pseudomonad Plasmids," in **"Microbiology 1978,"** D. Schlessinger, Ed., American Society for Microbiology: Washington, D.C., 1978, p. 121.

[22] D. L. Clark, A. A. Weiss and S. Silver, "Mercury and Organomercurial Resistances Determined by Plasmids in Pseudomonas," **J. Bacteriol., 132** (1977) 186.

[23] A. A. Weiss, S. D. Murphy and S. Silver, "Mercury and Organomercurial Resistances Determined by Plasmids in Staphylococcus aureus," **J. Bacteriol., 132** (1977) 197.

[24] F. D. Porter, C. Ong, S. Silver and H. Nakahara, "Selection for Mercurial Resistance in Hospital Settings," **Antimicrob. Agents Chemother., 22** (1982) 852.

[25] P. S. Meissner and J. O. Falkinham, III, "Plasmid-Encoded Mercuric Reductase in Mycobacterium scrofulcaeum," **J. Bacteriol., 157** (1984) 669.

[26] G. J. Olson, F. D. Porter, J. Rubinstein and S. Silver, "Mercuric Reductase Enzyme from a Mercury-Volatilizing Strain of Thiobacillus ferrooxidans," **J. Bacteriol., 151** (1982) 1230.

[27] S. Silver, Biology Department, Washington University, St. Louis, Missouri, unpublished results.

[27a] H. Nakahara, J. L. Schottel, T. Yamada, Y. Miyakawa, M. Asakawa, J. Harville and S. Silver, "Mercuric Reductase Enzymes from Group B Streptococcus and Streptomyces Species," **J. Gen. Microbiol.,** (1984) submitted.

[28] I. Komura, T. Funaba and K. Izaki, "Mechanism of Mercuric Chloride Resistance in Microorganisms. II. NADPH-Dependent Reduction of Mercuric Chloride and Vaporization of Mercury from Mercuric Chloride by a Multiple Drug Resistant Strain of Escherichia coli," **J. Biochem., 70** (1971) 895.

[29] A. O. Summers and S. Silver, "Mercury Resistance in a Plasmid-Bearing Strain of Escherichia coli," **J. Bacteriol., 112** (1972) 1228.

[30] T. Tezuka and K. Tonomura, "Purification and Properties of an Enzyme Catalyzing the Splitting of Carbon-Mercury Linkages from Mercury-Resistant Pseudomonas K-62 Strain. I. Splitting Enzyme 1," **J. Biochem., 80** (1976) 79.

[31] T. Tezuka and K. Tonomura, "Purification and Properties of a Second Enzyme Catalyzing the Splitting of Carbon-Mercury Linkages from Mercury-Resistant Pseudomonas K-62," **J. Bacteriol., 135** (1978) 138.

[32] J. L. Schottel, "The Mercuric and Organomercurial

Detoxifying Enzymes from a Plasmid-Bearing Strain of *Escherichia coli*," **J. Biol. Chem.**, 253 (1978) 4341.

[33] B. Fox and C. T. Walsh, "Mercuric Reductase: Purification and Characterization of a Transposon-Encoded Flavoprotein Containing an Oxidation-Reduction-Active Disulfide," **J. Biol. Chem.**, 257 (1982) 2498.

[34] K. Furukawa and K. Tonomura, "Metallic Mercury-Releasing Enzyme in Mercury-Resistant *Pseudomonas*," **Agric. Biol. Chem.**, 36 (1972) 217.

[35] K. Izaki, Y. Tashiro and T. Funaba, "Mechanism of Mercuric Chloride Resistance in Microorganisms. III. Purification and Properties of a Mercuric Ion Reducing Enzyme from *Escherichia coli* Bearing R Factor," **J. Biochem.**, 75 (1974) 591.

[36] C. H. Williams, Jr., L. D. Arscott and G. E. Schulz, "Amino Acid Sequence Homology Between Pig Heart Lipoamide Dehydrogenase and Human Erythrocyte Glutathione Reductase," **Proc. Natl. Acad. Sci., USA**, 79 (1982) 2199.

[37] R. L. Krauth-Siegel, R. Blatterspiel, M. Saleh, E. Schiltz, R. H. Schirmer and R. Untucht-Grau, "Glutathione Reductase from Human Erythrocytes. The Sequences of the NADPH Domain and of the Interface Domain," **Eur. J. Biochem.**, 121 (1982) 259.

[38] N. L. Brown, S. J. Ford, R. D. Pridmore and D. C. Fritzinger, "Nucleotide Sequence of a Gene from the *Pseudomonas* Transposon Tn501 Encoding Mercuric Reductase," **Biochemistry, 22** (1983) 4089.

[38a] T. K. Misra, N. L. Brown, D. C. Fritzinger, R. D. Pridmore, W. M. Barnes, L. Haberstroh and S. Silver, "Mercuric Ion-Resistance Operons of Plasmid R100 and Transposon Tn501: The Beginning of the Operon Including the Regulatory Region and the First Two Structural Genes," **Proc. Natl. Acad. Sci., USA, 81** (1984) 5975.

[38b] T. K. Misra, N. L. Brown, L. Haberstroh, A. Schmidt, D. Goddette and S. Silver, "Mercuric Reductase Structural Genes from Plasmid R100 and Transposon Tn501: Functional Domains of the Enzyme," **Gene,** (1984) in press.

[39] K. Furukawa and K. Tonomura, "Enzyme System Involved in the Decomposition of Phenyl Mercuric Acetate by Mercury-Resistant *Pseudomonas*," **Agric. Biol. Chem.**, 35 (1971) 604.

[40] S. Silver and T. G. Kinscherf, "Genetic and Biochemical Bases for Microbial Transformations and Detoxification of Mercury and Mercurial Compounds," in **"Biodegradation and Detoxification of Environmental**

Pollutants," A. M. Chakrabarty, Ed., CRC Press: Boca Raton, Florida, 1982, p. 85.
[41] T. G. Kinscherf and S. Silver, Biology Department, Washington University, St. Louis, Missouri, unpublished results.
[42] S. J. Rinderle, J. E. Booth and J. W. Williams, "Mercuric Reductase from R-Plasmid NR1: Characterization and Mechanistic Study," **Biochem.**, **22** (1983) 869.
[43] P. M. Bennett, J. Grinsted, C. L. Choi and M. H. Richmond , "Characterization of Tn501, a Transposon Determining Resistance to Mercuric Ions," **Mol. Gen. Genet.**, **159** (1978) 101.
[44] T. J. Foster, H. Nakahara, A. A. Weiss and S. Silver, "Transposon A-Generated Mutations in the Mercuric Resistance Genes of Plasmid R100-1," **J. Bacteriol.**, **140** (1979) 167.
[45] H. Nakahara, S. Silver, T. Miki and R. H. Rownd, "Hypersensitivity to Hg^{2+} and Hyperbinding Activity Associated with Cloned Fragments of the Mercurial Resistance Operon of Plasmid NR1," **J. Bacteriol.**, **140** (1979) 161.
[46] N. NiBhriain, S. Silver and T. J. Foster, "Tn5 Insertion Mutations Derived from Plasmid R100: Identification of two new mer Genes," **J. Bacteriol.**, **155** (1983) 690.
[47] V. Stanisich, La Trobe University, Bundoora, Victoria 3083, Australia, personal communication.
[48] S. Halbach and T. W. Clarkson, "Enzymatic Oxidation of Mercury Vapor by Erythrocytes," **Biochim. Biophys. Acta,** **523** (1978) 522.
[49] F. E. Brinckman, G. J. Olson and W. P. Iverson, "The Production and Fate of Volatile Molecular Species in the Environment: Metals and Metalloids," in **"Atmospheric Chemistry,"** Dahlem Konferenzen, E. D. Goldberg, Ed., Springer-Verlag: Berlin (1982) 231.
[50] B. C. McBride and T. L. Edwards, "Role of the Methanogenic Bacteria in the Alkylation of Arsenic and Mercury," in **"Biological Implications of Metals in the Environment,"** Proceedings, 15th Annual Hanford Life Sciences Symposium, Sept. 29-Oct. 1, 1975, Richland, Washington, **ERDA Symposium Series,** **42** (1977) 1.
[51] J. M. Wood, A. Cheh, L. J. Dizikes, W. P. Ridley, S. Rakow and J. R. Lakowicz, "Mechanisms for the Biomethylation of Metals and Metalloids," **Fed. Proc.**, **37** (1978) 16.
[52] J. W. Vonk and A. K. Sijpesteijn, "Studies on the Methylation of Mercuric Chloride by Pure Cultures of

Bacteria and Fungi," **Anton. vanLeeuwenhoek J. Microbiol. Serol.**, **39** (1973) 505.

[53] M. K. Hamdy and O. R. Noyes, "Formation of Methylmercury by Bacteria," **Appl. Microbiol.**, **30** (1975) 424.

[54] R. W. Hedges and S. Baumberg, "Resistance to Arsenic Compounds Conferred by a Plasmid Transmissible Between Strains of *Escherichia coli*," **J. Bacteriol.**, **115** (1973) 459.

[55] R. P. Novick and C. Roth, "Plasmid-Linked Resistance to Inorganic Salts in *Staphylococcus aureus*," **J. Bacteriol.**, **95** (1968) 1335.

[56] H. W. Smith, "Arsenic Resistance in Enterobacteria: Its Transmission by Conjugation and by Phage," **J. Gen. Microbiol.**, **109** (1978) 49.

[57] S. Silver, K. Budd, K. M. Leahy, W. V. Shaw, D. Hammond, R. P. Novick, G. R. Willsky, M. H. Malamy and H. Rosenberg, "Inducible Plasmid-Determined Resistance to Arsenate, Arsenite and Antimony(III) in *Escherichia coli* and *Staphylococcus aureus*," **J. Bacteriol.**, **146** (1981) 983.

[58] R. P. Novick, E. Murphy, T. J. Gryczan, E. Baron and I. Edelman, "Penicillinase Plasmids of *Staphylococcus aureus*: Restriction-Deletion Maps," **Plasmid**, **2** (1979) 109.

[59] S. Silver and D. Keach, "Energy-Dependent Arsenate Efflux: The Mechanism of Plasmid-Mediated Resistance," **Proc. Natl. Acad. Sci., USA**, **79** (1982) 6114.

[60] H. L. T. Mobley and B. P. Rosen, "Energetics of Plasmid-Mediated Arsenate Resistance in *Escherichia coli*," **Proc. Natl. Acad. Sci., USA**, **79** (1982) 6119.

[61] F. H. Osborne and H. L. Ehrlich, "Oxidation of Arsenite by a Soil Isolate of *Alcaligenes*," **J. Appl. Bacteriol.**, **41** (1976) 295.

[62] S. E. Phillips and M. L. Taylor, "Oxidation of Arsenite to Arsenate by *Alcaligenes faecalis*," **Appl. Environ. Microbiol.**, **32** (1976) 392.

[63] A. Albert, **"Selective Toxicity,"** 5th Ed., Chapman and Hall: London, 1973, p. 392.

[64] K. Arima and M. Beppu, "Induction and Mechanisms of Arsenite Resistance in *Pseudomonas pseudomallei*," **J. Bacteriol.**, **88** (1964) 143.

[65] M. Beppu and K. Arima, "Decreased Permeability as the Mechanism of Arsenite Resistance in *Pseudomonas pseudomallei*," **J. Bacteriol.**, **88** (1964) 151.

[66] W. R. Cullen, C. L. Froese, A. Lui, B. C. McBride, D. J. Patmore and M. Reimer, "The Aerobic Methylation of Arsenic by Microorganisms in the Presence of L-Methionine-Methyl-d_3," **J. Organometal. Chem.**, **139**

(1977) 61.
[67] B. C. McBride, H. Meriless, W. R. Cullen and A. W. Pickett, "Anaerobic and Aerobic Alkylation of Arsenic," **ACS Symp. Ser.**, **82** (1978) 94.
[68] W. R. Cullen, B. C. McBride and M. Reimer, "Induction of the Aerobic Methylation of Arsenic by *Candida humicola* ," **Bull. Environ. Contam. Toxicol.**, **21** (1979) 157.
[69] W. R. Cullen, B. C. McBride and A. W. Pickett, "The Transformation of Arsenicals by *Candida humicola*," **Can. J. Microbiol.**, **25** (1979) 1201.
[70] A. W. Pickett, B. C. McBride, W. R. Cullen and H. Manji, "The Reduction of Trimethylarsine Oxide by *Candida humicola*," **Can. J. Microbiol.**, **27** (1981) 773.
[71] S. Jensen and A. Jernelov, "Biological Methylation of Mercury in Aquatic Organisms," **Nature, 223** (1969) 753.
[72] T. Barkay, Program of Social Ecology, University of California, Irvine, unpublished results.
[73] R. D. Perry and S. Silver, "Cadmium and Manganese Transport in *Staphylococcus aureus* Membrane Vesicles," **J. Bacteriol.**, **150** (1982) 973.
[74] Z. Tynecka, Z. Gos and J. Zajac, "Reduced Cadmium Transport Determined by a Resistance Plasmid in *Staphylococcus aureus*," **J. Bacteriol.**, **147** (1981) 305.
[75] A. A. Weiss, S. Silver and T. G. Kinscherf, "Cation Transport Alteration Associated with Plasmid-Determined Resistance to Cadmium in *Staphylococcus aureus*," **Antimicrob. Agents Chemother.**, **14** (1978) 856.
[76] T. J. Tetaz and R. K. J. Luke, "Plasmid-Controlled Resistance to Copper in *Escherichia coli*," **J. Bacteriol.**, **154** (1983) 1263.
[77] Z. Tynecka, Z. Gos and J. Zajac, "Energy-Dependent Efflux of Cadmium Coded by a Plasmid Resistance Determinant in *Staphylococcus aureus*," **J. Bacteriol.**, **147** (1981) 313.
[78] S. Silver, R. D. Perry, Z. Tynecka and T. G. Kinscherf, "Mechanisms of Bacterial Resistances to the Toxic Heavy Metals Antimony, Arsenic, Mercury and Silver," in **"Drug Resistance in Bacteria: Genetics, Biochemistry, and Molecular Biology,"** S. Mitsuhashi, Ed., Japan Scientific Societies Press: Tokyo, 1982, p. 347.
[79] L. McMurry, R. E. Petrucci, Jr., and S. B. Levy, "Active Efflux of Tetracycline Encoded by Four Genetically Different Tetracycline Resistance Determinants in *Escherichia coli*," **Proc. Natl. Acad. Sci., USA, 77** (1980) 3974.
[80] A. J. Corrigan and P. C. Huang, "Cellular Uptake of

Cadmium and Zinc," **Biol. Trace Element Res.**, 3 (1981) 197.
[81] C. L. Fox, Jr., "Silver Sulfadiazine. A New Topical for *Pseudomonas* in Burns Unit," **Arch. Surg.**, 96 (1968) 184.
[82] C. L. Fox, Jr. and S. M. Modak, "Mechanism of Silver Sulfadiazine Action on Burn Wound Infections," **Antimicrob. Agents Chemother.**, 5 (1974) 582.
[83] A. O. Summers, G. A. Jacoby, M. N. Swartz, G. McHugh and L. Sutton, "Metal Cation and Oxyanion Resistances in Plasmids of Gram-Negative Bacteria," in **"Microbiology 1978,"** D. Schlessinger, Ed., American Society for Microbiology: Washington, D.C., 1978, p. 128.
[84] D. I. Annear, B. J. Mee and M. Bailey, "Instability and Linkage of Silver Resistance, Lactose Fermentation and Colony Structure in *Enterobacter cloacae* from Burn Wounds," **J. Clin. Path.**, 29 (1976) 441.
[85] K. Bridges, A. Kidson, E. J. L. Lowbury and M. D. Wilkins, "Gentamicin- and Silver-Resistant *Pseudomonas* in a Burns Unit," **Brit. Med. J.**, 1 (1979) 446.
[86] A. T. Hendry and I. O. Stewart, "Silver-Resistant Enterobacteriaceae from Hospital Patients," **Can. J. Mirobiol.**, 25 (1979) 915.
[87] G. L. McHugh, R. C. Moellering, C. C. Hopkins and M. N. Swartz, "*Salmonella typhimurium* Resistant to Silver Nitrate, Chloramphenicol, and Ampicillin," **Lancet**, 1 (1975) 235.
[88] P. D. Bragg and D. J. Rainnie, "The Effect of Silver Ions on the Respiratory Chain of *Escherichia coli*," **Can J. Microbiol.**, 20 (1974) 883.
[89] W. J. A. Schreurs and H. Rosenberg, "Effect of Silver Ions on Transport and Retention of Phosphate by *Escherichia coli*," **J. Bacteriol.**, 152 (1982) 7.
[90] D. H. Smith, "R Factors Mediate Resistances to Mercury, Nickel, and Cobalt," **Science**, 156 (1967) 1114.
[91] L. H. Bopp, A. M. Chakrabarty and H. L. Ehrlich, "Plasmid-Determined Resistance to Cr(VI) and Reduction of Cr(VI) to Cr(III)," **J. Bacteriol.**, 155 (1983) 1105.
[92] M. Ishihara, Y. Kamio and Y. Terawaki, "Cupric Ion Resistance as a New Marker of a Temperature Sensitive R Plasmid, Rts1 in *Escherichia coli*," **Biochem. Biophys. Res. Comm.**, 82 (1978) 74.
[93] D. J. Groves and F. E. Young, "Epidemiology of Antibiotic and Heavy Metal Resistance in Bacteria: Resistance Patterns in Staphylococci Isolated from Populations not Known to be Exposed to Heavy Metals," **Antimicrob. Agents Chemother.**, 7 (1975) 614.

[94] W. P. Ridley, L. J. Dizikes and J. M. Wood, "Biomethylation of Toxic Elements in the Environment," **Science, 197** (1977) 329.
[95] L. E. Hallas and J. J. Cooney, "Tin and Tin-Resistant Microorganisms in Chesapeake Bay," **Appl. Environ. Microbiol., 41** (1981) 446.
[96] L. E. Hallas, J. C. Means and J. J. Cooney, "Methylation of Tin by Estuarine Microorganisms," **Science, 215** (1982) 1505.
[97] G. E. Parris and F. E. Brinckman, "Reactions which Relate to Environmental Mobility of Arsenic and Antimony. II. Oxidation of Trimethylarsine and Trimethylstibine," **Environ. Sci. Technol., 10** (1976) 1128.
[98] H. Nakahara and H. Kozukue, "Volatilization of Mercury Determined by Plasmids in *E. coli* Isolated from an Aquatic Environment," in **"Drug Resistance in Bacteria: Genetics, Biochemistry, and Molecular Biology,"** S. Mitsuhashi, Ed., Japanese Scientific Societies Press: Tokyo, 1982, p. 337.
[99] P. Gardner, D. H. Smith, H. Beer and R. C. Moellering, Jr., "Recovery of Resistance (R) Factors from a Drug-Free Community," **Lancet, 2** (1969) 774.
[100] I. J. Maré, "Incidence of R Factors among Gram Negative Bacteria in Drug-Free Human and Animal Communities," **Nature, 220** (1968) 1046.

DISCUSSION

P. J. Sadler (Birkbeck College, University of London): Do you know of any detailed studies on the metal (and anion) requirements of bacteria? You mentioned that Ca^{2+} and Cl^- are often not requirements but of course they are often present in laboratory grade reagents as impurities. We find the same situation with Zn^{2+} and Fe^{3+}. For example, there is no need to add any Zn or Fe to the medium. If you vigorously deplete the reagent of Fe or Zn using chelating agents then deficiency can be demonstrated.

S. Silver: Studies of this nature started in the 1930s, and were carried out element by element. There have been no general studies in recent years. The difficulty is always, as your question states, that there are trace impurities present which are adequate for most elements. In our own laboratory, the limit for the requirement for calcium was below 1 µM as measured by AAS. The limit for Cl^-, without special effort, could be measured with a chloride-specific electrode. For Fe, there is a large literature on the use of chelating agents to deplete media. I do not know of any study demonstrating a Zn^{2+} requirement in bacteria exactly for the reason you state: the requirement is so small that the impurity level is sufficient. If you have data on a demonstrated Zn^{2+} requirement in bacteria, then that may be the first time that the impurities have been successfully removed to this level.

P. J. Sadler: The handling of Hg^{2+} by the reductase enzyme you described is fascinating. It is interesting to note that although Hg-SR thiol bonds are thermodynamically very strong, they are also very labile. Therefore, presumably they are easily handled (kinetically) by the di-cysteine active site you mentioned.

S. Silver: I do not know the data for the lability of mercury-thiol bonds. It is comforting that the kinetics can explain what is observed.

P. J. Sadler: The mechanism you described for bacterial resistance to cadmium involved exclusion of Cd^{2+} from cells. However, our work on *P. putida* (described briefly in the current edition of Inorg. Chim. Acta; Proceedings of Florence meeting) shows that here Cd^{2+} is concentrated by cells and cadmium binding proteins are produced. We do observe gross changes in membrane morphology for Cd^{2+}-resistant cells.

S. Silver: The mechanism of Cd^{2+} resistance that I described has been found only in *S. aureus*, and has never been found in Gram negative bacteria such as *P. putida*. In fact, most *Pseudomonas* isolates are as resistant to Cd^{2+} in the absence of a plasmid as are *S. aureus* when a resistance plasmid is present. I would like to see more information on your Cd^{2+} concentrating mechanism. Many of us have been hoping for and looking for a metallothionein in bacteria for a few years now, but your report is perhaps as close as anyone has come. The changes in membrane morphology need more experimental work and definition before we can understand what they mean.

J. M. Wood (Gray Freshwater Biological Institute, University of Minnesota): Do you regard the acquisition of plasmids for resistance to toxic metals a recent event on the evolutionary time scale?

S. Silver: I asked you the same question after your talk. Until you suggested this, John, I had (with very little basis) assumed that plasmids were an ancient invention, that occurred shortly after the evolution of the bacterial cell with its single chromosome. Certainly, toxic heavy metals were around from that time. I do not know what you mean as "recent" and hope you will expand on that point. Plasmids exist in essentially all bacterial types where they have been sought. They exist in the strict anaerobes; and I have assumed without reason that they predate the oxygen atmosphere. How do you answer such a question?

XXXI

ENVIRONMENTAL HAZARDS FROM THE GENETIC TOXICITY OF TRANSITION METAL IONS

E. H. Abbott

Department of Chemistry, Montana State University
Bozeman, Montana 59717 USA

The increase of metal concentrations in the environment caused by natural processes and human activities poses hazards to humans and all other organisms. Although the biological effects ascribed to a metal are caused by specific chemical compounds containing this metal, acceptable upper limits for the concentrations of metals in water are still expressed in "total metal" concentrations. The potential of metal complexes to cause mutations was investigated with a differential lethality bioassay and the Ames test. Many metal complexes were found to be mutagenic. Relatively small changes in the structure of metal complexes caused significant differences in their mutagenic activities. The interactions leading to mutations must involve a metal species that has at least some of the original ligands still attached to the metal ion.

The environmental and health effects of metal ion pollution are important and complex problems. At least eight transition elements are essential to living systems at very low concentrations. However most, if not all elements, are toxic at sufficiently high concentrations. Concentrations of trace elements may rise to toxic levels through natural processes or through the perturbation of the environment by various human activities. The beneficial and detrimental effects ascribed to a metal are caused by specific chemical compounds containing this metal. In general, the nature of these compounds, the modes of transport and transformation of these compounds, the mechanisms of toxicity, and proper methods of preventing toxicity are not known. Because transition metal compounds are known to be potentially hazardous to humans and other organisms, industrial and other human activities must not increase metal concentrations in the environment to harmful levels.

Environmental Regulations Concerning Metals

In the United States permissible concentrations in water were promulgated for only fourteen elements [1] (Table 1).

Table 1. *Acceptable Upper Limits for the Concentrations of Elements in Water* [1]

Element	Permissible Concentration in Drinking Water, mg/L	Element	Permissible Concentration in Drinking Water, mg/L
Ag	0.050	Fe	0.3
As	0.050*	Pb	0.050
Ba	1.0	Mn	0.050
Be	0.011**	Hg	0.002
Cd	0.010	Ni	***
Cr	0.050	Se	0.010
Cu	1.0	Zn	5.0

*For irrigation water: 0.1 mg/L.
**For soft freshwater; the limit for hard freshwater is 1.1 mg/L.
***Ni level checked by bioassay.

These water quality criteria were established in many cases quite arbitrarily, for instance, as some fraction of the concentration, at which chronic toxicity becomes apparent. The chemical forms of metals were not considered in setting these criteria. These criteria were criticized by the American Fisheries Society after careful analysis of the current pertinent literature [2]. The permissible concentrations were judged to be too high for Cd, Se, and Hg, and too low for Zn. The promulgated criteria for Cd, Cr, and Hg are suspect, because the effects of particular compounds of these elements were not considered. It is known that the toxicities of Cd, Cr, and Hg are strongly affected by the chemical nature of the compounds that contain these elements.

Other elements, for which permissible concentrations were not established, are also known to be hazardous. Aluminum, for instance, is fairly innocuous as the insoluble hydroxide or hydrous oxide, but will create problems in an acidic environment caused by acid rain or the dumping of acid mine water. Under such conditions mononuclear and polynuclear aluminum complexes form that are quite soluble. These soluble aluminum species may be responsible for the destruction of aquatic life and may cause Altzheimer's disease.

Genetic Damage through Exposure to Metal Ions

Metal ions and their complexes have the potential to cause genetic damage. For many kinds of toxicants concentration limits exist, below which any biological effects are insignificant. Organisms exposed to toxicants at levels higher than these limits may recover after the exposure ceases. For toxicants causing genetic damage such a concentration limit may not exist, and the toxicant may be hazardous at any concentration. Once a mutation occurred, it might persist even if the mutagen is withdrawn.

It is well established that metals can be carcinogenic. Human epidemeological studies implicated cadmium, chromium, nickel, and arsenic as carcinogens [3]. Human and animal studies suggested lead [4], cobalt [5], iron [6], titanium [5], and zinc [5] to be carcinogenic also. We studied the potential of metal complexes to cause genetic damage using the microbial bioassays of differential lethality and reversion of mutants. The differential lethality assay compares the toxicity of a metal complex toward a wild strain of *Escherichia coli* with the toxicity toward mutant strains lacking at least one mode of DNA repair. When a toxicant causes the death of repair-deficient bacteria at lower toxicant concentration than required to destroy the repair-proficient wild bacteria, the toxicant is presumed to have caused genetic damage in the repair-deficient strain. Because strains differ in the mechanism of their repair deficiency, differential lethality tests provide evidence for the genetic toxicity of toxicants and for the mechanism by which toxicity occurs. These tests were performed with strains and methods developed by G. Warren at Montanta State University [7]. The second bioassay, the classic Ames test [8], determines the ability of a toxicant to induce mutations converting strains of *Salmonella* unable to produce histidine to mutants able to produce histidine.

Table 2 summarizes selected results of bioassays conducted with substitutionally inert transition metal

Table 2. Selected Results of Microbial Bioassays for the Genetic Toxicity of Transition Metal Complexes

Compound*	Differential Lethality**				Ames Test** Number of His+ Revertants per Nanomole Toxicant		
	GW802	GW801	AA34	AB1886	TA92	TA98	TA100
cis-[Cr(bipy)$_2$Cl$_2$]Cl	512	8	64	16	2.55	1.05	2.05
cis-[Cr(phen)$_2$Cl$_2$]Cl	256	8	32	16	0.21	1.77	3.08
cis-[Cr(bipy)$_2$ox]I	512	2	8	16	0.22	0.63	2.52
cis-[Rh(en)$_2$Cl$_2$]Cl	16	†	†	†	0.04	†	†
cis-[Rh(phen)$_2$Cl$_2$]Cl	32	†	†	†	0.01	†	†
cis-[Rh(bipy)$_2$Cl$_2$]Cl	8	†	†	†	0.08	†	†
[Rh(CH$_3$CN)$_3$Cl$_3$]	32	†	†	†	0.30	†	†
trans-[Rh(py)$_4$Cl$_2$]Cl	128	†	†	8	0.18	†	†
trans-[Rh(py)$_4$Br$_2$]Br	512	16	8	16	0.99	†	†
trans-[Rh(3-pic)$_4$Cl$_2$]Cl	250	32	16	32	0.70	†	†
+[Co(en$_3$)]I$_3$	4	†	†	†	†	†	†
+[Co(en)$_3$]Cl$_3$	4	†	†	†	†	†	†
-[Co(en)$_3$]I$_3$	16	†	†	†	†	†	†
Cis-[Co(phen)$_2$Cl$_2$]Cl	64	8	†	8	0.03	†	†

*en: ethylendiamine; phen: 1,10-phenanthroline; bipy: 2,2-bipyridyl; ox: oxalate; py: pyridine; 3-pic: 3-picoline. These ligands are not mutagenic in the absence of a metal ion.

**GW802: recombinationless and excision minus; GW801: recombinationless; AA34: double recombinationless; AB1886: excision minus; TA92: his⁻ base pair substituted; TA98: his⁻ frame shifted; TA100: his⁻ base pair substituted. The results are reported as (lethal concentration for the wild type/lethal concentration for the repair-deficient type). A number of 2 is 95 percent significant.

† No mutagenic activity observed.

complexes [9-11]. Hexaaquochromium(III) salts such as $[Cr(H_2O)_6]Cl_3$ were found to be non-mutagenic by many investigators and proved to be inactive in our bioassays. The complexes $[Cr(en)_3]Cl_3$, $[Cr(NH_3)_4ox]Cl$, and cis-$[Cr(en)_2Cl_2]Cl$ were also inactive. However, when aromatic ligands are present in the complexes, mutagenicity is often observed (Table 2). Bipyridyl and phenanthroline complexes of chromium(III) are examples of compounds quite active in the lethality assay and the Ames test. The complex cis-$[Cr(bipy)_2Cl_2]Cl$ affects GW802 and TA92 to a larger degree than cis$[Cr(phen)_2Cl_2]Cl$, whereas TA100 and TA98 are more affected by the phenanthroline than the bipyridyl complex. These patterns are inconsistent with reduced or oxidized forms of the complexes as the true mutagens. Differences in the transport of these complexes through membranes are very likely not responsible for the observed effects. It is more plausible, that chromium(III) complexes become mutagenic, when they contain the right combination of ligands. The critical reaction leading to a mutation might occur while at least some of the ligands are still attached to chromium [9].

The mutagenicity of rhodium(III) complexes varies widely with their structure. Tripositive rhodium(III) complexes, such as $[Rh(phen)_3]Cl_3$, and the complex $[Rh(py)_3Cl_3]$ were found to be inactive. Whereas cis-dichlorobis(ethylenediamine)rhodium(III) chloride is mutagenic to many bacterial strains, the trans-form is inactive. Relatively small changes in the structure of complexes in the series of trans-dihalotetrakis(pyridine)rhodium(III) such as a change of halide, or a different substituent in the pyridine ring cause significant differences in activities (Table 2).

Several cobalt(III) complexes were found to be mutagenic. It is interesting to note, that small but significant differences in mutagenicity exist between the tris(ethylenediamine)cobalt(III) complexes of different absolute configuration (Table 2). This observation is evidence that cobalt is mutagenic as cobalt(III) and causes mutations, when at least two of the original ligands are still attached to the metal ion.

Many other metal complexes that are relatively inert toward ligand substitution might be mutagenic. Many metals are bioaccumulated by organisms. If these metals are not excreted, they might form mutagenic complexes inside the cell and cause genetic damage.

References

[1] Environmental Protection Agency, **"Quality Criteria for Water,"** U.S. Government Printing Office, 0-222-904, Washington, D.C., 1977.

[2] R. V. Thurston, R. C. Russo, T. A. Edsall and Y. M. Barber, Eds., **"A Review of the EPA Red Book: Quality Criteria for Water,"** Water Quality Section, American Fisheries Society: Bethesda, MD., 1979.

[3] F. W. Sunderman, "Carcinogenic Effects of Metals," **Fed. Proc., 37** (1978) 40.

[4] W. C. Cooper, "Cancer Mortality Patterns in the Lead Industry," **Ann. N.Y. Acad. Sci., 271** (1976) 250.

[5] F. W. Sunderman "Metal Carcinogenesis," in **"Advances in Modern Toxicology,"** R. A. Goyer and M. Mehlman, Eds., Hemisphere Corp: Washington, D.C., 1977, p. 257.

[6] A. E. MacKinnon and J. Banceiviez, "Sarcoma After Injection of Intramuscular Iron," **Br. Med. J., 2** (1973) 277.

[7] G. Warren "Detection of Genetically Toxic Metals by a Microliter Microbial Repair Assay," in **"Short-Term Bioassays in the Analysis of Complex Environmental Mixtures,"** S. Sander, J. Husingh, L. Claxton and S. Nesnow, Eds., Plenum Press: New York, 1981, p. 101.

[8] B. R. Ames, J. McCann and E. Yamasaki, "Methods for Detecting Carcinogens and Mutagens with the *Salmonella/* Mammalian-Microsome Mutagenicity Test," **Mutat. Res., 31** (1975) 347.

[9] G. Warren, P. Schultz, D. Bancroft, K. Bennett, E. H. Abbott and S. J. Rogers, "Mutagenicity of a Series of Hexacoordinate Rhodium(III) Compounds," **Mutat. Res., 88** (1981) 165.

[10] G. Warren, E. H. Abbott, P. Schultz, K. Bennett and S. J. Rogers, "Mutagenicity of a Series of Hexacoordinate Chromium(III) Compounds," **Mutat. Res., 90** (1981) 111.

[11] P. N. Schulz, G. Warren, S. Kosso and S. Rogers, "Mutagenicity of a Series of Hexacoordinate Cobalt(III) Compounds," **Mutat. Res., 102** (1982) 393.

XXXII

ENVIRONMENTAL INORGANIC ANALYTICAL CHEMISTRY

Kurt J. Irgolic

Department of Chemistry, Texas A&M University
College Station, Texas 77843 USA

The methods for the determination of total trace element concentrations in samples of interest to bioinorganic chemists are reviewed with special emphasis on multi-element techniques including mass spectrometry coupled with an inductively coupled plasma as an ion source. The importance of knowledge about the types of trace element compounds in environmental samples is emphasized and techniques for the identification and determination of kinetically or thermodynamically stable trace element compounds are presented. Examples of the analysis of samples containing nonvolatile trace element compounds (selenite, selenate; trialkyl tin species; arsenite, arsenate, organic arsenic compounds) by high pressure liquid chromatography and ion chromatography with graphite furnace atomic absorption spectrometers (Perkin Elmer, Hitachi Zeeman) or inductively coupled argon plasma emission spectrometers serving as single-element- or multi-element-specific detectors are given.

Without analytical chemistry most branches of chemistry would have been severely retarded in their development or would not have developed at all. Environmental chemistry would certainly be only of minor importance today without the accurate, precise, sensitive, and sophisticated analytical methods, which call attention to the presence of detrimental substances and the absence of essential elements in our environment. The statement "what we do not know, still hurts us" is certainly correct with respect to environmental concerns. Such a statement raises, however, only anxieties and does not provide guidelines for appropriate and effective countermeasures. Analytical chemistry delivers the information which is needed to assess an environmental situation, identify its causes, suggest preliminary action, and point toward a permanent solution.

Wilhelm Ostwald, wrote in 1894 in the preface to his classical book on the scientific fundamentals of analytical chemistry: "Analytical chemistry, or the art of recognising different substances and determining their constituents, takes a prominent position among the applications of the science, since the questions it enables one to answer arise wherever chemical processes are employed for scientific or technical purposes. Its supreme importance has caused it to be assiduously cultivated from a very early period in the history of chemistry, and its records comprise a large part of the quantitative work which is spread over the whole domain of the science" [1]. Aaron J. Ihde [2] in his book "The Development of Modern Chemistry", assesses the importance of analytical chemistry in the following manner: "It is not incorrect to ascribe to analytical chemistry a position of primary importance since only through chemical analysis can matter in its variety of forms be dealt with intelligently. The stimulus given to chemistry by new analytical approaches, either qualitative or quantitative, has been repeatedly observed. In general, however, analytical chemistry has never achieved recognition in keeping with its importance because the application of new techniques has resulted in new descriptive or theoretical knowledge that completely overshadows the technique which made the knowledge possible".

Analytical chemistry and specially environmental analytical chemistry has gained in importance during the past few decades. The results generated by analytical chemists have tremendous impact on the industrial and governmental sectors of our society. Suffice it here to mention only the concept of zero pollution, a highly debatable political and regulatory issue, which is directly connected to analytical instruments, detection limits, and the skills of analytical chemists [3].

Just as chemistry has been divided into organic and inorganic chemistry, environmental analytical chemistry has been separated into environmental analytical inorganic and environmental analytical organic chemistry. The different analytical methodologies applicable to these two branches of environmental analytical chemistry are one of the reasons for this division. This paper will outline some aspects of environmental analytical inorganic chemistry which shall include organometallic and organometalloidal chemistry by first discussing the determination of "total element" concentrations and then the determination of trace element compounds.

Determination of "Total Trace Element" Concentrations

When total trace element concentrations are determined, no attention is being paid to the chemical nature of the compounds, which contain the element of interest. The art and science of total trace element determinations have a long, distinguished history and have achieved great success. Analytical inorganic chemistry began as a "handmaiden" to the mining industry in the fifteenth century. The fire assay of ores was improved three centuries later through wet chemical analyses, which were introduced by the Swedish Chemist Torbern Bergman and subsequently improved and successfully applied to the analysis of minerals by Klaproth and Berzelius [4]. Titrimetric methods, electroanalytical methods, and spectroscopic techniques slowly replaced wet-chemical, gravimetric techniques. The instrumental period of analysis, which gained momentum after 1945, placed at the disposal of the analyst automated, computerized techniques for the rapid determination of metallic and non-metallic constituents of matter. These developments in analytical chemistry were guided by the desire for speed, selectivity, and sensitivity. Today the analyst is in the fortunate position to be able to determine any element in any sample at part per billion, part per trillion, and even sub-part per trillion levels provided the required expensive instruments, the skilled technical personnel to operate the instruments, and the necessary funds to buy the first and pay the latter are available.

The detection limits for many elements have reached levels assuring that these elements will be found in any sample for the simple reason that nature put them there. Although analytical methods used in environmental work should have detection limits one or two orders of magnitude below natural background concentrations, further efforts to increase sensitivity will not benefit environmental chemistry. Life evolved and successfully developed in a natural environment, which contained almost all known elements. Even if an ubiquitous element were found to be undesirable at the low natural background concentration, its large-scale removal to prevent exposure of a population would probably not be feasible.

Not too long ago analyses of samples gave results generally expressed in percent or fractional percent. The analysis of mineral waters - a matter of great importance for health spas - required a close look at trace elements, which were quantitated prior to the era of instrumental techniques by conventional methods after enriching the dissolved substances by evaporation of large volumes of water samples. Improvements in analytical techniques and introduction of new reagents pushed the detection limits toward ppth (mg/mL, g/L), ppm (µg/mL, mg/L), and ppb (ng/mL, µg/L). Now we are exploring the domain of ppt (pg/mL, ng/L) concentrations. The number of elements, that can be determined at these concentrations, and the number of methods, that are available to accomplish these tasks [5,6] are a testimonial to the hard and successful work of the analytical chemists.

Only a brief survey of the most common methods for the determination of total trace element concentrations can be given (Table 1). Most of the methods listed in Table 1 have detection limits in the region of µg/mL, and certain elements can be determined at even lower levels. Unless an investigation is limited to only one or a few elements, techniques and instrumentation capable of determining many elements simultaneously or sequentially in an automated manner are most desirable. The analysis of many samples for a large number of elements is not an uncommon task. Such tasks require automated, computer-controlled, sensitive instruments. Presently, plasma emission spectrometry with direct current or inductively coupled plasmas, atomic fluorescence spectrometry, X-ray fluorescence spectrometry and neutron activation analysis are the most commonly applied multi-element techniques. The high cost of the required instruments prevents many laboratories from using these techniques.

An exciting and promising new instrument has recently become commercially available. It combines an argon plasma known for its high ionization efficiency with a mass spectrometer, which serves as a sensitive detector. The "spectra" are much simpler than those obtained by optical plasma emission spectrometry and are claimed to be matrix-independent. At least 30 elements can be determined per minute with detection limits in the range of 0.1 to 10 µg/L. This inductively coupled plasma-mass spectrometry (ICP-MS) system seems to avoid the spectral interferences and background problems, that are rather common in optical plasma emission spectrometry.

These multi-element techniques do not make the "single-element" methods useless, outdated, unnecessary or impracticable. Fluorometry, flame emission and absorption, polarography, and graphite furnace atomic absorption have

Table 1. Selected Methods for the Determination of Total Trace Element Concentrations

Method	Type of Sample[†] Vol./Mass Required	Quantity of Element Required for Detection*, – log gram	Detection* Limits mg/L (kg)	Non-Destructive	Multi-Element Capability	Instrument Cost $\$ \times 10^{-3}$
Activation Analysis	Sol, S mg–g	12	0.001	Yes	Yes	high
Anodic Stripping Voltammetry	Sol few mL	11	<0.001	Yes	Yes	5
Atomic Fluorescence	Sol few mL	8	0.5 –0.001	No	Yes	70
Atomic Absorption						
Electrothermal	Sol, 0.1 mL	12	0.001	No	No	30
Flame	Sol, few mL	8	10. –0.001	No	No	10
Atomic Emission						
Flame	Sol, few mL	8	100. –0.0002	No	No	20
DC-Plasma	Sol, few mL	10	0.1 –0.0005	No	Yes	80
IC-Plasma	Sol, few mL	10	0.2 –0.0002	No	Yes	60–200
Fluorometry	Sol, S few mL	7	0.1 –0.001	Yes	No	10
Inductively Coupled Plasma-Mass Spectrometry	Sol few mL	7	0.1 –0.0001	No	Yes	170
Polarography	Sol, few mL	7	1.0 –0.001	Yes	Yes	5
X-Ray Fluorescence	S 0.5 g	7	1.0	Yes	Yes	100

*The quantities of an element required for detection and the detection limits vary considerably with the element to be determined.

[†] Sol: solution; S: solid

their rightful place in the arsenal of the analytical chemist and serve a very important function. Analytical results, especially those for trace elements obtained at levels approaching the detection limits of multi-element instruments, need to be verified by independent methods. Several of the single-element techniques are well suited for checking selected results [6].

The multi-element techniques - perhaps with exception of neutron activation analysis - are not yet sensitive enough for the determination of background levels of many trace elements in an unpolluted environment. Special techniques with or without preconcentration of the elements of interest are still needed to determine, for instance, mercury, lead, chromium, and several other elements in sea water.

The tasks of the chemical oceanographers, of scientists concerned with inorganic pollutants, and of analytical chemists would be much easier, if the detection limits of the most widely used multi-element techniques were one hundredth of those presently achievable. The development of detectors for plasma emission spectrometers with better sensitivity and improvements on the ICP-MS system justify the belief, that detection limits might recede to low ppt levels within the next few years. In the meantime, multi-element analyses of samples containing elements below current detection limits must rely on preconcentration employing extraction, chelating resins or immobilized chelates [5,7].

Determination of Trace Element Compounds

Knowledge of the composition of a sample in terms of total trace element concentrations is a first, important and necessary step in an evaluation of environmental impact, but does certainly not provide all or even the most important information. Beneficial or detrimental interactions of life-sustaining, biologically important molecules in living organisms occur in most cases not with elements or their free ions but with specific compounds of these elements. A solution, which has been analyzed by a helium microwave plasma system and has been found to contain C, H, and N, may serve as an extreme example. Knowledge about the C, H and N concentrations does not allow to declare this sample fit for human consumption. It is necessary to know the types of C, H, N-compounds present in the solution before pronouncing it safe. Whether the solution contains HCN or an amino acid, is certainly of utmost importance.

The idea, that the compounds of trace elements need to be identified, is not new. It never was doubted that identification of compounds or species is necessary in environmental organic chemistry. However, this idea has been only slowly accepted in the area of inorganic and organometallic environmental chemistry. Nobody questions the propriety of setting legal limits for organic pollutants in terms of compounds such as DDT or PCB. However, most of the legal limits for inorganic and organometallic substances are still expressed as total element concentrations.

An editorial in Analytical Chemistry [8] stated in 1971: "Legal limits for pollutants, particularly the heavy metals, are set in terms of the total concentration of each element rather than in terms of specific chemical forms. As our knowledge of the chemistry and toxicology improves, it is increasingly evident that such limits may have little meaning without a characterization as to chemical species". Analytical chemistry has made considerable progress toward the determination of chemical compounds of trace elements.

Environmental samples (water, organic matter, soil, rocks, particulates from the air) are generally very complex. Before a particular trace element compound can be determined, it must in most cases be separated to reduce the complexity of the sample before the compound can be measured by a suitable compound-specific or at least element-specific detector. Figure 1 presents such an analysis system in schematic form.

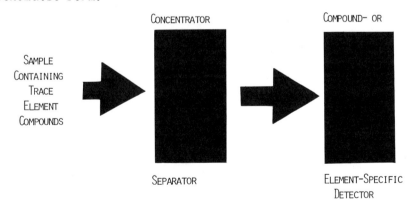

Figure 1. Schematic of a system for the separation and element-specific detection of trace element compounds.

Many methods are available for the concentration and separation of trace elements and trace-element compounds [9, 10]. Solvent extraction aided by chelating agents, electrolysis, ion exchange, and the many forms of chromatography may serve as examples. Element-specific and compound-specific detection systems are not offering so many choices.

The inorganic and organometallic compounds to be detected in the environment can be divided into three groups:

- Compounds with sufficient volatility to allow gas-chromatographic separation [$(CH_3)_2Se$, $(CH_3)_2AsH$, H_2S, $(C_2H_5)_4Pb$].
- Compounds that can be converted to volatile derivatives by reaction with suitable reagents, e.g., esterification, silylation, reduction to hydrides [$CH_3AsO_3H_2$].
- Compounds that are involatile and thus are not suitable for gas chromatography or cannot be converted to volatile derivatives [$(CH_3)_3AsCH_2COO$, arsenolipids, $(CH_3)_3SnX$].

The best compound-specific detector might be a mass-spectrometer (MS) monitoring the signal at a specific m/e value or operating over the entire mass range of the spectrometer in the electron impact or preferably in the chemical ionization mode. This detection system is used widely for volatile compounds and compounds convertible to volatile derivatives. Computer-controlled gas chromatography-mass spectrometry systems are available commercially. Non-volatile compounds cannot be identified by such a system. The separation of non-volatile compounds is best accomplished by liquid chromatography (LC). Although an LC-MS system is on the market, the transfer of involatile compounds from the effluent into the mass spectrometer is problematic. With the advent of commercial ICP-MS instruments, the use of the mass spectrometer as an element-specific detector (although not as a compound-specific detector) for liquid chromatography is now possible.

Any device that measures a unique property of an element or a molecule with sufficient sensitivity may serve as an element- or compound-specific detector. An important additional requirement for such detectors is insensitivity toward sample matrices. Detectors measuring atomic transitions are element-specific and relatively unaffected

Environmental Inorganic Analytical Chemistry   555

by the matrix. Therefore, flame emission and absorption spectrometers, graphite furnaces atomic absorption spectrometers (GFAA) [11,13], atomic fluorescence spectrometers and plasma atomic emission spectrometers [14,15] have been interfaced with gas chromatographs [11-14], high pressure liquid chromatographs (HPLC) and ion chromatographs (IC) [12-16].

Space limitations do not permit to present a comprehensive review of the various element-specific detection systems and their application to analytical problems. I will restrict myself to a few examples from the recent literature, which describe the use of GFAA or plasma emission spectrometers as element-specific detectors for high pressure liquid chromatography.

Two HPLC-GFAA systems were developed. The first system uses a Perkin-Elmer GFAA with a specially adapted autosampler [17]. The effluent passes through a well sampler with a volume of 50 µL. The autosampler arm dips into the well and transfers an aliquot of the effluent into the GF tube (Fig. 2). The HPLC-GFAA chromatogram (Fig. 3)

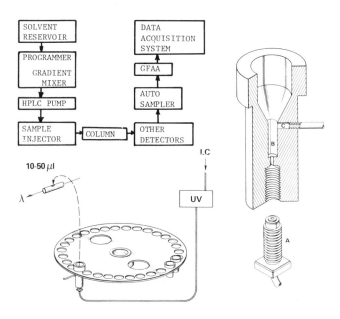

Figure 2. *The high pressure liquid chromatography-Perkin Elmer graphite furnace atomic spectrometer system. Redrawn from ref. 17.*

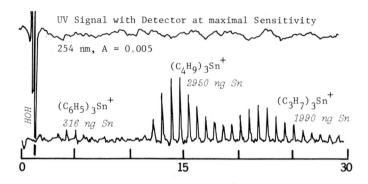

Figure 3. *Chromatogram of triorganyltin cations obtained with the HPLC-Perkin Elmer GFAA system on a Lichrosorb C_2 reverse-bonded phase column with methanol as mobile phase. Redrawn from ref. 17.*

obtained with a solution of triorganyltin ions clearly shows the advantage of this element-specific detector over a UV-detector. The second system employs a Hitachi-Zeeman GFAA with a sample valve, an injector and associated electronics to control the analysis sequence [18,19].

These systems were used to determine arsenite, arsenate, methylarsonic acid, dimethylarsinic acid [20,21,22] phenylarsonic acid [21], 4-aminophenylarsonic acid [20], arsenobetaine, arsenocholine [19], arsenic-containing lipids [23], lead compounds [24], tin derivatives [25-28], and selenite and selenate [29] employing either high pressure liquid chromatography [18-28] or ion chromatography [29]. The matrices analyzed include fresh water, motor oils, extracts from bacteria, lipids extracted from algae, fossil fuel precursors and products [30], oil shale [31], and shale oils [32]. Figure 4 gives an example of a chromatogram obtained with an IC-GFAA system.

Inductively coupled plasma atomic emission spectrometers (ICP-AES) possess several characteristics that make them useful HPLC detectors. ICP-AES is multi-element-specific, has detection limits in the ng/mL range, is relatively free from interferences and is easy to interface with a liquid chromatograph. ICP-AES/LC systems have thus far been used to separate Cu(II) complexes [33], iron and molybdenum carbonyls, alkylmercury compounds, lead compounds in gasoline, ferrocene derivatives [34], inorganic and organic arsenic compounds [34,35] and metal-ion-containing protein mixtures [36].

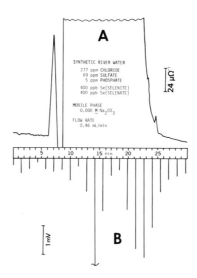

Figure 4. Ion chromatogram for selenite and selenate with conductivity detector (A) and graphite furnace atomic absorption detector (B). Reproduced with permission from ref. 29.

A computerized HPLC/ICP-AES system was developed which allows the simultaneous determination of five elements in the column effluent. The separation and detection of five arsenic compounds, of phosphate and selenite is shown in Fig. 5 [37]. The advantages of an HPLC/ICP-AES system are its multi-element specificity, the requirement that only compounds containing the same element must be separated by the chromatograph, the variability of the sampling rate, almost total sample consumption (if desired), which assures that sharp peaks will not be missed, and the independence of the signal from the nature of the compounds.

In order to successfully determine trace element compounds in a sample, the compounds must remain unchanged during dissolution, concentration or extraction procedures, and remain intact during the chromatographic separation. Not all compounds have such characteristics. Most of the trace element compounds that have thus far been separated and identified were organometallic, organometalloidal or inorganic anionic derivatives. These compounds are stable in solution and do not change when the solvent is changed, for instance, during chromatography.

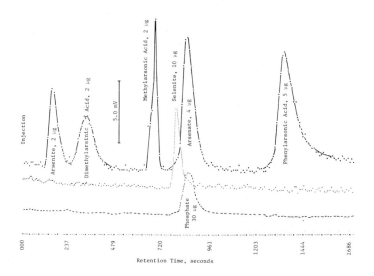

Figure 5. Chromatogram of five arsenic compounds, phosphate, and selenite obtained with a HPLC ICP-AES system. Reproduced with permission from ref. 37.

Many trace elements are methylated and transformed by organisms and emerge from their journey through living cells as compounds with carbon-element bonds [38]. Such compounds are easily determined. Figure 6 identifies the elements that form hydrolytically stable carbon-element bonds. Most of these elements are metalloids although metals such as Hg, Au, Pt, and Co also form organometallic compounds stable toward water. All of the other elements (excluding those claimed by the organic chemists) exist in the environment as ions. In aqueous solutions, these ions are present as complexes with water, inorganic anions, low molecular mass organic compounds, or functional groups of humic acid as ligands. Complexes of these ions can be separated, identified and quantitatively determined only when they are kinetically or thermodynamically stable under the conditions of the analytical procedure. If such stability does not exist, the presence of complexes must be inferred from the composition of the sample by calculation using appropriate stability constants.

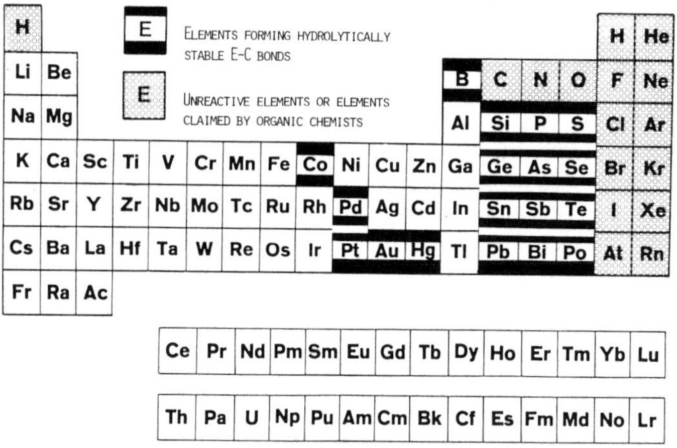

Figure 6. Elements forming hydrolytically stable organic compounds.

Conclusions

A variety of "hyphenated" analytical techniques are available which make it possible to identify and quantitatively determine trace element compounds at the µg/L levels. For compounds to be identified in this manner they must be kinetically or thermodynamically stable under the conditions of the analytical procedure. Volatile compounds and compounds convertible to volatile derivatives can be easily identified by GC-MS, GC-GFAA and similar techniques such as hydride generation in conjunction with element-specific detectors. For non-volatile compounds HPLC-GFAA, IC-GFAA and LC/ICP-AES are available. The detection limits of these systems are in the 100 µg/L range. The ICP-MS system promises to be a sensitive and multi-element-specific detection system for liquid chromatography. These techniques, specially those for non-volatile, non-reducible compounds, deserve wider application and much more attention with respect to their further development. Close cooperation between analytical chemists and bioinorganic chemists is very desirable and will bring important results. Unfortunately, HPLC-GFAA systems or appropriate attachment or conversion kits are not commercially available, and multi-channel plasma emission spectrometers are much too expensive to be acquired as a detector for a rather inexpensive chromatograph. However, progress in bioinorganic chemistry and in the area of transformation of trace element compounds within biological systems will be greatly accelerated by the application of these techniques.

The determination of total trace element concentrations does not pose any insurmountable problems, although it would be desirable to lower the detection limits of the currently available multi-element instruments by one to two orders of magnitude. Such an increased sensitivity would allow the determination of background levels in a routine manner and increase the confidence in the results that are close to the current detection limits.

References

[1] W. Ostwald, **"The Scientific Foundations of Analytical Chemistry"**, 3rd English Ed., McMillan & Co.: London, 1908, p. VII.
[2] A. J. Ihde, **"The Development of Modern Chemistry"**, Harper & Rowe, New York, 1964, p. 277.
[3] H. A. Laitinen, "Analytical Chemistry in Environmental Science I. The Meaning of Zero". **Anal. Chem., 43** (1971) 497.
[4] H. A. Laitinen and G. W. Ewing, Eds., **"A History of Analytical Chemistry"**, The Division of Analytical Chemistry of the American Chemical Society: Washington, D.C., 1977, p. 8.
[5] M. Pinta, **"Detection and Determination of Trace Elements"**, Ann Arbor Science Publishers, Inc.: Ann Arbor, Michigan, 1975.
[6] M. Pinta, **"Modern Methods for Trace Analysis"**, Ann Arbor Science Publishers, Inc.: Ann Arbor, Michigan, 1978.
[7] R. E. Sturgeon, S. S. Berman, S. N. Willie and J. A. H. Desaulniers, "Preconcentration of Trace Elements from Seawater with Silica-Immobilized 8-Hydroxyquinoline", **Anal. Chem., 53** (1981) 2337.
[8] H. A. Laitinen, "Analytical Chemistry in Environmental Science. II. The Need for Identification of Species", **Anal. Chem., 43** (1971) 809.
[9] Ref. 5, pp. 36-98.
[10] J. Minczewski, J. Chwastowska and R. Dybczynski, **"Separation and Preconcentration Methods in Inorganic Trace Analysis"**, John Wiley & Sons, New York, 1982.
[11] G. E. Parris, W. R. Blair and F. E. Brinckman, "Chemical and Physical Considerations in the Use of Atomic

Absorption Detectors Coupled with a Gas Chromatograph for Determination of Trace Organometallic Gases", **Anal. Chem.**, 49 (1977) 378.

[12] Y. K. Chan and P. T. S. Wong, "An Element- and Speciation-Specific Technique for the Determination of Organometallic Compounds", in **"Environmental Analysis"**, G. W. Ewing, Ed., Academic Press: New York, 1977, p. 215.

[13] Y. K. Chan and P. T. S. Wong, "Direct Speciation Analysis of Molecular and Ionic Organometals", in **"Trace Element Speciation in Surface Waters and It Ecological Implications"**, G. G. Leppard, Ed., NATO Conference Series. I: Ecology, Vol. 6, Plenum Press: New York, 1983, p. 87.

[14] P. C. Uden, "Plasma Emission Spectroscopic Detectors in Chromatography: A Review of a Developing Field", in **"Developments in Atomic Plasma Spectrochemical Analysis"**, R. M. Barnes, Ed., Heyden & Son, Ltd.: London, 1981, p. 302.

[15] K. J. Irgolic, R. A. Stockton and D. Chakraborti, "Simultaneous Inductively Coupled Argon Plasma Emission Spectrometer as a Multi-Element-Specific Detector for High Pressure Liquid Chromatography: The Determination of Arsenic, Selenium and Phosphorus Compounds", **Spectrochim. Acta, Part,** 38B (1983) 437, and references cited therein.

[16] K. L. Jewett and F. E. Brinckman, "The Use of Element-Specific Detectors Coupled with High-Performance Liquid Chromatographs", in **"Detectors in Liquid Chromatography"**, T. M. Vickrey, Ed., Marcel Dekker Publishers: New York, 1983, p. 205.

[17] F. E. Brinckman, W. R. Blair, K. L. Jewett and W. P. Iverson, "Application of a Liquid Chromatograph Coupled with a Flameless Atomic Absorption Detector for Speciation of Trace Organometallic Compounds", **J. Chromatogr. Sci.,** 15 (1977) 493.

[18] K. J. Irgolic, "Speciation of Arsenic Compounds in Water Supplies", Report 1982, EPA/600/1-82/010, Order No. PB82-257817, NTIS, 125 pp., **Chem. Abstr.,** 98, 113378.

[19] R. A. Stockton and K. J. Irgolic, "The Hitachi Graphite Furnace-Zeeman Atomic Absorption Spectrometer as an Automated, Element-Specific Detector for High Pressure Liquid Chromatography: The Separation of Arsenobetaine, Arsenocholine and Arsenite/Arsenate", **Intern. J. Environ. Anal. Chem.,** 6 (1979) 313.

[20] F. E. Brinckman, K. L. Jewett, W. P. Iverson, K. J. Irgolic, K. C. Ehrhardt and R. A. Stockton, "Graphite

Furnace Atomic Absorption Spectrophotometers as Automated Element-Specific Detectors for High Pressure Liquid Chromatography: The Determination of Arsenite, Arsenate, Methylarsonic Acid and Dimethylarsinic Acid," **J. Chromatogr., 191** (1980) 31.

[21] R. H. Fish, F. E. Brinckman and K. L. Jewett, "Fingerprinting Inorganic Arsenic and Organoarsenic Compounds in in-Situ Oil Shale Retort and Process Waters Using a Liquid Chromatograph Coupled with an Atomic Absorption Spectrometer as a Detector", **Environ. Sci. Technol., 16** (1982) 174.

[22] E. A. Woolson and N. Aharonson, "Separation and Detection of Arsenical Pesticide Residues and Some of Their Metabolites by High Pressure Liquid Chromatography-Graphite Furnace Atomic Absorption Spectrometry", **J. Assoc. Off. Anal. Chem., 63** (1980) 523.

[23] N. R. Bottino, E. R. Cox, K. J. Irgolic, S. Maeda, W. J. McShane, R. A. Stockton and R. A. Zingaro, "Arsenic Uptake and Metabolism by the Alga *Tetraselmis chui*", **ACS Symp. Ser., 82** (1978) 116.

[24] F. E. Brinckman and W. R. Blair, "Speciation of Metals in Used Oils: Recent Progress and Environmental Implications of Molecular Lead Compounds in Used Crankcase Oils", in "Proc. Workshop on Measurements and Standards for Recycled Oil-II", D. Becker, Ed., **NBS Special Publication, 556** (1979) 25.

[25] E. J. Parks, F. E. Brinckman and W. R. Blair, "Application of a Graphite Furnace Atomic Absorption Detector Automatically Coupled to a High-Performance Liquid Chromatograph for Speciation of Metal-Containing Macromolecules", **J. Chromatogr., 185** (1979) 563.

[26] W. R. Blair, G. J. Olson, F. E. Brinckman and W. P. Iverson, "Accumulation and Fate of Tri-n-butyltin Cation in Estuarine Bacteria", **Microb. Ecol., 8** (1982) 241.

[27] K. L. Jewett and F. E. Brinckman, "Speciation of Trace Di- and Triorganotins in Water by Ion-Exchange HPLC-GFAA", **J. Chromatogr. Sci., 19** (1981) 583.

[28] F. E. Brinckman, J. A. Jackson, W. R. Blair, G. J. Olson and W. P. Iverson, "Ultratrace Speciation and Biogenesis of Methyltin Transport Species in Estuarine Waters", in **"Trace Metals in Sea Water"**, C. S. Wong, E. Boyle, K. W. Bruland, J. D. Burton and E. D. Goldberg, Eds., NATO Conference Series IV: Marine Sciences, Vol. 9, Plenum Press: New York, 1983, p. 39.

[29] D. Chakraborti, D. C. Hillman, K. J. Irgolic and R. A. Zingaro, Hitachi Zeeman Graphite Furnace Atomic Absorption Spectrometer as a Selenium-Specific Detector

for Ion Chromatography: Separation and Determination of Selenite and Selenate", **J. Chromatogr., 249** (1982) 81.

[30] F. E. Brinckman, C. S. Weiss and R. H. Fish, "Speciation of Inorganic Arsenic and Organoarsenic Compounds in Fossil Fuel Precursors and Products", in **"Chemical and Geochemical Aspects of Fossil Fuel Extraction"**, T. F. Yen, Ed., Ann Arbor Science Publishers: Ann Arbor, Michigan, 1982, p. 197.

[31] R. H. Fish, R. S. Tannous, W. Walker, C. S. Weiss and F. E. Brinckman, "Organometallic Geochemistry. Isolation and Identification of Organoarsenic Compounds from Green River Formation Oil Shale", **J. Chem. Soc. Chem. Commun.,** (1983) 490.

[32] T. F. Degnan, C. S. Weiss, F. E. Brinckman and L. A. Rankel, "The Speciation and Removal of Arsenic Present in Whole and Processed Shale Oils", **Fuel,** submitted.

[33] D. M. Fraley, D. Yates and S. E. Manahan, "Inductively Coupled Plasma Emission Spectrometric Detection of Simulated High Performance Liquid Chromatographic Peaks", **Anal. Chem., 51** (1979) 2225.

[34] D. H. Gast, J. C. Krack, H. Poppe and F. J. M. J. Maessen, "Capabilities of On-Line Element-Specific Detection in High-Performance Liquid Chromatography Using an Inductively Coupled Argon Plasma Emission Detector", **J. Chromatogr., 185** (1979) 549.

[35] M. Morita, T. Vehiro and K. Fuwa, "Determination of Arsenic Compounds in Biological Samples by Liquid Chromatography with Inductively Coupled Argon Plasma-Atomic Emission Spectrometric Detection", **Anal. Chem., 53** (1981) 1806.

[36] M. Morita, T. Vehiro and K. Fuwa, "Speciation and Elemental Analysis of Mixtures by High Performance Liquid Chromatography with Inductively Coupled Argon Plasma Emission Spectrometric Detection", **Anal. Chem., 52** (1980) 351.

[37] K. J. Irgolic, R. A. Stockton, D. Chakraborti and W. Beyer, "Simultaneous Inductively Coupled Argon Plasma Emission Spectrometer as a Multi-Element-Specific Detector for High Pressure Liquid Chromatography: The Determination of Arsenic, Selenium and Phosphorus Compounds", **Spectrochim. Acta, Part B, 38B** (1983) 437.

[38] J. S. Thayer and F. E. Brinckman, "The Biological Methylation of Metals and Metalloids", in **"Advances in Organometallic Chemistry", Vol. 20,** F. G. A. Stone and R. West, Eds., Academic Press: New York, 1982, p. 313.

DISCUSSION

R. E. Sievers (University of Colarado, Boulder): You said that ICP coupled with a quadrupole mass spectrometer was more sensitive and less susceptible to interferences than detectors based on light emission. Because quadrupoles are low-resolution mass sorters, interferences from other species with the same nominal mass to charge ratio are to be expected.

K. J. Irgolic: ICP/MS certainly is not interference-free, but the interferencs are much easier to predict and less numerous than with optical systems. Interferences will be caused by the plasma gas, components of the atmosphere, and perhaps elements possessing several naturally occurring isotopes.

XXXIII

NEUTRON ACTIVATION AND RADIOTRACER METHODS APPLIED TO RESEARCH ON TRACE METAL EXPOSURE AND HEALTH EFFECTS

E. Sabbioni, R. Pietra, F. Mousty and M. Castiglioni*

Radiochemistry and Nuclear Chemistry, and Physics Division
Commission of the European Communities, Joint Research Center
Ispra Establishment
I-21020 Ispra, Varese, Italy*

Environmental biochemical toxicology research on trace elements requires the use of sensitive and sophisticated analytical techniques to determine µg/kg concentrations of trace metals in tissues and intracellular components of laboratory animals and humans. The results provided by these techniques are needed for an understanding of the biochemical mechanisms and the biotransformations involving trace metals. Nuclear and radioanalytical methods were developed and are currently applied to metallobiochemical research at the JRC-Ispra. Typical applications of these techniques to the study of the metabolism of inorganic arsenic in rats and rabbits show different levels of arsenic in the tissues of the two species. These levels are established by different degrees of interaction of the element with the intracellular components of the two species. These interactions are responsible for the detoxification of arsenic. Detoxification is closely related to the rate of methylation of arsenic in the tissues.

The risk, that large population groups could be exposed to trace metals in a form and dose incompatible with good health, has opened new problems in the area of enviromental biochemical toxicology of trace metals [1]. These problems require new analytical methods for their solutions. Interest in the determination of trace elements in environmental and tissue samples has shifted from concentrations of mg/kg or greater for elements such as Ca, P, Mg, Na, K, Zn, and Cu to ng/kg and lower concentrations for toxic elements such as V, As, Sb, Tl, Cd, Hg, and Pb. In addition, the determination of "total" trace metal concentrations in tissues is no longer sufficient and must be followed by the determination of trace element compounds. The study of the metabolic pathways of trace metals and the biochemical mechanisms responsible for the retention and the toxic effects of trace metals is essential for the evaluation of dose-effect relationships [2]. These investigations require knowledge of the distribution of trace metals in the cellular compartments of different tissues, biological fluids, and protein fractions [3]. Thus, very sensitive analytical techniques for ultramicro determinations of trace metals at µg/kg and lower concentrations in biological samples in combination with biochemical techniques of cellular fractionation are necessary to carry out environmental biochemical research in toxicology at the molecular level [4].

This paper presents nuclear and radioanalytical techniques developed at the JRC-Ispra for studies related to the metallobiochemistry of trace metal pollution [5]. In particular, significant advances in such research have been made possible through the use of neutron activation analysis and application of radiotracers with very high specific radioactivities. These radiotracers were prepared in the cyclotron and employed as labels for the study of metabolic and biochemical mechanisms associated with trace elements [6]. The results obtained in the field of environmetal biochemical toxicology during recent years suggest some research priorities which are briefly outlined.

Neutron Activation Analysis

Neutron activation analysis (NAA) is a method of elemental analysis that has excellent sensitivity for many elements. In addition, NAA is a multi-element technique offering the possibility of sensitive and simultaneous determination of many elements in a single sample. It is not surprising, therefore, that one of the major applications of NAA is the study of trace elements in human tissues and body fluids [7]. NAA has also the additional advantage of insensitivity to the biological matrix. The elements are determined with very high specificity through their characteristic radioactive emissions. Table 1 reports typical absolute detection limits for neutron activation analysis in the absence of interferences in the counting step. Typical concentrations of trace metals in human blood

Table 1. *Calculated Approximate NAA Detection Limits* for Trace Metals and Typical Trace Metal Content in Human Plasma and Lungs [8-10]*

Element	NAA Sensitivity (ng)	Metal Concentration (µg/kg)	
		Plasma	Lung
Ag	0.4	2.0 - 7.5	2. - 100.
As	0.06	1.0 - 10.	10. - 100.
Cd	0.4	3.	500.
Co	0.2	0.1 - 1.0	2. - 25.
Cr	2.	0.7 - 1.0	3. - 10.
Cu	0.01	1200.	1500. - 3500.
Hg	0.6	2.0 - 3.30	10. - 100.
Mn	0.002	0.5 - 1.0	400.
Mo	0.1	0.6	60. - 440.
Ni	20.	1.5 - 5.0	15. - 470.
Rb	4.	80. -200.	1000. - 3000.
Sb	0.02	3. - 50.	60. - 1000.
Se	2.	0.05	40. - 500.
Sn	120.	30. -100.	150. - 1750.
Th	0.08	<0.04	10.
U	0.04	0.1	1.
V	0.02	0.05- 1.0	4. - 20.
Zn	10.	1150.	6000. -15,000.

*Neutron flux: 10^{13} neutrons cm^{-2} sec^{-1}; irradiation time: 100 hours

and lungs are also listed.

In our laboratory procedures for the simultaneous determination of up to thirty elements were developed including preparation of the samples, neutron activation followed by radiochemical separation of the induced radiotracers, counting of the radioactivity by high resolution γ-ray spectroscopy using Ge(Li) detectors, and computer analysis of the γ-ray spectra [11]. In many cases the very low levels of some trace elements that must be determined in the biochemical samples (Table 1) necessitate the treatment of the samples before neutron irradiation. This treatment is a critical step. Possible accidental contamination or loss of elements must be carefully evaluated [12]. Although the treatment before irradiation requires special precautions, it brings some advantages. Treatment allows the elimination of a large fraction of the elements such as Na, Br, Cl, and P that strongly interfere with NAA. The elements to be determined can be preconcentrated making the concentrations of the blanks negligible.

Figure 1 shows the scheme of analysis for biological

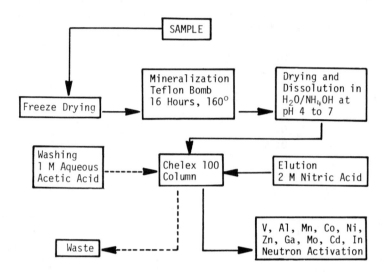

Figure 1. *Scheme for the preparation of biological samples preceding the determination of elements via their short-lived radioisotopes by neutron activation analysis.*

materials including a pre-separation step. The solution of the mineralized sample is passed through a Chelex-100 resin and the retained elements are eluted with nitric acid. The eluted elements are then determined by neutron activation analysis via their short-lived radioisotopes.

Radiotracers with High Specific Radioactivity

The production of many radiotracers with a very high specific radioactivity for the preparation of isotopically metal-labelled chemical species is the "key-tool" for in-vivo and in-vitro studies of metabolic patterns and biochemical mechanisms of toxicity [13]. These radiotracers can now be prepared with the new JRC MC-40 cyclotron by activation of metal samples or powder targets with charged particles. The cyclotron can accelerate protons to 10 to 38 MeV, deuterons to 5 to 19 MeV, and helium-4 to 10 to 38 MeV.

Table 2 lists the radioisotopes that can be produced in the

Table 2. *Radiotracers Produced by Proton Irradiation in the Cyclotron and Their Use for Metallochemical Studies*

Radioisotope Produced	$T_{1/2}$	Metallic Target Bombarded	Application
V-48	16.0d	Ti	Labelling of different chemical forms of V (V^{5+}, V^{4+}, V^{3+})
Cr-51	27.8d	V	Labelling of different inorganic chemical forms of Cr (Cr^{3+}, CrO_4^{2-})
As-74	17.5d	Ge	Labelling of inorganic and organic compounds of As (AsO_2^-, AsO_4^{3-}; arsenobetaine, arsenocholine)
Cd-109	1.3y	Ag	Labelling of Cd^{2+} and metallothioneins
Hg-197	2.7d	Au	Simultaneous labelling of organic and inorganic mercury compounds (Hg^{2+}; CH_3HgCl)
Tl-201	3.0d	Hg	Labelling of different chemical forms of Tl (Tl^+, Tl^{3+}, dimethyl thallium)
Pb-203	2.17d	Tl	Labelling of Pb^{2+} and tetraethyl lead
Bi-206	6.4d	Pb	Incorporation and labelling of metallothioneins

cyclotron and identifies their current use for metallobiochemical purposes. For many of these radiotracers prepared by (p,xn) reactions the excitation functions (formation rate vs. proton energy) were measured in recent years in collaboration with the Cyclotron Laboratory of Milan University as part of a program related to enviromental studies of trace metals [14, 15]. Figure 2 shows a typical excitation function for the preparation of Hg-(195m+197) by proton activation of a gold target. Whereas Hg-197 can be produced without significant contamination by gold radioisotopes using protons of 10 to 12 MeV energy, Hg-195m is contaminated by Au-196, Au-196m,

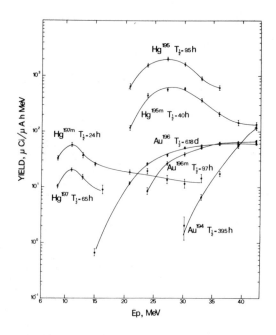

Figure 2. Yields of Hg-(197+197m), Hg-(195+195m), Au-(196+196m), and Au-194 as a function of proton energy.

and Au-194. The difficulties and the problems associated with the preparation of radiotracers and their use were discussed earlier [6].

Applications

Neutron activation analysis and radiotracers prepared at the cyclotron are currently used in our laboratory in two main areas of environmental biochemical toxicology research:

- In-vivo comparative metabolic studies of levels of trace elements in different chemical forms and in-vitro investigations on isolated biochemical systems and tissues of human origin.
- Studies related to analytical quality assurance in the microdetermination of trace metals in biological tissues.

Table 3 summarizes some main investigations related to metabolic studies of environmental levels of trace elements in different animal species. The following studies were

Table 3. *Investigations of the Metabolic Patterns of Trace Metals in Mammals in the Context of the JRC Project "Trace Metal Pollution" Using NAA and Radiotracers with Very High Specific Radioactivity*

Element Investigated	Metabolic Studies
As	Metabolism of inorganic As(III) and As(V), and organic As compounds such as arsenobetaine and arsenocholine in different animal species
V	Metabolic patterns in rats; relation with Fe metabolism; effects on enzymatic activities; transplacental transport; identification of V-biocomplexes in humans
Pb	Identification of Pb-binding components in the intra-cellular fractions of hepatic cell in the rat
Cd and others	Biosynthesis of metallothionein and incorporation into rat liver; long-term metabolic behavior in rats
Tl	Disposition in rat tissues; transplacental transport; intestinal absorption; long-term toxicity studies; influence of oxidation state on Tl metabolism
Sb	Influence of the oxidation state on Sb metabolism; metabolic pathways of organic Sb (stibophen)
Hg	Incorporation of Hg into rat liver and kidney metallothionein
Many elements	Long-term behavior in rats

selected to show the type of results one can expect.

Distribution of Copper and Europium in Laboratory Animals of Different Age

Knowledge about the "normal metabolism" of trace metals in laboratory animals is necessary for the identification of organs important for the accumulation of metals and required for the recognition of parameters that may be used in practice as indicators of exposure for the general population and occupational workers.

A group of thirty rats was housed in special cages for at least two years. The animals were fed well-characterized mineral water and rat pellets. The rats were killed after different periods. Tissues were removed and submitted to NAA. The results show that the elements can be classified with respect to the long-term variation of their concentrations in the tissues [16] (Fig. 3). The copper

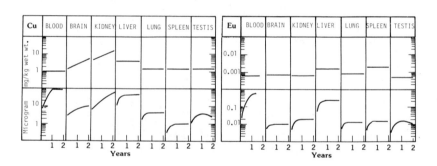

Figure 3. Long-term variation of Cu and Eu concentrations in tissues of rats.

concentrations in the tissues are in the range 1 to 10 mg/kg. Copper accumulates in the brain and the kidneys at different postnatal periods. In all other tissues, including blood, the concentrations are independent of age.

The concentration of europium in the tissues is in the order of µg/kg wet weight. No accumulation of this element was observed during the lifetimes of the rats. This observation tends to support the hypothesis that europium is controlled by a homeostatic mechanism in the body.

Metabolism of Arsenic in Different Animal Species

The study of the metabolism of arsenic is an example of the advantages accruing from the combined use of NAA and radiotracer methods. This methodology is capable of considerably advancing our understanding of the biochemical mechanisms involving trace metals (Table 4). NAA was used

Table 4. Some Results Related to Biochemical Mechanisms Responsible for the Different Metabolic Behavior of Arsenic in Rats and Rabbits

Rat	Rabbit
Neutron activation analysis showed	
mg/kg level of As in the tissues	µg/kg levels of As in the tissues
accumulation of As with age	no accumulation of As with age
Radiotracer experiments showed	
high affinity of As for red blood cell proteins and strong interaction with haemoglobin very little As in the plasma	poor affinity of As for red blood cells and plasma proteins
Arsenic Not Available For Methylation	**Arsenic Available for Methylation**
The absence of methylation leads to the accumulation of arsenic in the tissues and its very low excretion.	Formation of dimethylarsinic acid, which does not interact with proteins and is rapidly excreted (detoxification process).

to determine the long-term retention patterns of dietary arsenic in unexposed normal rats and rabbits [17]. The results show marked qualitative and quantitative differences among the arsenic concentrations in blood, brain, kidney, liver, lung, and spleen of the two animal species at postnatal periods ranging from 20 days to 15 months. In rats arsenic accumulates in all tissues analyzed but especially in blood. Blood contained 4,600 µg As/kg wet weight at 5 months. At this time the level of As in all other tissues was in the low mg/kg or sub-mg/kg range. In contrast, the concentration of arsenic in the tissues of rabbits was nearly the same at a few µg/kg in adult and young rabbits. Radiotracer experiments with As-(71+74) [18] lead to the following main conclusions [19]:

- Arsenic has a high affinity for red blood cell proteins of rats. Arsenic associates with haemoglobin and accumulates in red blood cells. In contrast, arsenic has a poor affinity for the red blood cells and the plasma proteins of rabbits. "A highly mobile pool" of arsenic in rabbits was identified as dimethylarsinic acid.
- Arsenic in the form of dimethylarsinic acid cannot interact with proteins and is rapidly excreted.

These findings show the existence of different biochemical mechansims for arsenic in the two animal species. In particular, the interaction of arsenic with intracellular components depends on the biotransformation of inorganic arsenic to dimethylarsinic acid. In rats arsenic is slowly methylated and excreted because the haemoglobin of this species has a peculiarly high affinity for arsenic. In rabbits, arsenic interacts fairly weakly with cellular components and is rapidly excreted as dimethylarsinic acid. When the organism detoxifies inorganic arsenic by methylation in the tissues followed by excretion, inorganic arsenic is not available to interact with cellular components.

Research Priorities

The results obtained during recent years in environmental biochemical toxicology research on trace metals suggest several important areas for further research requiring the use of novel analytical techniques.

- The determination of the subtle adverse effects induced by chronic exposure of living organisms to trace metals require sensitive, interference-free, element-specific analytical methods. Efforts are necessary to develop analytical procedures to study the biochemical effects of trace metals in laboratory animals at levels present in the environment.

- In nature, living organisms are rarely exposed to a single element. Therefore interactive effects must often be considered. Better knowledge of daily intake of different elements, of the ratios between them, and their chemical forms in the diet is necessary. Analytical methods are needed for the simultaneous determination of many elements to provide basic information for the evaluation of antagonistic and synergistic effects among trace metals.

- Different chemical forms of the same element may cause remarkably different biochemical and toxicological effects in mammals. A great need exists for analytical methods to study chemical species and the oxidation state of trace elements to which man is exposed. The recently discovered different metabolisms and different analytical responses of inorganic arsenic and of biotransformed compounds hint at the present serious lack of knowledge in this field.

- Knowledge of the biochemical mechanisms responsible for trace metal retention and toxic effects is essential for the evaluation of the dose-response relationships. Analytical procedures are necessary to relate total element concentrations in tissues to the results of investigations at the intracellular and molecular levels with the goal of identifying the biochemical components that interact with trace metals. A better understanding of the biochemical mechanisms may also be obtained by comparative studies with different animal species.

- Literature data on trace metals in human tissues [20] are of limited value for assessment studies, because the data vary considerably and are not correlated with any other parameters. Better evaluation of results to improve the practical use of trace element data by relating the levels determined in tissues to parameters such as body burden is necessary.

References

[1] K. Nogawa, A. Ishizaki, I. Shibata and J. Hagino, "Studies on the Women with Acquired Fanconi Syndrome Observed in the Ichi River Basin Polluted by Cadmium," **Environ. Res.**, **10** (1975) 280.

[2] E. Sabbioni and F. Girardi, "Metallobiochemistry of Heavy Metal Pollution. Nuclear and Radiochemical Techniques for Long-Term Low-Level Exposure (LLE) Experiments," **Sci. Total Environ.**, **7** (1977) 145.

[3] E. Sabbioni, E. Marafante, R. Pietra, L. Goetz, F. Girardi and E. Orivini, "The Association of Vanadium with the Iron Transport System in Human Blood as Determined by Gel Filtration and Neutron Activation

Analysis," **Proceedings, Intern. Symposium on Nuclear Activation Techniques in Life Sciences,"** May 22-26, 1978, International Atomic Energy Agency: Vienna, 1979, p. 179.
[4] E. Sabbioni, "Metallobiochemical Research at the JRC-Ispra as Carried Out by Nuclear and Radiochemical Methods," in **"Proceedings Intern. Workshop Trace Element Analytical Chemistry in Medicine and Biology,"** Neuherberg (FRG), 26-29 April 1980, P. Bratter and P. Schramel, Eds., Walter de Gruyter: Berlin, 1980, p. 407.
[5] E. Sabbioni, "Trace Metal Pollution Research at the JRC-Ispra, **Int. Symposium on Elements in Health and Disease,** Karachi, Pakistan, 13-15 February 1983, in press.
[6] E. Sabbioni, L. Goetz, C. Birattari and M. Bonardi, "Environmental Biochemistry of Current Environmental Levels of Heavy Metals: Preparation of Radiotracers with Very High Specific Radioactivity for Metallobiochemical Experiments on Laboratory Animals," **Sci. Total Environ., 17** (1981) 257.
[7] E. Sabbioni, "Neutron Activation Analysis: General Principles and Application to the Analysis of Biological Fluids," Ispra-Courses, **"Analytical Techniques for Heavy Metals in Biological Fluids,"** Ispra, November 27 - December 1, 1978.
[8] C. Vanoeteren, R. Cornelis, J. Versieck, J. Hoste and J. De Roose, "Trace Element Patterns in Human Lung Tissues," **J. Radioanal. Chem., 70** (1982) 219.
[9] J. Versieck and R. Cornelis, "Normal Levels of Trace Elements in Human Blood Plasma and Serum," **Anal. Chim. Acta, 116** (1980) 217.
[10] E. I. Hamilton, M. J. Minski and J. J. Cleary, "The Concentration and Distribution of some Stable Elements in Healthy Human Tissues for the United Kingdom," **Sci. Total Environ., 1** (1972/1973) 341.
[11] E. Sabbioni, "Radiotracer and Nuclear Techniques: Application in Metallobiochemical Investigations of Environmental Levels of Trace Elements," **Proceedings, Intern. Workshop on Trace Elements in New Zealand,** Dunedin, N.Z., May 20-21, 1981, J. V. Dunckley, Ed., Univ. Otago, Dept. of Nurition, Dunedin, N.Z., p. 322.
[12] R. Pietra, E. Sabbioni, A. Springer and L. Ubertalli, "Analytical Problems Related to the Preparation of Samples Used in Studies on Metallobiochemistry of Heavy Metal Pollution Using Neutron Activation Analysis," **J. Radioanal. Chem., 69** (1982) 365.
[13] E. Sabbioni, F. Girardi and E. Marafante, **"A Systematic**

Study of Biochemical Effects of Heavy Metal Pollution," Report EUR 5333.EN, Commission of the European Communities, Office for Official Publication of the European Communities, Luxembourg, 1975.

[14] L. Goetz, E. Sabbioni, E. Marafante, C. Birattari and M. Bonardi, "Cyclotron Production of Cd-107, 109 for Use in the Metallobiochemistry of Heavy Metal Pollution," **Radiochem. Radioanal. Letters, 45** (1980) 51.

[15] E. Sabbioni, E. Marafante, L. Goetz and C. Birattari, "Cyclotron Production of Carrier-Free V-48 and V-48 Compounds for Metabolic Studies in Rats," **Radiochem. Radioanal. Letters, 31** (1977) 39.

[16] F. Girardi, E. Marafante, R. Pietra, E. Sabbioni and A. Marchesini, "Application of Neutron Activation Analysis to Metallobiochemistry," **J. Radioanal. Chem., 37** (1977) 427.

[17] R. Pietra, E. Sabbioni and E. Marafante, "Comparative Metallobiochemical Studies on Present Environmental Levels of Arsenic in Mammals. I. Neutron Activation Analysis of Nanogram Levels of Arsenic in Tissues of Laboratory Animals," **J. Radioanal. Chem., 62** (1980) 41.

[18] D. Basile, C. Birattari, M. Bonardi, L. Goetz, E. Sabbioni and A. Salomone, "Excitation Functions and Production of Arsenic Radioisotopes for Environmental Toxicology and Biomedical Purposes," **Intern. J. Appl. Radiat. Isotopes, 32** (1981) 403.

[19] E. Marafante, F. Bertolero, J. Edel, R. Pietra and E. Sabbioni, "Intracellular Interaction and Biotransformation of Arsenite in Rats and Rabbits," **Sci. Total Environ., 24** (1982) 27.

[20] V. Iyengar, W. E. Kollmer and H. J. M. Bowen, **"The Element Composition of Human Tissues and Body Fluids,"** Verlag Chemie: New York, 1978.

XXXIV

TRACE ELEMENT-ORGANIC LIGAND SPECIATION IN OIL SHALE WASTEWATERS

John S. Stanley and Robert E. Sievers

Department of Chemistry, and Cooperative Institute for Research in Environmental Sciences, University of Colorado Campus Box 449, Boulder, Colorado 80309 USA

Experiments to speciate trace elements and organic ligands were conducted on oil shale wastewaters from in situ, modified in situ, and above-ground oil shale retorts. The complexing capacities of the wastewaters were determined. The concentrations of As, Se, Sr, Mg, Ca, Pb, Fe, Mo, Mn, Si, Cu, V, Ba, Co, Li, and Zn in samples passed through 0.45 µm filters and subjected to ultrafiltration were obtained by graphite furnace atomic absorption spectrometry and inductively coupled plasma emission spectrometry. The complexing capacities of the wastewaters were estimated by studies of the solubilization of copper at pH 10. The organic substances were separated into hydrophobic and hydrophilic fractions using Amberlite XAD-8 and ion exchange resins. The dissolved organic carbon concentration was measured to estimate the distribution of the acid, base, and neutral hydrophobic and hydrophilic organic compounds. Organic ligands were identified by high resolution, fused-silica gas chromatography with flame ionization, nitrogen-selective, and quadrupole mass spectrometry detectors. The results indicated that the complexing ability is related to the concentrations of basic organic compounds, particularly of alkyl-substituted pyridines, anilines, and quinolines that were identified in the hydrophobic base fraction.

The attempts to develop an oil shale industry in the United States have been accompanied by research to minimize adverse environmental impacts. These environmental research projects have focused on averting potential environmental hazards. The research has characterized a number of different operations for the recovery of oil shale and determined the composition of solid and aqueous wastes that must be safely disposed of in future commercial-scale activities. One aqueous waste, "retort water," is produced in the heating step required to pyrolyze the polymeric organic material in the shale. The retort waters and related process waters contain a large number of toxic trace elements and inorganic salts. The aqueous wastes are usually highly colored, have an obnoxious odor resulting from high concentrations of dissolved ammonia and volatile organic compounds, and contain dissolved organic compounds at 500 to 45,000 μg C/mL. It has been estimated that for every gallon of shale oil produced, 0.1 to 22 gallons of contaminated water will be generated depending on the type of retorting operations [1]. Thus, the commercial production of shale oil will have to control and dispose of a significant amount of waste. The treatment and reuse of wastewaters will influence the economic viability of shale oil plants.

Retort waters from a number of processes have been characterized with respect to total toxic trace elements and organic compounds. Traditionally, "total concentrations" of particular toxic elements and organic compounds were used in attempts to define environmental impacts. However, more recently the effect of the chemical nature of species on toxicity, bioavailability [2-5], and transport of many elements in the environment has been recognized. Much more detailed information is needed about the chemical forms of trace elements present in the environment and particularly about the identity of ligands and metal complexes that may be present. Recent reports issued by the U.S. Department of Energy and the Federal Interagency Committee on the Health and Environmental Effects of Energy Technologies expressed the need for the speciation of toxic elements in process and waste streams from synfuels activities. Earlier research concerning the speciation of trace elements in connection with oil production separated organic compounds in crude oils and products into several fractions and analyzed these fractions for trace elements [5, 6].

Speciation of toxic elements in oil-related aqueous wastes is limited to the identification of several arsenic compounds in retort waters by high performance liquid chromatography/graphite furnace atomic absorption spectrometry [7]. Many of the metal ions in oil shale wastewaters form quite labile complexes. The identification of these complexes is aided by information about the complexing agents in the water. Oil shale wastewaters have high concentrations of dissolved organic compounds, some of which are relatively strong complexing agents containing oxygen and nitrogen functionalities [7-16]. Stumm and Brauner [17] suggested that physical mechanical separations such as membrane filtration, dialysis, gel chromatography, anodic stripping voltammetry, solvent extraction, and chromatography are viable techniques for identifying coordinated or complexed species in environmental matrices. Florence and Batley [18] also discussed these techniques and the application of measurements of complexing ability. Several researchers used combinations of these procedures to investigate species distributions and complexation in aqueous systems much less contaminated than oil shale wastewaters [17-28].

In this work, three oil shale waters from different retort operations were studied. Because of the complexity of the oil shale wastewaters, several analytical techniques were employed for the identification of toxic trace element-organic ligand interactions. The techniques included physical separations by membrane filtration and pre-separation of components in complex wastewaters on macroreticular and ion exchange resins. The trace elements were determined by atomic absorption spectrometry and inductively coupled plasma emission spectrometry. The organic compounds were separated by high resolution gas chromatography with flame ionization and nitrogen-selective detectors and then identified by mass spectrometry. Major emphasis was placed on the identification of nitrogen bases, because these compounds are strong complexing agents, are biorefractory in oil shale wastes, and are responsible for a large percentage of the mutagenic activity of oil shale products [29, 30].

Experimental

Samples. The oil shale wastewaters were obtained from above-ground, modified in-situ, and true in-situ pilot plant operations. The aqueous waste of the above-ground process was taken from an oil-separator tank at the Paraho demonstration plant near Rifle, Colorado, in September 1978. This wastewater had accumulated during more than one year of operation of the facility because the tank was never completely emptied. The sample was taken near the oil/water interface. The retort was operated in a direct mode [31-33].

The modified in-situ retort water was obtained from the Occidental Corporation. The water was collected as a field composite of pilot plant condensate produced during March 1979 from Occidental's Retort 6 operation at Logan Wash, near De Beque, Colorado. The Oil Shale Task Force [34] presented a preliminary characterization and assessment of waste products from the Occidental Retort 6 operation.

The true in-situ retort water was an aliquot of a composite of 12,450 gallons of water produced at the Rock Springs, Wyoming, Site 9, experimental retort. The collection and handling of this wastewater is well-documented. Trace elements and major inorganic constituents were determined in this wastewater [35, 36]. The samples were received in dark amber bottles or Teflon containers. All samples were cooled following collection, kept cool during shipment, and stored at approximately 4°. Aliquots of each of these wastewaters were filtered through 0.45 µm type HA Millipore filters prior to ultrafiltration or chromatographic separation. Before use the filters were rinsed with 10 percent nitric acid in distilled, deionized water, and washed with several liters of distilled, deionized water.

Ultrafiltration. Amican ultrafiltration membranes XM50, PM30, UM20, UM10, UM2, and UM05 were used to fractionate the true in-situ water after filtration through a 0.45 µm Millipore filter. The membranes were chosen to separate solutes ranging in diameter from <5.1 to <1.0 nm. Each membrane was flushed extensively with distilled, deionized water. Twenty-milliliter aliquots of filtered retort water were passed through each membrane under 15 to 35 psi of nitrogen pressure. The ultrafiltered samples were collected in polyethylene bottles, acidified, and analyzed for trace elements by inductively coupled plasma emission and atomic absorption spectrometry.

Adsorbent separation. The organic compounds in the oil shale wastewaters were separated into hydrophobic and hydrophilic compounds according to the adsorbent separation scheme of Leenheer and Huffman [37, 38]. Hydrophobic bases and neutral solutes were removed by passing the 0.45 µm filtrates of the retort waters at their intrinsic pH of 8.5 to 8.7 through a column (20 mL bed volume) of Amberlite XAD-8 resin (Supelco Inc., 100-200 mesh). The eluent from this column was acidified to approximately pH 2 with concentrated nitric acid and then passed through an additional column of clean XAD-8 resin (20 mL bed volume). Compounds classified as hydrophobic acids were retained by the column and subsequently eluted by 0.1 M NaOH. The hydrophilic bases and acids were separated by subsequent passage of the acidified eluent through cation (AG-MP-50) and anion (AGMP-1) exchange resins (Biorad Laboratories, 100-200 mesh), respectively. Neutral hydrophilic compounds remained in the eluent from the anion exchange column. Hydrophobic bases and neutral components on the XAD-8 resin were eluted with 0.1 M HCl solution. The eluate was adjusted to pH 10 with 1.0 M NaOH and then passed through a clean XAD-8 column (20 mL bed volume). The hydrophobic bases were retained. The resin was air-dried and then placed in a Soxhlet. Analytical reagent grade methylene chloride (100 mL) (Fisher Scientific Company) was used to extract the nitrogen bases from the resin. The extracts were concentrated by roto-evaporation and flowing nitrogen to a final volume of 2 mL. These extracts were analyzed by high resolution capillary gas chromatography with flame ionization and nitrogen-selective detectors and mass spectrometry.

Dissolved organic carbon. A Beckman Model 915B total organic carbon analyzer was used for the determination of dissolved organic carbon. Each sample was diluted to contain 0 to 100 µg carbon per mL. The samples were acidified to pH 2 with concentrated phosphoric acid and then purged for at least 15 min with purified nitrogen to remove traces of dissolved carbon dioxide. The concentrations of organic carbon in the hydrophobic base and acid fractions were determined directly. The organic carbon related to hydrophobic neutral compounds and the hydrophilic acids, bases, and neutrals were obtained by difference between the dissolved organic carbon concentrations of the eluents from each of the four columns.

Complexing ability. The complexing ability was determined according to the procedures of Kunkel, Manahan, and Jones [39, 40]. Twenty-five milliliters of each of the three wastewater samples were acidified to pH 2 with nitric acid and filtered through a 0.45 µm filter. The samples were passed through a short column of purified Dowex 50-X8 in the sodium form. One milliliter of 0.05 M copper sulfate was added to each solution. The pH was adjusted to be between 9.8 and 10.2 by addition of 1 M NaOH. Additional copper sulfate solution was added if no precipitate was observed. The samples were heated almost to boiling with the pH maintained above 9.8. The samples were cooled, diluted to 100 mL, filtered through a 0.45 µm filter, acidified with nitric acid, and analyzed by flame atomic absorption spectrometry. The copper content of the filtrate was used as a relative measure of complexing ability.

Gas chromatography. A Hewlett-Packard Model 5830A gas chromatograph with a flame ionization detector modified for capillary columns and equipped with a Grob-type injection system was used for matching retention times of substituted pyridines, anilines, and quinolines in standards and samples. An HP 5730A gas chromatograph equipped with an HP NPD-18789A nitrogen-phosphorus selective detector was employed to screen samples for the presence of nitrogen-containing compounds. The detector was operated at 300° with a hydrogen flow of 3 mL/min, an air flow of 50 mL/min, makeup nitrogen at 30 mL/min, and a bead voltage of 16 to 20 V. This chromatograph was also equipped with a Grob-type capillary injection system. Fused silica capillary columns such as Hewlett-Packard SP-2100 [poly-(dimethylsiloxane), deactivated with Carbowax 20M], 25 m in length, J&W Scientific SE-52 [poly-(methylphenylsiloxane), deactivated by silylation], 30 m in length, and J&W SE-52 wide bore (0.42 mm i.d.), deactivated by silylation, 30 m in length were used. A typical temperature program for these analyses started with an initial temperature of 30 to 40°, had a temperature rise of 4°/min, and a final temperature of 270°.

Gas chromatography/mass spectrometry. Gas chromatography/mass spectrometry analyses were conducted using the Hewlett-Packard 5890A/5933A GC/MS/DS system. In most cases, the gas chromatographic separations were accomplished with a 12 m fused silica column coated with SP-2100, or a 30 m J&W fused silica column coated with SE-54 [1% vinyl in poly(methylphenylsiloxane), previously deactivated by silylation]. Mass spectra were obtained every 2 to 3 sec over the range of 50 to 400 atomic mass units. The mass spectra were interpreted through comparison with references in the EPA/NIH Mass Spectral Data Base (NSRDS-NBS 63) or with mass spectra of pure compounds.

Trace element analysis. The elements Fe, Mo, Zn, Cu, Mn, Si, V, Ba, Sr, Mg, Ca, Cd, and Pb were determined in the samples generated by ultrafiltration and adsorbent separation with an ARL 36000C inductively coupled plasma emission spectrometer by R. Meglen and Associates at the Center for Environmental Studies, Denver, Colorado. Arsenic (193.6 nm) and selenium (196.0 nm) were quantitated with the Perkin-Elmer 5000 atomic absorption spectrometer (0.7 nm slit) equipped with an HGA-500 high temperature graphite furnace and deuterium background correction using the matrix modification technique with 1 percent nickel solutions [41].

A Perkin-Elmer 360 atomic absorption spectrometer with a conventional air-acetylene flame was used for the determination of copper in the complexing ability measurements. Copper absorption was monitored at 324.7 nm with a slit width of 0.7 nm.

Results and Discussions

The wastewater samples should not be regarded to be entirely representative of the wastewaters generated in a commercial shale oil plant. Ultrafiltration is a unique method for the separation of trace elements and organic ligands in these samples. This separation technique exploits the differences in the size of compounds present in solution. This method was used previously to investigate the complex formation of humic and fulvic acids and to study the distribution of trace elements in marine environments. Florence and Batley [18] recently reviewed the application of ultrafiltration and the importance of determining the size of organic and inorganic complexes of metals in natural water systems. This report is the first describing the application of ultrafiltration to the study of the distribution of trace elements in oil shale wastewaters. The Amicon ultrafiltration membranes separate compounds of molecular weight between 500 and 50,000. The true in-situ retort water previously filtered through a 0.45 µm cellulose acetate Millipore filter was introduced into the ultrafiltration cell. The ultrafiltrates were analyzed for 12 elements (Table 1).

Large percentages of strontium, magnesium, and iron are present as fairly large molecules or are adsorbed on particulate matter that cannot pass through the pores of a 0.45 µm Millipore filter. Only iron was appreciably fractionated by ultrafiltration. Approximately 50 percent of the iron that passed through the membrane retaining substances with molecular weight >20,000 did not pass through the membrane with 1.9 nm pore size that retains molecules with molecular weight >10,000. The ultrafiltrates became lighter in color by passage through membranes of successively smaller pore size. The filtrate from the membrane of smallest (1.0 nm) pore size was nearly colorless but still had a strong ammonia odor. These results indicate that most of the trace elements studied existed in the retort water as small, simple ions and did not associate strongly with high molecular weight compounds. Iron, magnesium, silicon, and strontium that are retained on the

Table 1. Trace Element Concentrations in Unfiltered and Filtered Wastewater Samples from a True In-Situ Oil Shale Retort

Element	Emission Wavelength nm	Trace Element Concentration, mg/L			% Metal* as Apparent Simple Ion
		Unfiltered	0.45 μm Millipore	Ultra-filtration	
Sr	407.77	1.084	0.898	0.819	91
Mg	279.08	20.06	15.47	15.47	100
Ca	317.93	11.3	12.6	11.6	92
Pb	220.35	0.012	0.012	0.009	75
As	193.6**	0.82	0.69	0.53	77
Se	196.0**	0.22	0.16	0.08	50
Fe	259.94	1.09	0.57	0.27	47
Mo	281.61	0.49	0.41	0.31	76
Mn	257.61	0.06	0.07	0.07	100
Si	251.61	11.3	10.6	9.1	89
Cu	324.75	<0.02	<0.02	<0.02	-
V	311.07	<0.04	<0.04	<0.02	-

*Calculated from total dissolved concentration after filtration through 0.45 μm Millipore filter.

**Arsenic and selenium were determined by graphite furnace atomic absorption spectrometry, all other elements by inductively coupled argon plasma emission spectrometry.

0.45 μm filter were possibly present as mineral aggregates. A likely form for iron is a hydrous oxide [42]. The concentrations of arsenic and selenium were also lower in the 0.45 μm filtrate than in the unfiltered samples and decreased further on ultrafiltration. The significance of these concentration changes is difficult to assess particularly for the low concentration and concentration changes observed for selenium. The change in concentration for arsenic between the raw water and the 0.45 μm filtrate was significant.

Adsorbent separation was employed to better characterize the constituents of the wastewaters, identify ligands, and possibly isolate metallo-organic complexes. Leenheer and Huffman [37] suggested that this method might separate organic from inorganic solutes in natural waters and wastewaters. The XAD-8 resins remove organic solutes that are not very soluble in aqueous media, and perhaps neutral trace element-organic ligand complexes. The first pass through the XAD-8 column at the intrinsic pH of the retort water removed solutes classified as hydrophobic bases and neutrals. After the pH of the eluent was adjusted to approximately 2, it was passed through a second XAD-8 column. Compounds classified as hydrophobic acids were retained. Subsequent passage through the cation and the anion exchange resin removed hydrophilic solutes and also trace elements present either in cationic or anionic forms. The above-ground, modified in-situ, and true in-situ retort waters were fractionated by this method. The trace elements were determined in all the eluents (Table 2).

The largest changes in concentration occurred for most of the trace elements during passage of the acidified samples through the second XAD-8 column. Leenheer and Huffman [37] observed similar changes in natural waters and proposed that trace elements exist as metal-organic cations and anions and are absorbed on the macroreticular XAD-8 resins. However, the exchange capacities of these resins are low. The results of these experiments do not necessarily indicate that metal-organic ligand complexes present in the original wastewaters were isolated. Acidification of the retort water caused drastic changes. Thiosulfate, a major sulfur species in oil shale wastewaters, precipitated copious amounts of elemental sulfur on acidification. Organic acids are also precipitated as evidenced by their removal on the XAD-8 resin, and may possibly act as ion exchange sites at the adsorbent surface. Distinct differences in the concentration of zinc, calcium, magnesium, barium, cobalt, lithium, copper, and manganese were noted at this point of the separation scheme. Selenium, boron, and molybdenum were also removed by the second XAD-8 column together with the organic solutes classified as hydrophobic acids. The cation exchange resin effectively retained zinc, calcium, barium, manganese, and magnesium but not molybdenum. Molybdenum was removed by the anion exchange resin.

Arsenic did not behave the same in the three wastewaters. Ninety percent of the arsenic in the true in-situ retort water was removed by the anion exchange resin. The arsenic concentration in the modified in-situ retort water hardly changed at all. The above-ground wastewater lost almost 50 percent of the arsenic after acidification and passage through the second XAD-8 resin, but 3.3 mg/L of arsenic remained in the final solution.

Table 2. Trace Element Concentrations in Oil Shale Wastewaters After Passage through Amberlite XAD-8 and Ion Exchange Columns

Sample	As	Ba	Co	Cu	Li	Mg	Mn	Mo	Se	Zn
						Trace Element Concentration, mg/L				
Above-Ground	5.5	0.58	1.00	0.26	4.1	1,110	0.22	0.38	5.77	0.03
XAD-8(A)*	6.1	0.59	0.91	0.28	4.1	1,040	0.20	0.32	5.94	0.09
XAD-8(B)**	3.3	0.25	0.31	<0.01	1.4	410	0.11	<0.02	0.31	<0.01
AGMP-5***	3.7	0.01	0.22	0.05	<1.0	340	0.07	<0.02	0.33	<0.01
AGMP-1†	3.3	<0.002	0.08	<0.01	1.4	340	0.08	<0.02	0.19	<0.01
Modified In-Situ	1.02	0.44	0.16	0.04	5.7	108	<0.005	0.21	<0.01	0.02
XAD-8 (A)	1.02	0.43	0.14	0.03	6.1	102	<0.005	0.19	<0.01	<0.01
XAD-8 (B)	1.30	0.21	<0.01	0.03	1.2	35.9	<0.005	0.09	<0.01	<0.01
AGMP-50	1.02	0.002	<0.01	<0.01	<1.0	<0.3	<0.005	0.10	<0.01	<0.01
AGMP-1	0.90	0.004	<0.01	<0.01	1.3	<0.3	<0.005	<0.02	<0.01	<0.01
True In-Situ	0.73	0.31	<0.01	0.01	2.0	21.8	0.04	0.68	0.16	0.19
XAD-8 (A)	0.76	0.48	0.02	0.06	3.0	26.2	0.05	0.65	0.12	0.21
XAD-8 (B)	0.77	0.10	<0.01	0.02	<1.0	20.5	0.06	0.17	0.03	0.14
AGMP-50	0.73	<0.002	0.01	0.02	2.0	11.8	0.01	0.13	0.05	0.08
AGMP-1	0.06	<0.002	<0.01	<0.01	3.0	14.8	0.01	<0.02	<0.01	0.02
Stand. Dev. ††	0.01	0.002	0.01	0.01	1.0	0.3	0.005	0.02	0.01	0.01
Emission Wavelength nm	455.40	286.61	324.75	670.78	279.08	257.61	281.61	196.09	213.85	

*Concentrations after passage through the styrene-divinylbenzene column at the intrinsic pH of the wastewater.
**Concentrations after acidification with nitric acid to pH 2 and passage through a second XAD column.
***Concentrations after passage through an AGMP-50 cation exchange column.
†Concentrations after passage through an AGMP-1 anion exchange resin.
††Standard deviations were determined for standards and blanks only.

The selenium concentration was reduced to 20 percent of the original concentration in the true in-situ retort water following acidification and passage through the second XAD-8 column. This decrease in concentration may be caused by association of selenium with the hydrophobic acids that are also removed by the second XAD-8 column or by oxidation of selenium species, in which selenium is in a formal -2 oxidation state, to elemental selenium upon addition of nitric acid. The selenium concentration in the above-ground wastewater was reduced to 5 percent of its original value during the same separation step. The selenium concentration in the modified in-situ retort water was too low to detect any significant differences.

The observation that some of the trace elements such as magnesium, barium, cobalt, lithium, copper, and manganese were removed with the hydrophobic acid fraction required additional investigations into the extent of complexation of trace elements in the wastewaters. The modified in-situ and true in-situ retort waters after filtration through 0.45 μm filters were passed through Chelex-100, a weakly acidic chelating resin. The concentrations of a number of trace elements were determined in the eluents (Table 3). Chelex-100 effectively removed Ba, Sr, Mg, Ca, Cd, Li, Zn, and Mn from these two filtered retort waters. This procedure was claimed to identify stable complexes in solutions [26, 27]. The concentrations of As, Se, Fe, Mo, Si, and V did not change. These elements are not expected to be sorbed on a cation chelating resin. The behavior of iron might indicate, that iron is present as a ferrous hydroxide at the pH of the retort water or as a soluble organic complex. The Chelex-metal complexes of the retained elements must be more stable than the complexes that exist in solution. The selectivity of Chelex-100 for divalent over monovalent ions is approximately 5,000 to 1. This resin has a very strong affinity for transition elements even in highly concentrated salt solutions.

The oil shale wastewaters were further characterized by the distribution of the total organic carbon content in the fractions separated by the adsorption technique (Table 4). The above-ground process water had the highest total organic carbon concentration of the three wastewaters, a fact easily recognizable by visual inspection.

The true in-situ and modified in-situ waters contained hydrophobic and hydrophilic solutes in approximately equal concentrations. The solutes in the above-ground water

Table 3. Trace Element Concentrations in Oil Shale Wastewaters After Filtration Through 0.45 μm Filters and Passage of the Filtrates Through a Chelex-100 Column

	Concentration, mg/L			
	True In-Situ		Modified In-Situ	
Element	0.45 μm Filtrate	Chelex	0.45 μm Filtrate	Chelex
Ba	0.422	0.024	0.435	<0.002
Sr	0.898	0.014	ND*	<0.001
Mg	15.47	0.30	108.0	0.8
Ca	12.6	0.2	5.4	0.1
Cd	0.016	<0.001	0.02	0.01
Pb	<0.01	<0.01	0.02	<0.01
As	0.69	0.84	1.02	1.20
Se	0.16	0.12	<0.01	<0.01
Fe	0.57	0.44	0.07	0.10
Mo	0.41	0.39	0.21	0.18
Zn	4.32	0.11	ND	ND
Mn	0.08	0.01	<0.005	<0.005
Si	10.6	8.4	17.6	16.1
Cu	<0.02	<0.02	0.04	0.03
V	0.04	0.04	<0.02	0.03
Li	ND	ND	57.0	0.7
Co	ND	ND	0.16	<0.01

*Not determined

Table 4. Total Dissolved Organic Carbon in Fractions of Oil Shale Wastewaters Obtained by Adsorbent Separation

| Oil Shale | Dissolved Organic Carbon, mg/L | | | |
Wastewater	Total	Acids	Bases	Neutrals
Above-ground	41,900	23,740	7,970	10,490
		56%*	19%	25%
Modified In-Situ	2,300	1,020	75	1,230
		44%	3%	53%
True In-Situ**	1,000	480	250	270
		48%	25%	27%

*Percentage of total dissolved organic carbon.
**From reference 37.

consisted of 71 percent hydrophobic materials and 29 percent hydrophilic substances. The preponderance of hydrophobic solutes may be the result of the longer contact of the water with the crude oil in the separating tank. The acid and neutral solutes were the major components of the organic materials in these waters. The hydrophobic acid fraction of above-ground process water represents 52 percent of the total organic carbon content. In the modified in-situ and true in-situ retort waters, the concentration of hydrophobic acids was larger than the concentration of hydrophobic neutrals. The total concentration of hydrophobic acids and neutrals was highest in the above-ground waters, intermediate in the modified in-situ samples, and lowest in the true in-situ water. The same order was observed for the total organic carbon concentrations. The concentration of the basic solutes of the hydrophobic fraction is highest in the above-ground retort water, but lowest in the modified in-situ samples.

The concentrations of hydrophilic base solutes followed the same order as the hydrophobic bases. The modified in-situ water had approximately 5 mg/L of hydrophilic bases, a concentration at the detection limit of the analytical method. The modified in-situ water contains, therefore, probably little pyridine or aliphatic amines. Basic solutes are the major components of the hydrophilic substances in above-ground process water. The hydrophilic neutrals are more abundant than the acids in the above-ground and true in-situ wastewaters.

The complexing abilities of three oil shale wastewaters are a measure of their capacities to mobilize trace elements. The complexing abilities were determined via the solubilization of copper hydroxide at pH 10. High concentrations of divalent or trivalent cations, dissolved carbon as carbonate, ammonium salts, and dissolved ammonia interfered with the determination of complexing abilities in all untreated retort waters. To avoid interferences the retort waters were acidified and purged to remove inorganic carbon from solution. The water was passed through a cation exchange resin to reduce the concentrations of divalent cations such as Mg and Ca and heated to avoid interferences from ammonium salts and dissolved ammonia. Although copper is not a trace element of environmental concern in oil shale wastes, it does preferentially bind to nitrogen-containing ligands in aqueous solutions [43] and has been used as a model for complexing metal ions in environmental studies [16, 18, 19, 22-24, 39, 40, 44, 45]. The complexing abilities were found to be 570±30 mg Cu/L for the aboveground, 8.6±0.4 mg Cu/L for the modified in-situ, and 42±2 mg Cu/L for the true in-situ wastewaters. No apparent correlation exists between complexing abilities and the concentrations of total organic carbon. The total organic carbon content (Table 4) decreases in the order above-ground >> modified in-situ > true in-situ, whereas the complexing ability is in the sequence above-ground >> true in-situ > modified in-situ. However, the basic fractions of hydrophobic and hydrophilic solutes show the same trend as the complexing abilities. The acidic and neutral fractions of the hydrophobic and hydrophilic solutes are correlated with the complexing abilities.

The organic solutes from the oil shale wastewaters were separated into subclasses and the fractions analyzed by high resolution gas chromatography/mass spectrometry. The nitrogen bases in the hydrophobic base extracts were identified by the fragmentation patterns of each peak in the chromatograms and by their retention times. All retention times were measured using a fused silica capillary column. Novotny et al. [46] recently reported that the determination of quinoline derivatives in a recycle oil from a solvent-refined coal operation by gas chromatography is often difficult because of irreversible column absorption and peak asymmetry. These phenomena are especially pronounced in glass capillary gas chromatography. Problems with such nitrogen compounds, particularly aniline compounds, were observed in our earliest work with glass capillary columns. However, the peak shapes (Figs. 1, 2) for the nitrogen compounds in the base fractions of the oil shale wastewaters indicate that fused silica capillary columns are not plagued by these difficulties. The peaks obtained in these chromatograms are symmetrical and do not tail. The reproducibility of retention times for standard compounds varied from ± 0.10 min for pyridine to ± 0.05 min for acridine.

The chromatograms obtained with the nitrogen-phosphorus selective (NP) and the flame ionization (F) detectors (Figs. 1, 2) show that the nitrogen base fraction consists principally of nitrogen-containing compounds. The mass spectra of this sample had parent ions of odd molecular weight indicating the presence of an odd number of nitrogen atoms in the molecules. Figure 2 compares the FID chromatograms of the hydrophobic base fraction of the three different oil shale wastewaters. The numbered peaks are identified in Table 5. The compounds were identified by matching retention times and mass spectral patterns with

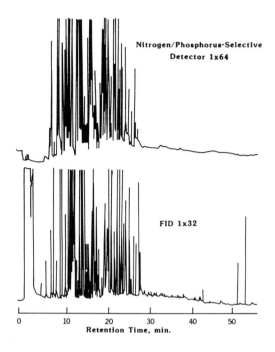

Figure 1. Gas chromatograms of the hydrophobic base extract from a true in situ oil shale retort water [16].

those of pure compounds. Each of the wastewaters contained 2,4,6-trimethylpyridine. This compound was positively identified. Other trimethylpyridines might also be present. Van Meter et al. [47] identified 2,3,5-, 2,3,6-, 2,4,5-, and 2,4,6-trimethylpyridines, and 2,3,-, 2,4-, 2,5-, 2,6-, 3,4-, and 3,5-dimethylpyridines in crude shale oils. The 2,3-, 2,4-, and 2,5-dimethylpyridines were also found in shale oil wastewaters [47]. Riley et al. [48] identified alkyl substituted pyridines in surface waters and groundwaters adjacent to an oil shale facility. Other alkylated pyridines were tentatively identified by mass spectrometry. The fragment ions at M-15 suggest that these compounds are tetramethyl-, dimethyl-, ethyl-, or diethylpyridines or similar derivatives. Propyl-, butyl-, and pentylpyridines were reported to be present in crude shale oils [49].

Anilines were previously reported to be present in crude shale oils [49]. Several anilines were separated from retort waters with the XAD-8 resin [13]. The quinoline

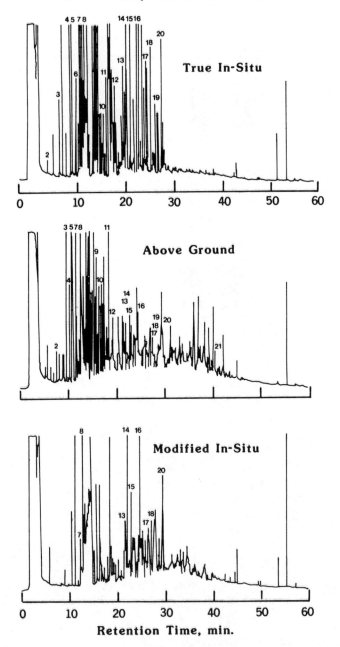

Figure 2. High resolution fused silica capillary chromatogram of the hydrophobic base fractions from oil shale wastewaters [16].

Table 5. *Nitrogen Base Compounds in the Hydrophobic Base Fractions of Three Oil Shale Wastewaters Identified by Gas Chromatography and Mass Spectroscopy*

Peak* No.	Compound	Compounds Identified**					
		True In-Situ		Modified In-Situ		Above-Ground	
		GC	MS	GC	MS	GC	MS
1	pyridine	X				X	
2	2-methylpyridine	X	X			X	
3	3-and/or 4-methylpyridine	X	X			X	X
4	2,4-dimethylpyridine	X	X			X	X
5	2,5-dimethylpyridine	X	X			X	X
6	2,3-dimethylpyridine	X	X				
7	aniline	X	X	X	X	X	X
8	2,4,6-trimethylpyridine	X	X	X	X	X	X
9	N-methylaniline					X	X
10	N,N-dimethylaniline	X				X	X
11	N-ethylaniline	X				X	X
12	2,4-dimethylaniline	X	X			X	X
13	N,N-diethylaniline	X	X	X	X	X	X
14	quinoline	X	X	X	X	X	X
15	isoquinoline	X	X	X	X	X	X
16	2-methylquinoline	X	X	X	X	X	X
17	7-methylquinoline	X	X	X	X	X	X
18	3-methylquinoline	X	X	X	X	X	X
19	2,6-and/or 2,7-dimethylquinoline	X	X			X	X
20	2,4-dimethylquinoline	X	X	X	X	X	X
21	acridine					X	X

*See Figure 2.
**GC: Gas chromatography; the compounds were identified by matching their retention times with those of standards. All compounds produced a response with the nitrogen-phosphorus selective detector.
MS: Mass spectrometry; the compounds were identified by comparison of their spectra with those of standards.
X: Denotes that evidence for the presence of a compound was obtained.

compounds were identified by mass chromatograms via their ions at 129, 143, and 157 that are representative of quinoline, isoquinoline, methylquinoline, and C_2-alkylquinolines. The mass chromatograms generated by monitoring these three ions were similar for the three waters.

Several organic nitrogen bases that are known to be excellent complexing agents are present in these wastewaters (Table 5). Pyridine, quinoline, and their alkyl derivatives are known to form stable complexes with a variety of metals. Because these compounds are biorefractory and resist degradation by conventional activated sludge wastewater treatment, they will be present in wastewaters for a long time and may mobilize metal ions particularly when these wastewaters are allowed to remain in contact with retorted shale.

Acknowledgements. The authors thank Dr. Robert Meglen and his staff at the Center for Environmental Sciences Analytical Laboratory for assistance in trace element analysis, Dr. Robert M. Barkley for mass spectrometry assistance, the Colorado Energy Fellowship Program, and the Department of Energy for assistance under DOE Contract numbers DE-AC02-79EV10298 and DE-AC02-83ER60121.

References

[1] D. S. Farrier, J. E. Virgona, T. E. Phillips and R. E. Poulson, "Environmental Research for In-Situ Oil Shale Processing," **11th Oil Shale Symposium Proceedings,** Colorado School of Mines Press, Golden, Colorado, 1978.

[2] E. A. Jenne and S. M. Luoma, "Forms of Trace Elements in Soils, Sediments, and Associated Waters: An Overview of Their Determination and Biological Availability," in **"Biological Implications of Metals in the Environment,"** R. E. Wildung and H. Drucker, Eds., Conf-750929, U.S. Department of the Interior, Geological Survey, 1977, p. 110.

[3] H. H. M. Bowen, **"Trace Elements in Biochemistry,"** Academic Press: New York, 1966.

[4] E. A. Jenne, "Trace Element Sorption by Sediments and Soils--Sites and Processes," in **"Molybdenum in the Environment,"** Vol. 2, W. R. Chappell and K. Peterson, Eds., Marcel Dekker: New York, 1976, p. 425.

[5] T. M. Florence and G. E. Batley, "Determination of the Chemical Forms of Trace Metals in Natural Waters with Special Reference to Copper, Lead, Cadmium and Zinc," **Talanta, 24** (1977) 151.

[6] G. G. Leppard, **"Trace Element Speciation in Surface Waters and Its Ecological Implications,"** Plenum Press: New York, 1983.

[7] R. H. Fish, "Speciation of Trace Organic Ligands and Inorganic and Organometallic Compounds in Oil Shale

Process Waters," **Proceedings,** 13th Annual Oil Shale Symposium, Colorado School of Mines Press, Golden, Colorado, 1980, p. 385.
[8] T. F. Yen, "Distribution of Short Chain n-Fatty Acids in Retort Water from Green River Oil Shale," **J. Am. Oil Chem. Soc.,** 54 (1977) 567.
[9] C. S. Wen and T. F. Yen, "Studies of Soluble Organics in Simulated In Situ Oil Shale Retort Waters by Electron Impact and Chemical Ionization from a Gas Chromatograph-Mass Spectrometer System," **Am. Chem. Soc., Div. Fuel Chem.,** 21 (1976) 290.
[10] E. W. Cook, "Organic Acids in Process Water from Green River Oil Shale," **Chem. Ind.,** 1 (1971) 485.
[11] J. Stanley, M. Conditt and R. E. Sievers, "Trace Elements and Organic Ligands in Oil Shale Wastes," in **"Trace Elements in Oil Shale,"** Progress Report Contract No. EV-10298, U.S. Department of Energy (DOE-10298-2) 1981, p. 217.
[12] R. E. Poulson, "Stationary Phases for Separation of Basic and Nonbasic Nitrogen Compounds or Hydrocarbons by Gas-Liquid Chromatography," **J. Chromatogr. Sci.,** 7 (1969) 152.
[13] H. A. Stuber, **"Selective Concentration and Isolation of Aromatic Amines from Water,"** Ph.D. Dissertation, University of Colorado, Boulder, Colorado, 1980.
[14] H. S. Hertz, J. M. Brown, S. N. Chesler, F. R. Guenther, L. R. Hilpert, W. E. May, R. M. Parris and S. A. Wise, "Determination of Individual Organic Compounds in Shale Oil," **Anal. Chem.,** 52 (1980) 1650.
[15] R. G. Riley, K. Shiosaki, R. M. Bean and D. M. Schoengold, "Solvent Solubilization, Characterization, and Quantitation of Aliphatic Carboxylic Acids in Oil Shale Retort Water Following Chemical Derivitization with Boron Trifluoride in Methanol," **Anal. Chem.,** 51 (1979) 1995.
[16] R. E. Sievers, M. K. Conditt and J. S. Stanley, "Speciation of Organic Compounds and Trace Elements in Oil Shale Wastewaters," in "Environmental Speciation and Monitoring Needs for Trace Metal-Containing Substances from Energy Related Processes," F. E. Brinckman and R. H. Fish, Eds., **NBS Special Publication 618** (1981) 93.
[17] W. Stumm and P. A. Brauner, "Chemical Speciation," in **"Chemical Oceanography,"** J. P. Riley and G. Skirrow, Eds., Academic Press: New York, 1975, p. 173.
[18] T. M. Florence and G. E. Batley, "Chemical Speciation in Natural Waters," **CRC Critical Reviews in Analytical Chemistry,** 9 (1980) 219.

[19] R. G. Smith, "Evaluation of Combined Application of Ultrafiltration and Complexation Capacity Technique to Natural Waters," **Anal. Chem.**, 48 (1976) 74.
[20] T. A. O'Shea and K. H. Nancy, "Characterization of Trace Metal-Organic Interactions by Anodic Stripping Voltammetry," **Anal. Chem.**, 48 (1976) 1603.
[21] I. Sinko and J. Dolezal, "Simultaneous Determination of Copper, Cadmium, Lead, and Zinc in Water by Anodic Stripping Polarography," **J. Electroanal. Chem.**, 25 (1970) 299.
[22] M. L. Crosser and H. E. Allen, "Determination of Complexation Capacity of Soluble Ligands by Ion Exchange Equilibrium," **Soil Science**, 123 (1977) 176.
[23] R. F. C. Mantoura and J. P. Riley, "The Use of Gel Filtration in the Study of Metal Binding by Humic Acids and Related Compounds," **Anal. Chim. Acta.**, 78 (1975) 193.
[24] C. M. G. Vanden Berg and J. P. Kramer, "Determination of Complexing Capacities of Ligands in Natural Waters and Conditional Stability Constants of the Copper Complexes by Means of Manganese Dioxide," **Anal. Chim. Acta.**, 106 (1979) 113.
[25] T. M. Florence and G. E. Batley, "Removal of Trace Metals from Sea-Water by a Chelating Resin," **Talanta**, 23 (1976) 179.
[26] G. E. Batley and T. M. Florence, "A Novel Scheme for the Classification of Heavy Metal Species in Natural Waters," **Anal. Letters**, 9 (1976) 379.
[27] P. Figura and B. McDuffle, "Determination of Labilities of Soluble Trace Metal Species in Aqueous Environmental Samples by Anodic Stripping Voltammetry and Chelex Columns and Batch Methods," **Anal. Chem.**, 52 (1980) 1433.
[28] T. M. Florence and G. E. Batley, "Exchange of Comments on Scheme for Classification of Heavy Metals Species in Natural Waters," **Anal. Chem.**, 52 (1980) 1960.
[29] M. R. Guerin, I. B. Rubin, T. K. Rao, B. R. Clark and A. L. Epler, "Distribution of Mutagenic Activity in Petroleum and Petroleum Substitutes," **Fuel**, 60 (1981) 281.
[30] B. W. Wilson, M. R. Peterson, R. A. Pelroy and J. T. Cresto, "In-Vitro Assay for Mutagenic Activity and Gas Chromatographic-Mass Spectral Analysis of Coal Liquefaction Material and the Products Resulting from Its Hydrogenation," **Fuel**, 60 (1981) 289.
[31] J. S. Fruchter, C. L. Wilkerson, J. C. Evans and R. W. Sanders, "Elemental Partitioning in an Above Ground Oil Shale Retort Pilot Plant," **Environ. Sci. Tech.**, 14

(1980) 1374.
[32] J. S. Fruchter, C. L. Wilkerson, J. C. Evans, R. W. Sanders and K. W. Abel, **"Source Characterization Studies at the Paraho Semiworks Oil Shale Retort,"** DOE/EY-76-C-06-1830, May 1979.
[33] T. R. Wildeman, A. M. Laffoon, K. Dahlin and D. Ramsey, "Chemical Analysis of Inputs and Outputs of Oil Shale Retorts," in **"Trace Elements in Oil Shale,"** COO-10298-1, 1980, p. 113.
[34] J. G. Flair, C. G. Goldstein, G. Grua, D. Smith and R. Franklin, **"Environmental Research on a Modified In-Situ Oil Shale Process: A Progress Report from the Oil Shale Task Force,"** DOE/EV-0078, May 1980.
[35] D. S. Farrier, R. E. Poulson, O. D. Skinner, J. C. Adams and J. P. Bower, "Acquisition, Processing and Storage for Environmental Research of Aqueous Effluents Derived from In-Situ Oil Shale Processing," **Proceedings, 2nd Pacific Chemical Engineering Congress,** Denver, Colorado, American Institute of Chemical Engineers: New York, 1977, p. 1031.
[36] D. S. Farrier, R. E. Poulson and J. P. Fox, "Interlaboratory, Multimethod Study of an In-Situ Produced Oil Shale Process Water," in **"Oil Shale Symposium: Sampling, Analysis, and Quality Assurance,"** EPA-600/9-80-023, June 1980, p. 182.
[37] J. A. Leenheer and E. W. D. Huffman, "Classification of Organic Solutes in Water by Using Macroreticular Resins," **J. Res. US Geol. Surv.,** 4 (1976) 737.
[38] E. W. D. Huffman, "Isolation of Organic Materials from In-Situ Oil Shale Retort Water Using Macroreticular Resins, Ion Exchange Resins, and Activated Carbon," in **"Measurement of Organic Pollutants in Water and Wastewater,"** ASTM STP 686, C. E. Van Hall, Ed., American Society for Testing and Materials: Philadelphia, Pennsylvania, 1979, p. 275.
[39] R. Kunkel and S. E. Manahan, "Atomic Absorption Analysis of Strong Heavy Metal Chelating Agents in Water and Wastewater," **Anal. Chem.,** 45 (1973) 1465.
[40] D. R. Jones and S. E. Manahan, "Elimination of Copper-Carbonate Complex Interference in Chelating Agent Analysis by Copper Solubilization," **Anal. Chem.,** 49 (1977) 10.
[41] T. D. Martin, J. F. Kopp and R. D. Ediger, "Determining Selenium in Water, Wastewater, Sediment and Sludge by Flameless Atomic Absorption Spectroscopy," **At. Absorption Newsl.,** 14 (1975) 109.
[42] R. A. Bailey, H. M. Clarke, J. P. Ferris, S. P. Krause and R. L. Strong, **"Chemistry of the Environment,"**

Academic Press: New York, 1976.
[43] M. Tanaka, "The Mechanistic Consideration on the Formation Constants of Copper(II) Complexes," **J. Inorg. Nucl. Chem., 36** (1974) 151.
[44] S. E. Manahan and D. R. Jones, "Atomic Absorption Detector for Liquid-Liquid Chromatography," **Anal. Lett., 6** (1973) 745.
[45] D. R. Jones and S. E. Manahan, "Measurement of Chelating Agent Levels in Media with High Concentrations of Interfering Metal Ions," **Anal. Lett., 8** (1975) 421.
[46] M. Novotny, R. Kump, F. Merli and L. J. Todd, "Capillary Gas Chromatography/Mass Spectrometric Determination of Nitrogen Aromatic Compounds in Complex Mixtures," **Anal. Chem., 52** (1980) 401.
[47] R. A. Van Meter, C. W. Bailey, J. R. Smith, R. T. Moore, C. S. Allbright, I. A. Jacobson, V. M. Hilton and J. S. Ball, "Oxygen and Nitrogen Compounds in Shale-Oil Naphtha," **Anal. Chem., 24** (1952) 1758.
[48] R. G. Riley, T. R. Garland, K. Shiosaki, D. C. Mann and E. Wilding, "Alkylpyridines in Surface Waters, Groundwaters, and Subsoils of a Drainage Location Adjacent to an Oil Shale Facility," **Environ. Sci. Technol., 15** (1981) 697.
[49] P. Brown, D. G. Earnshaw, F. R. McDonald and H. B. Jensen, "Gas-Liquid Chromatographic Separation and Spectrometric Identification of Nitrogen Bases in Hydrocracked Shale Oil Naphtha," **Anal. Chem., 42** (1970) 146.

XXXV

ENVIRONMENTAL CONDITIONS IN THE BAY OF AUGUSTA, SICILY: APPLICABILITY OF SIMULTANEOUS INDUCTIVELY COUPLED ARGON PLASMA EMISSION SPECTROMETRY FOR THE ANALYSIS OF SEAWATER

G. Magazzu, G. C. Pappalardo* and K. J. Irgolic**

Dipartimento di Biologia Animale ed Ecologia Marina
Universita di Messina, 98100 Messina, Italy
2^a Cattedra di Chimica Generale ed Inorganica, Facolta di Farmacia, and Centro Universitario di Studio per la Tutela Ambientale e la Medicina Sociale e Preventiva, Universita di Catania, I-95125 Catania, Italy
**Department of Chemistry, Texas A&M University*
College Station, Texas 77843 USA

Water samples from the Bay of Augusta, Sicily, were analyzed for major and minor elements by simultaneous inductively coupled argon plasma emission spectrometry (ICP) and for salinity, temperature, ammonium-nitrogen, cell population, chlorophyll a, and primary productivity. The results indicate that the bay receives pollutants through wastewaters from industrial and urban sources. Many samples of bay water had concentrations of arsenic, cadmium, copper, iron, nickel, lead, and zinc above the average concentrations characteristic of unpolluted seawater. The average Ca concentration of 521 mg/L, which is almost 100 mg/L higher than in the open ocean, is probably caused by the influx of Ca-rich freshwater. ICP was found to be a very convenient method for the simultaneous determination of many elements in seawater. Although moderate to severe pollution by trace elements can be easily detected by ICP, only eleven elements can be determined in this manner in unpolluted seawater.

The Bay of Augusta is a non-pristine, half-closed basin on the east coast of Sicily, 15 km north of Siracusa. The bay has an area of approximately 25 km^2, has a maximal depth of 30 m, and contains approximately 350 million m^3 of water.

During the last 25 years, land adjacent to the bay has become an important industrial area. Nine thousand ships transported 30 million tons of products into and out of the bay during the period 1978-1980. The very favorable position of Augusta Bay in the Mediterranean Sea led to the establishment of plants and refineries of considerable size along the west coast of the bay. These plants discharge seven million cubic meters of waste water per hour into the bay. Part of the urban waste water from the town of Augusta also flows into the bay.

During the last decade many phenomena, easily attributable to eutrophication, occurred in the Bay of Augusta. The color of the water changed from dark-green to brownish-red. Many marine organisms died. Algal blooms of *Peridinium trochoideum* [1] were found to be responsible for many of these phenomena. After the heavy bloom in August and September 1977, lesser blooms occurred in subsequent years.

The present investigation of the water in the Augusta Bay is part of a more comprehensive study undertaken in the framework of the UNEP MED-POL Phase II Program to determine profiles of major and minor elements in the Mediterranean Sea. The study of Augusta Bay deals with the special case of a polluted coastal zone of small size. In addition to nutrients, chlorophyll, temperature and salinity, the major and minor elements were determined by simultaneous inductively coupled argon plasma emission spectrometry (ICP). The use, advantages, potentialities, and limitations of ICP in monitoring the pollution of seawater are discussed.

Experimental

Sample collection and analytical methods. The samples were collected in Niskin bottles during the cruise of the vessel "Algesiro" (University of Messina) on February 8 and 9, 1983. The samples for the analysis of nutrients were filtered through Millipore HA (0.45 µm) and kept frozen at -20°. The pH values were determined potentiometrically on board. The samples for the chlorophyll determinations (800 mL) were filtered through Watman glass-fiber filters and stored frozen at -20°. Primary productivity was evaluated on board by incubation of surface water samples after addition of 4 µC of 14-C sodium bicarbonate. The analytical methods were described earlier [2-4]. The samples for analyses by ICP were filtered through Millipore HA (0.45 μm) immedately after collection. The samples were poured into 25 mL polyethylene containers, mixed with 0.05 mL of concentrated nitric acid (Suprapur Merck), and stored in a refrigerator at 4°.

Determination of major and minor elements by ICP. An ARL model 34000 simultaneous inductively coupled argon plasma emission spectrometer with 48 elements on the array was used for the determination of major and minor elements. After a preliminary analysis of a few samples to establish which elements are present, standards were prepared containing the elements in the range of the concentrations found in the preliminary experiments. The standards for the minor elements were prepared in a seawater matrix [5] (422 mg/L Ca, 436 mg/L K, 1326 mg/L Mg, 11,057 mg/L Na, 928 mg/L S as sulfate). Blanks of distilled water and a standard were analyzed at the beginning and after each set of 15 samples. The normalization program was employed to correct for instrumental drift using two standards with the elements in the concentration ranges characteristic of the seawater samples. The raw data from the standards and the samples were corrected for background employing the off-peak correction routine. The background intensities at +13 and -13 dial divisions from the peak maxima as determined by the "Scanning Accessory for Multielement Instrumentation" were used for background correction.

Results and Discussion

The sampling stations and their coordinates are shown in Figure 1.

Surface temperatures, salinity, ammonium-nitrogen, biological activity and pH. The surface temperatures in the Augusta Bay are shown in Figure 2. The lowest temperatures of less than 13.5° were found in the north-west section, which is also the region of lowest salinity (Fig. 2). Temperatures in the range 13.5° to 14.0° were encountered in the central zone of the bay, which coincides approximately with the intermediate salinity region. The water in the southern half of the bay and outside the bay (stations L2 to L8) had temperatures higher than 14.0°. The water with the highest salinity (>38.25‰) is on the seaward side of the bay. The average salinity at stations L2 to L8 outside the bay is 38.4‰. The effects of the discharge of process and cooling waters from industrial plants into the bay is

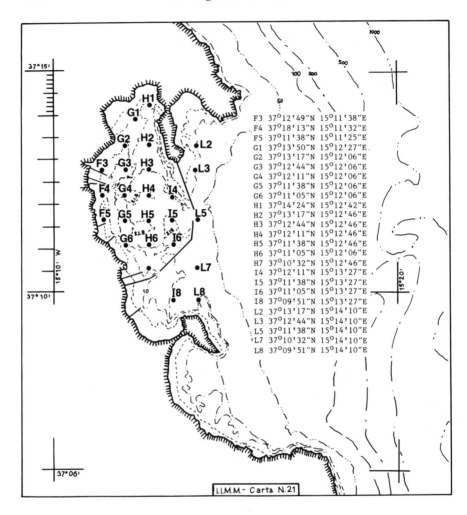

Figure 1. Location of sampling sites in the Bay of Augusta.

evident at station G6, where a wastewater pipeline ends. The temperature in this region is approximately one degree higher than in other sections of the bay.

The distribution of ammonium ion, the primary productivity, the distribution of chlorophyll a, and the cell population are shown in Figure 3. Because of their extremely low concentrations, nitrite, nitrate, and phosphate were not determined. A large amount of ammonium

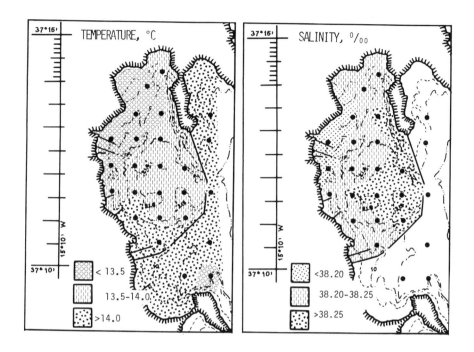

Figure 2. Temperature of surface waters and salinity in the Bay of Augusta, Sicily.

ion is brought into the bay with the waste water discharged underwater near station G6. This ion may be used to trace the dispersion of industrial waste water in the bay. The area bordered by a line from station F5 to station I6 and the southern portion of the levee had nitrogen-ammonium concentrations in the range from 1 µM to 6 µM. The average nitrogen concentration at the other stations was found to be 0.3 µM. Biological activity is strongly depressed in the southern coastal regions. The phytoplankton cell number, the concentration of chlorophyll, and the uptake of C-14 had their lowest values in these regions. The biological activity increased toward the northern part of the bay. The maximal concentration of chlorophyll was found in a region extending from the north-end of the bay along the coast of the peninsula, on which the town of Augusta is located, to a line between stations H6 and I6. The pH values and their averages determined for the Augusta Bay waters are the same as in unpolluted waters.

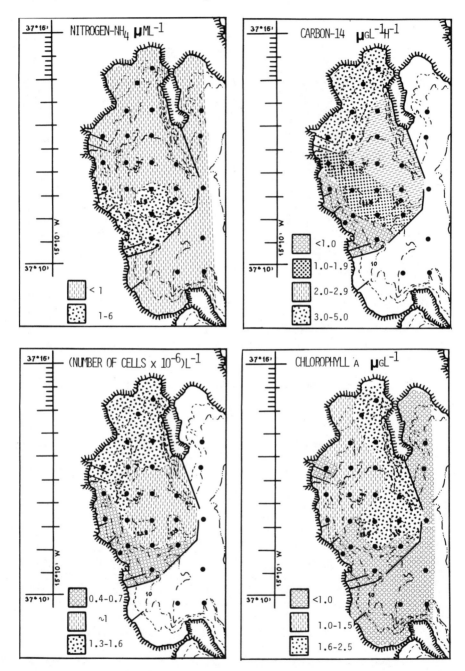

Figure 3. Ammonium-nitrogen, primary productivity, chlorophyll, and cell population in the Bay of Augusta.

Major and minor elements by ICP. The water samples from the stations in Figure 1 were analyzed for the 48 elements, for which detection limits are given in the first footnote to Table 1, by simultaneous inductively coupled argon plasma emission spectrometry. Only the elements listed in the body of Table 1 were present in the water samples at concentrations above the detection limits of the instrument.

Difficulties were encountered with the subroutines for data treatment. To obtain correct results, normalization and blank subtraction had to be used for the minor elements, whereas normalization without blank substraction was required for the major elements. The intercepts calculated by the subroutines for the normalized calibration curves are probably responsible for this situation. The detection limits for the elements present in the samples from Augusta Bay were determined for a seawater matrix and were found not to be drastically different from those for solutions in distilled water (Table 1). The detection limit of 7.2 mg/L for Na is the result of a change in the attenuator setting for the Na channel, which was required to move the 11,000 mg/L Na concentration characteristic for seawater from the flattened to the linear region of the calibration curve.

The averages of the results obtained by repeated analysis of a standard (Table 1) have relative standard deviations of a few percent unless the concentrations approach the detection limits. At concentrations of ten-times the detection limits, relative standard deviations of approximately ten percent are expected [6]. These results prove that the instrument remained stable during the period of four hours required for the analysis of the seawater samples.

The results of the analysis of the water samples from Augusta Bay by ICP are summarized in Table 1. The major elements [K 426±9, Mg 1343±27, Na $(1.11\pm0.02)\times10^4$, S 989±18 mg/L] are present in concentrations close to their average concentrations in seawater, whereas the Ca concentrations (521±7 mg/L) are approximately 80 to 100 mg/L higher than the norm. The observed calcium/chloride ratio (0.024) is much higher then the ratio of 0.02258 found for the surface waters of the Ionian Sea [7]. The increased calcium concentration in the Augusta Bay can be attributed to the large influx of freshwater - known to be calcium-rich - which is discharged by the land-based small water courses and industrial sources. A major element such as calcium may be a better "tracer" than a minor element for quick

Table 1. *Major and Minor Elements in Water Samples from the Bay of Augusta, Sicily, Determined by Simultaneous Inductively Coupled Argon Plasma Emission Spectrometry*

#	Sample Station Depth(m)	As μg/L	B mg/L	Ba μg/L	Ca mg/L	Cd μg/L	Cu μg/L	Fe μg/L	K mg/L	Li μg/L	Mg mg/L	Mn μg/L	Mo μg/L	Na mg/Lx10⁻⁴	Ni μg/L	Pb μg/L	S mg/L	Si μg/L	Sr mg/L	Zn μg/L
2	C3/0	*	5.1	3	531	9	*	17	447	261	1402	*	*	1.15	*	119	1022	*	8.08	20
3	F3/0	11	5.2	5	504	16	10	12	415	255	1308	*	18	1.09	*	147	967	27	8.38	14
4	F4/0	*	5.0	6	520	103	17	95	413	256	1353	3	14	1.12	*	139	997	32	7.93	59
5	F5/0	23	4.9	7	527	*	9	5	435	257	1361	6	29	1.11	10	146	1002	17	7.83	*
6	I4/0	75	5.2	17	522	25	48	33	434	256	1358	63	50	1.12	41	117	1000	95	8.12	24
7	I5/0	22	5.2	5	523	7	*	7	431	255	1350	*	*	1.12	*	105	1001	54	8.20	*
8	I6/0	*	5.0	4	519	*	*	9	429	255	1339	*	11	1.10	*	160	991	94	7.90	17
9	I6/5	18	5.1	6	525	9	*	22	432	256	1350	*	*	1.16	*	131	1001	24	7.95	15
10	I6/10	19	5.0	6	521	14	7	54	424	255	1332	*	*	1.09	16	156	991	18	7.87	27
11	I8/0	*	4.9	5	519	*	*	4	429	258	1346	32	*	1.11	*	157	996	42	7.95	*
12	L2/0	18	5.0	8	519	*	36	34	426	255	1348	*	21	1.11	110	124	993	79	7.87	44
13	L3/0	12	4.8	7	525	*	7	10	433	258	1375	6	*	1.13	14	105	1004	69	7.46	*
14	L5/0	13	4.8	6	531	*	5	*	438	261	1392	*	29	1.14	41	135	1015	19	7.59	22
16	L7/9	12	5.1	8	536	*	5	3	443	260	1402	*	25	1.14	17	122	1021	38	8.07	15
17	L8/0	28	4.9	9	520	*	9	16	425	256	1350	5	*	1.12	15	119	990	51	7.76	*
18	G1/0	18	5.1	8	513	13	*	7	418	254	1328	*	*	1.10	*	123	977	109	8.04	*
19	G2/0	24	5.0	7	526	*	5	4	432	256	1370	*	*	1.12	*	121	1003	77	7.89	15
20	G4/0	*	5.1	6	522	*	5	4	426	255	1348	3	25	1.12	*	115	992	18	8.15	*
21	G4/5	*	5.1	6	525	*	*	4	428	255	1353	3	14	1.12	*	113	994	15	7.97	23
22	G4/10	*	5.0	3	524	*	*	*	428	255	1347	*	*	1.10	*	146	993	30	7.93	17
23	G4/13	*	4.8	5	524	*	*	37	427	256	1342	*	*	1.10	*	138	996	8	7.54	25
24	G5/0	*	5.2	6	524	*	5	8	427	255	1340	*	*	1.11	*	111	987	32	8.07	15
25	G6/0	18	5.0	6	526	*	11	12	425	254	1330	3	19	1.10	*	118	981	47	7.76	32
26	G6/5	*	4.9	5	514	*	28	17	421	253	1313	*	*	1.09	*	134	971	21	7.74	29
27	G6/10	12	4.9	7	507	*	*	6	412	250	1290	*	*	1.07	*	128	957	11	7.68	14
28	H1/0	17	5.0	8	522	*	*	11	432	257	1346	*	*	1.11	*	119	994	43	7.78	*
29	H2/0	*	4.8	7	505	*	*	4	410	250	1290	*	*	1.07	*	140	956	254	7.52	*
31	H3/0	*	5.2	4	523	*	*	5	438	257	1358	*	13	1.12	*	107	994	108	8.30	*

Table 1. Cont'd

#	Sample Station Depth(m)	As µg/L	B mg/L	Ba µg/L	Ca mg/L	Cd µg/L	Cu µg/L	Fe µg/L	K mg/L	Li µg/L	Mg mg/L	Mn µg/L	Mo µg/L	Na mg/L×10^{-4}	Ni µg/L	Pb µg/L	S mg/L	Si µg/L	Sr mg/L	Zn µg/L
32	H3/2.5	*	5.1	5	520	*	*	*	430	255	1340	*	*	1.11	*	133	988	38	8.01	*
33	H3/5	22	5.0	6	526	5	5	3	436	256	1361	*	14	1.12	*	139	996	47	7.76	*
34	H3/8	25	5.0	9	511	5	7	21	415	252	1319	*	17	1.09	*	131	966	57	7.82	*
35	H3/12	12	5.0	4	521	*	*	*	423	254	1348	*	*	1.12	*	131	987	*	7.81	*
36	H3/15	*	4.8	5	530	*	*	9	432	260	1371	*	*	1.13	*	139	1005	19	7.51	*
37	H4/0	21	5.0	4	533	*	*	12	437	257	1385	*	*	1.14	*	98	1010	24	7.78	*
38	H5/0	*	5.1	6	503	7	9	30	403	250	1282	*	*	1.08	23	130	946	29	7.97	28
39	H6/0	*	5.0	5	513	5	*	11	414	254	1311	6	12	1.10	*	104	963	*	7.81	26
40	H6/5	*	5.1	6	518	*	*	7	418	253	1322	*	13	1.10	*	114	970	13	7.94	15
41	H6/10	*	4.8	5	523	5	*	8	425	255	1339	*	*	1.11	*	133	983	10	7.53	*
42	H7/0	*	4.9	6	523	*	8	14	409	252	1291	3	25	1.08	*	139	956	100	7.13	*
43	H7/5	*	5.0	5	525	*	*	6	427	254	1343	5	10	1.12	*	134	986	46	7.92	*
1	Standard	108	4.62	56	438	43	22	49	423	309	1307	18	18	1.09	46	235	926	359	10.2	61
15	Standard	116	4.80	59	460	43	27	52	438	318	1359	19	29	1.10	43	239	960	371	10.7	58
30	Standard	107	4.60	57	449	49	27	50	429	313	1314	19	15	1.06	47	223	932	357	10.3	63
44	Standard	131	4.54	55	452	49	22	50	432	314	1325	19	44	1.08	51	223	941	351	10.3	63
	Av./Std.Dev.†	116/11	4.6/1	57/2	450/9	46/3	25/3	50/1	430/6	313/4	1326/23	18.8/5	27/13	1.08/1	47/3	230/8	940/15	360/8	10.4/2	61/2
	RSD	9.5%	2.2%	3.5%	2.0%	6.5%	12.0%	2.0%	1.4%	1.3%	1.7%	2.7%	48.1%	0.9%	6.4%	3.5%	1.6%	2.2%	1.9%	3.3%
	Stand.Calc.	90	4.6	61	422	40	24	41	436	315	1326	16	20	1.10	52	200	928	253	10.45	60
	Av.Conc.Sea	3	4.6	30**	400	0.1	3**	10**	380	170	1350	2**	10	1.05	?	0.03	885	3000**	8.0	10
	DL Seawater	11	0.017	2	0.14	5	5	2	0.3	5	0.5	2	10	7.2 ppm	6	90	—	70	0.004	13

* At or below detection limits. The detection limits for distilled water solutions in µg/L are: Ag 6, Al 20, As 10, Au 16, B 7, Ba 4, Be 0.2, Bi 80, Ca 6, Cd 6, Co 10, Cr 8, Cu 4, Fe 4, Ge 60, Hf 20, Hg 40, I 40, K 120, La 10, Li 4, Mg 40, Mn 1, Mo 8, Na 40, Nb 10, Ni 18, P 100, Pb 60, Pd 60, Pt 80, S 80, Sb 60, Sc 2, Se 160, Si 16, Sn 40, Sr 0.2, Ta 6, Te 160, Th 100, Ti 16, U 200, V 8, W 30, Zn 10, Zr 4.

** The concentrations of these elements vary considerably. The average concentrations of elements in the sea were taken from ref. 5.

† Averages are given in the form "average/standard deviation", e.g., 10.4/2 = 10.4±0.2.

monitoring of industrial waste discharges in coastal seawaters. Changes in seawater quality and charateristics could in this way be quickly recognized by a rapid, precise and accurate analysis of seawater samples by ICP spectrometry.

Half the stations have arsenic concentrations approximately twice the detection limit of 11 µg/L. The average concentration of As in the ocean is 2.3 µg/L. The boron concentrations (5.0±0.1 mg/L) in the range 4.8 to 5.2 mg/L are close to the average for the ocean of 4.5 mg/L. The Ba concentrations are rather low at less than 10 µg/L. Only station I4/0 has 17 µg/L Ba.

Most of the stations produced samples with cadmium concentrations below the detection limit of 5 µg/L. A few stations had Cd concentrations in the range of 5 to 25 µg/L; station F4/0 had 103 µg/L Cd. The average Cd concentration in the sea is 0.05 µg/L.

Among the forty samples, nineteen had Cu concentrations in the range 5 to 48 µg/L. The highest concentrations were found in the samples coming from the stations along the coastal zone at the extreme west section of the bay where the most important industrial plants are located. Almost all stations had Fe in the range 3 to 95 mg/L. The Li concentration in the samples is almost constant at 225±3 µg/L, approximately 70 µg/L higher than the ocean average.

Few stations had Mn levels above the detection limit of 3 µg/L and Mo concentrations above 10 µg/L. Only nine samples had Ni concentrations above the detection limit of 6 µg/L in the range 14 to 110 µg/L. Five of these samples came from stations in section L outside the bay basin.

All samples appear to have lead at an average concentration of 128±15 µg/L ranging from 98 to 160 µg/L, a level exceeding the ocean average by a factor of approximately 4000. However, the results for Pb are tentative because the concentrations are less than twice the detection limit of 90 µg/L. Similarly, the results for Si are tentative. The poor agreement between the calculated and experimental concentrations of the standard indicate improper functioning of the channel, problems with the stability of the standard, or presence of interferences. The Sr concentrations in all the samples averaged 7.9±0.3 mg/L, close to the average ocean concentration of 8.5 mg/L. Most of the samples have zinc concentrations in the range 14 to 59 µg/L.

Many samples from the bay appear to have concentrations of arsenic, cadmium, copper, iron, nickel, lead, and zinc above the average concentrations characteristic for unpolluted seawater. Additional analyses by methods with better sensitivity are needed to confirm these results. Among the stations sampled, I4/0 has much higher As, Ba, Cd, Cu, Mn, Mo, and Ni concentrations than the other stations.

Augusta Bay appears to be a polluted area in which eutrophic phenomena frequently occur. Ammonium-nitrogen is added to the bay water through discharge of industrial wastes. The population of benthic organisms is depleted [8]. The results of the ICP analyses indicate also some contamination of the bay caused by land-based industrial and urban pollution sources. Additional investigations directed toward the identification of trace element compounds in the water and the sediments and the potential release of trace elements from the sediments [9] should be undertaken.

Inductively coupled argon plasma emission spectrometry is a very convenient method for the simultaneous determination of many elements in a large number of samples. However, ICP is sufficiently sensitive for the determination of only eleven elements (Li, Na, K, Mg, Ca, Sr, Ba, Mn, C, Si, S) at concentrations characteristic of unpolluted seawater. The combination of hydride-generation with ICP, which is claimed to lower detection limits for As, Se, Sn, Sb, Hg, Pb and Bi as much as 100-fold, might make these elements determinable in unpolluted seawater. The detection limits of ICP for metals of interest in pollution studies are in the 1-10 µg/L range. Moderate to severe pollution can easily be detected by ICP and certainly by hydride generation-ICP for the elements responsive to this technique.

In order to fully utilize the advantages of a multi-element spectroscopic system, preconcentration methods applicable to seawater must be developed which permit the utilization of aqueous standards. Many organic reagents are known to extract metal ions into organic solvents with enrichment factors of several hundred. However, metal standards soluble in organic solvents are not easily available. Therefore, re-extraction of the metal ions from the organic phase into an aqueous phase is very desirable. Such a system is now being investigated in our laboratory [10].

Acknowledgements. K. J. I. and G. C. P. gratefully acknowledge the financial support of their cooperative research endeavors by Ministero Della Pubblica Istruzione (Italy), NATO Research Grant No. 1659, the U.S. National Science Foundation (Grant No. INT-7921544), and the Robert A. Welch Foundation of Houston, Texas.

References

[1] S. Genovese, Dipartimento di Biologia Animale ed Ecologia Marina, Universita di Messina, private communication, 1983.
[2] S. Genovese and G. Magazzu, **"Manuale d'Analisi per le Acque Salmastre,"** Editrice Universitaria, Messina, 1969, 135 pp.
[3] J. D. H. Strickland and T. R. Parsons, "A Practical Handbook of Seawater Analysis," **Fish. Res. Board Canada,** Ottawa, 1972, 310 pp.
[4] G. Magazzu, **"Metodi per lo Studio del Plancton e della Produzione Primaria,"** Edizioni G. M., Vibo Valentia, 1979, 68 pp.
[5] R. A. Horne, **"Marine Chemistry,"** J. Wiley & Sons: New York, 1969, p. 152.
[6] G. C. Pappalardo, R. A. Stockton and K. J. Irgolic, "Inductively Coupled Argon Plasma Emission Spectrometry, The Method for the Determination of Trace Elements," **La Chimica e L'Industria, 64** (1982) 550.
[7] G. Magazzu and F. Corigliano, "Rapporti Ionici Ca/Cl e Mg/Cl con la Profondita nel Mar Ionia," **Boll. Pesca Piscic. Idrobiol., 29** (1974) 169.
[8] I. Di Geronimo, "Influence de la Pollution sur les Peuplements a Mollusques de la Baie d'Auguste (Sicile)," **Proceedings VIth ICSEM/IOC/UNEP Workshop on Pollution of the Mediterranean,** Cannes, Dec. 2-4, 1982, p. 715.
[9] S. Sciacca and R. Fallico, "Presenza e Concentrazione di Sostanza Inquinanti di Origine Industriale nei Faghi della Rada di Augusta (Siracusa)," **Inquinamento, 20** (1978) 1.
[10] D. Chakraborti, K. J. Irgolic and F. Adams, "Atomic Absorption Spectrometric Determination of Arsenite in Water Samples by Graphite Furnace After Extraction with Ammonium sec-Butyldithiophosphate," **J. Assoc. Off. Anal. Chem., 67** (1984) 277.

APPENDICES

```
List of Participants . . . . . . . . . . . .  615
Author Index . . . . . . . . . . . . . . . .  621
Subject Index  . . . . . . . . . . . . . . .  623
List of Abbreviations  . . . . . . . . . . .  653
```

LIST OF PARTICIPANTS
U.S.-ITALY WORKSHOP
ON
ENVIRONMENTAL INORGANIC CHEMISTRY

San Miniato, Italy, June 6-10, 1983

CHINA

Jennie Ching-I Liu
Director
Institute of Environmental
  Chemistry
Academia Sinica
P.O. Box 934
Beijing, China

Lu Zongpeng
Research Scientist
Institute of Environmental
  Chemistry
Academia Sinica
P.O. Box 934
Beijing, China

Peng An
Associate Research Professor
Institute of Environmental
  Chemistry
Academia Sinica
P.O. Box 934
Beijing, China

ITALY

Giuseppe Arena
Associate Professor of
  Analytical Chemistry
Istituto Departimentale Di
  Chimica e Chimica
  Industriale
Universita di Catania
Viale A. Doria 8
I-95125 Catania, Italy

Michael Bernhard
Scientific Adviser
Centro Studi Ambiente Marino
ENEA, P.O.B. 316
I-19100 La Spezia, Italy

Franco Baldi
Research Scientist
Istituto di Biologia
  Ambientale
Universita di Siena
Via delle Cerchia, 3
I-53100 Siena, Italy

Ivano Bertini
Professor of Inorganic
  Chemistry
Universita di Firenze
Via Gino Capponi
I-50132 Firenze, Italy

List of Participants

Vera Bianchi
Associate Professor
Istituto Biologia Animale
Universita di Padova
I-35000 Padova, Italy

Ugo Croatto
Professor of Chemistry
Istituto di Chimica Generale
Universita di Padova
Via Loredan 4
I-35100 Padova, Italy

Arnaldo Liberti
Professor
Istituto Sull'Inquinamento
  Atmosferico
Consiglio Nazionale delle
  Ricerche
Via Salaria Km. 29.300-C.P.10
I-00016 Roma, Italy

Vincenzo Minganti
Research Scientist
Istituto di Chimica Generale
Universita di Genova
Viale Benedetto XV, 3
I-16100 Genova, Italy

Giorgio Ostacoli
Professor of Analytical
  Chemistry
Universita di Turnio
Via G. Bidone 36
I-10125 Tornio, Italy

Renzo Capelli
Professor of General and
  Inorganic Chemistry
Istituto di Chimica Generale
Universita di Genova
Viale Benedetto XV, 3
I-16132 Genova, Italy

Innocente Franchini
Associate Professor of
  Occupational Medicine
Istituto Semeiotica Medica
Via Gramsci 14 Cattedra
  Medicina del Lavoro
Universita di Parma
I-43100 Parma, Italy

Giuseppe A. Magazzu
Professor of Oceanography
Dipartimento di Biologia
Animale ed Ecologia Marina
Universita di Messina
Via dei Verde 75
I-98100 Messina, Italy

Veronica Novelli
Assistente
Istituto di Chimica
Facolta Ingegneria
Universita di Udine
Viale Ungheria 43
I-33100 Udine, Italy

Giuseppe C. Pappalardo
Professor
Istituto Dipartimentale di
  Chimica
Universita di Catania
Viale A. Doria 8
I-95125 Catania, Italy

List of Participants

Giulio Queirazza
Chemist
ENEL - CRTN
Via Rubattio 54
I-20100 Milano, Italy

Ernst-Hermann Schulte
Biologist
ENEA - Centro Studi
Ambiente Marino
Casella Postale 316
I-19100 La Spezia, Italy

Gianni Tomassi
Scientist
Istituto Nazionale Della
  Nutrizione
Via Ardeatina 546
I-00100 Roma, Italy

Pietro A. Vigato
Research Scientist
Istituto di Chimica e
  Technologia die
  Radioelementi
Consiglio Nazionale Delle
  Ricerche
Area Delle Ricerca
I-35100 Padova, Italy

Enrico Sabbioni
Head, Project Heavy Metal
  Pollution
Nuclear and Radiochemistry
  Division
Commission of the European
  Communities
Joint Research Center
I-21020 Ispra, Varese, Italy

Pietro Scoppa
Senior Scientific Officer
Commission of the European
  Communities
Centro Ricerche Energia
  Ambiente
Casella Postale 316
I-19100 La Spezia, Italy

Ottavio Tubertini
Associate Professor
Istituto Chimico G. Ciamician
Universita di Bologna
Via Selmi, 2
I-40126 Bologna, Italy

OTHER EUROPEAN COUNTRIES

Sten Ahrland
Professor, Inorganic
  Chemistry I
Chemical Center
University of Lund
P.O.B. 740
S-220 07 Lund 7, Sweden

David E. Fenton
Reader in Chemistry
Department of Chemistry
The University of Sheffield
Sheffield S3 7HF
United Kingdom

List of Participants

Peter J. Sadler
Lecturer, Department of
  Chemistry
University of London,
  Birkbeck College
Malet Street
London WC1E 7HX
United Kingdom

S. C. Tam
Ph.D. Student
Inorganic Chemistry
  Laboratory
University of Oxford
South Parks Road
Oxford OX1 3QR
United Kingdom

Werner Stumm
Professor
Swiss Federal Institute of
  Technology
EAWAG
CH-8600 Dubendorf
Switzerland

P. A. Williams
Lecturer
University College
P.O. Box 78
Cardiff CF1 1XL, Wales
United Kingdom

U.S.A.

Edwin H. Abbott
Professor and Head,
Department of Chemistry
Montana State University
Bozeman, MT 59717

Owen Bricker
Water Resources Division
U.S. Geological Survey
Mail Stop 432
Reston, VA 22092

Gregory R. Choppin
Professor, Department of
  Chemistry
Florida State University
Tallahassee, FL 32306

Arthur E. Martell
Distinguished Professor
Department of Chemistry
Texas A&M University
College Station, TX 77843

Jon M. Bellama
Professor, Department of
  Chemistry
University of Maryland
College Park, MD 20742

F. E. Brinckman
Group Leader
Chemical & Biodegradation
  Process Group
Center for Material Science
National Bureau of Standards
Washington, D.C., 20234

Kurt J. Irgolic
Professor, Department of
  Chemistry
Texas A&M University
College Station, TX 77843

James J. Morgan
Professor, Environmental
  Engineering Science
California Institute of
  Technology
Pasadena, CA 91125

List of Participants

Kenneth N. Raymond
Professor, Department of
 Chemistry
University of California,
 Berkeley
Berkeley, CA 94720

Simon Silver
Professor, Department of
 Biology
Washington University
St. Louis, MO 63130

John M. Wood
Professor, Department of
 Biochemistry
University of Minnesota
P.O. Box 100
Navarre, MN 55392

Robert E. Sievers
Professor, Department of
 Chemistry
University of Colorado
Boulder, CO 80309

Oren F. Williams
Program Director
Synthetic Inorganic and
 Organometallic Chemistry
National Science Foundation
1800 G Street, N.W.
Washington, D.C., 20550

AUTHOR INDEX

Abbott, E.H.	541	Minoia, C.	463	
Acampora, M.	431	Morgan, J.J.	167	
Ahrland, S.	65	Motekaitis, R.	89	
Allegrini, I.	419	Mousty, F.	565	
Amico, P.	285	Mutti, A.	463,	473
Arena, G.	285	Myttenaere, C.	299	
Arfini, G.	463			
		Nicholson, J.K.	249	
Bellama, J.M.	239	Nies, J.D.	239	
Bernhard, M.	349			
Bianchi, V.	447	Ostacoli, G.	285	
Bricker, O.P.	135			
Brinckman, F.E.	195	Pappalardo, G.C.	601	
		Pedroni, C.	463	
Capelli, R.	155	Peng, An	393	
Casellato, U.	273	Pietra, R.	565	
Castiglioni, M.	565	Piovan, G.	431	
Cavalleri, A.	463	Possanzini, M.	419	
Choppin, G.R.	307	Raymond, K.N.	331	
Croatto, U.	59, 431	Rizzarelli, E.	285	
Daniele, P.G.	285	Sabbioni, E.	565	
		Sadler, P.J.	249	
Febo, A.	419	Sammartano, S.	285	
Fenton, D.E.	273	Schulte, E.H.	321	
Franchini, I.	463, 473	Scoppa, P.	299	
		Sievers, R.E.	579	
Hancock, R.D.	117	Silver, S.	513	
Higham, D.P.	249	Smith, R.M.	89	
		Stanley, J.S.	579	
Irgolic, K.J.	1, 547, 601	Stone, A.	167	
		Stumm, W.	401	
Jewett, K.L.	239	Sung, W.	167	
Levis, A.G.	447	Tang, H.-X.	359	
Liberti, A.	419			
Liu, Ching-I	359	Vigato, P.A.	273,	431
Lu Zongpeng	373			
		Wang, H.-K.	487	
Magazzu, G.	601	Wang, Z.-J.	393	
Mao, Meizhou	373	Wei, J.-X.	373	
Martell, A.E.	1, 89	Williams, P.A.	185	
Micoli, G.	463	Wood, J.M.	487	
Minganti, V.	155			

SUBJECT INDEX

Acceptors
  hard, soft 71, 491
Acetate
  methyl donor 210
Acetic Acids
  protonation constants 104
Acid-Base Reactions
  in the atmosphere 403
Acid Mine Water
  composition 505
Acid Precipitation 402
Acid Rain 403
  genesis 404
  in Switzerland 405
Acidity, atmospheric
  determination 419
Acids
  hard, soft 101, 121
Acids, hydrophobic in
  oil shale retort water 580
Activity Coefficient, for
  M(II) ions in seawater 72
Activity Rate Theory 200, 241
Actinide(IV) Ions
  sequestering agents for 331
Actinides, releases
  from reprocessing 308
Adsorption Sites 4
S-Adenosylmethionine
  methyl donor 497
  methylation of As 523
Aerosols 420
Air Pollutants
  sampling of 421
Albumin
  human, Pb complex 381
  bovine, Pb complex 381
Alcaligenes
  oxidation of arsenite 522
Algal Bloom
  in Augusta Bay 602

Algae
  green, 494
  ligand release by 13
  binding of Ni 502
Alkali Metals
  cryptand complexes 277, 278
  in seawater 72
Alkaline Earth Metals
  carbonate complexes 72
  fluoride complexes 72
  in seawater 72
  sulfate complexes 72
Aluminum
  Al-U citrate complexes 285
  immune response 256
  in environment 489, 542, 543
Aluminum Concentration
  effect on fish 405
  in water 405
Aluminum Hydroxides
  dissolution by acid rain 405
Altzheimer's Disease 543
Americium 333
  accumulation by organisms 327
  redistribution 327
Americium-241
  in ecosphere 308
  species in water 317
Ames Test, for
  genotoxicity of metals 544
Amines, aliphatic
  basicity in gas phase 120
  protonation constants 118
  protonation enthalpies 105
Amino Acids
  methylation of Hg 210
Ammonia
  metal complexes 102
  in the atmosphere 209, 403
Ammonia, atmospheric determination with denuders 429

Ammonium Ion
 in Augusta Bay 604, 606
Ammonium Salts, in air 423
Analytical Chemistry
 environmental 42, 547
Analytical Results
 reporting of 50
Anchovies
 Hg in 351, 158
 Cd, Pb, Cu, Mn, Zn in 158
Anilines, in oil shale
 retort waters 593
Anion Concentrations
 in freshwater 314
 in seawater 314
Anodic Stripping Voltammetry
 determination of
 –As 381
 –formation constants 380
 –metals 363, 410
 –metals in seawater 374
Anoxic Zones 70
Antibiotics, resistance to
 by S. aureus 526, 528
Anticancer Drugs 32, 257
Antimony
 methyl compounds 201
 study of metabolism 571
 valence states 382
Antimony(III), resistance to
 by E. coli 501, 520, 522
 by S. aureus 501, 520
Aragonite 68
Arsenate 197
 efflux from cells 501, 522
 in Dalian Bay 38
 in seawater 382
 stability 49
 transport system 501, 521
Arsenate, resistance to
 by E. coli 501, 520
 by S. aureus 501, 520
 resistance mechanism 521
Arsenic 199
 genotoxicity 452
 in Augusta Bay 608-610
 in oil shale water 586/7
 in seawater 368

Arsenic (contd.)
 metabolism in rat 573
 metabolism in rabbit 573
 methylation 523
 methyl compounds 201
 resistance to 520
 study of metabolism 571
 detn. of valence states 382
 valence state in waters 383
Arsenic Compounds
 detn. by HPLC-GFAA 556
 detn. by HPLC-ICP 557, 558
Arsenic-Containing
 Lipids 18, 23
Arsenic Cycle 18
Arsenic Species
 determination 381, 556/8
 in Dalian Bay 381
 in soils 381
 inorganic 381
 organic 381
Arsenite 197
 in Dalian Bay 381
 in seawater 382
 membrane transport 501
 methylation 20
 oxidation to As(V) 522
 stability 49
 toxicity 19
Arsenite, resistance to
 by E. coli 501, 520
 by P. pseudomallei 522
 by S. aureus 501, 520
Arsenobetaine 18
Arsenocholine 18
Arsenolicithins 23
Aspartylalanylhistidyl-
 N-metylamide
 Cu-Zn dinuclear compl. 290
 Cu-Cd dinuclear compl. 290
Atlantic
 metals from atmosphere 409
 Hg in water 355
 Hg in tuna 351
Atmosphere 196, 209
 acid-base reactions in 403
 acidity of 419
 analytical methods 58, 419

Atmosphere (contd.)
  element cycles 402
  pollution of 401
  redox reactions in 403
Atomic Absorption Spectrometry
  detn. of Cr 466
  element-specific detector 555
ATP
  heterobinuclear complexes 287
  V content 256
Augusta Bay, Sicily
  algal bloom 602
  ammonium ion in 606
  cell population 606
  chlorophyll a 606
  element concentr. 601, 607
  primary productivity 606
  salinity 603, 605
  surface temperature 603, 605
Azurite 73
  solubility product 74
  ion product in seawater 74

Bacillus
  Hg resistant 516
Background Concentrations
  of metals 384
Bacteria
  detn. of metals by NMR 258/9
  iron transport 491
  ligand release 13
  methylation of Sn 528
  mineral solubilization 212
  oxidation of arsenite 523
  oxidation of Fe(II) 213
  oxidation of Hg(0) 520
  sulfate reduction 70
  transformation of Hg 514
  transformation of metals 513
Bacteria, resistance to
  As, Sb 501, 520-522
  Cd 14, 501
  Ag 527
  Hg 516
  Sn 528
  methyltin 528
  Zn 528
Bakers Yeast 7

Barium
  EDTA complexes NMR 258
  NTA complex 95
Bases
  hard, soft 101, 121
  in oil shale water 587
Benthic Organisms 321/2
  influence on sediments 321
Bentonite 16, 215
Bertrand Diagram 254/5, 495
Beryllium
  genotoxicity 452
  immune response 256
Bicarbonate Complexes
  in seawater 72
Bicarbonte Concentrations
  in natural waters 93
Binuclear Complexes 8, 285
Bioaccumulation 196
  of elements 61
  of Hg 369
Bioavailability 25, 295
  of radionuclides 30, 301, 305
Bioconcentration
  of metals 500
Biofilms 209
Bioindicators 157, 163
Biological Materials
  detn. of elements by NAA 568
Biological Treatment
  of tannery wastes 435
Biomass 60
Biomethylation
  of As 523
  of Cd 208
  of Hg 499
  of metals 496
  of Pb 207
Bioturbation 145-149, 321/2
  mobilization of
  radionuclides 326
Bismuth
  deposition in cells 267
  detn. by ASV in seawater 374
  in coastal areas of China 375
  in Pacific 275
Bismuth(III) Complexes
  free energy relations 129

Bismuth(III), Resistance to
  by S. aureus 521
1,4-Bis(2,5,5-tris(carboxy-
  methyl)2,5-diazapentyl)-
  benzene 290
Blood, human
  trace metals in 567
Bluefin Tuna
  Hg in 350
Bohai Bay
  Hg species in fish 383
  metal concentrations in 376
Boleite 187
Bonitos
  metal concentr. in 160, 164
  methylmercury in 160, 164
Branching Index 220

Cadmium
  accumulation 257
  absorption in GI tract 438
  adsorption on particles 364
  binding by glutathione 262
  biomethylation 208
  biotransformation 257
  calcd. intake, Italians 444
  detn. by ASV in seawater 374
  essentiality 257
  excretion from cells 525
  foodsources for 441
  genotoxicity 452
  methyl compounds 201
  methylation 207
  polarography 79
  removal by EDTA 262, 264
  renal damage by 257
  retention 526
  species distrib. in sea 79
  study of metabolism 571
  tolerable intake 441
  transport system for 524
Cadmium
  in anchovies 158
  in coastal area, China 375/6
  in mussels 162/3
  in Nanhai Sea 376/7
  in Pacific 375
  in red cells 262

Cadmium (contd).
  in seawater 78
  in shrimp 161
  rivers 384
  in Xiang Riv. 379, 362/3, 359
Cadmium Bromide Chloride 78
Cadmium Complexes
  bromide 78
  calculation of formation
    constants 107/8
  carbonate 78/9
  chloride 78
  EDTA 258
  labile 80
  NTA 79, 80, 95
  sulfate 78
Cadmium Complexes
  mononuclear with
  -alanine 290
  -catechol 290
  -isoserine 291
  -labile 80
Cadmium Complexes
  homobinuclear with
  -isoserine 291
Cadmium Complexes
  heterobinuclear with
  -Co/isoserine 291
  -Cu/carnosine 288/9
  -Cu/L-DOPA 289
  -Cu/histamine 287
  -Cu/histidine 288
  -Cu/tripeptide 290
  -Mn/citrate 292
  -Ni/citrate 292
  -Zn/citrate 292
  -Zn/isoserine 291
Cadmium Cysteine Complex
  release by microbes 14
Cadmium Deposition
  from atmosphere into
  -Atlantic 409
  -Lake Constance 409
  -Swiss Rivers 409
Cadmium Intoxication
  signs of 437
Cadmium Metallothionein
  Cd-113 NMR 263

Cadmium Metal. (contd.)
H-1 NMR 263
from plaice 263
in mice 260
Cadmium-113 NMR 261, 263
rat liver metallothionein 262
Cadmium Oxide
cause of sarcomas 267
Cadmium Proteins 260
Cadmium Resistance 538
by S. aureus 501, 523, 525
Cadmium Species
in freshwater 414/5
in seawater 78
Cadmium Sulfide 257
Cadmium Transport
in plants 286
Calcite 68
Calcium Complexes
bicarbonate 68
carbonate in seawater 68
EDTA, NMR 258
NTA 95
Calcium Concentrations
in Augusta Bay 607/9
Calcium EDTA
absorption by man 91
treatment of Pb, Pu
poisoning 7, 33
Calcium Hydroxyapatite 266
Calcium Pyrophosphate 266
Candida humicola
methylation of As 523
Capacity Factors
correlation with molecular
surfaces 220
Carbon-14 300
Carbon Dioxide 402
Carbon Oxysulfide 403
Carbonate Complexes
alkal. earths in seaw. 72
Cd,Cu,Pb in freshwater 414/5
Co,Cu,Ni,Zn in seawater 73
Carbonate System
in seawater 68
Carbonates
and acid rain 405
adsorption of metal ions 364

Carcinogenesis 447/8
Carcinogenicity
of Cr 456, 541, 473, 447
of metals 41, 543
Carcinogens
tests for 449, 450
Cardanol
precursor, Cu extractant 274
L-Carnosine Complexes
Cu-Zn heterobinuclear 288/9
Cu-Cd heterobinuclear 288/9
Cashew Nuts
source of Cu extractant 274
Catalase
oxidation of Hg(O) 499, 520
Catalytic Wave Polarogr. 384
Catechoyl Amides
as chelating agents 337/8
Cell Membranes 47, 492
interactions with organo-
metallic compounds 25
transport of elements 38, 41
Cell Population
in Augusta Bay 606
Cell Surface
metal binding 37, 492
Cerium(III) Complexes
heterobinuclear with Cr 286
Cesium, radioactive
in ocean 300
Chelation Therapy 335
for Pb 7, 33, 91
for Pu 333, 342/6
Chelators, specific
for Pu, synthesis 339, 342
Chemical Extraction 361
of metal species 363
of Hg species 360/1
Chemistry
analytical 42, 547
aquatic 168
Chemotherapeutics 219, 257
Chesapeake Bay 142, 136, 213
box cores 148/9
composition of interstitial
water 141/3
fluxes of solids 146/7
sedimentation rates 139

Chesapeake Bay (contd.)
  sediment transport 137
  water circulation 136
Children
  exposure to Hg 438
China
  metal poll. 359, 373, 393
  metal pollution, coastal
    zone 374/5
Chloride Complexes
  Co,Cu,Ni,Zn in seawater 73
Chlorophyll a
  in Augusta Bay 608
Choline
  methyl donor 252
Chromates
  carcinogenicity 476
  effects on kidney, liver 475
  genotoxicity 453-455
  effects on exposed workers 476
  immune response 256
  lesions 475
  mechanism of genotoxicity 453
  mutagenicity 7, 452
  reduction in cells 453, 464
  resistance to 527
  toxicity 464
  transport into cell 453
Chrome Plating Workers
  exposure to Cr(VI) 464/5
  urinary Cr conc. 467
Chromite
  production 452
  use in U.S. 452
Chromium
  absorption 474
  accumulation in man 478
  body burden 464
  carcinogenicity 456, 544
  distribution in man 474
  essentiality 475
  excretion 464, 475, 478
  exposure of workers 452, 464
  genotoxicity 447, 452-7, 544
  half-life in man 478
  in air particulates 466
  in human plasma 474
  in man 474

Chromium (contd.)
  in urine 463, 466
  interaction with DNA 453
  mechanism, genotoxicity 453/5
  metabolism 466, 473, 478
  synergism with Zn 458
  toxicity 464, 473, 475, 544
  valence state detn. 382
Chromium(VI)
  see Chromate
Chromium Complexes
  genotoxicity 544
  heterobinuclear Ce(III) 286
  nicotinic acid/glutathione 7
  toxicity 91
Chromium Compounds
  recycling, tanneries 431/3
  use in industry 473
Chromium Excretion 464, 475/8
  mechanism 464
Chromium Oxides
  cause of sarcomas 267
Chromium, urinary
  as exposure test 469
  as measure of exposure 478
Chromosomes 448/9
  damage as toxicity test 14
Chromyl Chloride
  genotoxicity 457
Chrysotile 266
Citrate System
  with excess Cd,Cu,Pb,Zn 294
Citric Acid Species
  in Cu-Ni system 293
Citric Acid Complexes
  heterobinuclear 291/2
  mononuclear 293
Clays
  adsorption of Cd 364
  adsorption of Hg 365/6
  as reaction templates 214
Clostridium cochlearum
  methylation of Hg 497
Cobalt-60
  in seawater 301
Cobalt Complexes
  in seawater 72/3
  -species distribution 73

Cobalt Complexes (contd.)
 -stability constants 73
 with isoserine 291
 -mononuclear 291
 -homobinuclear 291
 -heterobinuclear, Cd,Zn 291
Cobalt(III) Complexes 120
 genotoxicity 544
Colloids 10
Complexes
 see also: formation constants
 chlorohydroxy 202
 formation gas phase 117
 formation in solution 117,123
Complexing Capacity
 of oil shale waters 591
 of water 381, 410
Concentration Units 51
Connectivity Index 220
Coordination Compounds 4, 89,
 65, 117, 249, 273, 285, 307,
 331, 541
Copper(I)
 chloride complexes 75
 in anoxic ocean 13
Copper(II)
 detn. by ASV in seawater 374
 detn. of free ion 409
 distribution in rats 572
 extraction 273/4
 in anchovies 158
 in bonitos 160, 164
 in coastal areas of China 375
 in mussels 162
 in Nanhai Sea 376/7
 in Pacific 375
 in shrimp 161
 in striped mullets 159
 in Tibet rivers 384
 in Tibet snow 385
 in Xiang River 362/3, 379
Copper(II) Complexes
 amines 109
 benzoic acids 99
 carboxylic acids 109
 diketones 100
 carcinogenicity 92
 dipeptides 110

Copper(II) Complexes (contd).
 extracellular ligands 502
 formation constants 109
 MECAMS 338
 NTA 95
Copper(II) Complexes
 homobinuclear
 -citrate 293
 -macrocycles 281
 -Schiff base 280
 mononuclear
 -alanine 290
 -catechol 290
 -citrate 293
 heterobinuclear
 -Cd, Ni, Zn/histidine 287
 -Cd, Zn/carnosine 288
 -Cd, Zn/L-DOPA 289
 -Cd, Zn/Histamine 288
 -Cd, Zn/tripeptide 290
 -Ni, Zn/citrate 292
 in seawater 72/3
 stability 490
Copper(I/II) Redox Couple 75
Copper Deposition
 from atmosphere into
 -Atlantic 409
 -Lake Constance 409
 -Swiss Rivers 409
Copper Minerals
 chloride-containing 187/8
 secondary 186
Copper Ores
 weathering
Copper(II), Resistance to
 by E. coli 527
 by S. aureus 527
Copper Species
 carbonates 415
 in natural waters 72, 415
Copper(I) Sulfide 75
Copper(II) Sulfide 75
Coronands
 extractants for Cu 274
Corrinoids 213
Covalent Parameter
 metal-H coupling const. 124
 for Lewis acids 125

Covalent Parameter (contd).
for Lewis bases 126
Crab Larvae
LC-50, triorganotins 223
18-Crown-6
dimethylthallium compd. 217
dimethyltin compd. 217
Crustaceans
uptake of transuranics 328
Cyanidium caldarium 37, 503
growth at low pH 503
metal removal, solution 504/5
sulfides on cell surface 506
Cyanobacteria
Ni at cell surface 502
Cyclam 119
Ni complex 120
Cysteine 262
in metallothionein 260

Dalian Bay, China
As in 381
Deep Sea 31, 70
and radioactive wastes 31
Dehydrogenase
arsenite 523
lipoamide 518
Demethylation, of
methylmercury 498, 514, 520
Denuders
behavior of
particles in 427/8
for detn. of SO_2 426, 429
for detn. of NH_3,
HNO_3, HCl 429
operation 424
theory 423
Dermatitis
from Cr(VI) 256, 476
Desferrichrome 335
Desferrioxamine B
affinity for Fe(III) 261
treatment of Fe overload 335
Desulfovibrio
sulfate reduction 497/9
Deposit Feeders 322/3, 326
Deposition Sites
of metals in organisms 267

Detection Limits 42, 55
analytical 549, 550
for NAA 567
in reporting analyses 50
Detectors
element-specific for LC 554
Detoxification
of inorganic compounds 61
of Hg compds., mechanism 516
2,6-Diacetylpyridine
synthesis of Schiff bases 276
Diamines, primary
Co,Cu,Ni complexes 105
-calc. formation enthalp. 105
enthalp. of protonation 105/6
Diaminoethers/thioethers
synthesis of Schiff bases 276
1,3-Diaminopropanol
synthesis of Schiff bases 277
Dichromate
see Chromates
mutagenicity 7
workers exposed to 465
Diet
metal intake through 437
Diethylenetriaminepentaacetate
Ca, Zn salt 333
-Pu removal from man 333
-effect on rats 91
-toxicity 91
Diffusion
between water/sediment 145/6
Fick's law 196
2,6-Diformylpyridine
synthesis of Schiff bases 277
1,2-Dihydroxybenzene
see catechol
2,5-Dihydroxybenzoic Acid
reduction of Mn oxide 178
**3-(3,4-Dihydroxyphenyl)-
L-alanine,** see L-DOPA
β-Diketones
metal complexes 97
Dimethylarsinic Acid 18, 240
identification by
HPLC-ICP 558
in rats 574

Dimethylcadmium
  stability in water 207
N,N-Dimethyl-2,3-dihydroxy-
  benzamide
  deprotonation constant 336
Dimethyl Diselenide 201
Dimethyl Disulfide 214
Dimethyl Ether
  proton basicity 121
Dimethyl(hydroxyethyl)arsine
  oxide 18
Dimethyl Mercury 18, 81, 523, 497
  dec. by light 499, 210
  from Hg(II) and acetate 210
  solubility 221
  surface area of molecule 221
Dimethyl(ribosyl)arsine
  Oxide 18
Dimethyl Selenide 18, 201
2,2-Dimethyl-2-silapentane-
  5-sulfonte, Na salt
  methylation of Hg 242
Dimethyl Sulfide 214, 403
Dimethyl Telluride 201
Dimethyl Thallium Compounds
  201/5
  compl. with crown ether 217
  compl. with porphyrin 217
  methyl donor 240
  methylation of Hg 205
Dimethyltin(IV) 18
  compl. with crown ether 217
  species in solution 240
DNA
  adsorption on mica 266
  damage 449
  interactions with Cr 453/5
  repair 449
DNA Replication
  and toxicity tests 14
Dolomite 68
Donors
  hard, soft 71, 491
Dogs
  Pu in plasma of 333
  Pu removal by chelators 340
L-DOPA
  heterobinuclear complexes 289

Dose
  definition 477
  optimization 302
Dose-Response Relationships
  and metal species 39
  studies of 8, 218
Drago Equation 101, 123

Ebullition 145
EDTA
  absorption by man 91
  effect on rats 91
  for Pu-contaminated wounds 333
  intravenous lethality 91
EDTA Complexes
  Ca, toxicity 91
  Cd 258
  -formation const. by ASV 380
  exchange with free EDTA 258
  in cells, H-1 NMR 258/9
Edwards Equation 100, 121, 124
Effects
  biological, definition 477
  electronic, in complex
    formation 117
Efflux Pumps 37, 501, 522
E_h-pH Diagram
  for C, N, S species 140
Electron Spectroscopy 46
Electronic Effects
  in complex formation 117
Element-Carbon Bonds
  hydrolytic stability 19, 559
Element Concentrations
  in Augusta Bay 607
  in man 252/3
  in earth's crust 253, 489
Element-Specific Detectors
  for HPLC, IC 554
Elements
  bonds with C 558/9
  essential 1, 253/4, 495, 541
  -cycling 2
  -for microbes 488
  -mobilization 2
  -toxicity 7
  nonessential 495
  tolerable intake 441

Elements (contd.)
 toxic 541
 -compounds of 373
 use by cells 488
Energy Sources
 nonrenewable 59
 renewable 60
Engraulis encrasicholus 156
 metal concentrations in 158
Enrichment 45
Enterobactin 335
 Fe(III) complex 336
 Pu(IV) complex 339
Erythrocytes
 accumulation of vanadate 255
Escherichia coli
 organomercurial lyases 517
 resistance to
 -arsenite, arsenate 520
 -As 501, 520
 -Cu 527
 -Hg 514/5
 -Sb 501, 520
 toxicity of metal ions 543
Estuaries 135
 toxic metals in 136, 150
Ethylenediamine
 protonation enthalpy 105
 Schiff base formation 277
Ethylenediamines
 C-methyl substituted 118
 -complexes of 119
Ethyl Mercury
 detoxification 516/7
Europium
 distribution in rats 572
Eutrophication 61
Exposure
 definition 477
 occupational to Cr 463, 473
Extraction, sequential 394
 of Hg species 393
 of Hg humates 396

Factor Analysis 51
Fauna, benthic 145
Ferrichrome
 Fe(III) complex 336

Ferrioxamine B
 Fe(III) complex 336
Ferritin 267
 for Pu storage 333
Fick's Diffusion Law 146
Fish
 Hg species in 383
Fishermen
 exposure to Hg 442/3
Flame Emission Spectrometry 555
Flow Rates, Air Sampling
 measurement of 422
Force Field Calculations
 empirical 120
Fluoride Complexes
 alkaline earths 72
 metals 102
Fluxes, Dissolved Materials
 in Chesapeake Bay 146
 measurement of 147
Foods
 Cd content 444
 Pb content 444
 radionuclides in 304
Fossil Fuels
 combustion 403, 420
Formation Constants
 calculations of 124
 chloromercury(II) species 243
 heterobinuclear complexes
 -Cd-Co/isoserine 291
 -Cd-Cu/carnosine 288/9
 -Cd-Cu/histamine 288
 -Cd-Cu/histidine 287
 -Cd-Cu/L-DOPA 289
 -Cd-Zn/isoserine 291
 -Co-Zn/isoserine 291
 -Cu-Ni/histidine 287
 -Cu-Zn/carnosine 288/9
 -Cu-Zn/histamine 288
 -Cu-Zn/histidine 287
 -Cu-Zn/L-DOPA 289
 homobinuclear complexes
 -citrate, Cd,Cu,Ni,Mn,Zn 293
 -isoserine, Cd,Co,Zn 291
 methyl mercury chloride 243
 mononuclear complexes
 -Cd/alanine 290

Formation Constants (contd.)
 -Cd/catechol 290
 -Cd/citrate 293
 -Cd/isoserine 291
 -Co/isoserine 291
 -Cu/alanine 290
 -Cu/catechol 290
 -Cu/citrate 293
 -Ni/citrate 293
 -Mn/citrate 293
 -Zn/alanine 290
 -Zn/catechol 290
 -Zn/citrate 293
Freshwater
 distribution of Cu, Cd, Pb
  species in 414/5
 pollution of 401
 valence of Se, As in 383
**Freundlich Adsorption
 isotherm** 364
Fungi
 ligand release by 13
 precipitation of U 503
Fulvic Acid 15, 70
 metal complexes 366
 standards for 16

Gastroenterocolitis
 from exposure to Cr(VI) 476
Gammaglobulin, human
 Pb complex 381
Genes 448
Genetic Engineering 506
Genotoxicity 447, 541
 Co(III) complexes 543
 Cr 454
 Cr(III) complexes 543
 Cr compounds 457
 metal complexes 14, 541, 451
 Rh(III) complexes 543
Germanium
 methyl compounds 201
Gliocladium roseum
 methylation of As 523
Glucose Tolerance Factor 475
Glutathione 262
 Cd binding 262, 265/6
 Cd, Zn exchange 263

Glutathione (contd.)
 Cd, Zn, H-1 NMR 264
Glutathione Reductase 518
Glycyl-L-histidine Complexes
 heterobinuclear 287
Gold
 deposition in cells 267
 methylation 206
**Graphite Furnace Atomic
 Absorption Spectrometry** 56
 for element-specific
  detection in HPLC, IC 555

Haemoglobin
 affinity for As 573/4
 binding of V 255
Haemosiderosis 99
Hammett Equation 99
Hancock Equations 101/2
Hardness Parameter
 and oxidation potential for
  bases 124, 127
 for Lewis acids 125
 for Lewis bases 126
Heats of Complex Formation 9
Heavy Metals
 as pollutants 186
 dietary intake 440
 exposure of humans 437
 in aquatic environment of
  China 373, 359, 393
 microbial resistance
  to 487, 496
Heterobinuclear Complexes
 see also: formation
  constants, individual metals
 Al-U 285
 Cu-Cd, -Ni, -Zn 286/8
 conditions for formation 294
Hexamethylcyclam
 Ni complex 121
Histamine Complexes
 heterobinuclear 288
L-Histidine Complexes
 heterobinuclear 287
HPLC-GFAA Systems 555
Humates
 of Hg 396

Humic Acids 15, 70, 93
  adsorption of Cd 364
  –of Hg 365/6
  dissolution of HgS 362
  concentration in natural
    water 93
  Hg complexes 360, 366
  influence on Pu
    solubility 309
  molecular weight 381
  of Ji Yun River 365
  reduction of Pu(VI) 314
  standards 16
Hydrocarbons, polyaromatic
  atmospheric deposition of 408
Hydrochloric Acid
  detn. with denuder 429
  in atmosphere 420
Hydrocyanic Acid 121
Hydrogen Chloride
  detn. in air by denuders 429
  in atmosphere 403
Hydrogen Sulfide
  dissociation 75
  in atmosphere 403
  proton basicity 121
  react. with methyl Hg 499
  react. with trimethyl Pb 499
Hydrolase
  organomercurial 516/7
Hydroxide as Ligand
  in metal complexe 73, 102
Hydroxo Complexes
  in seawater 73
2-Hydroxyaryloximes
  for Cu extraction 274
Hydroxycarboxylate Complexes
  Ce-Cr heterobinuclear 286
Hydroxyl Groups
  on surfaces 168
  on surface of FeOOH 171
p-Hydroxymercuribenzoate
  detoxification of 517
Hydroxypyridones
  acid-base properties 342
  in multidentate ligands 344
Hydrophobic Regions
  in biological systems 13

Hydroquinone
  reduction of Mn oxides 177/9
  –influence, org. compds. 179
Hypocalcemic Tetany 91

ICP-MS 550, 564
Illite
  absorption of Hg 365/6
Immune Response
  to metals 256
Inductive Effects 118, 123
  hidden 118
**Inductively-Coupled Argon
Plasma Emission
Spectrometry** 550, 601
  in element-spec. detect. 556
  –computerized 557
  detn., As, P, Se compds. 557
  seawater analysis 601/3/7/11
Inflammation
  of joints 266
Inorganic Compounds
  as therapeutic agents 32/5
  interactions with biota 3
  insoluble, toxicity tests 14
  transformations in
    environment 24
Insulin Metabolism
  regulaton by Cr 7
Intercalibration 160
Intercalation Compounds 216
Interstitial Water
  composition 141/5
Iodide/Iodate Redox System 69
Ion Association Model 9
Ion Selective Electrodes 409
Ionic Parameters 124
  for Lewis Acids 125
  for Lewis Bases 126
Ionophores 13
Iridium
  methylation 206
Irving-Rossotti Equation 96/7,
  103
Irving-Williams Order 72
Iron
  biological use 488
  deposition site in biota 267

Iron (contd.)
toxicity 7
Iron-59
in seawater 301
Iron(II) Carbonate 77
in sediments 142, 144
Iron(III) Complexes
calc., formation
 constants 107/8
hydroxy 76
mesoporphyrin 100
Iron Concentrations
in oil shale waters 585
Iron Cycles 169
Iron(II) Ion, hydrated 77
acid dissociation const. 76
concentration in seawater 76
Iron(III) Ion
hydrated
 -acid dissociation const. 76
 -conc. in seawater 76
sequestering agents
 for 331, 335
Iron(III) Hydroxide 76
Iron(III) Oxides
hydrous 491
 -adsorption of Cd on 364
 -adsorption of Hg on 365/6
 -adsorption of metals 415/6
 -catalyst, Mn oxidation 174
 -surface reactions 169
gamma-FeOOH 170
 -catalyst, Fe oxidation 171
 -solubility product 76
 -synthesis 171
Iron Overload Disease 262
treatment 335
Iron(II) Phosphate
in sediments 142, 144
Iron(II/III) Redox System 76
diagram 77/8
Iron(II) Silicates
in seawater 77
Iron(II) Sulfides
in sediments 142
Iron Transport
in aerobic bacteria 491

Isoserine Complexes
Cd, Co, Zn, 291

Jiaozhou Bay
Pb species in 377
Zn species in 377
Ji Yun River
Hg accumulation factors 369
Hg in sediments 365, 393, 398
Hg pollution of 360
Hg species in 362
Hg transport 367
Jin Sha River
metal pollution 367

Kaolinite
adsorbent for metals 363
adsorbent for Hg 366
Kidney
sink for Cr 478
Kinetics
in heterogeneous systems 16
of complex formation 6
of environm. reactions 57
of metal transport 491-493
of methylation of metals 200
radioelements in sea 301, 303
Kroeger-Drago Equation 102
Krypton
radioactive 300
Labile Metal Complexes
determination 46, 56
Lakes, alpine
metal concentrations in 408
pH of 405
water composition 406
acidity of 406-8
Lake Constance
atmospheric deposition
 of metals 409
Langmuir Adsorption
Isotherm 364
Lanthanide Ions, templates
for Schiff base formation 277
LC-50, crab larvae
triorganotin compounds 223
LD-50
toxicity test for metals 14

Lead
absorption in GI tract 438
biomethylation 207
calculated intake, Italy 444
determination by ASV 374
foodsources for 444
genotoxicity 452
in anchovies 158
in Chinese coastal
 areas 375/6
in Jiaozhou Bay 377
in mussles 162/3
in Nanhai Sea 376/7
in Pacific 375
in seawater 79-83
in shrimp 161
in striped mullets 159
in wine, limit for Italy 444
in Xiang River 362/3
methyl compounds 201, 207
methylation by methyl
 iodide 213
release from sediments 214
species in seawater 79
study of metabolism 571
tolerable intake 441
Lead(II) Acetate
toxicity 6
Lead Carbonates
solubility products 82
Lead Chloride 79, 83
Lead Chromates 456
Lead Compounds
detn. by HPLC-GFAA 556
Lead Complexes
carbonate in seawater 83
albumin, human, bovine 381
EDTA
–absorption by man 91
–H-1 NMR 258
–toxicity 6
humic acid 381
in seawater 83
in Tibet rivers 384
in Tibet snow 385
nitrilotriacetic acid 95
Lead Deposition, atmospheric
into Atlantic 409
Lead Deposition (contd.)
into Lake Constance 409
into Swiss Rivers 409
Lead Deposits
in cell nucleus 268
Lead Intoxication
chronic 438
Lead(II) Ion, hydrated
acid dissociation const. 82
Lead Minerals
chloride-containing 187/8
secondary 186
Lead Poisoning, treament
with BAL 33
with Ca-EDTA 33, 91
Lead Redox Systems 82
Lead Species
in freshwater 414/5
in seawater 79
Lead(II) Sulfate
in seawater 79, 83
Lepidocrocite 170
Lethality Assay 543
for metal complexes 544
Lifetimes
of chemical species 203
of metal-C bonds 203, 209
Ligands
hard, soft 131, 490/1
in environment 5
macrocyclic
–as extractants 275
–Ni, Cu, Zn complexes 281
organic
–for metal ions 411
–release by organisms 13, 492
therapy with 7, 32, 91,
 333, 342
transformation 9
Ligurean Sea
metals in organisms from 155
Limepits
tannery operations 433
**Linear Free Energy
Relations** 127
Ag(I)/Hg(II) complexes 128
Bi(III)/Hg(II) complexes 129
Hg(II)/H(I) complexes 129

Linear Free Energy Relations (contd.)
 for metal complexes 96
Liquid Chromatography 43
 detn. of trace element compounds with 554
 element-specific detectors for 554
Lung, human
 trace metals in 567
Lysosomes 13

Mackerel
 Hg in 350
Mackinawite
 in sediments 142
Macrofauna
 in sediments 323
 -population densities 324
Macrocycles
 metal transfer, membranes 217
 methylmetal complexes 217
 Schiff bases
 -metal ion templates for 277
 -Cu complexes 279/80
Macrocyclic Effect 119
Macrophages 266
Magnesium
 in oil shale waters 585
 template for macrocycles 216
Magnesium Complexes
 in seawater 68
 -bicarbonate 68
 -carbonate 68
 EDTA, H-1 NMR 258
 nitrilotriacetic acid 95
Magnetite 265
Malachite 73
 ion product, seawater 74
 solubility product 74
Malic Acid
 heterobinuclear complexes 287
Main Group Elements
 methylation 195
 transport by ligands 217
Man
 and marine radionuclides 304

Man (contd.)
 Pu contamination 333
Manganese
 adsorption on FeOOH 172
 in anchovies 158
 in bonitos 160, 164
 in mussels 162
 in seawater 77
 in shrimp 161
 in striped mullets 159
 oxidation 170/1
 -mechanism/FeOOH 174
 oxidation rate 172, 174
 oxidation state in sea 97
 release from sediments 214
 transport system, biota 524
 uptake 526
Manganese-54
 in seawater 301
Manganese(II) Carbonate
 in seawater 78
Manganese Complexes, citrate
 heterobinuclear Cd 293
 homonuclear 293
Manganese(I) Complexes
 LFER H(I) 130
 with methanethiol 130
Manganese cycle 169
Manganese Dust
 and carcinogenicity 266
Manganese(II) Ion, free
 concn. in seawater 77
Manganese Nodules 62, 77
Manganese Oxides
 hydrous
 -adsorbent for Cd 364
 -adsorbent for Hg 365/6
 -adsorbent for metals 415/6
 -on river particulates 141
 -reductive dissolution 170/6
 -red. by org. compds. 176, 181
 -solubility 176
 $MnO(OH)$ 78, 169, 176
 -in seawater 78
 -solubility 176
 -stability range 78
 -synthesis 177
 $MnO_{1.65}$

Manganese Oxides (contd.)
 -red. by org. compds. 177/8
 Mn_3O_4, stability range 78
Manganese(IV) Oxide
 catalyst Mn(II) oxid. 170/5
 in seawater 77
 reduction dissolution
  170, 176
 reduction by org. compds. 176
 stability range 78
Manganese(IV) Oxide/Mn(II)
 redox system 77
Marine Environment
 radionuclides in 299
Marine Food
 as Hg source 349
Mass Spectrometry
 fast atom bombardment 44
 for high molecular weight
  compounds 56
Meiofauna
 in sediments 323
 -population densities 324
Membrane Transfer
 of metals 25, 492
 of organometallic compds. 217
Mercury
 adsorption on particulates
  365/6
 bacterial transformation 514
 bioaccumulation 369
 biological half-life 352
 glutathione-bound in red
  cells 265
 immune response 256
 in anchovies 158, 351
 in bonitos, 160, 164
 in fish 349
 in Ji Yun River 359, 362
 -assoc. with humic acids 360
 -on particulates 363
 -separation of species 360
 -species distribution 363
 in mackerel 351
 in mussels 162/3
 in sardines 351
 in shrimp 161
 in striped mullets 159

Mercury (contd.)
 in tuna 350
 in sediments, anoxic 145
 -assoc. with humic acid 396
 -assoc. with org. compds. 397
 in seawater
 -Atlantic 335
 -halide complexes 81
 -Mediterranian 335/6
 -methylation 81
 -reactive, nonreactive 81
 -react. with S-ligands 82
 -species distribution 79, 81
 intake, tolerable 441
 reaction with organyl-
  bis(dimethylglyoximato)-
  cobalt(III) 245
 study of metabolism 571
 transport, Ji Yun River 367
 uptake by tuna, model 352
Mercury(195m, 197)
 production from Au 569/70
Mercury(II) Chloride
 methylation by acetate 210
 rate of dissociation 203
 react. with methyl-Cd 207
 react. with trimethyl-As 205
 reduction to Hg 208
Mercury Complexes, chloro
 adsorption on sediments 365
 formation constants 243
 LFER 128/9
Mercury Cycle 498
Mercury, elemental
 in air 210
 oxidation to Hg(II) 499
 -bacterial 514, 520
 -by catalase 520
Mercury(II) Haides
 reduction by $SnCl_2$ 81
Mercury Humates
 extraction 396
 stability 366
 thermal decomposition 396
Mercury, inorganic
 in seawater 355
 -Atlantic 355
 -Mediterranian 355

Mercury, inorganic (contd.)
 in Songhua River 359
 in tuna 350/1
 -half-life 352
Mercury(II) Iodide
 reaction with
  methylsilatrane 244
 -phenylsilatrane 244/5
Mercury(II) Ion
 in Cl-contg. solution 46
 intracellular traps for 499
 methylation 203-5,210,499,520
 reduction 499
 -bacterial 514
Mercury(II) Methylation by
 acetate 210
 amino acids 210
 bacteria 145
 Chlostridium cochlearum 497
 dimethylthallium 205
 methylsilicons 236,242,247
 silatranes 244
 trimethyllead 205
 trimethyltin 203, 240
 -chloride dependence 204, 241
 -ionic strength dept. 204,241
 -kinetics 241
 -mechanism 204
 -rate constants 204
Mercury Poisoning 199, 438
Mercury Reductase 499, 518
 antibiotics against 519
 composition 518
 DNA sequence in 518
 from soil Pseudomonad 518
 from Thiobacillus
  ferrooxidans 519
Mercury Resistance
 in E. Coli 514/5
 in Pseudomonas 515
 in S. aureus 515
Mercury Species
 chloride 241
 distribution model 363
 in aquatic environment 382
 in Chinese waters 382
 in fish of Bohai Bay 383
 in river sediments 396

Mercury Species (contd.)
 in seawater 45, 79, 368
 in tuna 349
 uptake by rice 382
Mercury(II) Sulfide 20, 145
 amorphous
 -dissolution by humics 362
 -stability 362
 dissolution 397
**Metabolism of trace
 elements** 570/1
Metals
 environmental regulations 542
 genotoxicity 451, 456, 541
 in active sites of enzymes 38
 oxidation state
 -influence on toxicity 7
 recovery from wastes 62, 431
Metal-Carbon Bonds
 lifetimes 203
 hydrolytically stable 559
Metal Complexes
 carcinogenicity 543
 determination of 44
 genotoxicity 541/3
 heterobinuclear 285
 homobinuclear 285
 in seawater 70
 released by organisms 13
 toxicity 89, 94, 541
Metal Concentrations
 in lakes 408
 in rain 408
 in seawater 65
Metal Deposition
 from atmosphere 409
Metal-H Coupling Constants
 and covalent parameter 124
Metals, heavy
 in China 359, 373, 393
 in Jin Sha River 367
 in marine environm. 367
Metal Ions
 as buffers 185
 biomethylation 207, 496
 electronegativities 250/1
 exchange in biol. systems 494

Metal Ions (contd.)
  hard, soft 71, 127, 131, 250, 251, 275, 491
  on particulates 411/3
  removal, C. caldarium 503/5
  surface binding 502
  surface complexes 15
  transport 196, 492, 524
  –into cells 491/2
  –kinetics 491/3
  uptake by cells 489
Metal-Metal Stimulation 286
Metal Oxides, chemistry
  in natural waters 167
Metal Species
  detn., need for 90, 409
  detn. methods for 409/10
  labile
  –detn. by ASV 374
  –in freshwater 378
  –in seawater 374
  in freshwater 378
  in organisms 164
  in seawater 376
  separation procedures 360
  toxicity 542
Metal Storage
  in organisms 267
Metal Sulfides 190
  bacterial dissolution 213
  on cell wall of C. caldarium 506
Metal Transport 15, 92, 254
Metalloenzymes 494
Metallothionein 526
  as metal trap 499
  binding of metals 495
  biosynthesis 261
  cadmium 260
  Cd-binding sites 260/1
  Cd removal 491
  Cd, Zn affinity 258, 261
  plaice 261
  rat liver 261
Methanobacterium 200
Methylamines
  correlation of protonation constants 104

Methylarsenic Acids 201
Methylarsines 18, 24, 198, 201
Methylarsonic Acid 18, 201
Methylase 208
Methylation
  of As 523
  of Cd 207
  of elements 18, 37, 195, 200, 206, 528
  of main group elements 195, 200
  of Hg 20, 203, 210, 240, 244, 514, 520
  of Tl 206, 240/2
Methyl Bromide
  solubility 221
  surface area 221
Methylcadmium 201
  NMR in water 208
  reaction with Hg(II) 207
Methylcadmium Hydroxide
  half-life in water 208
  methylation of Hg(II) 208
  NMR in water 208
Methylcobalamin 198
  methylation of As 523
  methylation of Hg 520
  methyl donor 497
  solubilization of oxides 213
Methylgermanes 201
Methyl Iodide
  methylation of Sn, Pb 213
  methyl donor 213
  production by seaweed 213
  reaction with sediments 214
  solubility 221
  surface area 221
Methylmercury 18, 45, 124, 201
  detoxification 516/7
  demethylation 498
  disproportionation 497
  formation 520, 497
  from Hg(II)/acetate 201, 210
  in bonitos 160, 164
  in fish 382
  in seawater 355, 352
  in Songhua River 359
  in tuna 349, 352

Methylmercury (contd.)
-biological half-life 352
in Xiang River 382
membrane transport 497
release from sediments 368
separation from inorg. Hg 382
species in solution 242
tolerable intake 441
toxicity 20
uptake 498
Methylmercury Chloride 79, 81
formation constant 243
Methylmercury Cyanide
CN-Cl exchange 203
Methylplumbanes 201
Methylsilanes 201
Methylsilatrane, reactions with Cd, Fe, Pb, Sn, Zn salts 245
Methylsilicon Compounds
methyl donors 240, 247
Methylstannanes 198, 201
Methylthallium(III) 242
Methyltin Hydrides
oxidation 24
Mica
absorption of DNA 266
Mice
distribution of Pu in 341
exposure to Cr(VI) 458
Pu removal, chelation 340/1
Microflotation 62
Microorganisms
essential elements for 488
in sediments 326
resistance to toxic metals 36, 487, 496, 513
Mineralization
in biology 265
in organic materials 157
Minerals
as metal buffers 185
solubilization 212
Mixed-Ligand Complexes 5
Models
for Hg in tuna 349
for metal-complex equil. 411
Molecular Topology 26, 219

Molybdenum
in Tibet rain, rivers, snow 384
interference with Cu 12
Monitoring of Exposure
to pollutants 477
Monohydrogenphosphate
in water 93
Montmorillonite
absorption of Hg 366
Mullets, striped 156
Mullus barbatus 156
metals in 159
Multi-Element Techniques
for trace metal analysis 43, 550, 556
Mussels 156
as bioindicators 163
metals in 162/3
Mutagens
tests for 449/50
Mutagenesis 447
Mutagenicity
of metal complexes 544/5
Mutations 448
Mycobacterium
Hg-resistant 516
Mytilus galloprovincialis 61, 156
metals in 162/3

Nanhai Sea
metal species in 376
Narragansett Bay 149
Nematodes
in sediments 323/4
Nephrops norvegicus 156
metals in 161
Neptunium
in water 317
Nervous System
effect of Pb on 438
Neurotoxicity
of methylelement compds. 41
Neutron Activation Analysis 550/1, 565/6
detection limits 567
sample preparation for 568

Nickel
 bioconcentration 502/3
 compounds 7
 deposition on cell walls 512
 genotoxicity 452
 immune response 256
 intracellular trap for 500
Nickel Complexes
 amines, Ni(I) 118/9
 formation constants
 -calculation of 107/8, 124
 heterobinuclear
 -Cd/citrate 292
 -Cu/citrate 292
 -Cu/histidine 287
 homonuclear 293
 in seawater 72/3
 macrocycles 281
 nitrilotriacetic acid 95
 toxicity 91
Nickel Itch 256
Nickel Oxides
 cause of sarcomas 267
Nickel Sulfide, Ni_3S_2
 carcinogen 266
Nickel Tetracarbonyl
 toxicity 7
Niebor-McBryde Equation 96/8
Nitrate/Nitrogen
 redox system 69
Nitric Acid
 in atmosphere 420
 -detn. with denuders 429
 in acid rain 403
Nitrilotriacetic Acid
 Ca salt, toxicity 91
 carcinogenicity 8
 concentration in water 93
 calculation of species
  distribution 93-5
 Cu salt, source of oral Cu 91
 Cd complexes 79
 Fe chelate 91
 in the environment 8
 influence on Cd species
  distribution 79, 80
 intake by man 92

Nitrilotriacetic Acid (contd.)
 metal complexes in water
  93, 95, 70
 oral LD-50 91
 sodium salts 92
 toxicity 8
 Zn chelate, wound repair 91
Nitrogen
 photoreduction on TiO_2 209
Nitrogen Bases
 in oil shale waters 402
Nitrogen Oxides 402, 420
Nonvolatile Compounds
 determination of 43, 554
North America
 acidity of lakes 406/7
Nuclear Energy
 and environment 300
 release of Pu 307
 waste from generation of 332
Nuclear Magnetic Resonance
Spectrometry 45, 47, 240, 257
 Cd-113 261
 multinuclear 56
Nuclear Tests
 atmospheric 299
Nuclear Waste 332

Oil Concentrations
 in coastal zones of China 376
Oil Shale
 methylelement compds. in 496
Oil Shale Wastewater 579
Organic Carbon, total
 in oil shale waters 589/90
Organic Compounds
 as ligands 5
 in oil shale waters 580, 587,
  592
Organic Matter
 mineralization of 157
 recycling by bioturbation 326
Organisms
 and marine radionuclides 301
 influence, element distrib. 36
Organomercurials
 bacterial resistance to
  498, 515

Organometallic Compounds 3, 18, 36
 coordination chemistry 19
Organylbis(dimethylglyoximato) cobalt(III)
 reaction with Hg(II) 245
Organylsilatranes
 reactions with HgX$_2$ 244
Organyltrialkoxysilanes 244
Oxidation
 in aqueous media 168
Oxidation Potential
 of bases
 -and hardness 124
 of seawater 65, 69
Oxygen Concentration
 in seawater 69
Oxygen/Water Systems 69

Pacific Ocean
 Hg in 382
 metal concentrations 375
Painters
 exposure to chromates 465/7
Palladium
 methylation 206
Particulates 15
 behavior in denuders 427/8
 in estuaries 135
 metal scavengers, lakes 415
 regulators
 -of metal concn. 411/3
Particulate Standards
 development of 16
Partition Coefficients
 water/octanol 218
Pattern Recognition 51
Pearson Equation 101
Penicillium putida 268
 accumulation of Cd 257
 Cd-EDTA complex in, NMR 259
 Cd resistant 257
4-Pentadecylsalicylaldehyde Oxime
 Cu extractant 274
Phenylmercuric Acetate
 detoxification 517

Phenylsilatrane
 reaction with HgX$_2$ 244
Phosphate
 detn. by HPLC-ICP 557/8
 transport system, cell 521
Phosphorus
 tolerable intake 441
Photomethylation 198, 208
 of Hg, Pb, Sn 211
Photosynthesis 209
Plaice
 metallothionein 261
Plankton
 Hg concentration in 354
Plants
 uptake of Cd 286
 uptake of metals 295
Plasma, animal
 metal detn. in, by NMR 258
Plasma Emission Spectrometry 550/1
 as element-spec. detector 555
 for seawater analysis 601
Plasmids
 As resistance 520
 Cd resistance 501, 513
 demethylation of CH$_3$Hg 498
 Hg resistance 514
 in genetic engineering 506
 metal resistance 496, 514, 519, 539
 Sb resistance 520
 Zn resistance 524/5
Platinum
 complexes 32, 124
 -anticancer drugs 257
 immune response 256/7
 methylation 206
Plutonium
 accumulation, organisms 327
 adsorption on particles 309
 biological properties 332
 chelation therapy for 333, 342/6
 chemistry of 310, 322/4
 concentration, dog plasma 333
 discharge at Windscale 308
 distribution in mice 341, 346

Plutonium (contd.)
from atomic weapons 308
from U-238 308
isotopes in environment 307
oxidation states 310
quantity in environment 307
redistribution 327
removal from mice 342, 346
solubility in seawater 309
storage in organisms 333
transferrin-bound 333
transport
-in organisms 333
-in environment 309
Plutonium-239 332
Plutonium(VI)
reduction by humic acid 314
Plutonium Cations
free 313
Plutonium(IV) Complexes
with enterobactin 339
Plutonium(III) Complexes
with humic acid
Plutonium Complexes
stability of 310
Plutonium Concentrations
at reprocessing facilities 308
in lakes, rivers, ocean 309
in Mediterranean Sea 310
in sediments 309
in soils 309
near test sites 308
Plutonium Contamination
of man 333
Plutonium Hydroxides 332
in the environment 8
polymeric 333
Plutonium Poisoning 33
Plutonium Species
cationic and pH 311
in freshwater, seawater 307, 309, 315, 316
model calculations 313
redox potentials 311
Plutonium Tetrahydroxide 310, 315
solubility product 334

Polarography
of Cd-glutathione 263/5
Pollutants
chemical 61
toxic effects of 447
Polyaminocarboxylic Acids
metal complexes 106/7
-calculation of formation constants 106
Polychaetes
Pu in 327
uptake of transuranics 328
Polymers, intracellular
as metal traps 37, 499
Porphyrins 207, 219
complexes with
-dimethylthallium 217
-gallium 217
-methylmetals 217
strapped 280
Posnjakite 189
as Cu buffer 189
Preservation of Samples
for detn. of trace element compounds 49
Primary Productivity
in Augusta Bay 605/6
Proteins
binding of Cr 474
Proton
basicity 121
hardness 125/6
LFER-Hg(II) 129
LFER-Mn(I) 130
Protonation Constants
carboxylate ligands 109
correlations with
-formation constants of metal complexes 98
correlation between organic acids and bases 104
N-contg. ligands 109
Pseudogout Syndrome 266
Pseudomonas
chromate resistant 527
mercuric reductase 518
organomercurial lyases 517

Pseudomonas aeruginosa
  resistance to organic Hg 515
Pseudomonas pseudomallei
  arsenite resistant 522
Pycnocline 147
Pyridines
  in oil shale waters 593
Pyrite 77
Pyrolusite 77

Quality Criteria
  for water, metal ions 542
Quality Control
  analytical laboratories 51
Quinolines
  in oil shale waters 595

Rabbit
  As metabolism 573
  response to arsenate 37
Radioactive Wastes
  disposal 31
Radioecology 299
Radionuclides 27, 61
  and organisms 301
  bioavailability 301
  dispersion in sea 303
  distribution, environment 27
  equilibrium distribution 300
  from fuel cycle 27
  from nonpower sources 28
  in marine environment 299
  kinetics in sea 301/3
  long-term behavior 30
  remobilization from sediment 322, 326
  risk to man 304
Radiotracers
  in environm. analysis 565
  high spec. activity 568/9
Rat
  As metabolism 573
  distribution of Cu in 572
  distribution of Eu in 572
  response to arsenate 37
Rate Constants 16
  for transmethylations 202

Rate Constants (contd.)
  for Hg(II) methylation 204, 241-245
Reactions
  at surfaces 168
  at very low concentrations 7
Reactions Rates, mercury(II)
  with trimethyltin 202/4, 241
  with methylsilicons 242
  with silatranes 244
Red Cells
  Cd in 262
  glutathione in,
  -Cd, Hg, Zn, binding 265
  lysed
  -detn. of metals in, NMR 258
Redox Potentials
  in cells 254
  of Pu species 311
Redox Reactions 198
  in atmosphere 403
  in estuaries 140/1
Reductase
  for Hg(II) 499, 516/7
  for Ni deposition 503
  glutathione 518
Regulations
  metals in water 542
Research Recommendations 12, 23, 35, 41, 55
Replicate Analysis 51
Resistance, microbial
  to antibiotics 526/8
  to As, Cd, Sb 501
  to metals 487, 496, 513, 527
Resonance Raman Spectrometry 45
Retention Indices 220, 221
Retort Waters
  from oil shale 580
  trace elements in 580, 588/90
  organic compds. in 580, 592
Rhinopathy
  from Cr(VI) exposure 476
Rhithropanopeus harrisii 222
  LC-50, triorganotins 223
Rhodium(III) Complexes
  genotoxicity 544

Rice
uptake of Hg 382
Risk Assessment 477
for Cr exposure 480/1
Risk Evaluation
for heavy metals 441
River Waters, China
metals in 362, 367, 379-84,
384/5

Salicylaldehydes
metal complexes 97
Salinity
in Augusta Bay 603/5
Salmonella
metal genotoxicity test 543
Sample Preparation
for neutron activation 568
Sample Preservation 57
Sampling 57
in aqueous environment 48
methodology 47
of air pollutants 47, 421
of organisms 48
Sarcomas
caused by Cr,Cd,Ni oxides 267
Sarda sarda 156
metals in 160, 164
methylmercury in 160, 164
Sardines
Hg in 350, 354
Scandinavia
acidity of lakes 406
Schiff Bases, macrocyclic
formation 276
metal complexes 279/80
Scorpion Fish 156
Scopulariopsis brevicaulis
methylation of As 523
Seawater 65
buffer system in 65
carbonate system 68
constituents of 65/6
inorganic chemistry of 65
ionic strength of 67
pH of 67
reactions, radioisotopes 300
salinity of 67

Seawater (contd.)
Te in 385
trace elements in 66
valence states of As, Se 383
Sediments
absorption of
–Cd 364
–chloromercury species 365/6
–Hg 365
–metal ions 363
anoxic 139
benthic 135, 139
Chesapeake Bay 137, 142,
148, 149
–mineral assemblages in 142
disturbances of 145
mercury in 393
mercury species in 362
–sequ. extraction of 395
mercury transport 367
organic 145
organisms in 323
separation of components 364
Pu in 309
release of metals after
treatment with MeI 214
sampling of 48
solubilization of metals 212
turnover rates 324
Sediment Standards 16
Selenate
detn. by IC-GFAA 556/7
in seawater 84
Selenoamino Acids 18
Selenite
detn. by IC-GFAA 556/7
in seawater 84
protonation constant 84
separation by HPLC-ICP 557/8
Selenium
in seawater 84
–Se(VI/IV) ratio 84
in oil shale waters 585, 589
in Tibet rivers, snow 384
interaction with Hg 38
methyl derivatives 201
valence states
–determination 382,556,557

Selenium (contd.)
 -in fresh-, seawater 383
Separation Methods 43
 for trace element compounds
 360, 373, 554
Sequential Extraction
 of Hg species 393
Serratia 518
Shrimp 156
 metals in 161
Siderite 142/4
Siderophores 33, 212, 334/5,
 491, 501
 synthesis 35
 treatment of Pu poisoning 33
Siderophilin
 Cr binding 474
Silatranes
 reactions with HgX_2 244
Silicates 214
Silicic Acid
 in water 405
Silicon
 in oil shale waters 585
 methyl derivatives 201
 reaction with haloalkanes 211
Silicon Dioxide
 absorption of Hg 365/6
Silver
 in Tibet rivers, rain 384
 sulfide on surface of cell 503
Silver Complexes
 amines 109, 118/9
 -formation constants 109
 anilines 100
 carboxylic acids 109
 halides 250/2
 LFER-Hg(II) 128
 phenyl-S,Se-acetic acids 100
 sulfide 76
Silver Halide Complexes
 stability in nonaqueous
 medium 250/2
Silver, microbial
 resistance 527
Small-Angle X-Ray
 Technique 47

Smithsonite 73
 ion product in seawater 74
 solubility product 74
Snow
 Ag in 384
 metals in 384/5
 Se in 384
Sodium EDTA 7
Sodium Sulfate Aerosol
 behavior in denuder 427
Sodium tetrachloromercurate
 adsorbent for SO_2 429
Soil
 Pu content 309
Solid Matter 4
 buffer system for metals 4,
 185
 surface active materials 4
Solid Surfaces
 absorption of metal ions 10
Solubility
 of Pu in seawater 309
Solubility Products 10
 see individual compounds
 of Pu tetrahydroxide 334
Solubilization
 of minerals in sediments 212
Solution-Particle Interphase
 15
Solvation Effects
 and steric effects 123
 in complex formation 121
Sorpaena porcus 156
Speciation 43
 of elements 18
Species Distribution 5, 6, 8
 determination of 8, 20/1
 in heterogeneous systems 10
 in homogeneous systems 9, 13
Specific Ionic Interaction
 Model 9
Springs, submarine
 proton source 67
Stability
 of trace element compds. 48
Stability Constants
 see also formation constants
 alkali metals/cryptands 277/8

Stability Constants (contd.)
and species distribution 6
calculation of 95
determination of 346
Fe(III) complexes,
-enterobactin, ferrichrome,
-ferrioxamine B 336
macrocycles, Cu, Zn, Ni 281
metal complexes 8
-and metal ion/ligand
 properties 96
-in heterogeneous systems 10
-in homogeneous systems 9, 12
-in seawater 72-85
Pu complexes 310/2
Standards, environmental
for analyses 50
for particulates 16
Staphylococcus aureus
mercuric reductase 519
resistance to
-antibiotocis 526
-arsenite, arsenate 520
-As 501, 520
-Bi(III) 521
-Cd 501, 523/5
-Cu 527
-mercurials 515
-Sb(III) 501, 520
-Zn 525
Steric Effects
and solvation effect 123
in complex formation 123
Steric/Specific-Solvation Parameter 124/5
correlation with
 Van der Waals radii 124
for Lewis Acids 125
for Lewis Bases 126
Stratosphere
bomb tests in 308
Streptomyces
Hg-resistant 516
mercuric reductase 518
reduction of Hg(II) 517
Structure-Activity Relationships 6, 218/9

Structure-Stability Relationships
for metal complexes 95
Structure-Toxicity Relations
for organometallic compds. 20
Strontium
EDTA complex, NMR 258
in oil shale waters 585
NTA complex 95
radioactive, in sea 300
Sulfates
in atmosphere 404
Sulfate Complexes, Sea
alkaline earths 72
Co, Cu, Ni, Zn 73
Sulfides
biological oxidation 432
concentration in anoxic
 zones 75
in tannery operations 432
Sulfide/Sulfur Redox Systems
 70, 75
Sulfhydryl Compounds 203
Sulfur Compounds
as ligands 19
oxidation of 169
Sulfur Dioxide
detn. with denuder 424/6/9
residence time, atmos. 403
Sulfur Oxides 402
in atmosphere 420
Sulfuric Acid
in atmosphere 404, 420
Superoxide Dismutase 278
Suspension Feeders 322/3
Susquehanna River 136, 142
Surfaces
influence on reactions 7
Surface Analysis 46
in-situ 57
Surface Areas, molecular
and solubility 219/21
correlation with
-LC-50 220
-retention indices 221
Surface Chemistry 211
Surface Complexes 169, 414
in freshwater 414

Surface Complexes (contd.)
 -of Cd, Cu, Pb 414/5
Surface Films 62
Surface Properties
 of crystals 266
Surface Reactions 198
Surveys, dietary 438/9
Swiss Rivers
 metals from atmosph. 409
Switzerland
 acid rain 404
Swordfish 156
 source of Hg intake 443
Synechococcus
 Cu-tolerant 500
 Ni-tolerant 500

Taft Equation 99
Taft-Hammett Sigma Values 218
Tannery Industry 431
 recycling of materials 433
 waste treatment 432
Technetium 30, 302, 328
 uptake by organisms 328
Tellurium
 detn. of valence states 382
 in fresh-, seawater 385
 methyl derivatives 201
Temperatures, surface
 in Augusta Bay 603/5
Template Reactions 214
Teratogenesis 447
Tests for mutagens,
 carcinogens 449/50
Tetramethylarsonium Salts 18
**2,2,3,3-Tetradeutero-
 3-(trimethylsilyl)propionate**,
 Na salt, methylation of
 Hg(II) 242
Tetramethyllead 1, 207
 half-life in day light 210
 in air 210
Tetramethylmethane
 solubility 221
 surface area 221
Tetramethyltin
 solubility in water 221
 surface area 221

Tetraorganyltin Compounds
 HPLC capacity factors 222
Tetrathiomolybdate 12
1,4,8,11-Tetrazaundecane
 Ni complex 120
 protonation constant 119
Thallassemia 7
Thallium
 methylation 206, 240
 -by trimethyltin 242
 methyl derivatives 201
 study of metabolism 571
Thallium(I) Acetate
 photoreaction 211
Therapeutic Agents 32, 257
Thermodynamic Cycles
 for metal complexes 103
Thermodynamic Equilibrium
 among complexes 6
Thermal Decomposition
 of Hg compds. 393/5, 399
Thiobacillus ferrooxidans
 Hg-resistant 516
 mercuric reductase 519
 reduction of Hg(II) 208
Thimerosal
 bacterial resistance to 515/6
Thunnus thynnus 156
 growth of 353
 Hg in 349, 350
 migration pattern 350
 model of Hg uptake by 352
Thiols
 as ligands 19
 as reducing agents 20
Tin
 bacterial resistance to 528
 detn. by ASV, seawater 374
 in coastal areas, China 375
 in Pacific 375
 methylation 528
 -by methyl iodide 213
 methyl compounds 201
 release from sediments 214
Tin(II) Acetate
 photoreaction 211
Tin Compounds
 detn. by HPLC-GFAA 556

Tin, Inorganic
  tolerable intake 441
Titanium Dioxide
  catalyst in photoreduction
    of nitrogen 209
Titanium Disulfide 216
Topology 26, 219
Toxicity
  of metals 5, 437, 447, 473
  of metal complexes 90
  tests for 7, 541
Toxicity Tests
  development, inexpensive 14
Trace Elements
  determination of
  -compounds of 547, 552, 555, 566
  -total concentrations 547,
    549, 566
  distribution in organisms 566
  in oil shale waters 580,
    585/6, 588
Trace Metals
  complexes in seawater 70
  concentrations
  -in marine organisms 155
  -problems with data 160
  in estuaries 135, 150
  mobilization 135, 150
Transducer
  for flow rate detn. 422
Transferrin 33, 261
  affinity for Fe(III) 261
  interaction with Pu 33
  Pu complex 333
  V transport 255
Transformations
  of Hg 514
  of metals by bacteria 513
  of organometallic, inorg.
    compds. in environment 24
Transmethylation 199, 203,
    240-245
  rate constants for 202, 240
Transport
  of metal ions 196, 524
  -across membranes 203, 492
  -kinetic studies 200, 493
  of Pu in environment 309

Transport Systems for
  arsenate 521
  Cd, Mn(II) 524
  phosphate 521
Transposons
  mercury resistance 519
  metal resistance 514
Transuranium Nuclides 300
  uptake by organisms 328
Traps, intracellular
  for Ni 500
  to remove metals 499
Threshold Levels 477
Trialkylammonium Cations
  deprotonation 122/3
Trialkylphosphonium Cations
  deprotonation 122
**1,4,7-Triazacyclononane
  N,N',N"-Triacetate**
  Ni(III) complex 120
Tributylchlorostannane 222
Trimethylamine
  proton basicity 121
Trimethylarsine 523
  reaction with $HgCl_2$ 205
Trimethylfluorosilane 244
Trimethyllead Compounds
  as methyl donors 240
  methylation of Hg(II) 205
  reaction with H_2S 499
Trimethylphosphine
  proton basicity 121
Trimethylstibine
  solubility in water 221
  surface area 221
**Trimethylsulfonium
  Chloride** 240
Trimethyltin Chloride
  as methyl donor 198, 240
  methylation of
  -Au, Ir, Pd, Pt, Tl 206
  -Hg(II) 203, 205, 240
  -Tl(III) 242
Trimethyltin Species
  in solution 241
Trinickel Disulfide
  carcinogenicity 7

Triorganyltin Compounds
  detn. by HPLC-GFAA 556
  surface areas 222
  toxicity to crab larvae 222
Tripeptide
  metal complexes 290
Trisodium Nitrilotriacetate 8
Tritium 300
Tuna 156, see Thunnus thynnus
Tunicates, marine
  V accumulation by 256
Tunichrome
  V complex 256
Tyrrhenian Sea, northern
  trace metals in organisms 155
Ultrafiltration
  of oil shale waters 582
Ultraviolet-Visible Spectrometry 45
Urea
  source of atmospheric ammonia 403
Urine
  Cr in 463/4, 467
  detn. of Cr in 465
Uranium
  extraction from seawater 63
  heterobinuclear complex
  -Al/citrate 285
  in seawater 59, 62, 83
  precipitation by fungi 503
Uranyl Ion
  complexes with
  -carbonate 83
  -Schiff bases 280
  -sulfate 83
  extraction from seawater 84

Valence States
  of elements in the environment 382
Valinomycin 277
Van der Waals Radii
  and steric/solvation effects 124
Vanadate
  in erythrocytes 255

Vanadium
  in ATP 256
  inhibitor of enzymes 256
  study of metabolism 571
Vanadocytes 256
Vanadyl Ion 255
Vivianite
  in sediments 142/4
Volatile Compounds
  of trace elements, detn. 43
Waste Disposal 60
Wastes
  in tannery industry 432
Wastewater, tanneries
  composition of 435
Water
  exchange with sediments 139
  interstitial 135, 143/5
Water Quality
  criteria for metals 542
Welders, stainless steel
  exposure to Cr 464/5
  urinary Cr 467
Willemite 191
Windscale
  Pu discharge 308
Wishbone Ligands 286, 290
Xia Wan River 394
  Hg in sediments 393, 398
Xiang River
  Cd, Cu, Pb species in 379
  methylmercury conc. 382
  sediments
  -Cd, Cu, Pb, Zn in 363
  -fractionation of 364
Xiphias gladius 156
X-Ray Fluorescence 550/1
  detn. of Cr in particles 466
Xizang, Tibet
  metals in river 384

Zeolites 215
Zhujiang River
  metal species in 376
Zhumulangmafeng, Tibet
  background conc., metals 384

Zinc
  binding to glutathione 263/5
  detn. by ASV, seawater 374
  in anchovies 158
  in bonitos 160, 164
  in coastal areas, China 375
  in Jiaozhou Bay 377
  in mussels 162
  in metallothionein 261
  in Nanhai Sea 377
  in Pacific 375
  in seawater 301
  in shrimp 161
  in striped mullets 159
  in Tibet rivers, snow 384/5
  in Xiang River 362/3
  tolerable intake 441
Zinc Carbonates 73
  ion products in seawater 74
  solubility products 74
  removal by EDTA from plaice metallothionein 261
Zinc Complexes
  calculation of formation constants 107/8
  EDTA
  -proton NMR 258
  -to treat Pu poisoning 33
  heterobinuclear

Zinc Complexes (contd.)
  -Cd, Co/isoserine 291
  -Cd, Cu, Ni/citrate 292
  -Cu/carnosine 288/9
  -Cu/L-DOPA 289
  -Cu/histamine 288
  -Cu/histidine 287
  -Cu/tripeptide 290
Zinc Complexes
  homobinuclear
  -citrate 293
  in seawater 72
  -carbonate 72
  -species 73
  -sulfate 72
  macrocycles 281
  mononuclear
  -alanine 290
  -catechol 290
  -citrate 293
  -isoserine 291
  nitrilotriacetic acid 95
Zinc Deposition, atmospheric
  in Atlantic 409
  in Lake Constance 409
  in Swiss rivers 409
Zinc Minerals
  secondary 186
Zinc Resistance
  S. aureus 524/5

LIST OF ABBREVIATIONS

aq	-aqueous
ASV	-anodic stripping voltammetry
atm	-atmosphere, unit of pressure
β	-stability constant
cm	-centimeter
contd.	-continued
cp	-cyclopentadienyl
CW	-continuous wave (laser)
E	-(redox) potential
EDTA	-ethylenediaminetetraacetic acid
FAB	-fast atom bombardment
g	-gram
GC	-gas chromatography
GC-MS	-gas chromatography-mass spectrometry
GFAA	-graphite furnace atomic absorption spectrometry
GI	-gastro-intestinal
HPLC	-high pressure liquid chromatography
IC	-ion chromatography
ICP	-inductively coupled argon plasma emission spectrometry
IR	-infrared
k	-unimolecular rate constant
k_2	-bimolecular rate constant
K	-equilibrium constant
kcal	-kilocalorie
kCi	-kilocurie
kg	-kilogram
kJ	-kilojoule
km	-kilometer
kP	-kilopascal, unit of pressure
K_{SO}	-solubility product constant
L	-liter
LC	-liquid chromatography
LC-50	-lethal concentration, 50%
LD-50	-lethal dose, 50%
LFER	-linear free energy relation
m	-meter
mm	-millimeter
M	-molar, moles per liter
mM	-millimolar
mg	-milligram
μg	-microgram
min	-minute
mL	-millileter

μL	—microliter
mol	—mole
MS	—mass spectrometry
mV	—millivolt
NBS	—National Bureau of Standards
nm	—nanometer
NMR	—nuclear magnetic resonance spectroscopy
NTA	—nitrilotriacetic acid
pH	—negative logarithm of H^+ concentration
pK	—negative logarithm of equilibrium constant
pK_w	—negative logarithm of ion product constant of water
pm	—picometer
pM	—negative logarithm of molarity
pOH	—negative logarithm of OH^- concentration
ppb	—part per billion, μg/L, μg/kg
ppm	—part per million, mg/L, mg/kg
ppt	—part per trillion, ng/L, ng/kg
R	—universal gas constant
s, sec	—second
S_E2	—bimolecular electrophilic substitution
S_N2	—bimolecular nucleophilic substitution
t	—time
T	—temperature
UV	—ultraviolet
V	—volt